Major American Universities Ph.D.
Qualifying Questions and Solutions – Mathematics

Problems and
Solutions in
Mathematics

Second Edition

Major American Universities Ph.D.
Qualifying Questions and Solutions – Mathematics

Problems and Solutions in Mathematics

Second Edition

Compiled by:

**Chen Ji-Xiu, Jiang Guo-Ying,
Pan Yang-Lian, Qin Tie-Hu,
Tong Yu-Sun, Wu Quan-Shui
and Xu Sheng-Zhi**

Edited by
Li Ta-Tsien
Fudan University, China

World Scientific

NEW JERSEY · LONDON · SINGAPORE · BEIJING · SHANGHAI · HONG KONG · TAIPEI · CHENNAI

Published by

World Scientific Publishing Co. Pte. Ltd.

5 Toh Tuck Link, Singapore 596224

USA office: 27 Warren Street, Suite 401-402, Hackensack, NJ 07601

UK office: 57 Shelton Street, Covent Garden, London WC2H 9HE

British Library Cataloguing-in-Publication Data
A catalogue record for this book is available from the British Library.

Major American Univ. Ph. D. Qualifying Questions and Solutions — Mathematics
PROBLEMS AND SOLUTIONS IN MATHEMATICS
Second Edition

ISBN-13 978-981-4304-95-5
ISBN-10 981-4304-95-6
ISBN-13 978-981-4304-96-2 (pbk)
ISBN-10 981-4304-96-4 (pbk)

Typeset by Stallion Press
Email: enquiries@stallionpress.com

Printed in Singapore.

Preface

This book covers six aspects of graduate school mathematics: Algebra, Topology, Differential Geometry, Real Analysis, Complex Analysis and Partial Differential Equations. It contains a selection of more than 500 problems and solutions based on the Ph.D. qualifying test papers of a decade of influential universities in North America. The mathematical problems under discussion are kept within the scope of the textbooks for graduate students. Finding solutions to these problems, however, involves a deep understanding of mathematical principles as well as an acquisition of skills in analysis and computation. As a supplement to textbooks, this book may prove to be of some help to the students in taking relevant courses. It may also serve as a reference book for the teachers concerned.

It has to be pointed out that this book should not be regarded as an all-purpose troubleshooter. Nor is it advisable to take the book as an exemplary text and commit to memory all the problems and solutions and make an indiscriminate use of them. Instead, the students are expected to make a selective survey of the problems, take a do-it-yourself approach and arrive at their own solutions which they may check against those listed in the book. It would be gratifying to see that the students can work out the problems on their own and come up with better solutions than those provided by the book. If the students fail to do so or their solutions may turn out to be incomplete, it may reveal the inadequacy of their knowledge or approach, thus spurring them to greater efforts to promote their skills. The very purpose of the authors in writing the book is just to help the students to discover the truth by trial and error.

This book was inspired by Professor K. K. Phua's proposals. We are particularly grateful to him for his support. We also wish to thank Dr. Xu Pei-jun, Professors Zhang Yin-nan, Hong Jia-xing and Chen Xiao-man for their painstaking efforts to collect test-oriented data. For selecting problems and providing solutions, we wish to acknowledge the following professors respectively: Wu Quan-shui (Part I), Pan Yang-lian (Part II), Jiang Guo-ying (Part III), Tong Yu-sun, Xu Sheng-zhi (Part IV), Chen Ji-xiu (Part V) and Qin Tie-hu (Part VI). We are also indebted to Professor Guo Yu-tao for carefully reading and correcting the manuscript. Finally, we pay tribute to Dr. Cai Zhi-jie for printing out the manuscript.

Li Ta-tsien

School of Mathematical Sciences

Fudan University

Shanghai 200433

China

Preface to the Second Edition

It has been twelve years since the first publication of this book. During the period, new mathematical problems, including many classical and interesting ones, kept emerging in the Ph.D. Qualifying Test of some prestigious universities in North America. For the improvement of this book and with the encouragement of World Scientific Publishing Co., we carefully selected more than one hundred new mathematical problems from the Ph.D. Qualifying Test papers in the past decade and provided solutions to them respectively, which can be taken as a necessary supplement to the first edition. Besides, we also made suitable corrections or amendments to the first edition regarding the solutions to a few problems which were not written in a very clear way or contained loopholes. We hope that the second edition of the book will be more helpful to readers.

Finally, on this occasion, we would like to express our gratitude to the readers for their concern and to the World Scientific for their support.

Li Ta-tsien
School of Mathematical Sciences
Fudan University
Shanghai 200433
China

Contents

PART 1
Algebra

Section 1

Linear Algebra

1101

Let V be a real vector space of dimension at least 3 and let $T \in \mathrm{End}_{\mathbb{R}}(V)$. Prove that there is a non-zero subspace W of V, $W \neq V$, such that $T(W) \subseteq W$.

(Indiana)

Solution.

Make V into an $\mathbb{R}[\lambda]$-module by defining $\lambda \cdot v = T(v)$ for all $v \in V$. Thus for $\sum_{i=0}^{n} a_i \lambda^i \in \mathbb{R}[\lambda]$ and $v \in V$

$$\left(\sum_{i=0}^{n} a_i \lambda^i \right) \cdot v = \sum_{i=0}^{n} a_i T^i(v).$$

It is clear that a subspace W of V is an $\mathbb{R}[\lambda]$-submodule of V if and only if $T(W) \subseteq W$.

Now suppose V is a simple $\mathbb{R}[\lambda]$-module. Then $V \simeq \mathbb{R}[\lambda]/I$ for some maximal ideal of $\mathbb{R}[\lambda]$. Since $\mathbb{R}[\lambda]$ is a P.I.D., there exists an irreducible polynomial $f(\lambda)$ of $\mathbb{R}[\lambda]$ such that $I = (f(\lambda))$. So

$$3 \leq \dim_{\mathbb{R}}(V) = \dim_{\mathbb{R}} \mathbb{R}[\lambda]/(f(\lambda)) = \deg f(\lambda).$$

This implies that we have an irreducible polynomial $f(\lambda)$ with degree ≥ 3 in $\mathbb{R}[\lambda]$. This is a contradiction. Hence V is not a simple $\mathbb{R}(\lambda)$-module, that is, there is a non-zero subspace W of V, $W \neq V$, such that $T(W) \subseteq W$.

1102

Let V be a finite-dimensional vector space over a field K.

Let S be a linear transformation of V into itself. Let W be an invariant subspace of V (that is, $SW \subseteq W$). Let $m(t)$, $m_1(t)$, and $m_2(t)$ be the minimal polynomial of S as linear transformation of V, W and V/W respectively.

(a) Prove that $m(t)$ divides $m_1(t) \cdot m_2(t)$.
(b) Prove that if $m_1(t)$ and $m_2(t)$ are relatively prime, then

$$m(t) = m_1(t) \cdot m_2(t).$$

(c) Give an example of a case in which $m(t) \neq m_1(t) \cdot m_2(t)$.

(*Indiana*)

Solution.

As usual, V can be viewed as a $K[t]$-module via the linear transformation S. Since W is an S-invariant subspace of V, W is a $K[t]$-submodule of V. Then, it is clear that $(m(t)) = \text{Ann}_{K[t]}V$, $(m_1(t)) = \text{Ann}_{K[t]}W$ and $(m_2(t)) = \text{Ann}_{K[t]}V/W$.

(a) Since

$$m_1(t) \cdot m_2(t) \cdot V \subseteq m_1(t) \cdot W = 0,$$

$$m_1(t) \cdot m_2(t) \in \text{Ann}_{K[t]}V = (m(t)).$$

Hence $m(t)$ divides $m_1(t) \cdot m_2(t)$.

(b) Since

$$m(t) \in \text{Ann}_{K[t]}V \subseteq \text{Ann}_{K[t]}W = (m_1(t)),$$

$m_1(t)$ divides $m(t)$. Similarly, $m_2(t)$ divides $m(t)$. Since $m_1(t)$ and $m_2(t)$ are relatively prime, $m_1(t) \cdot m_2(t)$ divides $m(t)$. Then we have $m(t) = m_1(t) \cdot m_2(t)$, since $m(t)$, $m_1(t)$ and $m_2(t)$ are all monic polynomials.

(c) Let W be a 2-dimensional vector space over the field $\mathbb{Q}$ of rational numbers and $S : W \to W$ be a linear transformation with minimal polynomial $t^2 + 1$. Let $V = W \oplus W$ and $S : V \to V$ be the natural extension of S to V. Then it is clear that $m(t) = m_1(t) = m_2(t) = t^2 + 1$. So $m(t) \neq m_1(t) \cdot m_2(t)$ in this example.

1103

Let V be a finite-dimensional vector space over $\mathbb{R}$ and $T : V \to V$ be a linear transformation such that (a) the minimal polynomial of T is irreducible and (b) there exists a vector $v \in V$ such that $\{T^i v | i \geq 0\}$ spans V. Show that V does not have proper T-invariant subspace.

(*Indiana*)

Solution.

V can be viewed as a module over the polynomial ring $\mathbb{R}[\lambda]$ via $f(\lambda) \cdot x = f(T) \cdot (x)$, for any $f(\lambda) \in \mathbb{R}[\lambda]$ and $x \in V$. Then we have $V = \mathbb{R}[\lambda] \cdot v$, a cyclic module, since $\{T^i v | i \geq 0\}$ spans V by (b). Let $m(\lambda)$ be the minimal polynomial of the linear transformation $T : V \to V$. Then $m(\lambda) \in \mathrm{Ann}_{\mathbb{R}[\lambda]}(v)$. Since $m(\lambda)$ is irreducible, we have

$$\mathbb{R}[\lambda]/(m(\lambda)) \simeq \mathbb{R}[\lambda] \cdot v = V$$

(we may assume that $V \neq 0$). So V is an irreducible $\mathbb{R}[\lambda]$-module. Thus, V does not have proper T-invariant subspace.

1104

Let A be an $n \times n$ matrix with entries in $\mathbb{C}$. Show that A has n distinct eigenvalues in $\mathbb{C}$ if and only if A commutes with no nonzero nilpotent matrix.

(*Indiana*)

Solution.

Necessity. Suppose that A has n distinct eigenvalues $\lambda_1, \lambda_2, \ldots, \lambda_n$ in $\mathbb{C}$. Then there exists an invertible $n \times n$ matrix P such that

$$PAP^{-1} = \text{diag}\{\lambda_1, \ldots, \lambda_n\}.$$

If A commutes with some nilpotent matrix B, we have to show $B = 0$. Since $\lambda_1, \lambda_2, \ldots, \lambda_n$ are distinct and

$$PAP^{-1} = \text{diag}\{\lambda_1, \ldots, \lambda_n\}$$

commutes with $P^{-1}BP$, $P^{-1}BP$ is a diagonal matrix. But the nilpontency of B implies that $P^{-1}BP$ is nilpotent. Hence we have $P^{-1}BP = 0$. So $B = 0$.

Sufficiency. Suppose that the characteristic polynomial of A has multiple roots. We have to show that A commutes with some nonzero nilpotent matrix. Let $\text{diag}(J_1, J_2, \ldots, J_t)$ be the Jordan canonical form of A and

$$PAP^{-1} = \text{diag}(J_1, J_2, \ldots, J_t),$$

where P is an $n \times n$ invertible matrix and J_i is a Jordan block of order e_i. Without loss of generality, we may assume that $e_1 > 1$ (If all the $e_i = 1$, then it is easy to see that A commutes with some nonzero nilpotent matrix).

Let B_1 be the Jordan block of order e_1, with 0 on the diagonal. Then $J_1 B_1 = B_1 J_1$ and $B_1 (\neq 0)$ is nilpotent. Let

$$B' = \text{diag}(B_1, B_2, \ldots, B_t)$$

where B_i $(i \geq 2) = 0 \in M_{e_i}(\mathbb{C})$. Then

$$B' \cdot PAP^{-1} = PAP^{-1} \cdot B'.$$

Taking $B = P^{-1}B'P$, we have $B \neq 0$, which is nilpotent, and $AB = BA$.

1105

Suppose V is a finite-dimensional vector space over a field F, $T : V \to V$ a linear map such that the minimal polynomial of T coincides with the characteristic polynomial, which is the square of an irreducible polynomial in $F[\lambda]$. Show that if $\vec{u}$, $\vec{v}$ and $\vec{w}$ are any three non-zero vectors in V, then at least two of the three subspaces spaned by the sets $\{T^i\vec{u}\}_{i\geq 0}$, $\{T^i\vec{v}\}_{i\geq 0}$ and $\{T^i\vec{w}\}_{i\geq 0}$ coincide.

(*Stanford*)

Solution.

V can be viewed as a module over the polynomial ring $F[\lambda]$ simply by $f(\lambda) \cdot x = f(T) \cdot (x)$ for any $x \in V$, $f(\lambda) \in F[\lambda]$. Let $\{u_1, u_2, \ldots, u_n\}$ be a base of V over F, $A = (a_{ij})_{n \times n}$ be the matrix of T relative to the base. In general, a normal form for $\lambda I - A$ in $M_n(F[\lambda])$ has the form

$$\mathrm{diag}\{1, \ldots, 1, d_1(\lambda), \ldots, d_s(\lambda)\}$$

where the $d_i(\lambda)$ are monic of positive degree and $d_i(\lambda)|d_j(\lambda)$ if $i \leq j$. By the structure theory of finitely generated modules over P.I.D., there exist z_i $(i = 1, 2, \ldots, s) \in V$ such that $V = F[\lambda] \cdot z_1 \oplus F[\lambda] \cdot z_2 \oplus \cdots \oplus F[\lambda] \cdot z_s$ where $\mathrm{Ann}(z_i) = (d_i(\lambda))$. Here, according to the assumptions, the minimal polynomial $m(\lambda)$ of T is $\det(\lambda I - A)$, so

$$m(\lambda) = d_s(\lambda) = \det(\lambda I - A).$$

Hence $s = 1$ and

$$V = F[\lambda] \cdot z_1 \cong F[\lambda]/(m(\lambda))$$

is cyclic. Since $m(\lambda)$ is the square of some irreducible polynomial, $V = F[\lambda] \cdot z_1$ has exactly two non-zero submodules. Obviously, the three subspaces generated by the sets $\{T^i\vec{u}\}_{i\geq 0}$, $\{T^i\vec{V}\}_{i\geq 0}$ and $\{T^i\vec{w}\}_{i\geq 0}$ are non-zero submodules of V over $F[\lambda]$. So at least two of them coincide.

1106

Let V be a finite-dimensional vector space over $\mathbb{C}$ with basis $\{v_1, \ldots, v_n\}$. Let σ be a permutation on $\{v_1, \ldots, v_n\}$ and thus induce a linear transformation A on V. Show that A is diagonalizable.

(*Harvard*)

Solution.

By re-ordering the elements $v_1, v_2, \ldots, v_n$, we assume that

$$\sigma = (v_1 \cdots v_{i_1})(v_{i_1+1} \cdots v_{i_2}) \cdots (v_{i_s+1}, \cdots v_n),$$
$$(1 \le i_1 < i_2 < \cdots < i_s < n),$$

when σ is expressed as the product of disjoint cycles. (This decomposition may have 1-cycles.) Let W_j be the subspace of V generated by $\{V_{i_{j-1}+1}, \ldots, v_{i_j}\}$ for $j = 1, 2, \ldots, s+1$ ($i_0 = 0$, $i_{s+1} = n$). Then the W_j's are invariant subspaces of A and $V = W_1 \oplus W_2 \oplus \cdots \oplus W_{s+1}$. Let M_j be the matrix of $A|_{w_j} : W_j \to W_j$ relative to the base $\{V_{i_{j-1}+1}, \ldots, v_{i_j}\}$ of W_j over $\mathbb{C}$. Then $M = \operatorname{diag}\{M_1, \ldots, M_{s+1}\}$ is the matrix of A relative to the base $\{v_1, v_2, \ldots, v_n\}$. So it suffices to prove that every M_j is diagonalizable.

Hence, without loss of generality, we may assume that σ is the n-cycle $(v_1 v_2 \cdots v_n)$. The matrix of A relative to the base $\{v_1, v_2, \ldots, v_n\}$ is

$$M = \begin{pmatrix} 0 & 1 & 0 & \cdots & \cdots & 0 \\ 0 & 0 & 1 & \cdots & \cdots & 0 \\ \cdots & \cdots & \cdots & \cdots & \cdots & \cdots \\ \cdots & \cdots & \cdots & \cdots & 0 & 1 \\ 1 & 0 & 0 & \cdots & 0 & 0 \end{pmatrix}.$$

It is easy to see that the minimal polynomial of M is $\lambda^n - 1$, and thus M is diagonalizable.

This completes the proof that A is diagonalizable.

1107

Let V be a finite-dimensional vector space over the field of rational numbers. Suppose T is a non-singular linear transformation of V such that $T^{-1} = T^2 + T$. Prove that 3 divides the dimension of V, and prove that if $\dim V = 3$, then all such $T's$ are similar.

(*Harvard*)

Solution.

Since $T^{-1} = T^2 + T$, T is annihilated by the polynomial $\lambda^3 + \lambda^2 - 1$. Obviously, $\lambda^3 + \lambda^2 - 1$ is irreducible over the field $\mathbb{Q}$ of rational numbers. Thus $\lambda^3 + \lambda^2 - 1$ is the minimal polynomial $m(\lambda)$ of T.

Now let n be the dimension of V over $\mathbb{Q}$, A be the matrix of T relative to some base of V, $\mathrm{diag}\{\overbrace{1,\ldots,1}^{n-s}, d_1(\lambda),\ldots,d_s(\lambda)\}$ be the normal form for $\lambda I - A$ where the $d_i(\lambda)$ are monic of positive degree and $d_i(\lambda)|d_j(\lambda)$ if $i \leq j$. By the irreduciblity of $d_s(\lambda) = m(\lambda) = \lambda^3 + \lambda^2 - 1$, we have

$$d_1(\lambda) = d_2(\lambda) = \cdots = d_s(\lambda) = \lambda^3 + \lambda^2 - 1.$$

Since $\det(\lambda I - A) = d_1(\lambda) \cdots d_s(\lambda)$,

$$3 \cdot s = \deg(\det(\lambda I - A)) = n.$$

Thus we have proved that 3 divides the dimension of V.

If $\dim V = 3$, then $\lambda I - A$ is equivalent to $\mathrm{diag}\{1, 1, \lambda^3 + \lambda^2 - 1\}$. The rational canonical form for A (or T) is $\begin{pmatrix} 0 & 1 & 0 \\ 0 & 0 & 1 \\ 1 & 0 & -1 \end{pmatrix}$.

It follows that all the T's are similar when $\dim V = 3$.

1108

Let F_q be a finite field with $q = p^n$ elements, where p is a prime. Let $\Pi : F_q \to F_q$ be the Frobenius automorphism

$\Pi(x) = x^p$. Prove that Π considered as a linear map over F_p is diagonalizable if and only if n divides $p - 1$.

(Harvard)

Solution.

It is well-known that $F_p \subseteq F_q$ is a Galois extension with

$$\mathrm{Gal}(F_q/F_p) = \{1, \Pi, \Pi^2, \ldots, \Pi^{n-1}\}.$$

By the Normal Base Theorem, there exists a $u \in F_q$ such that $\{u, \Pi(u), \Pi^2(u), \ldots, \Pi^{n-1}(u)\}$ is a base for F_q over F_p. When Π is considered as a linear map of F_q over F_p, the matrix of Π relative to the base $\{u, \Pi(u), \Pi^2(u), \ldots, \Pi^{n-1}(u)\}$ is

$$M = \begin{pmatrix} 0 & 1 & 0 & \cdots & 0 \\ 0 & 0 & 1 & \cdots & 0 \\ \cdots & \cdots & \cdots & \cdots & \cdots \\ 0 & 0 & 0 & \cdots & 1 \\ 1 & 0 & 0 & \cdots & 0 \end{pmatrix}.$$

The normal form for $\lambda E - M$ is $\mathrm{diag}\{1, \ldots, 1, \lambda^n - 1\}$ and the minimal polynomial $m(\lambda)$ of M is $\lambda^n - 1 = \det(\lambda E - M)$.

Suppose that Π is diagonalizable as a linear map over F_p. Then $m(\lambda) = \lambda^n - 1$ has no multiple root, and all the roots of $\lambda^n - 1$ are in F_p. On the other hand, all the root of $\lambda^n - 1$ forms a subgroup of $F_p^* = F_p\backslash\{0\}$. Thus n divides $p - 1$.

Conversely, if n divides $p - 1$, $\lambda^n - 1$ has no multiple root and

$$\lambda^n - 1 = (\lambda - 1) \cdot (\lambda - a^d)(\lambda - a^{2d}) \cdots (\lambda - a^{(n-1)d})$$

where $d = \frac{p-1}{n}$ and a is the generator of the group F_p^*. Hence M is similar to $\mathrm{diag}\{1, a^d, \ldots, a^{(n-1)d}\}$ in $M_n(F_p)$. So Π is diagonalizable as a linear map over F_p.

1109

Let $A(t)$ be a non-singular matrix whose elements are differentiable functions of real variable t. Let $A'(t)$ denote the matrix formed by the derivatives of the elements. Show that the derivative of the determinant $\det A$ satisfies

$$\frac{d}{dt}(\det A) = \det A \cdot \text{trace}\,(A' \cdot A^{-1}).$$

(Harvard)

Solution.

Let

$$A(t) = \begin{pmatrix} a_{11}(t) & a_{12}(t) & \cdots & a_{1n}(t) \\ a_{21}(t) & a_{22}(t) & \cdots & a_{2n}(t) \\ \cdots & \cdots & \cdots & \cdots \\ a_{n1}(t) & a_{n2}(t) & \cdots & a_{nn}(t) \end{pmatrix}.$$

Then

$$A'(t) = \begin{pmatrix} a'_{11}(t) & a'_{12}(t) & \cdots & a'_{1n}(t) \\ a'_{21}(t) & a'_{22}(t) & \cdots & a'_{2n}(t) \\ \cdots & \cdots & \cdots & \cdots \\ a'_{n1}(t) & a'_{n2}(t) & \cdots & a'_{nn}(t) \end{pmatrix}$$

and

$$A^{-1} = (\det A)^{-1} \begin{pmatrix} A_{11} & A_{21} & \cdots & A_{n1} \\ A_{12} & A_{22} & \cdots & A_{n2} \\ \cdots & \cdots & \cdots & \cdots \\ A_{1n} & A_{2n} & \cdots & A_{nn} \end{pmatrix}$$

where A_{ij} is the algebraic cofactor of $a_{ij}(t)$. Hence

$$\det A \cdot \text{trace}\,(A'A^{-1}) = \sum_{j=1}^{n} a'_{1j}(t)A_{1j} + \sum_{j=1}^{n} a'_{2j}(t)A_{2j}$$

$$+ \cdots + \sum_{j=1}^{n} a'_{nj}(t)A_{nj}$$

$$= \sum_{i}\sum_{j} a'_{ij}(t)A_{ij} = \sum_{j}\left(\sum_{i} a'_{ij}(t)A_{ij}\right).$$

For $1 \le k \le n$, let

$$
A_k = \begin{pmatrix}
a_{11}(t) & a_{12}(t) & \cdots & a_{1n}(t) \\
\cdots & \cdots & \cdots & \cdots \\
a'_{k1}(t) & a'_{k2}(t) & \cdots & a'_{kn}(t) \\
\cdots & \cdots & \cdots & \cdots \\
a_{n1}(t) & a_{n2}(t) & \cdots & a_{nn}(t)
\end{pmatrix}.
$$

So,

$\det A \cdot \operatorname{trace}(A'A^{-1}) = \det A_1 + \det A_2 + \cdots + \det A_n.$

On the other hand, by definition,

$$
\frac{d}{dt}(\det A) = \frac{d}{dt}\left(\sum_{(p_1 \cdots p_n)} (-1)^{\tau(p_1 \cdots p_n)} a_{1p_1}(t) a_{2p_2}(t) \cdots a_{np_n}(t) \right)
$$

$$
= \sum_{p_1 \cdots p_n} (-1)^{\tau(p_1 \cdots p_n)} \left(\sum_{k=1}^{n} a_{1p_1}(t) \cdots a'_{kp_k}(t) \cdots a_{np_n}(t) \right)
$$

$$
= \sum_{k=1}^{n} \left(\sum_{(p_1 \cdots p_n)} (-1)^{\tau(p_1 \cdots p_n)} a_{1p_1}(t) \cdots a'_{kp_k}(t) \cdots a_{np_n}(t) \right)
$$

$$
= \det A_1 + \det A_2 + \cdots + \det A_n.
$$

Hence we have proved that

$$
\frac{d}{dt}(\det A) = \det A \cdot \operatorname{trace}(A' \cdot A^{-1}).
$$

1110

Let V be the vector space of polynomials $p(x) = a + bx + cx^2$ with real coefficients a, b, and c. Define an inner product on V by

$$
(p, q) = \frac{1}{2} \int_{-1}^{1} p(x)q(x)\,dx.
$$

(a) Find an orthonormal basis for V consisting of polynomials $\phi_0(x)$, $\phi_1(x)$, and $\phi_2(x)$, having degree 0, 1, and 2, respectively.

(b) Use the answer to (a) to find the second degree polynomial that solves the minimization problem

$$\min_{p \in V} \int_{-1}^{1} (p(x) - x^3)^2 dx.$$

(Courant Inst.)

Solution. (a) Since $1, x, x^2$ is a base of V, we can orthonormalize $1, x, x^2$ to get an orthonormal basis

$$\phi_0(x) = 1, \quad \phi_1(x) = \sqrt{3}x, \quad \phi_2(x) = -\frac{\sqrt{5}}{2} + \frac{3\sqrt{5}}{2}x^2$$

of V with degree 0, 1 and 2 respectively as usual.
(b) Let

$$p_0(x) = c_0\phi_0(x) + c_1\phi_1(x) + c_2\phi_2(x) \in V \ (c_i \in \mathbb{R})$$

and

$$q_0(x) = x^3 - p_0(x)$$

such that $q_0(x)$ is orthogonal to V, that is

$$(x^3 - p_0(x), \phi_i(x)) = 0 \quad (i = 0, 1, 2).$$

Then for any $p(x) \in V$,

$$\int_{-1}^{1} (p(x) - x^3)^2 dx = 2|x^3 - p(x)|^2$$

$$= 2|x^3 - p_0(x) + p_0(x) - p(x)|^2$$

$$= 2|q_0(x) + p_0(x) - p(x)|^2.$$

Since $(q_0(x), p_0(x) - p(x)) = 0$,

$$\int_{-1}^{1} (p(x) - x^3)^2 dx = 2|q_0(x)|^2 + 2|p_0(x) - p(x)|^2.$$

It follows that

$$\min_{p \in V} \int_{-1}^{1} (p(x) - x^3)^2 dx = 2|q_0(x)|^2.$$

Since $(q_0(x), \phi_i(x)) = 0$ if and only if

$$(x^3, \phi_i(x)) = (p_0(x), \phi_i(x)) = c_i \quad (i = 0, 1, 2).$$

By an easy calculation, we get $c_0 = 0$, $c_1 = \frac{\sqrt{3}}{5}$ and $c_2 = 0$. Hence $p_0(x) = \frac{3}{5}x$ and $q_0(x) = x^3 - \frac{3}{5}x$. Obviously,

$$\int_{-1}^{1} \left(x^3 - \frac{3}{5}x \right)^2 dx = \frac{8}{175}.$$

Thus, when $p(x) = \frac{3}{5}x$,

$$\int_{-1}^{1} (p(x) - x^3)^2 dx$$

is minimal, and

$$\min_{p \in V} \int_{-1}^{1} (p(x) - x^3)^2 dx = \frac{8}{175}.$$

1111

Suppose that A is an $n \times n$ matrix and that

$$x = \begin{pmatrix} x_1 \\ \vdots \\ x_n \end{pmatrix}, \quad y = \begin{pmatrix} y_1 \\ \vdots \\ y_n \end{pmatrix}, \quad y^* = (y_1, \dots, y_n).$$

Suppose that all the entries of A, x, and y are real.

(a) Show that there exist numbers a and b so that

$$\det(A + sxy^*) = a + bs.$$

(b) Show that if $\det(A) \neq 0$ then

$$a = \det(A) \text{ and } b = \det(A) \cdot y^* A^{-1} x.$$

(c) Is it true that $a = 0$ if $\det(A) = 0$?

(*Courant Inst.*)

Solution.
(a) Directly,

$$\det(A + sxy^*) = \det \begin{pmatrix} a_{11} + sx_1y_1 & \cdots & a_{1n} + sx_1y_n \\ \cdots & \cdots & \cdots \\ a_{n1} + sx_ny_1 & \cdots & a_{nn} + sx_ny_n \end{pmatrix}$$

$$= \det \begin{pmatrix} a_{11} & \cdots & a_{1n} \\ \cdots & \cdots & \cdots \\ a_{n1} & \cdots & a_{nn} \end{pmatrix}$$

$$+ s \cdot \left(\det \begin{pmatrix} x_1y_1 & \cdots & x_1y_n \\ a_{21} & \cdots & a_{2n} \\ \cdots & \cdots & \cdots \\ a_{n1} & \cdots & a_{nn} \end{pmatrix} + \det \begin{pmatrix} a_{11} & \cdots & a_{1n} \\ x_2y_1 & \cdots & x_2y_n \\ a_{31} & \cdots & a_{3n} \\ \cdots & \cdots & \cdots \\ a_{n1} & \cdots & a_{nn} \end{pmatrix} \right.$$

$$\left. + \cdots + \det \begin{pmatrix} a_{11} & \cdots & a_{1n} \\ \cdots & \cdots & \cdots \\ a_{n-1,1} & \cdots & a_{n-1,n} \\ a_ny_1 & \cdots & x_ny_n \end{pmatrix} \right).$$

Hence $\det(A + sxy^*) = a + bs$ for some a and b, for any s.

(b) Since for any s, $\det(A + sxy^*) = a + bs$ as in (a), we have $a = \det(A)$ and

$$
b = \det \begin{pmatrix} x_1 y_1 & \cdots & x_1 y_n \\ a_{21} & \cdots & a_{2n} \\ \cdots & \cdots & \cdots \\ a_{n1} & \cdots & a_{nn} \end{pmatrix} + \det \begin{pmatrix} a_{11} & \cdots & a_{1n} \\ x_2 y_1 & \cdots & x_2 y_n \\ a_{31} & \cdots & a_{3n} \\ \cdots & \cdots & \cdots \\ a_{n1} & \cdots & a_{nn} \end{pmatrix}
$$

$$
+ \cdots + \det \begin{pmatrix} a_{11} & \cdots & a_{1n} \\ \cdots & \cdots & \cdots \\ a_{n-11} & \cdots & a_{n-1n} \\ x_n y_1 & \cdots & x_n y_n \end{pmatrix}
$$

$$
= \sum_{j=1}^{n} x_1 y_j A_{1j} + \sum_{j=1}^{n} x_2 y_j A_{2j} + \cdots + \sum_{j=1}^{n} x_n y_j A_{nj}
$$

$$
= \sum_{i=1}^{n} x_i \left(\sum_{j=1}^{n} y_j A_{ij} \right),
$$

where A_{ij} is the algebraic cofactor of a_{ij}. If $\det A \neq 0$,

$$
A^{-1} = \frac{1}{\det(A)} \begin{pmatrix} A_{11} & \cdots & A_{n1} \\ \cdots & \cdots & \cdots \\ A_{1n} & \cdots & A_{nn} \end{pmatrix}.
$$

Thus

$$
b = \sum_{i=1}^{n} x_i \left(\sum_{j=1}^{n} y_j A_{ij} \right)
$$

$$
= y^* \cdot \begin{pmatrix} A_{11} & \cdots & A_{n1} \\ \cdots & \cdots & \cdots \\ A_{1n} & \cdots & A_{nn} \end{pmatrix} \cdot x
$$

$$
= y^* (\det(A) \cdot A^{-1}) x
$$

$$
= \det(A) \cdot y^* A^{-1} x.
$$

(c) If $\det(A) = 0$, $\quad a = \det(A) = 0$.

1112

Let A be an $n \times n$ real matrix with distinct (possibly complex) eigenvalues, $\lambda_1, \lambda_2, \ldots, \lambda_n$, and corresponding eigenvector $v_1, v_2, \ldots, v_n$. Assume that $\lambda_1 = 1$ and that $|\lambda_j| < 1$ for $2 \leq j \leq n$. Prove that $\lim_{n \to \infty} A^n v$ exists. Define $T : \mathbb{C}^n \to \mathbb{C}^n$ by $T(v) = \lim_{n \to \infty} A^n v$. Find the dimensions of the kernel and image of T and give basis for both.

(Courant Inst.)

Solution.

Let $P = (v_1, v_2, \ldots, v_n)$. Since $\lambda_1, \lambda_2, \ldots, \lambda_n$ are distinct, P is an invertible matrix in $M_n(\mathbb{C})$ and $P^{-1}AP = \text{diag}\{\lambda_1, \lambda_2, \ldots, \lambda_n\}$.

For any $v \in \mathbb{C}^n$,

$$\lim_{n \to \infty} A^n v = \lim_{n \to \infty} P \, \text{diag}\{\lambda_1^n, \lambda_2^n, \ldots, \lambda_n^n\} \cdot P^{-1} \cdot v$$

$$= P \cdot \text{diag}\{1, 0, \ldots, 0\} \cdot P^{-1} \cdot v$$

(Since $\lim_{n \to \infty}(X_n Y_n) = \lim_{n \to \infty} X_n \cdot \lim_{n \to \infty} Y_n$.) Let $(e_1, e_2, \ldots, e_n)$ be the standard orthonormal basis of $\mathbb{C}^n$. Then the matrix of T with respect to this basis is $P'^{-1} \text{diag}(1, 0, \ldots, 0) \cdot P'$. Let

$$\begin{pmatrix} f_1 \\ \vdots \\ f_n \end{pmatrix} = P' \cdot \begin{pmatrix} e_1 \\ \vdots \\ e_n \end{pmatrix}.$$

Then $(f_1, f_2, \ldots, f_n)$ is a basis of $\mathbb{C}^n$ and

$$\text{diag}(1, 0, \ldots, 0) = P' \cdot P'^{-1} \cdot \text{diag}(1, 0, \ldots, 0) P' P'^{-1}$$

is the matrix of T with respect to the basis $(f_1, f_2, \ldots, f_n)$. Hence $\{f_1\}$ is a basis of $\text{Im}(T)$, $\{f_2, \ldots, f_n\}$ is a basis of $\ker(T)$, and $\dim(\text{Im}(T)) = 1$, $\dim(\ker(T)) = n - 1$.

1113

Let A be a matrix. Define

$$\sin(A) = A - \frac{1}{3!}A^3 + \frac{1}{5!}A^5 - \cdots .$$

For

$$A = \frac{\pi}{4}\begin{pmatrix} 7 & -3 \\ -3 & 7 \end{pmatrix},$$

express $\sin(A)$ in closed form.

(Courant Inst.)

Solution.

Denote

$$B = \begin{pmatrix} 7 & -3 \\ -3 & 7 \end{pmatrix}.$$

Then B is similar to

$$\begin{pmatrix} 4 & 0 \\ 0 & 10 \end{pmatrix}$$

and

$$\begin{pmatrix} 1 & 1 \\ 1 & -1 \end{pmatrix}\begin{pmatrix} 4 & 0 \\ 0 & 10 \end{pmatrix}\begin{pmatrix} 1 & 1 \\ 1 & -1 \end{pmatrix}^{-1} = \begin{pmatrix} 7 & -3 \\ -3 & 7 \end{pmatrix}.$$

Obviously,

$$\begin{pmatrix} 1 & 1 \\ 1 & -1 \end{pmatrix}^{-1} = \begin{pmatrix} \dfrac{1}{2} & \dfrac{1}{2} \\ \dfrac{1}{2} & -\dfrac{1}{2} \end{pmatrix}.$$

Hence

$$\sin(A) = \begin{pmatrix} 1 & 1 \\ 1 & -1 \end{pmatrix} \cdot \left[\frac{\pi}{4} \cdot \begin{pmatrix} 4 & 0 \\ 0 & 10 \end{pmatrix} - \frac{1}{3!}\left(\frac{\pi}{4}\right)^3 \cdot \begin{pmatrix} 4^3 & 0 \\ 0 & 10^3 \end{pmatrix} \right.$$

$$+ \frac{1}{5!}\left(\frac{\pi}{5}\right)^5 \cdot \begin{pmatrix} 4^5 & 0 \\ 0 & 10^5 \end{pmatrix} - \cdots \Bigg] \begin{pmatrix} 1 & 1 \\ 1 & -1 \end{pmatrix}^{-1}$$

$$= \begin{pmatrix} 1 & 1 \\ 1 & -1 \end{pmatrix} \cdot P \cdot \begin{pmatrix} \frac{1}{2} & \frac{1}{2} \\ \frac{1}{2} & -\frac{1}{2} \end{pmatrix},$$

where

$$P = \begin{pmatrix} \displaystyle\sum_{k=1}^{\infty} (-1)^k \cdot \frac{1}{(2k-1)!} \cdot \pi^{2k-1} & 0 \\ 0 & \displaystyle\sum_{k=1}^{\infty} (-1)^k \cdot \frac{1}{(2k-1)!} \left(\frac{5}{2}\pi\right)^{2k-1} \end{pmatrix}.$$

So,

$$\sin(A) = \begin{pmatrix} 1 & 1 \\ 1 & -1 \end{pmatrix} \cdot \begin{pmatrix} \sin \pi & 0 \\ 0 & \sin\frac{5\pi}{2} \end{pmatrix} \cdot \begin{pmatrix} \frac{1}{2} & \frac{1}{2} \\ \frac{1}{2} & -\frac{1}{2} \end{pmatrix}$$

$$= \begin{pmatrix} 1 & 1 \\ 1 & -1 \end{pmatrix} \cdot \begin{pmatrix} 0 & 0 \\ 0 & 1 \end{pmatrix} \begin{pmatrix} \frac{1}{2} & \frac{1}{2} \\ \frac{1}{2} & -\frac{1}{2} \end{pmatrix}$$

$$= \frac{1}{2} \begin{pmatrix} 1 & -1 \\ -1 & 1 \end{pmatrix}.$$

1114

Let A be the 9×9 tridiagonal matrix

$$A = \begin{pmatrix} -2 & 1 & & & & & \\ 1 & -2 & 1 & & & & \\ & 1 & \ddots & \ddots & & & \\ & & \ddots & \ddots & \ddots & & \\ & & & \ddots & -2 & 1 \\ & & & & 1 & -2 \end{pmatrix}.$$

All entries not shown are zero.

(a) Show that

$$x^T A x = -(x_1^2 + (x_2 - x_1)^2 + \cdots + (x_9 - x_8)^2 + x_9^2).$$

(b) Use part (a) to show that A has all negative eigenvalues.

(c) Let B be the following 10×10 tridiagonal matrix, which aggrees with A except in the first row and column:

$$B = \begin{pmatrix} 1 & -3 & & & & \\ -3 & -2 & 1 & & & \\ & 1 & -2 & 1 & & \\ & & 1 & \ddots & \ddots & \\ & & & \ddots & \ddots & \end{pmatrix}.$$

Let $\lambda_{\max}(B)$ be the largest eigenvalue of B. Show that $\lambda_{\max}(B) > 1$.

(d) Show that B has 9 negative eigenvalues.

(*Courant Inst.*)

Solution.

(a) A direct verification shows that (a) is true.

(b) Since $x^T A x \leq 0$ for all $x \in \mathbb{R}^9$ and $x^T A x = 0$ if and only if $x = 0$, all the eigenvalues of A are negative.

(c) Obviously, the symmetric matrix $\lambda_{\max} I - B$ is semi-definite positive. Then for any $Y \in \mathbb{R}^{10}$, $Y^T(\lambda_{\max} I - B)Y \geq 0$, this is, $Y^T B Y \leq \lambda_{\max} Y^T Y$. Taking $Y^T = (1, -1, 0, \ldots, 0)$, we have $Y^T B Y = 5 \leq \lambda_{\max} \cdot 2$. Thus $\lambda_{\max} > 1$.

(d) Let $V_1 = \{(0, x_1, \ldots, x_9) | x_i \in \mathbb{R}\} \subseteq \mathbb{R}^{10}$. Then for any $Y \in V_1$, $Y^T B Y \leq 0$ and $Y^T B Y = 0$ if and only if $Y = 0$. If B has more than two nonnegative eigenvalues. Let λ_1 and λ_2 be two of them, we have $y_1, y_2 \in \mathbb{R}^{10}$ such that $y_1 \perp y_2, y_1^T y_1 = y_2^T y_2 = 1$ and $B y_i = \lambda_i y_i$ $(i = 1, 2)$. Let $V_2 = \langle y_1, y_2 \rangle$, the subspace spaned by y_1 and y_2. Then for any $y = a_1 y_1 + a_2 y_2 \in V_2$, $y^T B y = a_1^2 \lambda_1 + a_2^2 \cdot \lambda_2 \geq 0$.

Since

$$\dim V_1 + \dim V_2 = 9 + 2 = 11 > 10,$$

there exists $0 \neq y \in V_1 \cap V_2$. Then we have $y^T B y < 0$ ($y \neq 0$, $y \in V_1$) and $y^T B y \geq 0$ ($y \in V_2$), a contradiction. Thus B has 9 negative eigenvalues.

1115

Let T be a real symmetric, positive-definite, $n \times n$ matrix with distinct eigenvalues $\lambda_1 > \lambda_2 > \cdots > \lambda_n$. Show that

$$\lambda_2 = \max_V \min_{x \in V - \{0\}} \frac{|Tx|}{|x|},$$

where V ranges over all two-dimensional subspaces of $\mathbb{R}^n$ and $|x|$ is the Euclidean norm of x.

(Hint. Show $=$ by showing $\leq$ and $\geq$ separately. You may wish to express T in a basis of eigenvectors.)

(Courant Inst.)

Solution.

By assumption, there exists an $n \times n$ orthogonal matrix P such that $P'TP = \mathrm{diag}(\lambda_1, \lambda_2, \ldots, \lambda_n)$. Since for any orthogonal matrix H and any $y \in \mathbb{R}^n$, $|Hy| = |y|$, we have

$$\min_{x \in V - \{0\}} \frac{|Tx|}{|x|} = \min_{x \in V - \{0\}} \frac{|P'^{-1}\mathrm{diag}(\lambda_1,\lambda_2,\ldots,\lambda_n) \cdot P^{-1}x|}{x}$$

$$= \min_{x \in V - \{0\}} \frac{|\mathrm{diag}(\lambda_1,\ldots,\lambda_n) \cdot P^{-1}x|}{|P^{-1}x|}.$$

If V is a two-dimensional subspace, $P^{-1}V$ is also two-dimensional. Hence

$$\max_V \min_{x \in V - \{0\}} \frac{|Tx|}{|x|} = \max_V \min_{x \in V - \{0\}} \frac{|\mathrm{diag}(\lambda_1,\ldots,\lambda_n) \cdot x|}{|x|}$$

when V ranges over all two dimensional subspaces of $\mathbb{R}^n$.

Now let $\{e_1, e_2, \ldots, e_n\}$ be the standard orthonormal basis of $\mathbb{R}^n$, and $V = (e_1, e_2)$. Then

$$
\min_{x \in V - \{0\}} \frac{|\operatorname{diag}(\lambda_1 \ldots, \lambda_n)x|}{|x|}
$$

$$
= \min \left\{ \frac{\sqrt{\lambda_1^2 a_1^2 + \lambda_2^2 a_2^2}}{\sqrt{a_1^2 + a_2^2}} \,\middle|\, 0 \neq (a_1, a_2) \in \mathbb{R}^2 \right\}
$$

$$
= \lambda_2 \ (\lambda_1 > \lambda_2 > 0).
$$

It follows that

$$
\max_{V} \min_{x \in V - \{0\}} \frac{|Tx|}{|x|} \geq \lambda_2.
$$

On the other hand, for any two-dimensional subspace V of $\mathbb{R}^n$, we can find an orthogonal basis $\{f_1, f_2, \ldots, f_n\}$ of $\mathbb{R}^n$ such that $V = (f_1, f_2)$. Let

$$
(f_1, f_2, \ldots, f_n)^T = Q(e_1, e_2, \ldots, e_n)^T.
$$

Then $Q = (q_{ij})_{n \times n}$ is an orthogonal matrix, and if $0 \neq x = a_1 f_1 + a_2 f_2 \in V$.

$$
\frac{|\operatorname{diag}(\lambda_1, \ldots, \lambda_n) \cdot x|}{|x|} = \frac{\sqrt{\displaystyle\sum_{k=1}^{n} (\lambda_1 a_1 q_1 k + \lambda_2 a_2 q_{2k})^2}}{\sqrt{a_1^2 + a_2^2}}
$$

$$
\leq \frac{\sqrt{\displaystyle\sum_{k=1}^{n} \lambda_1^2 a_1^2 q_{1k}^2} + \sqrt{\displaystyle\sum_{k=1}^{n} \lambda_2^2 a_2^2 q_{2k}^2}}{\sqrt{a_1^2 + a_2^2}}
$$

$$
= \frac{\sqrt{\lambda_1^2 a_1^2} + \sqrt{\lambda_2^2 a_2^2}}{\sqrt{a_1^2 + a_2^2}}.
$$

So,

$$\min_{x \in V - \{0\}} \frac{|\mathrm{diag}(\lambda_1, \ldots, \lambda_n)x|}{|x|}$$

$$\leq \min\left\{ \frac{\sqrt{\lambda_1^2 a_1^2} + \sqrt{\lambda_2^2 a_2^2}}{\sqrt{a_1^2 + a_2^2}} \middle| 0 \neq (a_1, a_2) \in \mathbb{R}^2 \right\} \leq \lambda_2,$$

and

$$\max_V \min_{x \in V - \{0\}} \frac{|Tx|}{|x|} \leq \lambda_2.$$

Thus

$$\max_V \min_{x \in V - \{0\}} \frac{|Tx|}{|x|} = \lambda_2$$

when V ranges over all two-dimensional subspace of $\mathbb{R}^n$.

1116

For any $n \times n$ matrix P, consider the sum

$$R(P) = \sum_{k=0}^{\infty} P^k.$$

(a) Prove that if $\sum_{k=0}^{\infty} \|P^k\| < \infty$, then $(I - P)^{-1}$ exist and $R(P) = (I - P)^{-1}$.

(b) Assume that $\|P\| < 1$ in a matrix norm induced by a vector norm. Prove that $\|(I - P)^{-1}\| \leq \frac{1}{1-\|P\|}$.

(c) Use part (a) to compute the inverse of

$$A = \begin{pmatrix} 1 & 1 & 0 & 0 \\ 0 & 1 & 1 & 0 \\ 0 & 0 & 1 & 1 \\ 0 & 0 & 0 & 1 \end{pmatrix}.$$

(*Courant Inst.*)

Solution.

(a) First, we claim that the norms of all eigenvalues of P are < 1. Let $\lambda_1, \lambda_2, \ldots, \lambda_n$ be the n eigenvalues of P and

$$Q^{-1}PQ = \begin{pmatrix} \lambda_1 & & * \\ & \ddots & \\ 0 & & \lambda_n \end{pmatrix}$$

be the Jordan form of P. Denote

$$\phi_m(x) = 1 + x + \cdots + x^m.$$

Then $\phi_m(\lambda_1), \phi_m(\lambda_2), \ldots, \phi_m(\lambda_n)$ are the n eigenvalues of

$$S_m = E + P + P^2 + \cdots + P^m$$

and

$$Q^{-1}S_mQ = \begin{pmatrix} \phi_m(\lambda_1) & & * \\ & \ddots & \\ 0 & & \phi_m(\lambda_n) \end{pmatrix}.$$

Since $\sum_{k=0}^{\infty} \|P^k\| < \infty$ and the norms over vector space are all equivalent,

$$R(P) = \sum_{k=0}^{\infty} P^k = \lim_{m \to \infty} S_m$$

is convergent. Since

$$\lim_{m \to \infty} Q^{-1}S_mQ = Q^{-1} \left(\lim_{m \to \infty} S_m \right) Q,$$

$\lim_{m \to \infty} \phi_m(\lambda_i)$ exist, that is, $\sum_{m=0}^{\infty} \lambda_i^m$ is convergent ($i = 1, 2, \ldots, n$). Hence $|\lambda_i| < 1$ ($1 \leq i \leq n$).

Thus all the eigenvalues of $I - P$ are non-zero, and $(I - P)$ is invertible.

Since

$$S_m(I - P) = I - P^{m+1},$$
$$\lim_{m \to \infty} S_m = \lim_{m \to \infty} (I - P^{m+1})(I - P)^{-1}$$
$$= \left(I - \lim_{m \to \infty} P^{m+1}\right)(I - P)^{-1}$$
$$= (I - 0)(I - P)^{-1}$$
$$= (I - P)^{-1}.$$

Thus

$$R(P) = \sum_{k=0}^{\infty} P^k = (1 - P)^{-1}.$$

(b) By definition,

$$\|P\| = \sup \left\{ \frac{\|Pv\|}{\|v\|} \,\middle|\, v \text{ non-zero vector} \right\}.$$

Since $\|P\| < 1$, for any non-zero vector u, $\|Pu\| \le \|P\| \cdot \|u\| < \|u\|$ for the corresponding vector norm.

Now suppose $(I - P) \cdot u = 0$ for some vector u. Then $\|u\| = \|I \cdot u\| = \|Pu\|$. We must have $u = 0$. The matrix $I - P$ is therefore invertible, and we can write $(I - P)^{-1} = I + P(I - P)^{-1}$. Then we have

$$\|(I - P)^{-1}\| \le \|I\| + \|P\| \cdot \|(I - P)^{-1}\| = 1 + \|P\| \cdot \|(I - P)^{-1}\|.$$

Thus $\|(I - P)^{-1}\| \le \frac{1}{1 - \|P\|}$, since $\|P\| < 1$.

(c) Denote

$$N = \begin{pmatrix} 0 & -1 & 0 & 0 \\ 0 & 0 & -1 & 0 \\ 0 & 0 & 0 & -1 \\ 0 & 0 & 0 & 0 \end{pmatrix}.$$

Then $A = I - N$. So

$$A^{-1} = (I - N)^{-1} = I + N + N^2$$
$$+ N^3 + \cdots = I + N + N^2 + N^3$$

$$= \begin{pmatrix} 1 & -1 & 1 & -1 \\ 0 & 1 & -1 & 1 \\ 0 & 0 & 1 & -1 \\ 0 & 0 & 0 & 1 \end{pmatrix}.$$

1117

Let G be a p-group and $G \times V \to V$ be a linear action of G on a finite vector space over a finite field F_{p^n}. Using Sylow theory, prove that there exist a basis of V in which all the transformation $v \to \sigma \cdot v$ ($\sigma \in G$) are unipotent matrices.

(Columbia)

Solution.

Let

$$G' = \{V \to V, v \mapsto \sigma \cdot v | \sigma \in G\}.$$

Then G' is a p-group, since there is a surjective homomorphism $G \to G'$ ($\leq \operatorname{End}_F(V)$). We fix a base of V over F_{p^n}. Then G' is isomorphic to a subgroup H of $GL(m, F_{p^n})$, where $m = \dim_{F_{p^n}} V$.

It is well known that

$$|GL(m, F_{p^n})| = (p^{nm} - 1)(p^{nm} - p^n) \cdots (p^{nm} - p^{n(m-1)})$$

$$= p^{\frac{nm(m-1)}{2}} \cdot \prod_{j=1}^{m} (p^{jn} - 1)$$

and

$$U = \left\{ \begin{pmatrix} 1 & & & * \\ & 1 & & \\ & & \ddots & \\ 0 & & & 1 \end{pmatrix} \right\}$$

is a Sylow p-subgroup of $GL(m, F_{p^n})$. By Sylow Theorem, there exist some $P \in GL_n(m, F_{p^n})$ such that $PHP^{-1} \subseteq U$. Thus, if we change the base of V via P, the corresponding matrices of the linear transformations in G' are in U, which are unipotent matrices.

1118

Let S be an endomorphism of a finite-dimensional vector space V over a field F whose characteristic polynomial is not equal to its minimal polynomial. Show that there is an endomorphism T of V so that T commutes with S but T is not a polynomial in S. (T is a polynomial in S if $T = a_0 I + a_1 S + a_2 S^2 + \cdots + a_k S^k$ for some k and $a_i \in F$.)

(*Stanford*)

Solution.
Let

$$F[S] = \{f(S) : V \to V | f(\lambda) \in F[\lambda]\}.$$

Then, $\dim_F F[S] = \deg m(\lambda)$, where $m(\lambda)$ is the minimal polynomial of S. Since the minimal polynomial of S is not equal to its characteristic polynomial, $\dim_F F[S] < \dim_F V$.

On the other hand, let $d_1(\lambda), d_2(\lambda), \ldots, d_s(\lambda)$ be the invariant factors $\neq 1$ of S and let $n_i = \deg d_i(\lambda)$, then by Frobenius Theorem, the dimension of the vector space over F of matrices commutative with the matrix of S is

$$N = \sum_{j=1}^{s} (2s - 2j + 1) n_j.$$

Obviously,

$$N \geq \sum_{j=1}^{s} n_j = \dim_F V.$$

So there is a linear transformation T of V such that T commutes with S but T is not a polynomial is S.

1119

Let A be an $n \times n$ real matrix, all of whose (complex) eigenvalues are real and positive. Show that for any integer $m \geq 1$, there exist at least one $n \times n$ real matrix B with $B^m = A$.

(Hint. Make use of the Jordan form $S + N$ of a conjugate of A, where S is diagonal and N is strictly triangular.)

(*Stanford*)

Solution.

Suppose first that

$$A = \begin{pmatrix} a & 1 & & & & 0 \\ 0 & a & 1 & & & \\ & & \ddots & \ddots & & \\ 0 & 0 & \cdots & a & 1 \\ 0 & 0 & \cdots & 0 & a \end{pmatrix}$$

be a Jordan block with a real and positive. For any integer $m \geq 1$, let $b = \sqrt[m]{a}$ and

$$B = \begin{pmatrix} b & 1 & & & & 0 \\ 0 & b & 1 & & & \\ & & \ddots & \ddots & & \\ 0 & 0 & \cdots & b & 1 \\ 0 & 0 & \cdots & 0 & b \end{pmatrix}.$$

It is easy to see that

$$B^m = \begin{pmatrix} b^m & C_m^1 b^{m-1} & C_m^2 b^{m-2} & \cdots & C_m^{n-1} b^{m-n+1} \\ 0 & b^m & C_m^1 b^{m-1} & \cdots & C_m^{n-2} b^{m-n+2} \\ \vdots & & \ddots & \ddots & \vdots \\ \vdots & & & \ddots & C_m^1 b^{m-1} \\ 0 & \cdots & & \cdots & b^m \end{pmatrix}.$$

Obviously, $\lambda E - A$ and $\lambda E - B^m$ have the same invariant divisors $\{1, \ldots, 1, (\lambda - a)^n\}$, A and B^m are similar. Thus

$$A = Q^{-1} B^m Q = (Q^{-1} B Q)^m$$

for some invertible $n \times n$ real matrix Q.

Now let A be an $n \times n$ real matrix, all of whose eigenvalues are real and positive. Then there exists an $n \times n$ invertible real matrix P such that

$$P^{-1} A P = \mathrm{diag}(J_1, J_2, \ldots, J_k)$$

where J_i $(1 \leq i \leq k)$ are Jordan blocks with diagonal elements real and positive. As proved in the above, for any integer $m \geq 1$, any $1 \leq i \leq k$, there exist a real matrix B_i such that $B_i^m = J_i$ $(1 \leq i \leq k)$. Hence

$$P^{-1} A P = (\mathrm{diag}(B_1, B_2, \ldots, B_k))^m$$

and

$$A = \left(P \cdot \begin{pmatrix} B_1 & & & 0 \\ & B_2 & & \\ & & \ddots & \\ 0 & & & B_k \end{pmatrix} P^{-1} \right)^m = B^m,$$

where

$$B = P^{-1} \cdot \mathrm{diag}(B_1, B_2, \ldots, B_k) \cdot P.$$

1120

Let V be a finite-dimensional vector space over an algebraically closed field K, and let $T : V \to V$ be a linear transformation.

(a) Show that there are T-invariant subspaces $V_i \subseteq V$ and elements of $\alpha_i \in K$ such that V is the direct sum of the V_i

and $(T - \alpha_i I) : V_i \to V_i$ is nilpotent. (A transformation N is nilpotent if $N^n = 0$, for some $n \geq 1$.)

(b) Show that there are polynomials $S(T), N(T) \in K[T]$ such that $T = S(T) + N(T)$, $S(T) : V \to V$ is diagonalizable, and $N(T) : V \to V$ is nilpotent.

(Hint. Use the Chinese Remainder Theorem.)

(Stanford)

Solution.

Viewing V as a module over the polynomial ring $K[\lambda]$ via T. Let $(\lambda - \alpha_1)^{e_{11}}, \ldots, (\lambda - \alpha_1)^{e_{1r_1}}; (\lambda - \alpha_2)^{e_{21}}, \ldots, (\lambda - \alpha_2)^{e_{2r_2}}; (\lambda - \alpha_n)^{e_{n1}}, \ldots, (\lambda - \alpha_n)^{e_{nr_n}}$ be the elementary divisors of $V_{K[\lambda]}$, where $a_1, a_2, \ldots, \alpha_n$ are distinct and

$$0 < e_{11} \leq e_{12} \leq \cdots \leq e_{1r_1}; \cdots ; 0 < e_{n1} \leq e_{n2} \leq \cdots \leq e_{nr_n}.$$

By the structure theorem of finitely generated modules over PID, we may write

$$V = K[\lambda]w_{11} \oplus \cdots \oplus K[\lambda]w_{1r_1} \oplus K[\lambda]w_{21} \oplus \cdots \oplus K[\lambda]w_{2r_2}$$
$$\oplus \cdots \oplus K[\lambda]w_{n1} \oplus \cdots \oplus K[\lambda]w_{nr_n},$$

where all the $w_{ij} \in V$ and

$$Ann(w_{ij}) = ((\lambda - \alpha_i)^{e_{ij}}).$$

Denote $V_{ij} = K[\lambda]w_{ij}$. Then

$$V = V_{11} \oplus \cdots \oplus V_{1r_1} \oplus \cdots \oplus V_{n1} \oplus \cdots \oplus V_{nr_n}.$$

The V_{ij}'s are T-invariant subspace and $(T - \alpha_i I) : V_{ij} \to V_{ij}$ is nilpotent. Thus (a) is proved.

Since

$$(\lambda - \alpha_1)^{e_{1r_1}}, (\lambda - \alpha_2)^{e_{2r_2}}, \ldots, (\lambda - \alpha_n)^{e_{nr_n}}$$

are pairwise relatively prime, there exists some $S(\lambda) \in K[\lambda]$ such that

$$S(\lambda) \equiv \alpha_i (\mathrm{mod}(\lambda - \alpha_i)^{e_i r_i}), \quad (1 \le i \le n),$$

by Chinese Remainder Theorem. Then

$$S(T) \cdot w_{ij} = \alpha_i \cdot w_{ij}$$

for all w_{ij} and it is easy to see that $S(T) : V \to V$ is diagonalizable. Let $N(\lambda) = \lambda - S(\lambda)$. Obviously, $T = S(T) + N(T)$ and $\lambda - \alpha_i | N(\lambda)$ for any $1 \le i \le n$. So the minimal polynomial $m(\lambda) | N(\lambda)^h$ for some positive integer h. Thus $N(T) : V \to V$ is nilpotent. Thus (b) is proved.

Section 2

Group Theory

1201

Let G be the group of real 2×2 matrices, of determinant one. Describe the set of conjugacy classes of elements of G.

(Harvard)

Solution.

Let $g \in G$, λ_1 and λ_2 be the eigenvalues of g viewed in $M_2(\mathbb{C})$. Then $\lambda_1 \cdot \lambda_2 = 1$.

(i) If $\lambda_1 = \lambda_2$, then $\lambda_1 = \lambda_2 = 1$ or $\lambda_1 = \lambda_2 = -1$. g is similar to

$$\text{(a) } \begin{pmatrix} 1 & 0 \\ 0 & 1 \end{pmatrix} \text{ or (b) } \begin{pmatrix} -1 & 0 \\ 0 & -1 \end{pmatrix} \text{ or (c) } \begin{pmatrix} 1 & 1 \\ 0 & 1 \end{pmatrix} \text{ or (d) } \begin{pmatrix} -1 & 1 \\ 0 & -1 \end{pmatrix}$$

in $GL_2(\mathbb{R})$.

In case (a), $g = \begin{pmatrix} 1 & 0 \\ 0 & 1 \end{pmatrix}$;

In case (b), $g = \begin{pmatrix} -1 & 0 \\ 0 & -1 \end{pmatrix}$;

In case (c), let A be an invertible 2×2 matrix with real entries such that

$$AgA^{-1} = \begin{pmatrix} 1 & 1 \\ 0 & 1 \end{pmatrix}.$$

33

We take $N = d^{-\frac{1}{2}} \cdot A$ if $d = \det A > 0$, or

$$N = (-d)^{-\frac{1}{2}} \begin{pmatrix} -1 & 0 \\ 0 & 1 \end{pmatrix} A$$

if $d = \det A < 0$, then $N \in G$ and

$$N g N^{-1} = \begin{pmatrix} 1 & 1 \\ 0 & 1 \end{pmatrix}$$

or

$$N g N^{-1} = \begin{pmatrix} 1 & -1 \\ 0 & 1 \end{pmatrix}.$$

Hence g is conjugate to $\begin{pmatrix} 1 & 1 \\ 0 & 1 \end{pmatrix}$ or $\begin{pmatrix} 1 & -1 \\ 0 & 1 \end{pmatrix}$ in G. But $\begin{pmatrix} 1 & 1 \\ 0 & 1 \end{pmatrix}$ is not conjugate to $\begin{pmatrix} 1 & -1 \\ 0 & 1 \end{pmatrix}$ in G, for if

$$A \cdot \begin{pmatrix} 1 & 1 \\ 0 & 1 \end{pmatrix} A^{-1} = \begin{pmatrix} 1 & -1 \\ 0 & 1 \end{pmatrix}$$

where $A = \begin{pmatrix} a_{11} & a_{12} \\ a_{21} & a_{22} \end{pmatrix} \in G$, then we have $a_{21} = 0$, $a_{11} = -a_{22}$ and $\det A = -a_{11}^2 = 1$ which is a contradiction. Thus in case (c), we have two conjugacy classes $\left[\begin{pmatrix} 1 & 1 \\ 0 & 1 \end{pmatrix}\right]$ and $\left[\begin{pmatrix} 1 & -1 \\ 0 & 1 \end{pmatrix}\right]$.

In case (d), g is similar to $\begin{pmatrix} -1 & 1 \\ 0 & -1 \end{pmatrix}$ in $GL_2(\mathbb{R})$. As in case (c), g is similar (conjugate) to $\begin{pmatrix} -1 & 1 \\ 0 & -1 \end{pmatrix}$ or $\begin{pmatrix} -1 & -1 \\ 0 & -1 \end{pmatrix}$ in G and $\begin{pmatrix} -1 & 1 \\ 0 & -1 \end{pmatrix}$ is not conjugate to $\begin{pmatrix} -1 & -1 \\ 0 & -1 \end{pmatrix}$ in G. Thus in case (d), we have two conjugacy classes $\left[\begin{pmatrix} -1 & 1 \\ 0 & -1 \end{pmatrix}\right]$ and $\left[\begin{pmatrix} -1 & -1 \\ 0 & -1 \end{pmatrix}\right]$.

(ii) If $\lambda_1 \neq \lambda_2$ and $\lambda_i \in \mathbb{R}$ $(i = 1, 2)$, then g is similar to $\begin{pmatrix} \lambda_1 & 0 \\ 0 & \lambda_2 \end{pmatrix}$ in $GL_2(\mathbb{R})$.

There exists $M \in GL_2(\mathbb{R})$ such that

$$M g M^{-1} = \begin{pmatrix} \lambda_1 & 0 \\ 0 & \lambda_2 \end{pmatrix}.$$

We take $N = d^{-\frac{1}{2}} \cdot M$ if $d = \det M > 0$ or

$$N = (-d)^{-\frac{1}{2}} \begin{pmatrix} -1 & 0 \\ 0 & 1 \end{pmatrix} M$$

if $\det M = d < 0$. Then $N \in G$ and

$$NgN^{-1} = \begin{pmatrix} \lambda_1 & 0 \\ 0 & \lambda_2 \end{pmatrix}.$$

Thus g is conjugate to $\begin{pmatrix} \lambda_1 & 0 \\ 0 & \lambda_2 \end{pmatrix}$ in G. So in this case, we have the conjugacy classes

$$\left\{ \left[\begin{pmatrix} \lambda & 0 \\ 0 & 1/\lambda \end{pmatrix} \right] \; \middle| \; \lambda \in \mathbb{R} \text{ and } \lambda \neq \pm 1 \right\}.$$

(iii) If $\lambda_1 \neq \lambda_2$ and $\lambda_i \in \mathbb{C}\backslash\mathbb{R}$ $(i = 1, 2)$, we denote $\lambda_1 = \cos\theta + i\sin\theta$ and $\lambda_2 = \cos\theta - i\sin\theta$ (since $\lambda_1 \cdot \lambda_2 = 1$ and trace $(g) = \lambda_1 + \lambda_2 \in \mathbb{R}$), where $|\theta| < \pi$, $\theta \neq 0$. g is similar to $\begin{pmatrix} \cos\theta + i\sin\theta & 0 \\ 0 & \cos\theta - i\sin\theta \end{pmatrix}$ in $GL_2(\mathbb{C})$, and also similar to $\begin{pmatrix} \cos\theta & \sin\theta \\ -\sin\theta & \cos\theta \end{pmatrix}$ in $GL_2(\mathbb{C})$.

As it is well known, that two matrices $A, B \in M_2(\mathbb{R})$ are similar in $M_2(\mathbb{C})$ implies that A and B are similar in $M_2(\mathbb{R})$. So there exists $M \in GL_2(\mathbb{R})$ such that

$$MgM^{-1} = \begin{pmatrix} \cos\theta & \sin\theta \\ -\sin\theta & \cos\theta \end{pmatrix}.$$

As in the above, we conclude that g is conjugate to

$$\begin{pmatrix} \cos\theta & \sin\theta \\ -\sin\theta & \cos\theta \end{pmatrix}$$

or

$$\begin{pmatrix} \cos\theta & -\sin\theta \\ \sin\theta & \cos\theta \end{pmatrix}$$

in G. But

$$\begin{pmatrix} \cos\theta & \sin\theta \\ -\sin\theta & \cos\theta \end{pmatrix}$$

and

$$\begin{pmatrix} \cos\theta & -\sin\theta \\ \sin\theta & \cos\theta \end{pmatrix}$$

are not conjugate in G, for if

$$A = \begin{pmatrix} a_{11} & a_{12} \\ a_{21} & a_{22} \end{pmatrix} \in G$$

such that

$$A\begin{pmatrix} \cos\theta & \sin\theta \\ -\sin\theta & \cos\theta \end{pmatrix} A^{-1} = \begin{pmatrix} \cos\theta & -\sin\theta \\ \sin\theta & \cos\theta \end{pmatrix},$$

then $a_{12} = a_{21}$, $a_{11} = -a_{22}$ and

$$\det A = -(a_{11}^2 + a_{21}^2) = 1,$$

which is a contradiction. So in this case, we have the conjugacy classes

$$\left[\begin{pmatrix} \cos\theta & \sin\theta \\ -\sin\theta & \cos\theta \end{pmatrix}\right]$$

and

$$\left[\begin{pmatrix} \cos\theta & -\sin\theta \\ \sin\theta & \cos\theta \end{pmatrix}\right].$$

To sum up, the set of conjugate classes of elements of G is

$$\left\{\left[\begin{pmatrix} \lambda_1 & 0 \\ 0 & \lambda_2 \end{pmatrix}\right]\,\bigg|\,\lambda_1, \lambda_2 \in \mathbb{R} \text{ and } \lambda_1 \cdot \lambda_2 = 1\right\}$$

$$\cup \left\{\left[\begin{pmatrix} \lambda & \pm 1 \\ 0 & \lambda \end{pmatrix}\right]\,\bigg|\,\lambda = \pm 1\right\}$$

$$\cup \left\{\left[\begin{pmatrix} \cos\theta & \sin\theta \\ -\sin\theta & \cos\theta \end{pmatrix}\right]\,\bigg|\,0 < |\theta| < \pi\right\}.$$

1202

Let $G = GL_n(F_q)$, the group of invertible $n \times n$ matrices over the finite field F_q with $q = p^r$, p a prime, $U = U_n(F_q)$, the subgroup of upper triangular matrices with 1's on the diagonal.

(a) Calculate the orders of G and U. Deduce that U is a Sylow p-subgroup of G.

(b) Deduce that every p-subgroup of G is conjugate to a subgroup of U.

(c) Determine the number of G-conjugates of U.

(d) Show that $g \in G$ has p-power order iff $g = I + N$ with $N^n = 0$.

(e) Show that G contains an element of order $q^n - 1$. (Hint. Make use of F_{q^n}.)

(*Columbia*)

Solution.

(a) When forming a matrix in $GL_n(F_q)$, we may choose the first row in $q^n - 1$ ways (a row of zeros not being allowed), the second row in $q^n - q$ ways (no multiple of the first row being allowed), the third row in $q^n - q^2$ ways (no linear combination of the first two rows being allowed), and so on. Thus we can conclude that the order of $GL_n(F_q)$ is

$$(q^n - 1)(q^n - q)(q^n - q^2)\cdots(q^n - q^{n-1}) = q^{\frac{n(n-1)}{2}} \cdot \prod_{i=1}^{n}(q^i - 1)$$

$$= p^{\frac{rn(n-1)}{2}} \cdot \prod_{i=1}^{n}(p^{ri} - 1).$$

When forming a matrix in $U_n(F_q)$, we may choose the first row in q^{n-1} ways, the second row in q^{n-2} ways: the third rows in q^{n-3} ways and so on. Hence the order of $U_n(F_q)$ is

$$q^{n-1} \cdot q^{n-2} \cdots q = q^{\frac{n(n-1)}{2}} = p^{\frac{rn(n-1)}{2}}.$$

So $U = U_n(F_q)$ is a p-subgroup of $G = GL_n(F_q)$. Since $p \nmid [G:U] = \prod_{i=1}^{n}(p^{ri} - 1)$, U is a Sylow p-subgroup of G.

(b) It follows directly from Sylow Theorem.

(c) By Sylow Theorem, the number of G-conjugates of U is equal to $[G : N_G(U)]$ where $N_G(U)$ is the normalizer of U in G.

Suppose $g = (g_{ij})_{n \times n} \in N_G(U)$, or $gUg^{-1} = U$. Then for any $1 \le i < j \le n$, there exists some

$$u = \begin{pmatrix} 1 & u_{12} & \cdots & \cdots & u_{1n} \\ 0 & 1 & u_{23} & \cdots & u_{2n} \\ 0 & 0 & 1 & \cdots & u_{3n} \\ \cdots & \cdots & \cdots & \cdots & \cdots \\ 0 & 0 & 0 & \cdots & 1 \end{pmatrix} \in U$$

such that $g(I + E_{ij}) = ug$ where E_{ij} is the matrix with a lone 1 in the (i, j) place, 0's elsewhere. Checking the last row of $g(I + E_{ij})$ and ug, we get $g_{ni} = 0$. So $g_{ni} = 0$ if $i < n$. Then checking the $(n-1)$th row, we get $g_{n-1,i} = 0$ if $i < n-1$. By this way, we get $g_{ij} = 0$ if $i > j$. Hence $g \in T = T_n(F_q)$, the set of matrices $(a_{ij}) \in G$ with $a_{ij} = 0$ $(0 \le j < i \le n)$.

Conversely, if $g \in T$, it is easy to see that $gUg^{-1} \subseteq U$ (by matrix multiplication). So $g \in N_G(U)$ and we have $N_G(U) = T$.

We can define a map $\theta: T \to \overbrace{F_q^* \times \cdots \times F_q^*}^{n}$ by mapping a matrix onto its principal diagonal, matrix multiplication shows that θ is a group epimorphism whose kernel is precisely $U = U_n(F_q)$. So the order of T is

$$(q-1)^n \cdot q^{\frac{n(n-1)}{2}} = (p^r - 1)^n \cdot p^{\frac{rn(n-1)}{2}}.$$

Hence the number of G-conjugates of U is $\prod_{i=1}^{n}(q^i - 1)/(q-1)^n$.

(d) Suppose $g \in G$ has p-power order. Then $\langle g \rangle$ is a p-subgroup and $\langle g \rangle$ is conjugate to a subgroup of U by (b). There exists an $h \in G$ such that $hgh^{-1} \in U$. Denote $hgh^{-1} = I + M$ where M is a matrix with zero on and below the diagonal.

Obviously $M^n = 0$ and $g = I + h^{-1}Mh = I + N$ where $N = h^{-l}Mh$ and $N^n = h^{-1}M^n h = 0$.

Conversely, if $g = I + N$ with $N^n = 0$, the Jordan canonical form for N is a matrix with zero on and below the diagonal. Thus g is similar (conjugate) to some matrix in U. So the order of g is p-power.

(e) Consider the finite field F_{q^n} as an extension of the field F_q and as a vector space over F_q. For any $a\ (\neq 0) \in F_{q^n}$, we have a linear map $a_l : F_{q^n} \to F_{q^n}$, $a_l(x) = ax$ (over F_q) which is invertible. We fix a base of F_{q^n} over F_q, then we can obtain a group homomorphism $\sigma : F^*_{q^n} \to G = GL_n(F_q)$, $\sigma(a) = a_l$. Obviously, σ is injective. It is well known that $F^*_{q^n}$ has an element with order $q^n - 1$. It follows that G contains an element of order $q^n - 1$.

1203

(a) Let G be a finite group and H a subgroup. Prove that if $G = \bigcup_{g \in G} gHg^{-1}$, then $H = G$.

(b) Recall that a subgroup M of a group G is said to be maximal if the only subgroups H satisfying $M \subseteq H \subseteq G$ are $H = M$ and $H = G$. Let G be a finite group with the property that all of its maximal subgroups are conjugate. Prove that G is cyclic of prime power order.

(Indiana)

Solution.

(a) Let S be the set of all subgroups of G. Then we have an action of G on the set S defined to be $g \cdot K = gKg^{-1}$ for any $g \in G$ and $K \in S$. Obviously, the stabilizer of H under this action is the normalizer $N(H)$ of H in G, and the length of the orbit defined by H is $[G : N(H)]$.

If $G = \bigcup_{g \in G} gHg^{-1}$, then

$$|G| = \left| \bigcup_{g \in G} gHg^{-1} \right| \leq [G : N(H)] \cdot (|H| - 1) + 1.$$

If H is not normal in G, then $[G : N(H)] > 1$. Hence

$$\begin{aligned}
|G| &\leq [G : N(H)] \cdot |H| - [G : N(H)] + 1 \\
&< [G : N(H)] \cdot |H| \\
&\leq [G : H] \cdot |H| = |G|.
\end{aligned}$$

This is a contradition. So H is normal in G. Hence

$$G = \bigcup_{g \in G} gHg^{-1} = H.$$

(b) First we claim that G is cyclic. Suppose that G is not cyclic. Then for any $a \in G$, $\langle a \rangle < G$. So $\langle a \rangle$ is contained in some maximal subgroup of G (since G is finite). Let H be a maximal subgroup of G. Then we have $G = \bigcup_{g \in G} gHg^{-1}$ by assumption. By (a) $G = H$. This contradicts the maximality of H. Hence G is cyclic. Now, it is easy to see that G is cyclic of prime power order since all of its maximal subgroup are conjugate.

1204

Suppose that H and K are normal subgroups of a finite group G and that $G/H \simeq K$.

(a) Give an example to show that G/K need not be isomorphic to H.

(b) Show that if H is simple, then $G/K \simeq H$.

(Columbia)

Solution.

(a) Let $G = Q_8 = \{\pm 1, \pm i, \pm j, \pm k\}$ be the Hamilton's Quaternion group and $H = \langle i \rangle = \{\pm 1, \pm i\}$, $K = \langle -1 \rangle = \{\pm 1\}$ be the subgroups of G generated by i and -1 respectively. Obviously, we have $H \lhd G$, $K \lhd G$ and G/H is cyclic of order 2, so it is isomorphic to K. But $G/K = \{\bar{1}, \bar{i}, \bar{j}, \bar{k}\} \simeq K_4$ (Klein four group), K is not isomorphic to H.

(b) Let $\{1\} = K_0 \lhd K_1 \lhd \cdots \lhd K_{n-1} \lhd K_n = K$ be a composition series of K. Since $G/H \simeq K$, there exist subgroups H_i such that $H_i \supseteq H$, $H_i/H \simeq K_i$ $(i = 0, 1, \ldots, n)$ and $H_i \lhd H_{i+1}$ $(i = 0, \ldots, n-1)$.

Suppose H is simple, then

$$\{1\} \lhd H \lhd H_1 \lhd \cdots \lhd H_{n-1} \lhd H_n = G$$

is a composition series of G with composition factors H and $H_i/H_{i-1} \simeq K_i/K_{i-1}$ $(i = 1, 2, \ldots, n)$. On the other hand,

$$\{1\} = K_0 \lhd K_1 \lhd \cdots \lhd K_{n-1} \lhd K \lhd G$$

is a normal series of G with factors K_i/K_{i-1} $(i = 1, 2, \ldots, n)$ and G/K. By the Jordan–Hölder Theorem, we have the isomorphism $G/K \simeq H$.

1205

Prove that if G is a finitely generated infinite group then for each positive integer n, G has only finitely many subgroups of index n.

(Hint. Let $H \leq G$ and define $H_G = \bigcap_{g \in G} g^{-1}Hg$. Show that if $H \leq G$ (finite index) then there exists a homomorphism of G/H_G into $S_{[G:H]}$, where S_n denotes the symmetric group on n letters.)

(Columbia)

Solution.

Let H be any subgroup of G such that $[G : H] = n$. Then $G \times G/H \to G/H$, $g \cdot (xH) = (gx) \cdot H$ defines an action of G on G/H, where G/H denotes the set of all left cosets xH $(x \in G)$. The kernel of this action is $H_G = \bigcap_{g \in G} g^{-1}Hg$. Hence the group G/H_G is isomorphic to a subgroup of S_n, the symmetric group on n letters. This induces a homomorphism $G \to S_n$ with H_G as its kernel.

Since G is finitely generated, there are only finite number of homomorphisms from G to S_n. It follows that the set

$$\{H_G | H < G, [G : H] = n\}$$

is finite. Since G/H_G is finite, G/H_G has only finitely many subgroups of index n. Thus G has only finitely many subgroups of index n.

1206

Let

$$G = \left\{ \begin{pmatrix} a & b \\ 0 & a^{-1} \end{pmatrix} \Big| a \in \mathbb{R}^*, b \in \mathbb{R} \right\} \subseteq GL(2, \mathbb{R}).$$

Show that G is solvable but not nilpotent.

(Columbia)

Solution.

Let θ be the map $G \to \mathbb{R}^* \times \mathbb{R}^*$ mapping $\begin{pmatrix} a & b \\ 0 & a^{-1} \end{pmatrix}$ to its diagonal (a, a^{-1}). Matrix multiplication shows that θ is a homomorphism, and

$$\ker \theta = \left\{ \begin{pmatrix} 1 & b \\ 0 & 1 \end{pmatrix} \Big| b \in R \right\}.$$

Obviously, $N = \ker \theta (\simeq (\mathbb{R}, +))$ is an abelian subgroup of G, and $G/N \simeq \mathbb{R}^* \times \mathbb{R}^*$ is also abelian. Hence G is a solvable group.

Suppose

$$\begin{pmatrix} a & b \\ 0 & a^{-1} \end{pmatrix} \in C(G),$$

the center of G. Then for any $\begin{pmatrix} x & y \\ 0 & x^{-1} \end{pmatrix} \in G$,

$$\begin{pmatrix} a & b \\ 0 & a^{-1} \end{pmatrix} \begin{pmatrix} x & y \\ 0 & x^{-1} \end{pmatrix} = \begin{pmatrix} x & y \\ 0 & x^{-1} \end{pmatrix} \begin{pmatrix} a & b \\ 0 & a^{-1} \end{pmatrix}.$$

So $ay + bx^{-1} = xb + ya^{-1}$ for any $x \neq 0$, $y \in \mathbb{R}$. It follows that $b = 0$, $a = \pm 1$. Thus

$$C(G) = \left\{ \begin{pmatrix} 1 & 0 \\ 0 & 1 \end{pmatrix}, \begin{pmatrix} -1 & 0 \\ 0 & -1 \end{pmatrix} \right\}.$$

A direct discussion as above shows that $C(G/C(G)) = \{1\}$. Hence

$$C_1(G) = C_2(G) = \cdots = C_n(G) = \cdots \nleq G$$

where $C_{i+1}(G)/C_i(G) = C(G/C_i(G))$ $(C_0(G) = 1)$. So G is not nilpotent.

1207

Let p be a prime and let V be an n-dimensional vector space over F_p. Let $G = GL_{F_p}(V)$. Recall that $|G| = (p^n - 1)(p^n - p) \cdots (p^n - p^{n-1})$.

(a) Recall that a linear transformation is called semisimple if its minimal polynomial is separable. Prove that a transformation $T \in G$ is semisimple if and only if $T^{p^m - 1} = 1$ for some positive integer m.

(b) Let H be a subgroup of G of order a power of p. Show that H can be simultaneously upper triangularized, that is, there is a basis of V with respect to which all of the elements of H are upper triangular.

(Hint. Find a Sylow p-subgroup of G.)

(*Indiana*)

Solution.

(a) Let $T \in G$ and $f(\lambda)$ be the minimal polynomial of T over F_p. Then $\lambda \nmid f(\lambda)$. If T is semisimple, that is, $f(\lambda)$ is separable (Remark. Here separability means that $f(\lambda)$ has no multiple roots), then $f(\lambda) = f_1(\lambda) f_2(\lambda) \cdots f_k(\lambda)$ for some distinct irreducible polynomials f_i $(1 \leq i \leq k)$. Denote $n_i = \deg f_i(\lambda)$ and $m = n_1 \cdot n_2 \cdots n_k$. Let E be a splitting subfield of $\lambda^{p^m} - \lambda$ over F_p and $E_i = F_p[\lambda]/(f_i(\lambda))$ $(1 \leq i \leq k)$. Then $|E| = p^m$ and E_i

is a field of order p^{n_i}. Since $n_i|m$, E_i is isomorphic to a subfield of E. So E contains an element r_i whose minimal polynomial over F_p is $f_i(x)$. Since $r_i \neq 0$ and $r_i^{p^m-1} = 1$, $f_i(\lambda)|(\lambda^{p^m-1} - 1)$, $(1 \leq i \leq k)$. Hence $f(\lambda) = f_1(\lambda) \cdot f_2(\lambda) \cdots f_k(\lambda)|(\lambda^{p^m-1} - 1)$. Thus $T^{p^m-1} = 1$.

Next suppose $T^{p^m-1} = 1$ for some positive integer m. Then $f(\lambda)|(\lambda^{p^m-1} - 1)$. Since $(\lambda^{p^m-1} - 1)' = -\lambda^{p^m-2} \neq 0$, $f(\lambda)$ has no multiple roots. Thus T is semisimple.

(b) Let $U = U_{F_p}(V)$, the subgroup of upper triangular matrices with 1's on the diagonal. Then

$$|U| = p^{n-1} \cdot p^{n-2} \cdots p = p^{\frac{n(n-1)}{2}}$$

(when forming a matrix in U, we may choose the first row in p^{n-1} ways, the second row in p^{n-2} ways and so on). Since

$$|G| = (p^n - 1)(p^n - p) \cdots (p^n - p^{n-1})$$
$$= p^{\frac{n(n-1)}{2}} \cdot (p^n - 1) \cdot (p^{n-1} - 1) \cdots (p - 1),$$

U is a Sylow p-subgroup of G.

Now let H be a subgroup of G of order a power of p. By Sylow Theorem, H is contained in some Sylow p-subgroup K of G and K is conjugate to U. Hence there exists an element g in G such that $gHg^{-1} \subseteq gKg^{-1} = U$. Thus we have proved that H can be simultaneously upper triangularized.

1208

Let p and q be distinct prime numbers. Let G be a group of order $p^3 \cdot q$ such that its commutator subgroup K is of order q. Let H be a p-Sylow subgroup of G.

(a) Show that H is abelian and $G = HK$.
(b) Show that there are elements $h \in H$ and $k \in K$ such that $hk \neq kh$.
(c) From (b) show that p divides $q - 1$.

(Indiana)

Solution.

(a) Since K is the commutator subgroup of G. K is normal in G and G/K is abelian. So $HK = KH$ is a subgroup of G. Since $|K| = q$, $|H| = p^3$, $H \cap K = \{e\}$ and $|HK| = p^3 \cdot q = |G|$. Hence $G = HK$ and $H \cong H/(e) = H/H \cap K \simeq HK/K \simeq G/K$ is abelian.

(b) Suppose that for any elements $h \in H$ and $k \in K$ holds $hk = kh$. Since K is abelian and H is abelian, $G = HK$ is abelian. This contradicts the fact that the commutator subgroup K of G is of order q. This proves (b).

(c) First we claim that H is not normal in G. Otherwise, for any $h \in H$ and $k \in K$, $hkh^{-1}k^{-1} \in H \cap K = \{e\}$ and so $hk = kh$. By Sylow Theorem, the number of p-Sylow subgroups of G is greater than 1 and divides q. So it is q. Again by Sylow Theorem, $p | q - 1$.

1209

(a) Let p, q be primes, $p > q > 2$. Let G be a group of order pq^2. Show G has a subgroup of order pq.

(b) What can you say if $q = 2$ (and $p > q$)?

(*Indiana*)

Solution.

(a) Let r_p be the number of Sylow p-subgroups of G. Then, by Sylow Theorem, $r_p | q^2$ and $p | r_p - 1$. Since p, q are primes and $p > q > 2$, it is easy to see that $r_p = 1$. So G has only one Sylow p-subgroup H, which is normal in G and of order p. By Cauchy Theorem, G also contains an element of order q. So G has a subgroup K of order q. Thus $H \cdot K$ is a subgroup of order $p \cdot q$ since H is normal in G.

(b) If $q = 2$, G may not contain a subgroup of order $p \cdot q$. For example, A_4, the alternating group of degree 4, is a group of order $3 \cdot 2^2$. A_4 does not have a subgroup of order 6.

1210

Let A be the abelian group on generators e, f, and g, subject to the relations

$$9e + 3f + 6g = 0,$$

$$3e + 3f = 0,$$

$$3e - 3f + 6g = 0.$$

Give a decomposition of A as a direct sum of cyclic groups of prime order or infinite order.

(Stanford)

Solution.

Let F be the free abelian group $\mathbb{Z}e_1 \oplus \mathbb{Z}e_2 \oplus \mathbb{Z}e_3$, K be the subgroup of F generated by $f_1 = 9e_1 + 3e_2 + 6e_3$, $f_2 = 3e_1 + 3e_2$ and $f_3 = 3e_1 - 3e_2 + 6e_3$. Obviously $A \simeq F/K$. Denote

$$M = \begin{pmatrix} 9 & 3 & 6 \\ 3 & 3 & 0 \\ 3 & -3 & 6 \end{pmatrix} \in M_3(\mathbb{Z}).$$

It is easy to get the normal form $\mathrm{diag}\{3, 6, 0\}$ of M in $M_3(\mathbb{Z})$ and to find

$$P = \begin{pmatrix} 1 & 0 & -1 \\ 0 & 0 & 1 \\ 0 & 1 & 1 \end{pmatrix}$$

and

$$Q = \begin{pmatrix} 0 & 1 & 0 \\ 0 & -1 & 1 \\ 1 & -2 & -1 \end{pmatrix}$$

such that $QMP = \mathrm{diag}\{3, 6, 0\}$ (P and Q are invertible matrices in $M_3(\mathbb{Z})$). Let

$$(e_1', e_2', e_3')' = P^{-1}(e_1, e_2, e_3)'$$

and

$$(f_1', f_2', f_3')' = Q \cdot (f_1, f_2, f_3)'.$$

Then

$$F = \mathbb{Z}e_1' \oplus \mathbb{Z}e_2' \oplus \mathbb{Z}e_3'$$

and

$$K = \mathbb{Z}f_1 + \mathbb{Z}f_2 + \mathbb{Z}f_3 = 3\mathbb{Z}e_1' \oplus 6\mathbb{Z}e_2'.$$

So

$$A \cong F/K = \mathbb{Z}e_1' \oplus \mathbb{Z}e_2' \oplus \mathbb{Z}e_3'/3\mathbb{Z}e_1' \oplus 6\mathbb{Z}e_2'$$
$$\simeq \mathbb{Z}/(3) \oplus \mathbb{Z}/(6) \oplus \mathbb{Z}$$
$$\simeq \mathbb{Z}_2 \oplus \mathbb{Z}_3 \oplus \mathbb{Z}_3 \oplus \mathbb{Z}.$$

1211

Let M be an $n \times n$ matrix of integers. Suppose that M is invertible when viewed as a matrix of rational numbers, i.e., that there exists an $n \times n$ matrix N with rational entries so that MN equals the $n \times n$ identity matrix. View M as an endomorphism of $\mathbb{Z}^n$.

(a) Show that $\mathbb{Z}^n/M\mathbb{Z}^n$ is finite.

(b) Show that the order of $\mathbb{Z}^n/M\mathbb{Z}^n$ is equal to the absolute value of the determinant of M.

(*Stanford*)

Solution.

(a) It follows directly from (b). It is also obvious from the facts that the map $g : \mathbb{Z}^n/M\mathbb{Z}^n \to \mathbb{Z}^n/|M|\mathbb{Z}^n$, $\delta(\bar{X}) = \overline{M^*X}$ is injective, where $|M|$ is the determinant of M, M^* is the adjoint of M, and the fact $|\mathbb{Z}^n/|M|\mathbb{Z}^n| = (|\det M|)^n$.

(b) Let $D = \mathrm{diag}\{d_1, d_2, \ldots, d_n\}$ be the normal form of M in $M_n(\mathbb{Z})$, where the $d_i \neq 0$ and $d_i | d_{i+1}$ $(i = 1, 2, \ldots, n-1)$, and $P, Q \in M_n(\mathbb{Z})$ be invertible matrices in $M_n(\mathbb{Z})$ such that $D = QMP$. Obviously

$$\det D = d_1 \cdot d_2 \cdots d_n = \det Q \cdot \det M \cdot \det P = \det M \text{ or } -\det M.$$

Let $\{e_1, e_2, \ldots, e_n\}$ be a base for the free module $\mathbb{Z}^n$. Denote

$$(e'_1, e'_2, \ldots, e'_n)' = P^{-1}(e_1, e_2, \ldots, e_n)'.$$

Then $\{e'_1, e'_2, \ldots, e'_n\}$ is another base of $\mathbb{Z}^n$. Let

$$(f_1, f_2, \ldots, f_n)' = M \cdot (e_1, e_2, \ldots, e_n)' = MP \cdot (e'_1, e'_2, \ldots, e'_n)'.$$

Then $M\mathbb{Z}^n$ is generated by $\{f_1, f_2, \ldots, f_n\}$. Since Q is invertible in $M_n(\mathbb{Z})$, $M\mathbb{Z}^n$ can be generated by $\{f'_1, f'_2, \ldots, f'_n\}$ where

$$(f'_1, f'_2, \ldots, f'_n)' = Q \cdot (f_1, f_2, \ldots, f_n)'.$$

Obviously,

$$\begin{aligned}
(f'_1, f'_2, \ldots, f'_n)' &= Q \cdot (f_1, f_2, \ldots, f_n)' \\
&= QMP \cdot (e'_1, e'_2, \ldots, e'_n)' \\
&= \text{diag}\{d_1, d_2, \ldots, d_n\} \cdot (e'_1, e'_2, \ldots, e'_n)'.
\end{aligned}$$

Hence

$$\begin{aligned}
\mathbb{Z}^n/M\mathbb{Z}^n &= \mathbb{Z}e'_1 \oplus \mathbb{Z}e'_2 \oplus \cdots \oplus \mathbb{Z}e'_n/(d_1 e'_1, d_2 e'_2, \ldots, d_n e'_n) \\
&\simeq \mathbb{Z}/(d_1) \oplus \mathbb{Z}/(d_2) \oplus \cdots \oplus \mathbb{Z}/(d_n).
\end{aligned}$$

It follows that the order of $\mathbb{Z}^n/M\mathbb{Z}^n$ is $|d_1| \cdot |d_2| \cdots |d_n| = |\det M|$.

1212

Let $D = \mathbb{Z}[\xi]$ with $\xi = \frac{-1+\sqrt{-3}}{2}$. Calculate the order of the additive group $G = D^2/K$ where K is the D-submodule of D^2 generated by $(2\xi + 1, \xi - 1)$, $(\xi + 2, \xi - 4)$ and $(21, 21)$. Then express G as a product of cyclic groups.

(Harvard)

Solution.

Since ξ is a root of the irreducible polynomial $x^2 + x + 1$, $D = \mathbb{Z}[\xi] = \mathbb{Z} \oplus \mathbb{Z}\xi$ (as additive group).

Let $\{e_1, e_2\}$ be a base of the free D-module D^2. Then

$$D^2 = De_1 \oplus De_2 = \mathbb{Z}e_1 \oplus \mathbb{Z}\xi e_1 \oplus \mathbb{Z}e_2 \oplus \mathbb{Z}\xi e_2,$$

$\{e_1, \xi e_1, e_2, \xi e_2\}$ is a base of D^2 as $\mathbb{Z}$-module, and $\{(1 + 2\xi)e_1 + (-1+\xi)e_2, (2+\xi)e_1 + (-4+\xi) \cdot e_2, 21 \cdot e_1 + 21 \cdot e_2\}$ is a generating subset of the D-submodule K of D^2. For any $a + b\xi \in D$,

$$(a + b\xi)[(1 + 2\xi)e_1 + (-1 + \xi)e_2]$$
$$= a(e_1 + 2\xi e_1 - e_2 + \xi e_2) + b(-2e_1 - \xi e_1 - e_2 - 2\xi e_2),$$
$$(a + b\xi)[(2 + \xi)e_1 + (-4 + \xi)e_2]$$
$$= a(2e_1 + \xi e_1 - 4e_2 + \xi e_2) + b(-e_1 + \xi e_1 - e_2 - 5\xi e_2),$$
$$(a + b\xi)(21e_1 + 21e_2)$$
$$= a(21 \cdot e_1 + 0 \cdot \xi e_1 + 21 \cdot e_2 + 0 \cdot \xi e_2)$$
$$+ b(0 \cdot e_1 + 21 \cdot \xi e_1 + 0 \cdot e_2 + 21 \cdot \xi e_2).$$

It is easy to see that $\{e_1 + 2\xi e_1 - e_2 + \xi e_2, -2e_1 - \xi e_1 - e_2 - 2\xi e_2, 2e_1 + \xi e_1 - 4e_2 + \xi e_2, -e_1 + \xi e_1 - e_2 - 5\xi e_2, 21e_1 + 21e_2, 21\xi e_1 + 21\xi e_2\}$ is a generating subset of K as $\mathbb{Z}$-module (additive group).

Denote

$$A = \begin{pmatrix} 1 & 2 & -1 & 1 \\ -2 & -1 & -1 & -2 \\ 2 & 1 & -4 & 1 \\ -1 & 1 & -1 & -5 \\ 21 & 0 & 21 & 0 \\ 0 & 21 & 0 & 21 \end{pmatrix}.$$

It is routine to get the normal form

$$N = \begin{pmatrix} 1 & 0 & 0 & 0 \\ 0 & 1 & 0 & 0 \\ 0 & 0 & 3 & 0 \\ 0 & 0 & 0 & 21 \\ 0 & 0 & 0 & 0 \\ 0 & 0 & 0 & 0 \end{pmatrix}$$

for A in $M_{6\times4}(\mathbb{Z})$ and to find invertible matrices $P \in M_6(\mathbb{Z})$ and $Q \in M_4(\mathbb{Z})$ such that $PAQ = N$. It follows that $D^2/K \simeq \mathbb{Z}/(3) \oplus \mathbb{Z}/(21)$ and the order of D^2/K is 63.

1213

Prove that $SL(2,\mathbb{Z})$ is generated by $\begin{pmatrix} 1 & \pm 1 \\ 0 & 1 \end{pmatrix}$ and $\begin{pmatrix} 0 & -1 \\ 1 & 0 \end{pmatrix}$ where $SL(2,\mathbb{Z})$ is the group of 2×2 matrices with integral coefficients and determinant $= 1$.

(Stanford)

Solution.

Suppose that $M = \begin{pmatrix} a & b \\ c & d \end{pmatrix} \in SL(2,\mathbb{Z})$.

If $a = 0$ or $b = 0$, then M has the form $\begin{pmatrix} 0 & 1 \\ -1 & d \end{pmatrix}$, $\begin{pmatrix} 0 & -1 \\ 1 & d \end{pmatrix}$, $\begin{pmatrix} 1 & b \\ 0 & 1 \end{pmatrix}$ and $\begin{pmatrix} -1 & b \\ 0 & -1 \end{pmatrix}$ where $b, d \in \mathbb{Z}$.

If both a and c are nonzero, we claim that $\begin{pmatrix} a & b \\ c & d \end{pmatrix}$ is a product of the matrices in $SL(2,\mathbb{Z})$ with the form

$$\begin{pmatrix} 1 & b' \\ 0 & 1 \end{pmatrix}, \begin{pmatrix} 1 & 0 \\ c' & 1 \end{pmatrix}, \begin{pmatrix} 0 & -1 \\ 1 & d' \end{pmatrix}$$

for some $b', c', d' \in \mathbb{Z}$.

Since $\det M = ad - bc = 1$, a and c are relatively prime. It is easy to know that if $a = 1$,

$$\begin{pmatrix} a & b \\ c & d \end{pmatrix} = \begin{pmatrix} 1 & 0 \\ c & 1 \end{pmatrix} \begin{pmatrix} 1 & b \\ 0 & 1 \end{pmatrix}$$

and if $c = 1$,

$$\begin{pmatrix} a & b \\ c & d \end{pmatrix} = \begin{pmatrix} 1 & a \\ 0 & 1 \end{pmatrix} \begin{pmatrix} 0 & -1 \\ 1 & d \end{pmatrix}.$$

If $a > c > 1$, there exist some q, $a_1 \in \mathbb{Z}$ such that $1 \le a_1 < c$ and $a = qc + a_1$. Hence

$$\begin{pmatrix} a & b \\ c & d \end{pmatrix} = \begin{pmatrix} 1 & q \\ 0 & 1 \end{pmatrix} \begin{pmatrix} a_1 & b' \\ c & d \end{pmatrix}$$

for some $b' \in \mathbb{Z}$.

Similarly, if $c > a > 1$, let $c = ha + c_1$, $1 \le c_1 < a$, then

$$\begin{pmatrix} a & b \\ c & d \end{pmatrix} = \begin{pmatrix} 1 & 0 \\ h & 1 \end{pmatrix} \begin{pmatrix} a & b \\ c_1 & d' \end{pmatrix}$$

for some $d' \in \mathbb{Z}$.

According to the above discussions, it can be derived that $M = \begin{pmatrix} a & b \\ c & d \end{pmatrix}$ is a product of the matrices in $SL(2,\mathbb{Z})$ with the form

$$\begin{pmatrix} 1 & b' \\ 0 & 1 \end{pmatrix}, \begin{pmatrix} 1 & 0 \\ c' & 1 \end{pmatrix}, \begin{pmatrix} 0 & -1 \\ 1 & d' \end{pmatrix}$$

for some $b', c', d' \in \mathbb{Z}$.

So to prove that $SL(2,\mathbb{Z}) = \left(\begin{pmatrix} 1 & \pm 1 \\ 0 & 1 \end{pmatrix}, \begin{pmatrix} 0 & -1 \\ 1 & 0 \end{pmatrix} \right)$, the subgroup generated by $\begin{pmatrix} 1 & \pm 1 \\ 0 & 1 \end{pmatrix}$ and $\begin{pmatrix} 0 & -1 \\ 1 & 0 \end{pmatrix}$, it suffices to prove that all $\begin{pmatrix} 1 & b' \\ 0 & 1 \end{pmatrix}$, $\begin{pmatrix} 1 & 0 \\ c' & 1 \end{pmatrix}$, $\begin{pmatrix} 0 & -1 \\ 1 & d' \end{pmatrix}$, $\begin{pmatrix} 0 & 1 \\ -1 & d' \end{pmatrix}$, $\begin{pmatrix} -1 & b' \\ 0 & -1 \end{pmatrix}$ are in $\left(\begin{pmatrix} 1 & \pm 1 \\ 0 & 1 \end{pmatrix}, \begin{pmatrix} 0 & -1 \\ 1 & 0 \end{pmatrix} \right)$ where $b', c', d' \in \mathbb{Z}$.

It is easy to check this by the property of the elementary matrices. This completes the proof.

1214

Let G be a finite group, K a normal subgroup of G and P a Sylow subgroup of K. If N is the normalizer of P in G (i.e., $N = \{g \in G \mid gPg^{-1} = P\}$ then show that $G = K \cdot N$.

(Indiana)

Solution.

For any $g \in G$, we have $gPg^{-1} \subseteq gKg^{-1} \subseteq K$ since K is normal in G. Hence gPg^{-1} is also a Sylow subgroup of K. By Sylow Theorem, there exists $h \in K$ such that $gPg^{-1} = hPh^{-1}$. Hence $h^{-1}gP(h^{-1}g)^{-1} = P$ and $h^{-1}g \in N$. So we have $g = h \cdot h^{-1}g \in K \cdot N$. Thus we have $G = K \cdot N$.

1215

Let P be a Sylow subgroup of a finite group G. Let $N = \{x \in G \,|\, xPx^{-1} = P\}$. Let H be a subgroup of G, $H \supseteq N$. Prove: If $y \in G$ such that $yHy^{-1} = H$, then $y \in H$.

(*Indiana*)

Solution.

Obviously, P is a Sylow subgroup of H. Since $yPy^{-1} \subseteq yHy^{-1} \subseteq H$, yPy^{-1} is also a Sylow subgroup of H. So there exists an $h \in H$ such that $yPy^{-1} = hPh^{-1}$. Hence $h^{-1}yP \cdot (h^{-1}y)^{-1} = P$ and $h^{-1}y \subseteq N$. So

$$y = h \cdot h^{-1}y \in H \cdot N \subseteq H.$$

1216

Let G be a finite group. The probability of G to be commutative is defined by

$$P(G) = \frac{|\{(a,b) \in G \times G \,|\, ab = ba\}|}{|G \times G|}.$$

(a) If G is not commutative, prove that $P(G) \leq 5/8$.
(b) Give an example of G such that $P(G)$ is exactly $5/8$.

(*Stanford*)

Solution.

(a) Let $C(G)$ be the center of G and $C_G(a) = \{b \in G \,|\, ab = ba\}$ be the centralizer of a in G for any $a \in G$. If G is not commutative, then, obviously, $|G/C(G)| \geq 4$.

Since

$$|\{(a,b) \in G \times G | ab = ba\}| = \sum_{a \in G} |C_G(a)|$$

$$= |C(G)| \cdot |G| + \sum_{a \in G - C(G)} |C_G(a)|,$$

$$P(G) = \frac{|C(G)|}{|G|} + \sum_{a \in G - C(G)} \frac{|C_G(a)|}{|G|^2}$$

$$= \frac{1}{|G|} \left(|C(G)| + \sum_{a \in G - C(G)} \frac{1}{[G : C_G(a)]} \right)$$

$$\le \frac{1}{|G|} \left[|C(G)| + \frac{1}{2} \cdot (|G| - |C(G)|) \right]$$

$$= \frac{1}{2} \cdot \left(1 + \frac{|C(G)|}{|G|} \right)$$

$$\le \frac{1}{2} \cdot \left(1 + \frac{1}{4} \right) = \frac{5}{8}$$

(b) Let G be the dihedral group $D_4 = \{x^i y^j \mid 0 \le i \le 1, 0 \le j \le 3, x^2 = y^4 = 1, xy = y^3 x\}$. Then $|G| = 8$ and $C(G) = \{1, y\}$. For any $a \in G - C(G)$, obviously, $3 \le |C_G(a)| < 8$, and so $|C_G(a)| = 4$. It is easy to see that $P(G) = 5/8$.

Section 3

Ring Theory

1301

(a) Prove the ring $\mathbb{Z}[\sqrt{-2}]$ is Euclidean.

(b) Using (a), find all integer solutions to the equation $y^2 + 2 = x^3$.

(*Harvard*)

Solution.

(a) It is readily known that $\mathbb{Z}[\sqrt{-2}]$ is a subring of $\mathbb{C}$, hence an integral domain. For any $a = m + n\sqrt{-2} \in \mathbb{Z}[\sqrt{-2}]$, we define $\delta(a) = m^2 + 2n^2$. So we have a map $\delta : \mathbb{Z}[\sqrt{-2}]^* \to \mathbb{N}$, $a \mapsto \delta(a)$.

Suppose that $a, b \neq 0 \in \mathbb{Z}[\sqrt{-2}]$. Then $ab^{-1} = \mu + \nu\sqrt{-2}$ where μ and ν are rational numbers. We can find integers u and v such that $|u - \mu| \leq 1/2$ and $|v - \nu| \leq 1/2$. Then

$$a = b(\mu + \nu\sqrt{-2})$$
$$= b[(u + \mu - u) + (v + \nu - v)\sqrt{-2}]$$
$$= bq + r$$

where $q = u + v\sqrt{-2}$ is in $\mathbb{Z}[\sqrt{-2}]$ and

$$r = a - bq = b(\mu - u) + b(\nu - v)\sqrt{-2}.$$

Obviously $r \in \mathbb{Z}[\sqrt{-2}]$ and

$$\delta(r) = \delta(b) \cdot (|\mu - u|^2 + 2|\nu - v|^2)$$

$$\leq \delta(b) \cdot \left(\frac{1}{4} + 2 \cdot \frac{1}{4}\right) < \delta(b).$$

Hence $\mathbb{Z}[\sqrt{-2}]$ is Euclidean.

(b) By (a), $\mathbb{Z}[\sqrt{-2}]$ is a unique factorization domain. $\{1, -1\}$ is the set of units of $\mathbb{Z}[\sqrt{-2}]$. Suppose (x_0, y_0) be an integer solution to the equation $y^2 + 2 = x^3$. In the ring $\mathbb{Z}[\sqrt{-2}]$, we have $(y_0 + \sqrt{-2})(y_0 - \sqrt{-2}) = x_0^3$. Since the integral divisor of $y_0 + \sqrt{-2}$ or $y_0 - \sqrt{-2}$ can only be ± 1, x_0 is not a prime element in $\mathbb{Z}[\sqrt{-2}]$. If $a = m + n\sqrt{-2}$ is an irreducible divisor of x_0, $\bar{a} = m - n\sqrt{-2}$ is also an irreducible divisor of x_0, and if $a | y_0 + \sqrt{-2}$, then $\bar{a} | y_0 - \sqrt{-2}$ from which it follows that $a^3 | y_0 + \sqrt{-2}$, $\bar{a}^3 | y_0 - \sqrt{-2}$. So without loss of generality, $y_0 + \sqrt{-2}$ is of the form $(m + n\sqrt{-2})^3$, $m, n \in \mathbb{Z}$. Hence

$$\begin{cases} y_0 = m^3 - 6mn^2 \\ 1 = 3m^2n - 2n^3 = (3m^2 - 2n^2) \cdot n. \end{cases}$$

Thus it is easy to conclude that the integer solution to the equation $y^2 + 2 = x^3$ are

$$\begin{cases} y = 5 \\ x = 3 \end{cases} \text{ and } \begin{cases} y = -5 \\ x = 3. \end{cases}$$

1302

Let p be a prime. Show that for any element $a \in \mathbb{Z}/p\mathbb{Z}$, there exist $b, c \in \mathbb{Z}/p\mathbb{Z}$ such that $a = b^2 + c^2$.

(Harvard)

Solution.

When $p = 2$, it is trivial.

Assume that $p \neq 2$. Let S be the set $\{b^2 | b \in \mathbb{Z}/p\mathbb{Z}\}$. Then $S = \{\bar{0}, \bar{1}, \bar{2}^2, \ldots, (\overline{\frac{p-1}{2}})^2\}$ has exactly $\frac{p+1}{2}$ elements. On one hand

for any $i \neq j, 0 \leq i \leq \frac{p-1}{2}, 0 \leq j \leq \frac{p-1}{2}, \bar{i}^2 - \bar{j}^2 = \overline{i+j} \cdot \overline{i-j} \neq \bar{0}$
($0 < i+j \leq p-1, 0 < i-j \leq \frac{p-1}{2}$ if $i > j$), $\{\bar{0}, \bar{1}, \bar{2}, \ldots, \overline{(\frac{p-1}{2})^2}\}$
are $\frac{p+1}{2}$ elements in S. On the other hand, for any $\bar{n}^2 \in S$ where
$\frac{p-1}{2} < n < p, 0 < p-n \leq \frac{p-1}{2}$ and $\bar{n}^2 = \overline{n^2} = \overline{(n-p)^2} = \overline{n-p}^2$.
Thus $S = \{\bar{b}^2 | b \in \mathbb{Z}/p\mathbb{Z}\} = \{\bar{0}, \bar{1}, \bar{2}^2, \ldots, \overline{(\frac{p-1}{2})}^2\}$ has exactly $\frac{p+1}{2}$
elements.

Now for any element $a \in \mathbb{Z}/p\mathbb{Z}$, the set $a - S := \{a - b^2 | b \in \mathbb{Z}/p\mathbb{Z}\}$ has exactly $\frac{p+1}{2}$ elements. Since $a - S \subseteq \mathbb{Z}/p\mathbb{Z}$ and $\mathbb{Z}/p\mathbb{Z}$ has only p elements, $(a - S) \cap S \neq \emptyset$. Hence there exist $b, c \in \mathbb{Z}/p\mathbb{Z}$ such that $a - b^2 = c^2$, that is $a = b^2 + c^2$.

1303

For a ring R, R^* denotes its multiplicative group of units, $M_n(R)$ the ring of $n \times n$ matrices over R, and $GL_n(R) = M_n(R)^*$.
   (a) If $a \in R$ is nilpotent ($a^m = 0$ for some m) then $1 - a \in R^*$.
   (b) Let J be a nilpotent ideal of R ($J^m = 0$ for some m).
Show that $GL_n(R) \to GL_n(R/J)$ is surjective, with kernel

$$I + M_n(J) = \{A \in M_n(R) | A \equiv I \bmod M_n(J)\}.$$

   (c) Let $\mathcal{F}_q$ denote the finite field with $q = p^r$ elements (p prime). What is the order of $GL_3(\mathcal{F}_q)$?
   (d) Show that $U = \left\{ \begin{pmatrix} 1 & a & b \\ 0 & 1 & c \\ 0 & 0 & 1 \end{pmatrix} \middle| a, b, c \in \mathcal{F}_q \right\}$ is a p-Sylow
subgroup of $GL_3(\mathcal{F}_q)$.
   (e) Show that $GL_3(\mathcal{F}_q)$ contains an element of order $q^3 - 1$.
   (f) Find the order of $GL_3(\mathbb{Z}/25\mathbb{Z})$, and describe a 5-Sylow subgroup.

(Columbia)

Solution.
   (a) Since $(1 - a)(1 + a + \cdots + a^{m-1}) = (1 + a + \cdots + a^{m-1})(1 - a) = 1 - a^m = 1, 1 - a \in R^*$.

(b) Let $\bar{A} = (\bar{a}_{ij})_{n \times n} \in GL_n(R/J)$ and $\bar{B} = (\bar{b}_{ij})_{n \times n} = \bar{A}^{-1}$, where $A = (a_{ij})_m$ and $B = (b_{ij})_{n \times n} \in M_n(R)$. $\overline{AB} = \overline{BA} = I = \mathrm{diag}(\bar{1}, \bar{1}, \ldots, \bar{1})$ implies that $I - AB \in M_n(J)$ and $I - BA \in M_n(J)$. So there exist integers m_1 and m_2 such that $(I - AB)^{m_1} = (I - BA)^{m_2} = 0$. Then it is easy to find some X and $Y \in M_n(R)$ such that $AX = I$ and $YA = I$. Thus $A \in GL_n(R)$ and $GL_n(R) \to GL_n(R/J)$ is surjective.

Obviously, the kernel of this map is

$$\{A = (a_{ij}) | \bar{A} = \mathrm{diag}\{\bar{1}, \bar{1}, \ldots, \bar{1}\}\}$$
$$= \{A \in M_n(R) | A = I \bmod M_n(J)\}$$
$$= I + M_n(J).$$

(c) By calculating the possibility of the row vectors of A in $GL_3(\mathcal{F}_q)$, it is easy to see that

$$|GL_3(\mathcal{F}_q)| = (q^3 - 1)(q^3 - q)(q^3 - q^2) = q^3(q^3 - 1)(q^2 - 1)(q - 1).$$

(d) $|U| = q^3$. Since U is a subgroup of $GL_3(\mathcal{F}_q)$ with order p^{3r} and

$$|GL_3(\mathcal{F}_q)| = p^{3r}(p^{3r} - 1)(p^{2r} - 1)(p^r - 1),$$

U is a p-Sylow subgroup of $GL_3(\mathcal{F}_q)$.

(e) We consider the field extension $\mathcal{F}_q \subseteq \mathcal{F}_{q^3}$. There exists $a \in \mathcal{F}_{q^3}$ such that $\mathcal{F}_{q^3}^* = (a)$. Obviously, $T_a : \mathcal{F}_{q^3} \to \mathcal{F}_{q^3}$, $x \mapsto ax$ is an invertible linear transformation over $\mathcal{F}_q$. Let $A \in GL_3(\mathcal{F}_q)$ be the matrix of T_a relative to some base for $\mathcal{F}_{q^3}/\mathcal{F}_q$. Since $o(a) = q^3 - 1$, the order of A in $GL_3(\mathcal{F}_q)$ is $q^3 - 1$.

(f) Let $R = \mathbb{Z}/25\mathbb{Z}$ and $J = 5\mathbb{Z}/25\mathbb{Z} < R$. Then $J^2 = 0$ and $\bar{R} = R/J \simeq \mathbb{Z}/5\mathbb{Z}$. By (b), the kernel of the surjective homomorphism

$$GL_3(\mathbb{Z}/25\mathbb{Z}) \to GL_3(\mathbb{Z}/5\mathbb{Z})$$

is $I + M_3(J)$. Obviously

$$|I + M_3(J)| = 5^9$$

and

$$|GL_3(\mathbb{Z}/5\mathbb{Z})| = 5^3(5^3 - 1)(5^2 - 1)(5 - 1) = 2^7 \cdot 3 \cdot 5^3 \cdot 31.$$

So

$$|GL_3(\mathbb{Z}/25\mathbb{Z})| = 2^7 \cdot 3 \cdot 5^{12} \cdot 31.$$

Let

$$P = \{A = (a_{ij})_{3 \times 3} \in GL_3(\mathbb{Z}/25\mathbb{Z}) | \sigma(A) \in U \subseteq GL_3(\mathbb{Z}/5\mathbb{Z})\}$$

where σ is the homomorphism $GL_3(\mathbb{Z}/25\mathbb{Z}) \rightarrow GL_3(\mathbb{Z}/5\mathbb{Z})$. Then $P/\ker \sigma \simeq U$ and $|P| = |\ker \sigma| \cdot |U| = 5^{12}$. Thus P is a 5-Sylow subgroup of $GL_3(\mathbb{Z}/25\mathbb{Z})$.

1304

Describe an infinite set of integral solutions (a, b) of the Pell's equation $x^2 - 2y^2 = 1$ (i.e., $a, b \in \mathbb{Z}$ such that $a^2 - 2b^2 = 1$).

(*Indiana*)

Solution.

Suppose that (a, b) is an integral solution of $x^2 - 2y^2 = 1$. Then a must be odd and b even. Let $a = 2a' + 1$ and $b = 2b'$. Then we have $b'^2 = \frac{a'(a'+1)}{2}$. So $\frac{a'(a'+1)}{2}$ must be a square number. Hence either both $\frac{a'}{2}$ and $a' + 1$ are square numbers or a' and $\frac{a'+1}{2}$ are square numbers.

If $\frac{a'}{2}$ and $a' + 1$ are square numbers, say $a' = 2b_0^2$, $a' + 1 = a_0^2$ for some integers a_0, b_0, then $b'^2 = b_0^2 \cdot a_0^2$, (a_0, b_0) is a solution of $x^2 - 2y^2 = 1$, and $a = 2a_0^2 - 1$, $b = \pm 2a_0 b_0$.

From the above consideration, it is easy to see that if (a_0, b_0) is a solution of $x^2 - 2y^2 = 1$, then $(a_1 = 2a_0^2 - 1, b = 2a_0 b_0)$ is another solution of $x^2 - 2y^2 = 1$.

Obviously, $(3, 2)$ is an integral solution of $x^2 - 2y^2 = 1$. So if we take $a_0 = 3$, $b_0 = 2$ and $a_i = 2a_{i-1}^2 - 1$, $b_i = 2a_{i-1}b_{i-1}$ $(i \geq 1)$, we get an infinite set of integral solutions $\{(a_i, b_i) | i \geq 0\}$ of the Pell's equation $x^2 - 2y^2 = 1$.

Remark. If a' and $\frac{a'+1}{2}$ are square numbers, say $a' = a_0^2$, $a'+1 = 2b_0^2$, then (a_0, b_0) is an integer solution of $x^2 - 2y^2 = -1$ and $a = 2a_0^2 + 1$, $b = \pm 2a_0 b_0$. Conversely, if (a_0, b_0) is a solution of $x^2 - 2y^2 = -1$, then $(a = 2a_0^2 + 1, b = \pm 2a_0 b_0)$ is an solution $x^2 - 2y^2 = 1$. For example, from the solution $(7, 5)$ of $x^2 - 2y^2 = -1$, we get a solution $(99, 70)$ of $x^2 - 2y^2 = 1$.

1305

Let p be a prime. Let R be a commutative ring in which $a^p = a$ for all $a \in R$.

(a) If R is an integral domain, prove that R is isomorphic (as a ring) to $\mathbb{Z}/p\mathbb{Z}$.

(b) In the general case (i.e., R not necessarily an integral domain), prove that R is isomorphic (as a ring) to a subring of a direct product (not necessarily finite) of rings each of which is isomorphic to $\mathbb{Z}/p\mathbb{Z}$.

(*Indiana*)

Solution.

(a) Suppose that R is an integral domain. For any $a\ (\neq 0) \in R$, since $a^p - a = a\ (a^{p-1} - 1) = 0$, $a^{p-1} = 1$. Hence R is a field, and for any $a \in R$, a is a root of the polynomial $x^p - x = 0$. It follows that R has at most p elements. Since p is a prime, $\mathrm{Char}(R) = p$ and $R \simeq \mathbb{Z}/p\mathbb{Z}$.

(b) Since for any $a \in R$, $a^p = a$, the nil radical of R is 0, that is, the intersection of all prime ideals of R is 0. Hence

$$R \hookrightarrow \prod_{\text{all prime } P} R/P.$$

For any prime ideal P of R, R/P is an integral domain and for any $\bar{a} \in R/P$, $\bar{a}^p = \overline{a^p} = \bar{a}$. By (a), $R/P \simeq \mathbb{Z}/p\mathbb{Z}$. Thus we have proved that R is isomorphic to a subring of a direct product of rings, each of which is isomorphic to $\mathbb{Z}/p\mathbb{Z}$.

1306

Let R be a commutative ring with 1. If R satisfies the a.c.c. (ascending chain condition) for finitely generated ideals then show that R satisfies the a.c.c. for all ideals. Give example (without proof) of such a ring which is an integral domain but not a p.i.d.

(*Indiana*)

Solution.

If R satisfies the a.c.c. for finitely generated ideals, then for any ideal I of R, I is finitely generated. Otherwise, we can construct a strictly ascending chain of finitely generated subideals of I. Hence R is a Noetherian ring, i.e., R satisfies the a.c.c. for all ideals.

$\mathbb{Z}[x]$, the polynomial ring over $\mathbb{Z}$, satisfies the a.c.c. for all ideals, but not a p.i.d.

1307

Let R be a commutative ring. Let A be an ideal of R.

(a) Show that $S = \{1 + a | a \in A\}$ is a multiplicatively closed subset of R.

(b) Show there is a one-to-one correspondence between the prime ideals of $S^{-1}R$ and those prime ideals P of R such that $P + A \neq R$.

(c) If A is contained in every maximal ideal of R, what is $S^{-1}R$?

(d) If A is a maximal ideal of R, what can you say about the structure of $S^{-1}R$?

(*Indiana*)

Solution.

(a) Obviously, $1 = 1 + 0 \in S$. For any $1 + a$ and $1 + b$, $a, b \in A$,

$$(1 + a)(1 + b) = 1 + (a + b + ab) \in S.$$

It follows that S is a multiplicatively closed subset of R.

(b) It is well known that there is a one-to-one correspondence between the prime ideas of $S^{-1}R$ and those prime ideals P of R such that $P \cap S = \emptyset$. Obviously, $P \cap S \neq \emptyset$ if and only if $P + A = R$ here. This is what we need to prove (b).

(c) If A is contained in every maximal ideal of R, then $A \subseteq J(R)$, the Jacobson radical of R. So every elements in S is invertible in R. Hence $S^{-1}R = R$.

(d) If A is a maximal ideal of R, then $S^{-1}A$ is the unique maximal ideal of $S^{-1}R$ by (b). So $S^{-1}R$ is a local ring, and it is easy to see that $S^{-1}R$ is isomorphic to R_A, the localization of R at the maximal ideal A.

1308

Let I be a nilpotent ideal in a ring R, let M and N be R-modules, and let $f : M \to N$ be an R-homomorphism. Show that if the induced map

$$\bar{f} : M/IM \to N/IN$$

is surjective, then f is surjective.

(*Harvard*)

Solution.

Since

$$\bar{f} : M/IM \to N/IN$$

is surjective, $f(M) + IN = N$. It follows that

$$I \cdot N/f(M) = IN + f(M)/f(M) = N/f(M).$$

Hence

$$N/f(M) = I \cdot N/f(M) = I^2 \cdot N/f(M) = \cdots =$$
$$I^n \cdot N/f(M) = \cdots = 0,$$

because I is nilpotent. So $N = f(M)$ and f is surjective.

1309

Let F be a finite field containing 5 elements. Let t be transcendental over F. Explicitly construct one non-archimedean absolute value $||$ on $F(t)$. If $\overline{F(t)}$ is the completion of $F(t)$ with respect to $||$, show that the set of $\alpha(t) \in \overline{F(t)}$ satisfying $|\alpha(t)| \leq 1$ is a local ring.

(Columbia)

Solution.

Let $p = p(t) = t - 1 \in F[t]$. We define $V_p : F(t) \to \mathbb{R}$ by $V_p(0) = \infty$ and $V_p(f(t)) = k$ if $0 \neq f(t) = p(t)^k \cdot b(t)/c(t)$, where $b(t), c(t) \in F[t]$ and

$$(b(t), p(t)) = 1 = (c(t), p(t)).$$

Obviously, we have

(i) $V_p(f(t)) = \infty$ if and only if $f(t) = 0$,

(ii) $V_p(f(t) \cdot g(t)) = V_p(f(t)) + V_p(g(t))$ and

(iii) $V_p(f(t) + g(t)) \geq \min(V_p(f(t)), V_p(g(t)))$.

Then, by defining $|f(t)|_p = 2^{-V_p(f(t))}$, we get an absolute value $||_p$ (called p-adic absolute value) on $F(t)$. Obviously, $||_p$ is non-archimedean $(|f(t) + g(t)|_p \leq \max\{|f(t)|_p, |g(t)|_p\})$.

Suppose $\overline{F(t)}$ is the completion with respect to $||_p$. Let $R = \{\alpha(t) | \alpha(t) \in \overline{F(t)}, |\alpha(t)|_p \leq 1\}$ and $\mathcal{M} = \{\alpha(t) \in R \mid |\alpha(t)| < 1\}$. Since

$$|\alpha(t) + \beta(t)|_p \leq \max\{|\alpha(t)|_p, |\beta(t)|_p\}$$

and

$$|\alpha(t)\beta(t)|_p = |\alpha(t)|_p |\beta(t)|_p,$$

R is a subring of the field $\overline{F(t)}$ and $\mathcal{M}$ is an ideal of R.

For any $\alpha(t) \in R \backslash \mathcal{M}$, we have $|\alpha(t)|_p = 1$ and $|\alpha(t)^{-1}|_p = 1$ where $\alpha(t)^{-1} \in \overline{F(t)}$ is the inverse of $\alpha(t)$. So, $\alpha(t)^{-1} \in R$ and

$\alpha(t)$ is invertible in R. Thus R is a local ring with maximal ideal $\mathcal{M}$.

1310

Prove that the ideal generated by $X_1, \ldots, X_n$ in the polynomial ring $\mathbb{C}[X_1, \ldots, X_n]$ cannot be generated by fewer than n elements.

(*Stanford*)

Solution.
Suppose that $\{Y_1, Y_2, \ldots, Y_m\}$ is another generating subset of the ideal $\langle X_1, X_2, \ldots, X_n \rangle$ of the ring $\mathbb{C}[X_1, X_2, \ldots, X_n]$. We have to show that $m \geq n$.

We can write

$$X_i = a_{i1}Y_1 + \cdots + a_{im}Y_m + f_i$$

for any X_i where $a_{i1}, a_{i2}, \ldots, a_{im} \in \mathbb{C}$, and f_i is a sum of monomials in $Y_1, \ldots, Y_m$ of degree two or greater. In the same way, we also have

$$Y_j = b_{j1}X_1 + b_{j2}X_2 + \cdots + b_{jn}X_n + g_j$$

for $j = 1, 2, \ldots, m$, where $b_{j1}, \ldots, b_{jn} \in \mathbb{C}$, and g_j is a sum of monomials in $X_1, X_2, \ldots, X_n$ of degree two or greater.

So

$$X_i = \left(\sum_{j=1}^{m} a_{ij}Y_j \right) + f_i = \sum_{j=1}^{m} a_{ij} \left(\sum_{k=1}^{n} b_{jk}X_k + g_k \right) + f_i$$

$$= \sum_{k=1}^{n} \left(\sum_{j=1}^{m} a_{ij}b_{jk} \right) X_k + (\text{terms in } X_1, \ldots, X_n \text{ of degree} \geq 2)$$

$(i = 1, 2, \ldots, n)$.

Since $X_1, X_2, \ldots, X_n$ are algebraically independent over $\mathbb{C}$,

$$\sum_{j=1}^{m} a_{ij}b_{jk} = \delta_{ik} \quad (i, k = 1, 2, \ldots, n).$$

It follows that

$$\begin{pmatrix} a_{11} & a_{12} & \cdots & a_{1m} \\ a_{21} & a_{22} & \cdots & a_{2m} \\ \cdots & \cdots & \cdots & \cdots \\ a_{n1} & a_{n2} & \cdots & a_{nm} \end{pmatrix} \begin{pmatrix} b_{11} & b_{12} & \cdots & b_{1n} \\ b_{21} & b_{22} & \cdots & b_{2n} \\ \cdots & \cdots & \cdots & \cdots \\ b_{m1} & b_{m2} & \cdots & b_{mn} \end{pmatrix} = I_{n \times n}.$$

Hence $m \geq n$.

1311

Let A be a commutative ring, and let I and J be two (proper) ideals such that every prime ideal of A contains either I or J but no prime ideal contains both I and J. Prove that

$$A \simeq A_1 \times A_2$$

for some (nontrivial) rings A_1 and A_2.

(*Stanford*)

Solution.

Since every prime ideal of A contains either I or J but no prime ideal contains both I and J, we have $IJ \subseteq N(A)$ (the nil radical of A) and $I + J = A$. There exist $a \in I$, $b \in J$ such that $a + b = 1$, and since $ab \in IJ \subseteq N(A)$, there exists an integer n such that $(ab)^n = a^n \cdot b^n = 0$. Let $I_1 = \langle a^n \rangle$ and $I_2 = \langle b^n \rangle$. Then I_1 and I_2 are proper and $I_1 + I_2 = A$, since

$$1 = (a + b)^{2n} \in \langle a^n \rangle + \langle b^n \rangle = I_1 + I_2.$$

By Chinese Remainder Theorem, we have

$$I_1 \cap I_2 = I_1 \cdot I_2 = \langle a^n \rangle \cdot \langle b^n \rangle = \langle a^n b^n \rangle = 0$$

and thus $A \simeq A/I_1 \times A/I_2$ where A/I_1 and A/I_2 nontrivial.

1312

Define $f : [0, 1) \to [0, 1)$ by

$$f(x) = \begin{cases} 2x, & \text{if } 0 \le 2x < 1, \\ 2x - 1, & \text{if } 1 \le 2x < 2. \end{cases}$$

Find all x such that

$$f(f(f(f(f(f(x)))))) = x.$$

(Indiana)

Solution.

For any real number a, $[a]$ denotes the largest integer $\le a$ (Gauss function). We denote $\{a\} = a - [a]$ here. Then it is clear that

$$f(x) = 2x - [2x] = \{2x\}$$

for any $x \in [0, 1)$. Hence

$$f(f(f(f(f(f(x))\cdots) = \{2 \cdot \{2 \cdot \{2 \cdot \{2 \cdot \{2 \cdot \{2x\}\cdots\}.$$

Since for any real number a and positive integer n, we have

$$\{n \cdot \{a\}\} = n\{a\} - [n \cdot \{a\}] = n(a - [a]) - [n(a - [a])]$$

$$= na - n \cdot [a] - [na - n[a]] = na - n[a] - ([na] - n[a])$$

$$= na - [na] = \{na\},$$

$$f(f(f(f(f(f(x)))))) = \{2^6 x\} = \{64x\}.$$

Notice that $x = \{64x\} = 64x - [64 \cdot x]$ if and only if $63x = [64x]$. Since $x \in [0, 1)$, $63x = [64x] = [63x + x]$ (to be an integer) if and only if

$$x \in \left\{0, \frac{1}{63}, \frac{2}{63}, \dots, \frac{62}{63}\right\}.$$

Thus $f(f(f(f(f(f(x))\cdots) = x$ if and only if $x = \frac{k}{63}$, $0 \le k \le 62$.

1313

Let T be the ring of all real trigonometric polynomials

$$f(x) = a_0 + \sum_{n=1}^{N} a_n \cos nx + b_n \sin nx.$$

Define $\deg f(x) = N$ where a_N or $b_N \neq 0$. Show that $\deg f \cdot g = \deg f + \deg g$. Use this result to prove that T is an integral domain which is not a unique factorization domain.

(*Columbia*)

Solution.

By using the orthogonality of the set $\{1, \cos nx, \sin nx | n \in \mathbb{N}\}$, it is easy to see that

$$f(x) = a_0 + \sum_{n=1}^{N} a_n \cos nx + b_n \sin nx \equiv 0$$

if and only if all the coefficients of $f(x)$, i.e., $a_0, a_1, \ldots, a_N$ and $b_0, b_1, \ldots, b_N$ are zero.

Let

$$f(x) = a_0 + \sum_{n=1}^{N} a_n \cos nx + b_n \sin nx$$

and

$$g(x) = c_0 + \sum_{m=1}^{M} c_m \cos mx + d_m \sin mx.$$

Suppose that $\deg f(x) = N$ and $\deg g(x) = M$. Since

$$(a_n \cos nx + b_n \sin nx) \cdot (c_m \cos mx + d_m \sin mx)$$

$$= \frac{a_n c_m - b_n d_m}{2} \cos(n+m)x + \frac{a_n d_m + b_n c_m}{2} \sin(n+m)x$$

$$+ \frac{a_n c_m + b_n d_m}{2} \cos(n-m)x + \frac{-a_n d_m + b_n c_m}{2} \sin(n-m)x,$$

$$f(x) \cdot g(x)$$

$$= \sum_{k=0}^{N+M} \left[\left(\sum_{n+m=k} \frac{a_n c_m - b_n d_m}{2} + \sum_{|n-m|=k} \frac{a_n c_m + b_n d_m}{2} \right) \cos kx \right.$$

$$+ \left(\sum_{n+m=k} \frac{a_n d_m + b_n c_m}{2} + \sum_{n-m=k} \frac{-a_n d_m + b_n c_m}{2} \right.$$

$$\left. \left. + \sum_{m-n=k} \frac{a_n d_m - b_n c_m}{2} \right) \sin kx \right].$$

The coefficients of $\cos(N+M)x$ and $\sin(N+M)x$ in $f(x) \cdot g(x)$ are $\frac{a_N c_M - b_N d_M}{2}$ and $\frac{a_N d_M + b_N c_M}{2}$ respectively. If both of them are zero, then

$$a_N \cdot c_M \cdot d_M = b_N \cdot d_M^2 = -b_N c_M^2,$$

and so $b_N(c_M^2 + d_M^2) = 0$. Since $c_M^2 + d_M^2 \neq 0$, $b_N = 0$. Then we have $a_N c_M = a_N d_M = 0$. Hence $a_N = 0$, which is contrary to a_N or $b_N \neq 0$.

Thus we have proved that

$$\deg(f(x) \cdot g(x)) = N + M = \deg f(x) + \deg g(x).$$

Now $f(x) \cdot g(x) = 0$ can happen only if either $f(x) = 0$ or $g(x) = 0$ by the above degree relation. So T is an integral domain.

If $f(x) \cdot g(x) = 1$, then the degree relation implies that $\deg f(x) = 0 = \deg g(x)$. Hence the units of T are $\{\pm 1\}$.

Obviously, in T, we have

$$\cos 2x = (\cos x + \sin x)(\cos x - \sin x)$$

$$= (1 + \sqrt{2}\sin x)(1 - \sqrt{2}\sin x).$$

Again by the degree relation, all the factors $\cos x \pm \sin x$, $1 \pm \sqrt{2}\sin x$ are irreducible, and $\cos x \pm \sin x$ and $1 \pm \sqrt{2}\sin x$ are not associates. Thus $\cos 2x$ in T does not have an essentially unique factorization into irreducible elements. Hence T is not a unique factorization domain.

1314

Let A be a Noetherian integral domain integrally closed in its field of fractions K. Let L be a finite separable field extension of K. If B is the integral closure of A in L, outline a proof that B is Noetherian.

(*Stanford*)

Solution.
First, we claim that the trace function $\mathrm{Tr}_{L/K} \neq 0$. Let $p = \mathrm{Char}(K)$ and $n = [L : K]$. If $p = 0$ or $(p, n) = 1$, it is easy to see that $\mathrm{Tr}_{L/K} \neq 0$ ($\mathrm{Tr}_{L/K}(b) = nb$ for any $b \in K$). Now, suppose that $p \neq 0$ and $p|n$. Since $L \supseteq K$ is a finite separable extension, $L = K[\alpha]$ for some $\alpha \in L$. Denote $\alpha = \alpha_1, \alpha_2, \ldots, \alpha_n$ be the set of the conjugates of α. For any $f(\alpha) \in L$, $\mathrm{Tr}_{L/K}(f(\alpha)) = \sum_{i=1}^n f(\alpha_i)$. Let $x^n - c_1 x^{n-1} + c_2 x^{n-2} - \cdots + (-1)^n \cdot c_n$ be the minimal polynomial of α over K and j be the minimal positive integer such that $p \nmid j$ and $c_j \neq 0$. Then by using Newton's identities on the elementary symmetric polynomials, we get

$$\mathrm{Tr}_{L/K}(\alpha^j) = (-1)^{n+j+1} \cdot j \cdot c_j \neq 0.$$

Thus $\mathrm{Tr}_{L/K} \neq 0$.

Let $u_1, u_2, \ldots, u_n$ be a basis of L over K contained in B. The bilinear function $(x, y) \to \mathrm{Tr}_{L/K}(xy)$ is non-degenerate since $\mathrm{Tr}_{L/K} \neq 0$. Let $v_1, v_2, \ldots, v_n$ be the dual basis of $u_1, u_2, \ldots, u_n$ (the elements in L satisfying $\mathrm{Tr}_{L/K}(u_i v_j) = \delta_{ij}$ for all i, j). Then, for any $x \in B$, x has the form $k_1 v_1 + k_2 v_2 + \cdots + k_n v_n$ ($k_i \in B$). Since $xu_i \in B$, $k_i = \mathrm{Tr}_{L/K}(xu_i) \in A$ for any i. Hence $B \subseteq Av_1 + Av_2 + \cdots + Av_n$. Since A is Noetherian, B is Noetherian.

1315

If A is a commutative ring and $A[[x]]$ is the ring of formal power series over A, show that if $f = \sum_{n=0}^{\infty} a_n x^n$ is nilpotent, then all the elements $a_i \in A$ are nilpotent. If A is Noetherian, prove the converse, i.e., that if all the elements a_i are nilpotent, then f is nilpotent.

(Stanford)

Solution.
Suppose that $f = \sum_{n=0}^{\infty} a_n x^n$ is nilpotent. It is easy to see that $f^m = 0$ implies that $a_0^m = 0$. So, first we get that a_0 is nilpotent. Assume that $a_0, a_1, \ldots, a_k$ is nilpotent. Since A is commutative, $a_0 + a_1 x + \cdots + a_k x^k$ is nilpotent and

$$f - (a_0 + a_1 x + \cdots + a_k x^k)$$

$$= \sum_{n=k+1}^{\infty} a_n x^n = x^{k+1} \left(\sum_{n=k+1}^{\infty} a_n x^{n-k-1} \right)$$

is nilpotent. Hence $\sum_{n=k+1}^{\infty} a_n x^{n-k+1}$ is nilpotent and so a_{k+1} is nilpotent. By induction, all the elements a_i are nilpotent.

Conversely, suppose that A is Noetherian and all the elements a_i are nilpotent. Let I be the ideal generated by $\{a_i | 0 \leq i < \infty\}$. Since A is commutative Noetherian, I is finitely generated and nilpotent. If $I^m = 0$, that is, $b_1 b_2 \cdots b_m = 0$ for any $b_1, b_2, \ldots, b_m \in I$, then it is easy to see that $f^m = 0$. Hence f is nilpotent.

1316

Let F be a field, and for $n \geq 1$ let R_n be the subring of the ring of polynomials $F[x]$ consisting of polynomials

$$f_0 + f_1 x + f_2 x^2 + \cdots + f_k x^k \in F[x]$$

such that $f_1 = f_2 = \cdots = f_n = 0$.

(a) Show that R_n is a Noetherian ring.

(b) Show that the field of fractions of R_n is $F(x)$, and determine the integral closure of R_n in its field of fractions.

(Stanford)

Solution.

(a) Since $F[x]$ is Noetherian and

$$F[x] = R_n + R_n x + R_n x^2 + \cdots + R_n x^n$$

is finitely generated as R_n-module, R_n is a Noetherian ring by Artin–Tate Lemma (or Eakin Theorem).

(b) Observe that

$$\frac{a(x)}{b(x)} = \frac{x^{n+1} a(x)}{x^{n+1} b(x)}$$

for any $a(x)/b(x) \in F(x)$. It is clear that the field of fractions of R_n is $F(x)$.

Obviously, x is integral over R_n, since x is a root of $X^{n+1} - x^{n+1} \in R_n[X]$. Hence $F[x]$ is integral over R_n. On the other hand, if $a(x)/b(x) \in F(x)$ is integral over R_n, $a(x)/b(x)$ is integral over $F[x]$. Since $F[x]$ is an integrally closed domain, $a(x)/b(x) \in F[x]$. Hence $F[x]$ is the integral closure of R_n in its field of fractions.

1317

Describe all subrings of $\mathbb{Q}$. (A subring contains 1 by definition.)

(Stanford)

Solution.

Let R be a subring of $\mathbb{Q}$. Then $R \supseteq \mathbb{Z}$ by definition. Let

$$S = \left\{ 0 \neq p \in \mathbb{Z} \,\middle|\, \frac{1}{p} \in R \right\}.$$

Obviously, S is a multiplicatively closed subset of $\mathbb{Z}$ containing 1. Let $\mathbb{Z}_S$ be the localization of $\mathbb{Z}$ at the multiplicatively closed subset S. For any $q/p \in \mathbb{Z}_S$ ($q \in \mathbb{Z}, p \in S$), $\frac{q}{p} = q \cdot \frac{1}{p} \in R$. Hence $\mathbb{Z}_S \subseteq R$.

On the other hand, for any $q/p \in R$, we may assume that $(p, q) = 1$ and $pl + qk = 1$ for some $l, k \in \mathbb{Z}$. Then

$$\frac{1}{p} = \frac{pl + qk}{p} = l + k \cdot \frac{q}{p} \in R.$$

So $p \in S$ and $q/p \in \mathbb{Z}_S$. Hence $R \subseteq \mathbb{Z}_S$. It follows that $R = \mathbb{Z}_S$ where

$$S = \left\{ 0 \neq p \in \mathbb{Z} \middle| \frac{1}{p} \in R \right\}.$$

Thus

$$\{\mathbb{Z}_S | S \text{ is a multiplicatively closed subset of } \mathbb{Z}\}$$

is the set of all subrings of $\mathbb{Q}$ (S can be choosed as the complement in $\mathbb{Z}$ of the union of some prime ideals of $\mathbb{Z}$).

Section 4

Field and Galois Theory

1401

(1) Let X be a finite set and G a subgroup of the group of permutations of X. Define a relation $\sim$ on X by requiring $x \sim y$ if either $x = y$ or the transposition (x, y) (which interchanges $x, y \in X$ and leaves all other elements fixed) is an element of G. Show the following.

(a) $\sim$ is an equivalence relation.

(b) If G acts transitively, then all equivalence classes are distinct and contain the same number of elements.

(c) If $\mathrm{Card}(X)$ is a prime number and if G acts transitively and contains at least one transposition then G must be the whole permutation group of X.

(2) Suppose $f \in \mathbb{Q}[x]$ is irreducible and has degree p, a prime number. If f has exactly $p - 2$ real roots and 2 complex roots, show that the Galois group of f over $\mathbb{Q}$ is the symmetric group S_p on p symbols. Show that the polynomial

$$(x^2 + 4) \cdot x \cdot (x^2 - 4)(x^2 - 16) - 2$$

is irreducible and determine its Galois group over $\mathbb{Q}$.

(*Columbia*)

Solution.

(1) (a) By the definition of $\sim$, $\sim$ is reflexive and symmetric. Since

$$(x, y)(y, z)(x, y)^{-1} = (x, z), \quad (x \neq y, y \neq z),$$

it is easy to see that $\sim$ is transitive. Hence $\sim$ is an equivalence relation.

(b) Let $\{x_1, x_2, \ldots, x_n\}$ and $\{y_1, y_2, \ldots, y_m\}$ be two equivalence classes of $\sim$ determined by $x = x_1$ and $y = y_1$ respectively. Since G acts transitively on X, there exists $\sigma \in G$ such that $\sigma(x) = y$. For any $1 < i \leq n$, we have

$$(\sigma(x_1), \sigma(x_i)) = \sigma(x_1, x_i)\sigma^{-1} \in G.$$

Thus $\{\sigma(x_1), \sigma(x_2), \ldots, \sigma(x_n)\}$ are n distinct elements belonging to the equivalence class determined by $y_1 = \sigma(x_1)$. Hence $m \geq n$. Similarly, we have $n \geq m$. Thus $m = n$. So all equivalence classes contain the same number of elements.

(c) Suppose G acts transitively and contains at least one transposition. Then, by (b), all the equivalence classes contain the same number of elements, say n, and $n > 1$. So $\mathrm{Card}(X) = m \cdot n$, where m is the number of distinct equivalence classes determined by $\sim$. Since $\mathrm{Card}(X)$ is a prime, $m = 1$ and $n = \mathrm{Card}(X)$. Thus for any $x, y \in X$, $x \sim y$, i.e., $(x, y) \in G$. It follows that G is the whole permutation group of X, which is generated by all the transpositions.

(2) Suppose

$$f(x) = \prod_{i=1}^{p}(x - r_i)$$

in $\mathbb{C}[x]$. So $E = \mathbb{Q}(r_1, \ldots, r_p)$ is a splitting field of $f(x)$ over $\mathbb{Q}$ contained in $\mathbb{C}$. Identify $G = \mathrm{Gal}(E/\mathbb{Q})$ with a permutation group of the set $X = \{r_1, \ldots, r_p\}$ of the (distinct) roots by $\eta \rightarrow \eta|_x$. For any $r_i, r_j \in X$, since $f(x)$ is irreducible and $f(r_i) = 0 = f(r_j)$, there exists an isomorphism of $\mathbb{Q}(r_i)/\mathbb{Q}$ into

$\mathbb{Q}(r_j)/\mathbb{Q}$ by sending r_i to r_j. Since $E = \mathbb{Q}(r_1, \ldots, r_p)$ is a splitting field of $f(x)$ over $\mathbb{Q}(r_i)$, and also over $\mathbb{Q}(r_j)$, this isomorphism can be extended to an automorphism η of $E/\mathbb{Q}$. Then $\eta \in$ Gal$(E/\mathbb{Q})$ and $\eta(r_i) = r_j$, which shows G acts transitively on X.

Consider the conjugation automorphism on $\mathbb{C}$. This maps $f(x)$ to itself. Let r_1 and r_2 be the two non-real roots of $f(x)$. Thus the conjugation interchanges r_1 and $r_2 = \bar{r}_1$ and fixed all other real roots. Hence the restriction of the conjugation to E is an element of G and it is a transposition. Thus the Galois group G of $f(x)$ over $\mathbb{Q}$ is the symmetric group S_p on $\{r_1, \ldots, r_p\}$.

By Eisenstein criterion,

$$f(x) = (x^2 + 4)\, x\, (x^2 - 4)(x^2 - 16) - 2$$

is irreducible over $\mathbb{Q}$. Let

$$g(x) = (x^2 + 4)\, x\, (x^2 - 4)(x^2 - 16).$$

The real roots of $g(x)$ are $0, \pm 2, \pm 4$, and the graph of $y = g(x)$ has the form

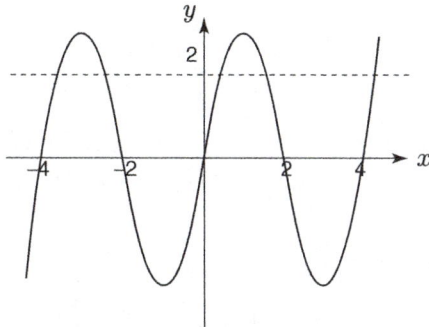

Fig. 1.1.

Since $|g(n)| > 2$ for any odd integer n, it is easy to see that $f(x) = g(x) - 2$ has five real roots and two non-real roots. Hence the Galois group of $f(x)$ over $\mathbb{Q}$ is S_7.

1402

Let p be an odd prime. Let ζ_p be a primitive p^{th} root of unity and g a primitive root (mod p) (i.e., g is a generator for $(\mathbb{Z}/p\mathbb{Z})^*$). Fix e, a divisor of $p-1$ and put $f = (p-1)/e$. Define

$$\eta_i = \sum_{j=0}^{f-1} (\zeta_p)^{g^{ej+i}}$$

Show that, for any i, η_i generates a subfield of $\mathbb{Q}(\zeta_p)$ of degree e over $\mathbb{Q}$.

(Hint. Use the generator $\sigma : \zeta_p \rightarrow (\zeta_p)^g$ of the Galois group of $\mathbb{Q}(\zeta_p)$ over $\mathbb{Q}$.)

(Columbia)

Solution.

Since g is a primitive root (mod p), $\sigma : \zeta_p \mapsto (\zeta_p)^g$ is an automorphism of $\mathbb{Q}(\zeta_p)$ over $\mathbb{Q}$ and

$$G = \mathrm{Gal}(\mathbb{Q}(\zeta_p)/\mathbb{Q}) = \langle\sigma\rangle = \{\sigma, \sigma^2, \ldots, \sigma^{p-2}, \sigma^{p-1} = 1\}.$$

Obviously, $\{\sigma(\zeta_p), \sigma^2(\zeta_p), \ldots, \sigma^{p-1}(\zeta_p) = \zeta_p\}$ is a base for $\mathbb{Q}(\zeta_p)/\mathbb{Q}$.

Let $H = \langle\sigma^e\rangle < G$. Then $|H| = f$,

$$[\mathbb{Q}(\zeta_p) : \mathrm{Inv}\, H] = |H| = f$$

and $[\mathrm{Inv}(H) : \mathbb{Q}] = e$. Since H is normal in G, $\mathrm{Inv}(H)/\mathbb{Q}$ is a Galois extension and $\mathrm{Gal}(\mathrm{Inv}(H)/\mathbb{Q}) \simeq G/H$.

Now for any i,

$$\sigma(\eta_i) = \sigma\left(\sum_{j=0}^{f-1}(\zeta_p)^{g^{ej+i}}\right)$$

$$= \sigma\left(\sum_{j=0}^{f-1}\sigma^{ej+i}(\zeta_p)\right)$$

$$= \sum_{j=0}^{f-1}\sigma^{ej+i+1}(\zeta_p)$$

$$= \eta_{i+1}$$

and

$$\sigma^e(\eta_i) = \sigma^e \left(\sum_{j=0}^{f-1} \sigma^{ej+i}(\zeta_p) \right)$$

$$= \sum_{j=0}^{f-1} \sigma^{e(j+1)+i}(\zeta_p)$$

$$= \sum_{j=1}^{f-1} \sigma^{ej+i}(\zeta_p) + \sigma^{ef+i}(\zeta_p)$$

$$= \sum_{j=1}^{f-1} \sigma^{ej+i}(\zeta_p) + \sigma^{i}(\zeta_p)$$

$$= \sum_{j=0}^{f-1} \sigma^{ej+i}(\zeta_p)$$

$$= \eta_i.$$

It follows that $\eta_i \in \mathrm{Inv}(H)$ and $\{\eta_0, \eta_1, \ldots, \eta_{e-1}\}$ is the orbit $\mathrm{Gal}(\mathrm{Inv}(H)/\mathbb{Q})(\eta_i)$ of η_i for any i. Since $\{\sigma(\zeta_p), \sigma^2(\zeta_p), \ldots, \sigma^{p-1}(\zeta_p)\}$ is a base for $\mathbb{Q}(\zeta_p)/\mathbb{Q}$,

$$\eta_0 = \sigma(\zeta_p) + \sigma^e(\zeta_p) + \cdots + \sigma^{e(f-1)}(\zeta_p),$$

$$\eta_1 = \sigma^2(\zeta_p) + \sigma^{e+1}(\zeta_p) + \cdots + \sigma^{e(f-1)+1}(\zeta_p),$$

$$\ldots$$

$$\eta_{e-1} = \sigma^{e-1}(\zeta_p) + \sigma^{2e-1}(\zeta_p) + \cdots + \sigma^{e(f-1)+e-1}(\zeta_p)(= \sigma^{p-2}(\zeta_p))$$

are distinct. Hence $\prod_{j=0}^{e-1}(x - \eta_j)$ is the minimal polynomial for η_i over $\mathbb{Q}$ (for any i). It follows that η_i is a primitive element for $\mathrm{Inv}(H)/\mathbb{Q}$, i.e., $\mathrm{Inv}(H) = \mathbb{Q}(\eta_i)$ for any i. Thus η_i generates a subfield of $\mathbb{Q}(\zeta_p)$ of degree e.

1403

Let K be a field and x be an element of an extension of K such that x is transcendental over K. Put $G = \text{Aut}(K(x)/K)$, and let H denote the subgroup of G consisting of the substitutions $x \to x + b$ with $b \in K$.

(a) Let $A, B \in K[X]$, $AB \notin K$, $\gcd(A, B) = 1$, and put $y = A(x)/B(x)$. Show in succession that the polynomial $A - yB \in K(y)[X]$ is not 0, that x is algebraic over $K(y)$, that y is transendental over K, that $A - yB$ is irreducible in $K(y)[X]$, and that

$$[K(x) : K(y)] = \max\{\deg A, \deg B\}.$$

(b) Show that the elements of G are given by the fractional linear substitutions $x \to ax + b/cx + d$ with $a, b, c, d \in K$ and $ad - bc \neq 0$.

(c) Show that when K is infinite, then G is infinite and the fixed field of G is K. Find the fixed field of H.

(d) Show that when $K = \mathbb{F}_q$ and put

$$z = (x^{q^2} - x)^{q+1}/(x^q - x)^{q^2+1},$$

then $\text{ord}\, G = q^3 - q$ and the fixed field of G is $\mathbb{F}_q(z)$. Conclude that $\mathbb{F}_q(x)$ is a Galois extension of a given field L with $\mathbb{F}_q \subseteq L \subseteq \mathbb{F}_q(x)$ if and only if $L \supseteq \mathbb{F}_q(z)$. Find the fixed field of H.

(Columbia)

Solution.

(a) Since $\gcd(A, B) = 1$, there exist $S, T \in K[X]$ such that $AS + BT = 1$ in $K[X]$. Suppose $A - yB = 0$ in $K(y)[X]$. Then

$$y = ASy + BTy = A(Sy + T).$$

It follows that $\deg A = 0$ and $\deg B = \deg(A) = 0$, which is contrary to $AB \notin K$. Thus $A - yB \neq 0$.

Since $A(x) - yB(x) = 0$, x is algebraic over $K(y)$. If y is algebraic over K, x must be algebraic over K, which is contrary to the assumption. Hence y is transendental over K.

Again, since $\gcd(A, B) = 1$ in $K[X]$ ($\subseteq K(y)[X]$), $A - yB$ is irreducible in $K[y][X] = K[X][y]$. Thus $A - yB$ is irreducible in $K(y)[X]$. Thus the minimal polynomial of x over $K(y)$ is the monic polynomial which is a multiple in $K(y)$ of $A - yB$ and

$$[K(x) : K(y)] = \deg(A - yB) = \max\{\deg A, \deg B\}.$$

(b) By (a), $y = A(x)/B(x)$ is a generator of $K(x)/K$ (i.e., $K(y) = K(x)$) if and only if $\max\{\deg A, \deg B\} = 1$, or if and only if y has the form $ax + b/cx + d$, $ad - bc \neq 0$.

(c) By (b), it is easy to see that $G \simeq GL_2(K)/K^* \cdot I_2$, where

$$K^* \cdot I_2 = \{\mathrm{diag}(a, a) | a \in K^*\}.$$

If K is infinite, so is G.

Obviously, $K \subseteq \mathrm{Inv}(G)$, the fixed subfield of G. On the other hand, if there exists some $y = f(x)/g(x) \in \mathrm{Inv}(G) \backslash K$, then by (a), $K(x)/K(y)$ is a finite-dimensional simple extension. Hence $\mathrm{Aut}_{K(y)} K(x)$ is finite, which is contrary to the facts that $G \subseteq \mathrm{Aut}_{K(y)} K(x)$ and G is infinite. Thus we have $\mathrm{Inv}(G) = K$.

Similarly, we have $\mathrm{Inv}(H) = K$, since H is also infinite.

(d) Suppose $K = \mathbb{F}_q$. Then

$$\mathrm{ord}\, G = \left| \frac{GL_2(\mathbb{F}_q)}{\mathbb{F}_q^+ \cdot I_2} \right| = \frac{(q^2 - 1)(q^2 - q)}{q - 1} = q^3 - q$$

and

$$[\mathbb{F}_q(x) : \mathrm{Inv}(G)] = q^3 - q.$$

For any

$$\sigma : x \mapsto ax + b/cx + d, \quad (ad - bc \neq 0)$$

in G, it is routine to check

$$\sigma(z) = \sigma((x^{q^2} - x)^{q+1}/(x^q - x)^{q^2+1}) = z.$$

Hence $\mathbb{F}(z) \subseteq \text{Inv}(G)$. On the other hand, we have

$$(X^{q^2} - X)^{q+1}/(X^q - X)^{q^2+1}$$

$$= \left(\frac{(X^{q^2} - X)}{(X^q - X)}\right)^{q+1} \cdot \frac{1}{(X^q - X)^{q^2-q}}$$

$$= \frac{(1 + X^{q-1} + X^{(q-1)2} + \cdots + X^{(q-1)q})^{q+1}}{(X^q - X)^{q^2-q}}.$$

Denote

$$A = (1 + X^{q-1} + X^{(q-1)2} + \cdots + X^{(q-1)q})^{q+1}$$

and

$$B = (X^q - X)^{q^2-q}.$$

Then $\gcd(A, B) = 1$ and

$$\max\{\deg A, \deg B\} = q^3 - q.$$

Hence we have $[\mathbb{F}_q(x) : \mathbb{F}_q(z)] = q^3 - q$. Thus $\text{Inv}(G) = \mathbb{F}_q(z)$ and $\mathbb{F}_q(x)$ is a Galois extension of a given field L with $\mathbb{F}_q \subseteq L \subseteq \mathbb{F}_q(x)$ if and only if $L \supseteq \mathbb{F}_q(z)$.

Similarly, we have $[\mathbb{F}_q(x) : \text{Inv}(H)] = q$ since $|H| = q$. Let $A = X^q - X$ and $B = 1$ in $\mathbb{F}_q[X]$ and let $z' = x^q - x \in \mathbb{F}_q(x)$. Then $z' \in \text{Inv}(H)$ and by (a), $[\mathbb{F}_q(x) : \mathbb{F}_q(z')] = q$. Thus

$$\text{Inv}(H) = \mathbb{F}_q(z') = \mathbb{F}_q(x^q - x).$$

1404

Let $E = \mathbb{C}(Y)$ with Y an indeterminate, $F = \mathbb{C}(Z)$ with $Z = Y^n + Y^{-n}$, and $\zeta = e^{2\pi i/n}$.

(a) Show that there are unique automorphisms σ and τ of $E/\mathbb{C}$ such that $\sigma(Y) = \zeta Y$ and $\tau(Y) = Y^{-1}$, and that the subgroup G of $\text{Aut}(E)$ which these generate is isomorphic to D_n, the Dihedral group of order $2n$.

(b) Show that Y is a root of a polynomial of degree $2n$ with coefficients in F.

(c) Show that E/F is a Galois extension with Galois group G.

(*Columbia*)

Solution.

(a) Since Y is a generator of E over $\mathbb{C}$, there are unique homomorphisms σ and τ of E to E over $\mathbb{C}$ such that $\sigma(Y) = \zeta Y$ and $\tau(Y) = Y^{-1}$. Obviously, $\sigma^n = 1 = \tau^2$. Hence σ and τ are automorphisms of $E/\mathbb{C}$. Since $\mathrm{ord}(\sigma) = n$, $\mathrm{ord}(\tau) = 2$ and $\tau\sigma = \sigma^{n-1}\tau$ in G, $G = \langle \sigma, \tau \rangle \simeq D_n$, the Dihedral group of order $2n$.

(b) Since

$$X^{2n} - Z \cdot X + 1 = X^{2n} - (Y^n + Y^{-n}) + 1$$
$$= (X^n - Y^n)(X^n - Y^{-n}) \in F[X],$$

Y is a root of $X^{2n} - ZX + 1$.

(c) Obviously, $E/\mathrm{Inv}(G)$ is a Galois extension with Galois group G and $[E : \mathrm{Inv}(G)] = 2n$. Since $\sigma(Z) = \sigma(Y^n + Y^{-n}) = Z$ and $\tau(Z) = Z$, we have $Z \in \mathrm{Inv}(G)$ and $F = \mathbb{C}(Z) \subseteq \mathrm{Inv}(G)$. It is readily verified that $X^{2n} - ZX + 1$ is irreducible over F (Z is transendental over $\mathbb{C}$ and $X^{2n} - ZX + 1$ is irreducible in $\mathbb{C}[Z][X]$, for example, using 1403). E is a splitting field of $X^{2n} - ZX + 1$ since

$$X^{2n} - ZX + 1 = (X - Y)(X - \zeta Y)\cdots(X - \zeta^{n-1}Y)$$
$$(X - Y^{-1})(X - \zeta Y^{-1})\cdots(X - \zeta^{n-1}Y^{-1})$$

in E. It follows that E/F is Galois and $[E : F] = 2n$. Hence we have $F = \mathrm{Inv}(G)$ and E/F is Galois with Galois group G.

1405

Let $F = \mathbb{Q}(x)$ be the field of rational polynomials in one variable x over $\mathbb{Q}$ (i.e., the quotient field of the polynomial ring

$\mathbb{Q}[x]$). Consider the elements σ, τ in $\mathrm{Aut}_{\mathbb{Q}}(F)$ (i.e., field auto-morphisms of F) given by $\sigma(x) = 2 - x$ and $\tau(x) = \frac{x}{x-1}$.

(a) Find the subgroup G (of $\mathrm{Aut}_{\mathbb{Q}}(F)$) generated by σ and τ.

(b) If K is the fixed field of G in F, find a finite subset S of F such that $K = \mathbb{Q}(S)$.

(c) How many subfields lie strictly between K and F? How many of these are Galois over K? Justify your answers.

(Indiana)

Solution.

(a) Since $\sigma^2(x) = x$, $\tau^2(x) = x$ and $\tau\sigma(x) = \frac{x-2}{x-1} = \sigma\tau(x)$, we have $\sigma^2 = 1$, $\tau^2 = 1$ and $\sigma\tau = \tau\sigma$. Hence

$$G = \langle \sigma, \tau \rangle \simeq K_4,$$

where K_4 is the Klein 4-group.

(b) Obviously $\sigma(x - 1) = 1 - x$ and $\tau(x - 1) = \frac{1}{x-1}$. It is easy to see that $(x - 1)^2 + \frac{1}{(x-1)^2} \in K$, the fixed field of G in F. Denote $\eta = (x - 1)^2 + \frac{1}{(x-1)^2}$. Then $\mathbb{Q}(\eta) \subseteq K \subseteq F$. Let

$$f(t) = t^4 - \eta t^2 + 1 \in \mathbb{Q}(\eta)[t].$$

Then $f(t)$ is irreducible in $\mathbb{Q}(\eta)[t]$ and F is a splitting field of $f(t)$ since

$$f(t) = (t - x + 1)(t + x - 1) \cdot \left(t - \frac{1}{x-1} \right)\left(t + \frac{1}{x-1} \right)$$

in $F[t]$. Hence $\mathbb{Q}(\eta) \subseteq F$ is a Galois extension and $[F : \mathbb{Q}(\eta)] = 4$. On the other hand, $K \subseteq F$ is a Galois extension and $[F : K] = |G| = 4$. It follows that

$$K = \mathbb{Q}(\eta) = \mathbb{Q}\left((x - 1)^2 + \frac{1}{(x-1)^2} \right).$$

(c) By the fundamental theorem of Galois theory, there are exactly 3 subfields lying strictly between K and F which correspond to the three proper subgroups of $G \simeq K_4$. Since all the subgroups of K_4 are normal, all these 3 subfields are normal over K (finite-dimensional and separable). Thus all these 3 subfields are Galois over K.

1406

Consider $\mathbb{Q}(t)$, the field of quotients of the polynomial ring $\mathbb{Q}[t]$. Let σ and τ be elements of $\mathrm{Aut}_{\mathbb{Q}}(\mathbb{Q}(t))$ given by $\sigma(t) = \frac{t-1}{t+1}$ and $\tau(t) = -t$. Let G be the subgroup of $\mathrm{Aut}_{\mathbb{Q}}(\mathbb{Q}(t))$ generated by σ and τ.

(a) Identify G.

(b) Let H be the subgroup of G generated by σ^2 and τ. Find $\alpha \in \mathbb{Q}(t)$ such that $\mathbb{Q}(\alpha)$ is the fixed field of H (Justify your answer).

(Indiana)

Solution.

(a) Obviously, $\tau^2(t) = t$, $\sigma^2(t) = \sigma\left(\frac{t-1}{t+1}\right) = -\frac{1}{t}$, $\sigma^3(t) = -\frac{t+1}{t-1}$ and $\sigma^4(t) = t$. It is easy to see that $r^2 = 1$, $\sigma^4 = 1$ and $\tau\sigma = \sigma^3\tau$. It follows that G is isomorphic to the Dihedral group D_4.

(b) Since

$$\tau\left(t^2 + \frac{1}{t^2}\right) = \sigma^2\left(t^2 + \frac{1}{t^2}\right) = t^2 + \frac{1}{t^2}, \quad t^2 + \frac{1}{t^2} \in \mathrm{Inv}\, H,$$

the fixed field of H. Hence,

$$\mathbb{Q}\left(t^2 + \frac{1}{t^2}\right) \subseteq \mathrm{Inv}\, H \subseteq \mathbb{Q}(t)$$

and obviously

$$[\mathbb{Q}(t) : \mathrm{Inv}\, H] = |H| = 4.$$

On the other hand,

$$f(x) = x^4 - \left(t^2 + \frac{1}{t^2}\right) \cdot x^2 + 1$$

$$= (x^2 - t^2)\left(x^2 - \frac{1}{t^2}\right)$$

$$= (x - t)(x + t)\left(x - \frac{1}{t}\right)\left(x + \frac{1}{t}\right)$$

is irreducible over $\mathbb{Q}(t^2 + \frac{1}{t^2})$ and $\mathbb{Q}(t)$ is a splitting field over $\mathbb{Q}\left(t^2 + \frac{1}{t^2}\right)$ of separable polynomial $f(x)$. So

$$\mathbb{Q}(t) \supseteq \mathbb{Q}\left(t^2 + \frac{1}{t^2}\right)$$

is a Galois extension and

$$\left[\mathbb{Q}(t) : \mathbb{Q}\left(t^2 + \frac{1}{t^2}\right)\right] = 4.$$

It follows that

$$\left[\mathrm{Inv}\,H : \mathbb{Q}\left(t^2 + \frac{1}{t^2}\right)\right] = 1,$$

that is,

$$\mathrm{Inv}\,H = \mathbb{Q}\left(t^2 + \frac{1}{t^2}\right).$$

1407

Let $\mathbb{Q}$ denote the field of rational numbers. Let $K = \mathbb{Q}(\alpha)$, where α is a root of $f(x) = x^3 - 3x + 1$.

(a) Prove that $f(x)$ is irreducible over $\mathbb{Q}$.

(b) Prove that $K/\mathbb{Q}$ is Galois.

(Hint. Consider $\alpha^2 - 2$.)

(c) Find a generator of the Galois group $\mathrm{Gal}(K/\mathbb{Q})$.

(*Indiana*)

Solution.

(a) If $f(x) = x^3 - 3x + 1$ is reducible over $\mathbb{Q}$, then $f(x)$ has a factor with degree one in $\mathbb{Q}[x]$, that is, $f(x)$ has a root in $\mathbb{Q}$. Since $f(x) = x^3 - 3x + 1$ is monic, the rational roots must be integral factors of 1. But ± 1 are not roots of $f(x)$. This is a contradiction. Thus $f(x)$ is irreducible over $\mathbb{Q}$.

(b) Let β, γ be the other two roots of $f(x) = x^3 - 3x + 1$ in a splitting field of $f(x)$ over K. Obviously, $\alpha + \beta + \gamma = 0$, $\alpha\beta\gamma = -1$ and the discriminant Δ of $f(x)$ is 3^4. Hence $\beta + \gamma = -\alpha$ and

$$\beta - \gamma = \frac{(\alpha - \beta)(\beta - \gamma)(\gamma - \alpha)}{(\alpha - \beta)(\gamma - \alpha)}$$

$$= \frac{\sqrt{\Delta}}{-\alpha^2 + (\beta + \gamma)\alpha - \beta\gamma}$$

$$= \frac{\sqrt{81} \cdot \alpha}{3(1 - 2\alpha)}$$

$$= \frac{\sqrt{81}}{3(1 - \alpha^2)}.$$

It follows that α, $\frac{1}{1-\alpha} = -\alpha^2 - \alpha + 2$ and $-\frac{1}{1-\alpha} - \alpha = \alpha^2 - 2$ are the roots of $f(x)$. So K is a splitting field of $f(x)$. Hence $K/\mathbb{Q}$ is Galois since $\mathrm{Char}(\mathbb{Q}) = 0$. (We may check $\alpha^2 - 2$ is a root of $f(x)$ directly by using the hint.)

(c) From (a) and (b), it is easy to see that the Galois group $\mathrm{Gal}(K/\mathbb{Q}) \simeq A_3$.

$$\sigma : K = \mathbb{Q}(\alpha) \to K$$

$$\alpha \mapsto \alpha^2 - 2$$

is a generator of $\mathrm{Gal}(K/\mathbb{Q})$.

1408

Let K be a field and x an indeterminate.

(a) Show that the rational functions $f_a = \frac{1}{x-a}$ $(a \in K)$ are linearly independent over K.

(b) As K-modules, $K[x]$ and $K(x)$ have what dimensions?

(c) Let G denote the additive group of K, acting on $K(x)$ by $a \in G$ sending x to $x + a$. Assume that K is infinite.

Let $f \in K(x)$. Show that the G-orbit of f spans a finite-dimensional K-module if and only if $f \in K[x]$.

(*Columbia*)

Solution.

(a) Suppose that $\{f_a = \frac{1}{x-a} \,|\, a \in K\}$ is linearly dependent over K. There exist some non-zero elements $\alpha_1, \ldots, \alpha_n \in K$ and some distinct elements $a_1, a_2, \ldots, a_n \in K$ such that

$$\sum_{i=1}^{n} \alpha_i f_{a_i} = 0.$$

Hence

$$\sum_i \alpha_i \left(\prod_{j \neq i} (x - a_j) \right) = 0$$

and

$$(x - a_i) \left| \prod_{j \neq i} (x - a_j), \right.$$

which is not true. Thus $\{f_a = \frac{1}{x-a} \,|\, a \in K\}$ is linearly independent.

(b) $\dim_K K[x] = \dim_K K(x) = \infty$.

(c) Suppose

$$f(x) = a_n x^n + a_{n-1} x^{n-1} + \cdots + a_1 x + a_0 \in K[x].$$

We claim that for any $n+1$ distinct elements $\alpha_1, \alpha_2, \ldots, \alpha_{n+1}$ in K (note that K is infinite), $f(x+\alpha_1), f(x+\alpha_2), \ldots, f(x+\alpha_{n+1})$ spans the subspace generated by the G-orbit of f.

For any $a \in G$, we take $(\beta_1, \beta_2, \ldots, \beta_{n+1}) \in K^{n+1}$ to be the solution of the equation system

$$\begin{pmatrix} 1 & 1 & \cdots & 1 \\ \alpha_1 & \alpha_2 & \cdots & \alpha_{n+1} \\ \alpha_1^2 & \alpha_2^2 & \cdots & \alpha_{n+1}^2 \\ \cdots & \cdots & \cdots & \cdots \\ \alpha_1^n & \alpha_2^n & \cdots & \alpha_{n+1}^n \end{pmatrix} \begin{pmatrix} x_1 \\ x_2 \\ x_3 \\ \vdots \\ x_n \end{pmatrix} = \begin{pmatrix} 1 \\ a \\ a^2 \\ \vdots \\ a^n \end{pmatrix}.$$

Then it is clear that

$$(x + a)^n = \beta_1(x + \alpha_1)^n + \beta_2(x + \alpha_2)^n + \cdots + \beta_{n+1}(x + \alpha_{n+1})^n$$

and also

$$(x + a)^i = \beta_1(x + \alpha_1)^i + \beta_2(x + \alpha_2)^i + \cdots + \beta_{n+1}(x + \alpha_{n+1})^i$$

for $1 \leq i < n$. Hence

$$f(x+a) = \beta_1 f(x+\alpha_1) + \beta_2 f(x+\alpha_2) + \cdots + \beta_{n+1} f(x+\alpha_{n+1}).$$

This shows that $\{f(x + \alpha_i) \mid 1 \leq i \leq n + 1\}$ spans the subspace generated by the G-orbit of f.

On the other hand, suppose $f(x) \in K(x)$, and the G-orbit of f, $\{f(x + a) \mid a \in K\}$ spans a finite-dimensional K-module. We write $f(x) = g(x)/h(x)$ where $g(x)$, $h(x) \in K[x]$ and $(g(x), h(x)) = 1$. Let $f(x+\alpha_1)$, $f(x+\alpha_2)$, ..., $f(x+\alpha_n)$ $(\alpha_i \in K)$ span the subspace $\langle \{f(x + a) \mid a \in K\} \rangle$. Then for any $a \in K$, there exist $\beta_1, \beta_2, \ldots, \beta_n$ in K such that

$$\begin{aligned}
f(x + a) &= \frac{g(x + a)}{h(x + a)} \\
&= \sum_{i=1}^{n} \beta_i f(x + \alpha_i) \\
&= \sum_{i=1}^{n} \beta_i \frac{g(x + \alpha_i)}{h(x + \alpha_i)} \\
&= \frac{\sum_{i=1}^{n} \beta_i g(x + \alpha_i) \prod_{j \neq i} h(x + \alpha_j)}{\prod_{i=1}^{n} h(x + \alpha_i)}.
\end{aligned}$$

Since $(g(x + a), h(x + a)) = 1$, we have

$$h(x + a) \Big| \prod_{i=1}^{n} h(x + \alpha_i)$$

for any $a \in F$. Since K is infinite, an easy discussion in a splitting field of $h(x)$ over F will lead to $\deg h(x) = 0$. Thus we have

$$f(x) = \frac{g(x)}{h(x)} \in F[x].$$

1409

Let E be a finite Galois extension of F and let $f(x)$ be an irreducible polynomial in $F[x]$. Show that all the irreducible factors of $f(x)$ over E are of the same degree.

(*Columbia*)

Solution.

For any $\sigma \in G = \text{Gal}(E/F)$, we still denote σ to be the isomorphism $E[x] \to E[x]$ which extends σ on E and maps x to x. Since $f(x) \in F[x]$, $\sigma(f(x)) = f(x)$. Let $e(x)$ be a monic irreducible factor of $f(x)$ over E. Then, for any $\sigma \in G$, $\sigma(e(x))$ is an irreducible factor of $f(x)$ over E.

We prove in the following that all monic irreducible factors of $f(x)$ over E arise in this way. Thus all the irreducible factors of $f(x)$ over E are of the same degree.

Suppose $e'(x)$ is another monic irreducible factor over E. Let α and α' be roots of $e(x)$ and $e'(x)$ in some extension field of E. Then, α and α' are roots of $f(x)$, which is irreducible over F. Hence we have an isomorphism $\eta : F(\alpha) \to F(\alpha')$ which sends a to $a(a \in F)$ and α to α'.

Since E/F is Galois, we can write $E = F(\beta)$, where β is a root of $g(x)$, a separable irreducible polynomial over F. Obviously E is a splitting field of $g(x)$. Then $E(\alpha) = F(\alpha)(\beta)$ and $E(\alpha') = F(\alpha')(\beta)$ are splitting fields of $g(x)$ over $F(\alpha)$ and $F(\alpha')$ respectively. Hence $\eta : F(\alpha) \to F(\alpha')$ can be extended to an isomorphism η of $E(\alpha)$ onto $E(\alpha')$. Note that $\eta(\beta)$ may not be β, but $\eta(\beta)$ is a root of $\eta(g(x)) = g(x)$. It follows that $\eta|_E : E \to E$ is in G.

Now, since α and α' are roots of the monic irreducible polynomial $e(x)$ and $e'(x)$ over E respectively, $\eta : E(\alpha) \to E(\alpha')$ is an isomorphism and $\eta(\alpha) = \alpha'$, we must have $(\eta|_E)(e(x)) = e'(x)$ and $\deg e(x) = \deg e'(x)$.

1410

(a) Let K be a field of characteristic $p > 0$. Show that the polynomial $t^p - t - c$ in $K[t]$ is either irreducible or splits completely into p linear factors over K.

(Hint. If u is a root of $t^p - t - c$ then so is $u + 1$.)

(b) Let F be the splitting field of the polynomial $t^{62} - 1$ over $\mathbb{Z}_5$. Show that $[F : \mathbb{Z}_5] = 3$.

(Hint. First prove that the zeroes of $t^{62} - 1$ form a cyclic group G of order 62.)

(Indiana)

Solution.

Suppose that $t^p - t - c = f(t) \cdot g(t)$ in $K[t]$ where $f(t)$ is a monic polynomial of degree n, $1 \le n \le p-1$. Let E be a splitting field of $t^p - t - c$ and let $u \in E$ be a root of this polynomial. Then for any $m \in \mathbb{Z}_p$, the prime field of K,

$$(u+m)^p - (u+m) - c = u^p + m^p - u - m - c = u^p - u - c = 0.$$

Hence we have

$$f(t) \cdot g(t) = \prod_{m \in \mathbb{Z}_p} (t - u - m)$$

and there exist $i_1, i_2, \ldots, i_n \in \mathbb{Z}_p$ such that

$$f(t) = (t - u - i_1)(t - u - i_2) \cdots (t - u - i_n).$$

Comparing the coefficients of the term of degree $n-1$, we obtain $n \cdot u + i_1 + i_2 + \cdots + i_n \in K$. So we have $n \cdot u \in K$. Since $p \cdot u = 0$

and there exist integers v and w such that $v \cdot n + wp = 1$, $u = (v \cdot n + wp) \cdot u = v(n \cdot u) \in K$. Thus we have

$$t^p - t - c = \prod_{m \in \mathbb{Z}_p} (t - u - m)$$

in $K[t]$.

(b) Let G be all the zeroes of $t^{62} - 1$ in F. Obviously, G is a subgroup of F^* and G is cyclic of order 62 since

$$(t^{62} - 1)' = 62t^{61} = 2t^{61} \neq 0.$$

It follows that

$$[F : \mathbb{Z}_5] \geq 3.$$

Let E be an extension field of $\mathbb{Z}_5$ such that $[E : \mathbb{Z}_5] = 3$. Then E^* is a cyclic group of order $5^3 - 1 = 124$. Let G' be its unique subgroup of order 62. Then all the elements of G' satisfies $t^{62} - 1$. So $t^{62} - 1$ splits in E and E is a splitting field of $t^{62} - 1$. Thus we have $E \simeq F$ and $[F : \mathbb{Z}_5] = 3$.

1411

(a) Suppose you are given a field L, $\mathbb{Q} \subseteq L \subseteq \mathbb{C}$, such that $L/\mathbb{Q}$ is algebraic and every finite field extension K/L, $K \subseteq \mathbb{C}$ is of even degree. Show that every finite field extension of L must in fact have degree equal to a power of 2.

(b) Show that such a field L actually exists.

(Indiana)

Solution.

(a) Let $K/L\,(K \subseteq \mathbb{C})$ be a finite field extension. We have to show $[K : L]$ is a power of 2. For this purpose, we may assume that K is Galois over L. Let $G = \operatorname{Gal} K/L$ and $|G| = 2^n \cdot m$ where m is odd. By Sylow Theorem, G has a subgroup H of order 2^n. If K' is the corresponding subfield of K/L, then

$[K : K'] = 2^n$ and $[K' : L] = m$. Since L has no proper odd-dimensional extension field, we must have $m = 1$, and so $K' = L$ and $[K : L] = 2^n$.

(b) Let L be the field of real algebraic numbers, that is, the subfield of $\mathbb{R}$ of numbers which are algebraic over $\mathbb{Q}$. Then $\mathbb{Q} \subseteq L \subseteq \mathbb{C}$ and $L/\mathbb{Q}$ is algebraic.

Now for any finite field extension K/L, $K \subseteq \mathbb{C}$, there exists some element $\alpha \in K$ such that $K = L(\alpha)$ by Primitive Element Theorem. Let $f(x)$ be the minimal polynomial of α over L. Then $f(x)$ has no real root since $f(x)$ is irreducible in $L[x]$. So, $f(x)$, when decomposed in $\mathbb{R}[x]$, is a product of irreducible polynomials of degree 2. Hence,

$$[K : L] = [L(\alpha) : L] = \deg f(x)$$

is even. By (a), it is in fact a power of 2.

1412

Let K be a field of characteristic $p \neq 0$. The set $\{x^p \mid x \in K\}$ is a subfield of K that is denoted by K^p (no proof required).

(a) Let L be an intermediate field between K^p and K. If $[L : K^p]$ is finite, prove that it is a power of p.

(b) A subset B of K is called p-independent if for any finite set $b_1, b_2, \ldots, b_m$ of distinct elements of B

$$[K^p(b_1, b_2, \ldots, b_m) : K^p] = p^m.$$

Prove that if $K^p \neq K$, then K contains a maximal p-independent subset B.

(c) Prove that the set B of part (b) satisfies $K^p(B) = K$.

(*Indiana*)

Solution.

(a) If $[L : K^p]$ is finite, there exists a finite set of elements $\{b_1, b_2, \ldots, b_n\}$ such that

$$K^p(b_1, b_2, \ldots, b_n) = L.$$

Without loss of generality, we can assume that $b_i \notin K^p(b_1, \ldots, b_{i-1})$ for any $1 \leq i \leq n$ ($K^p(b_1, \ldots, b_{i-1}) = K^p$ when $i = 1$). Since

$$b_i^p \in K^p \subseteq K^p(b_1, \ldots, b_{i-1})$$

and

$$b_i \notin K^p(b_1, \ldots, b_{i-1}),$$

$t^p - b_i^p$ is irreducible in $K^p(b_1, \ldots, b_{i-1})[t]$. Hence

$$[K^p(b_1, \ldots, b_i) : K^p(b_1, \ldots, b_{i-1})] = p.$$

It follows that

$$[L : K^p] = [K^p(b_1, b_2, \ldots, b_n) : L]$$
$$= \prod_{i=1}^{n} [K^p(b_1, \ldots, b_i) : K^p(b_1, \ldots, b_{i-1})]$$
$$= p^n.$$

(b) If $K^p \neq K$, there exist p-independent subsets. For example, if $b \in K \backslash K^p$, $B = \{b\}$ is a p-independent subset of K. Now suppose that $\{B_i \mid i \in I\}$ is a chain of p-independent subsets of K. Let $B = \bigcup_{i \in I} B_i$. Then for any finite set $\{b_1, \ldots, b_m\}$ of distinct elements of B, there exists some i, such that $\{b_1, b_2, \ldots, b_m\} \subseteq B_i$. Since B_i is a p-independent subset,

$$[K^p(b_1, b_2, \ldots, b_m) : K^p] = p^m.$$

It follows that B is p-independent. By Zorn's Lemma, K contains a maximal p-independent subset.

(c) Let B be a maximal p-independent subset of K. Suppose $K^P(B) \subset K$. Let $b \in K \backslash K^p(B)$. Then $B \cup \{b\}$ is p-independent. The reason is that, for any distinct elements $b_1, b_2, \ldots, b_m$ in $B \cup \{b\}$, if $b \notin \{b_1, b_2, \ldots, b_m\}$, then

$$\{b_1, b_2, \ldots, b_m\} \subseteq B$$

and

$$[K^p(b_1, \ldots, b_m) : K^p] = p^m,$$

and if $b \in \{b_1, b_2, \ldots, b_m\}$, say, $b = b_m$, then $\{b_1, \ldots, b_{m-1}\} \subseteq B$ and we still have

$$[K^p(b_1, \ldots, b_m) : K^p] = [K^p(b_1, \ldots, b_m) : K^p(b_1, \ldots, b_{m-1})]$$
$$\cdot [K^p(b_1, \ldots, b_{m-1}) : K^p]$$
$$= p \cdot p^{m-1} = p^m.$$

Since $B \subset B \cup \{b\}$, the independency of $B \cup \{b\}$ contradicts the maximality of B. Thus we have $K^p(B) = K$.

1413

Let K be a finite field with p^r elements (p a prime) and n be a positive integer. If m is an integer which divides n and $f(t) \in K[t]$ is an irreducible polynomial of degree m, show that f divides $t^{p^{rn}} - t$.

(*Indiana*)

Solution.

Let E be a splitting field of $t^{p^{rn}} - t$ over K. Then $|E| = p^{rn}$ and $[E : K] = n$. Let α be a root of the irreducible polynomial $f(t)$ in some extension field of K. Then $|K(\alpha)| = p^{rm}$ and $[K(\alpha) : K] = m$.

Since $m \mid n$, E contains a subfield L such that $K \subseteq L \subseteq E$ and $L \simeq K(\alpha)$. Hence there exists an element $\beta \in L \subseteq E$ such that $f(\beta) = 0$, that is, $f(t)$ is the minimal polynomial of β. Since $\beta^{p^{rn}} - \beta = 0$, we have $f(t) \mid (t^{p^{rn}} - t)$.

1414

Let K/F be a finite extension of fields and let L and E be intermediate fields, with E/F Galois and $[K : L] = p$, a prime. Prove that if p does not divide $[E : F]$ then $E \subseteq L$.

(*Indiana*)

Solution.

Let $f(x)$ be a separable irreducible polynomial over F such that E is its splitting field. Let

$$E \cdot L = E(L) = L(E)$$

be the composite of E and L in K. Then $E \cdot L = L(E)$ is a splitting field of $f(x)$ over L. Hence $E \cdot L$ is Galois over L. Let $\alpha \in E$ be a root of $f(x)$. Then $E = F(\alpha)$ and $E \cdot L = L(E) = L(\alpha)$. For any

$$\sigma \in \mathrm{Gal}(E \cdot L/L) = \mathrm{Gal}(L(\alpha)/L),$$

it is clear that $\sigma|_E \in \mathrm{Aut}(E/L \cap E)$. And further, we have

$$\mathrm{Gal}(E \cdot L/L) \simeq \mathrm{Gal}(E/L \cap E).$$

Now suppose $E \not\subseteq L$. Then $E \cdot L = K$, since $L \subseteq E \cdot L$ and $[K : L] = p$, a prime. It follows that

$$p = |\mathrm{Gal}(E \cdot L/L)| = |\mathrm{Gal}(E/L \cap E)|$$

dividers $|\mathrm{Gal}(E/F)| = [E : F]$, contrary to the assumption. Thus we have $E \subseteq L$.

1415

Let K_i be the subfields of $\mathbb{C}$ defined as follows: $K_0 = \mathbb{Q}$. If $i \geq 0$, K_{i+1} is the smallest subfield of $\mathbb{C}$ containing the set

$$\{\theta \in \mathbb{C} \,|\, \theta^n \in K_i \text{ for some } n > 0\}.$$

Let

$$K = \bigcup_{i=0}^{\infty} K_i.$$

(1) Prove K is a field.
(2) Let $f(x) \in K[x]$ be irreducible. Prove that $\deg(f) \geq 5$.

(Indiana)

Solution.

(1) Since

$$K_0 \subseteq K_1 \subseteq \cdots \subseteq K_i \subseteq K_{i+1} \subseteq \cdots$$

is a chain of subfields of $\mathbb{C}$. It is clear that $\bigcup_{i=1}^{\infty} K_i$ is a subfield of $\mathbb{C}$.

(2) (Remark. $\deg f(x)$ may be 1).

Let $f(x) \in K[x]$ be an irreducible polynomial with $\deg f(x) > 1$. There exists some i such that $f(x) \in K_i[x]$. Suppose $\deg f(x) \leq 4$. By the formulas for the roots of quadratic, cubic and quartic equations and

$$K_{i+1} = \{\theta \in \mathbb{C} \mid \theta^n \in K_i \text{ for some } n > 0\},$$

$f(x)$ splits in $K_{i+3}[x]$, hence in $K[x]$. Contradicts the irreducibility of $f(x)$. Hence $\deg f(x) \geq 5$.

1416

Let K be an extension field of F_p, the field with p elements. Let a be an algebraic element in K. Prove that $[F_p(a) : F_p]$ is the smallest positive integer m such that $a^{g(m)} \in F_p$, where $g(m) = \frac{p^m - 1}{p - 1}$.

(*Indiana*)

Solution.

Let $n = [F_p(a) : F_p]$. Then $|F_p(a)| = p^n$ and $a^{p^n - 1} = 1$. Since $(a^{(g(n))})^{p-1} = a^{p^n - 1} = 1$, $a^{g(n)} \in F_p$.

On the other hand, if $a^{g(m)} \in F_p$, for some positive integer m, then

$$a^{p^m - 1} = (a^{g(m)})^{p-1} = 1,$$

so a is a root of $x^{p^m} - x$. Let E be a splitting field of $x^{p^m} - x$ over F_p and $a \in E$. Then $[E : F_p] = m$ and $F_p \subseteq F_p(a) \subseteq E$. Hence

$$n = [F_p(a) : F_p] \mid [E : F_p] = m.$$

Thus $[F_p(a) : F]$ is the smallest positive integer m such that $a^{g(m)} \in F_p$.

1417

Let $F \supseteq K$ be a field extension of finite degree m. Let $f \in K[t]$ be an irreducible polynomial of degree n. If m and n are coprime then show that f remains irreducible in $F[t]$.

(Indiana)

Solution.

Suppose that $f(t)$ is reducible in $F[t]$ and let $f(t) = g(t) \cdot h(t)$ in $F[t]$ where $g(t)$ is an irreducible polynomial in $F[t]$ of degree k, $1 \le k < n$. Let $E = F(\alpha)$, where α is a root of $g(t)$ in some extension field of F. Then $[E : F] = k$ since $g(t)$ is irreducible in $F[t]$. So

$$[E : K] = [E : F][F : K] = k \cdot m.$$

On the other hand,

$$[E : K] = [E : K(\alpha)] \cdot [K(\alpha) : K] = [E : K(\alpha)] \cdot n,$$

since α is a root of $f(t)$ and $f(t)$ is irreducible in $K[t]$. It follows that $n \mid k \cdot m$, which contradicts $(m, n) = 1$ and $1 \le k < n$. Thus $f(t)$ is irreducible in $F[t]$.

1418

Find a Galois extension E over $\mathbb{Q}$ with $\mathrm{Gal}(E/\mathbb{Q})$ cyclic of order 16.

(Stanford)

Solution.

For any positive integer n, the cyclotomic field $\mathbb{Q}(z_n)$ over $\mathbb{Q}$ is a Galois extension and $[\mathbb{Q}(z_n) : \mathbb{Q}] = \phi(n)$ where z_n is an n-th primitive root of the unit, $\phi(n)$ is the Euler ϕ-function. It is easy to see that $|\mathrm{Gal}(\mathbb{Q}(z_i)/\mathbb{Q})| = \phi(n)$ and $\mathrm{Gal}(\mathbb{Q}(z_n)/\mathbb{Q}) \simeq \mathrm{Aut}(G)$

where G is the cyclic group of order n. When n is prime, $\mathrm{Aut}(G)$ is cyclic of order $n - 1$.

So if we take $n = 17$, $E = \mathbb{Q}(z_{17})$, then E is a Galois extension of $\mathbb{Q}$ with $\mathrm{Gal}(E/\mathbb{Q})$ cyclic of order 16.

1419

Find a Galois extension E over $\mathbb{Q}$ with $\mathrm{Gal}(E/\mathbb{Q})$ cyclic of order 32.

(*Stanford*)

Solution.
As in 1418, for any positive integer n, the cyclotomic field $\mathbb{Q}(z_n)$ over $\mathbb{Q}$ is a Galois extension and $[\mathbb{Q}(z_n) : \mathbb{Q}] = \phi(n)$ where z_n is an n-th primitive root of the unit, $\phi(n)$ is the Euler ϕ-function. It is easy to see that $|\mathrm{Gal}(\mathbb{Q}(z_i)/\mathbb{Q})| = \phi(n)$ and $\mathrm{Gal}(\mathbb{Q}(z_n)/\mathbb{Q}) \simeq \mathrm{Aut}(G)$ where G is the cyclic group of order n. When $n = 2^m$ and $m \geq 3$, it is well known that $\mathrm{Aut}(G) \simeq \mathbb{Z}_2 \oplus \mathbb{Z}_{2^{m-2}}$.

By the Fundamental Theorem of Galois Theory, if we take $E = \mathrm{Inv}(\mathbb{Z}_2)$, then $\mathbb{Q} \subseteq E$ is a Galois extension and $\mathrm{Gal}(E/\mathbb{Q}) \simeq \mathbb{Z}_{2^{m-2}}$.

Taking $m = 7$, then $\mathbb{Q} \subseteq E$ is a cyclic extension of order 32.

1420

Let E/F be a finite Galois extension, $G = \mathrm{Gal}(E/F)$ and $a \in E$. Consider the F-linear map $M_a : E \to E$, $M_a(x) = ax$. Show that its trace is given by $\mathrm{Tr}_F(M_a) = \sum \sigma(a)$ where σ varies over G.

(*Columbia*)

Solution.
Let z be a primitive element of E/F, $[E : F] = n$ and

$$G = \mathrm{Gal}(E/F) = \{\sigma_1, \sigma_2, \ldots, \sigma_n\}.$$

Then

$$f(x) = \prod_{i=1}^{n}(x - \sigma_i(z))$$

$$= x^n + a_1 x^{n-1} + \cdots + a_n$$

is the minimal polynomial of z over F, and $\sigma_1(z), \sigma_2(z), \ldots,$ $\sigma_n(z)$ are distinct.

Now, for any $a \in E$, a has the form

$$\alpha_0 + \alpha_1 z + \cdots + \alpha_{n-1} z^{n-1} \quad (\alpha_i \in F),$$

since $\{1, z, \ldots, z^{n-1}\}$ is a base for E/F. To prove that

$$\mathrm{Tr}_F(M_a) = \sum_{i=1}^{n} \sigma_i(a),$$

it suffices to prove that

$$\mathrm{Tr}_F(M_{z^k}) = \sum_{i=1}^{n} \sigma_i(z^k)$$

for any $1 \le k \le n - 1$.

Obviously, $f(x)$ is also the minimal polynomial of the F-linear map $M_z : E \to E$, $M_z(x) = z \cdot x$. Since $f(x)$ has distinct roots $\sigma_1(z), \sigma_2(z), \ldots, \sigma_n(z)$ in E, the matrix of M_z ($\in M_n(F)$), say, relative to the base $\{1, z, \ldots, z^{n-1}\}$, is similar to $\mathrm{diag}\{\sigma_1(z), \ldots, \sigma_n(z)\}$ in $M_n(E)$. Anyway, we have

$$\mathrm{Tr}_F(M_z) = \sum_{i=1}^{n} \sigma_i(z)$$

and for any $1 \le i \le n - 1$,

$$\mathrm{Tr}_F(M_{z^k}) = \mathrm{Tr}_F((M_z)^k)$$

$$= \sigma_1(z)^k + \sigma_2(z)^k + \cdots + \sigma_n(z)^k$$

$$= \sum_{i=1}^{n} \sigma_i(z^k).$$

This completes the proof.

1421

Show that the discriminant D of the polynomial $f(x) = x^n + ax + b$ (i.e., $D = \prod_{i \neq j}(\alpha_i - \alpha_j)$ if $f = \prod_{i=1}^{n}(x - \alpha_i)$) is given by the formula:

$$D = n^n b^{n-1} + a^n(1 - n)^{n-1}.$$

(Hint. One has $D = \prod_i f'(\alpha_i)$.)

(*Princeton*)

Solution.

By product rule, $f'(x) = \sum_{i=1}^{n} \prod_{j \neq i}(x - \alpha_j)$. Therefore

$$f'(\alpha_i) = \prod_{j \neq i}(\alpha_i - \alpha_j)$$

and

$$\prod_i f'(\alpha_i) = \prod_i \prod_{j \neq i}(\alpha_i - \alpha_j) = \prod_{j \neq i}(\alpha_i - \alpha_j) = D.$$

For any

$$f(x) = a_0 x^n + a_1 x^{n-1} + \cdots + a_n$$

and

$$g(x) = b_0 x^m + b_1 x^{m-1} + \cdots + b_m,$$

the Sylvester determinant of f and g is defined as the following determinant $R(f, g)$ of degree $m + n$.

$$R(f, g) = \begin{vmatrix} a_0 & a_1 & a_2 & \cdots & \cdots & a_n & 0 & \cdots & 0 \\ 0 & a_0 & a_1 & \cdots & \cdots & a_{n-1} & a_n & \cdots & 0 \\ 0 & 0 & a_0 & \cdots & \cdots & a_{n-2} & a_{n-1} & \cdots & 0 \\ \cdots & \cdots & \cdots & \cdots & \cdots & \cdots & \cdots & \cdots & \cdots \\ 0 & 0 & \cdots & \cdots & a_0 & \cdots & \cdots & \cdots & a_n \\ b_0 & b_1 & b_2 & \cdots & \cdots & \cdots & 0 & \cdots & 0 \\ 0 & b_0 & b_1 & \cdots & \cdots & \cdots & \cdots & \cdots & \cdots \\ \cdots & \cdots & \cdots & \cdots & \cdots & \cdots & \cdots & \cdots & \cdots \\ 0 & 0 & 0 & \cdots & b_0 & b_1 & \cdots & \cdots & b_m \end{vmatrix}$$

It is well known that $R(f, g) = a_0^m \prod_{i=1}^{n} g(x_i)$ where $x_1, x_2, \ldots, x_n$ are the roots of $f(x)$. Hence for $f(x) = x^n + ax + b$,

$$D = R(f, f') = \begin{vmatrix} 1 & 0 & 0 & \cdots & a & b & 0 & \cdots & 0 & 0 \\ 0 & 1 & 0 & \cdots & 0 & a & b & \cdots & 0 & 0 \\ 0 & 0 & 1 & \cdots & 0 & 0 & a & \cdots & 0 & 0 \\ \cdots & \cdots & \cdots & \cdots & \cdots & \cdots & \cdots & \cdots & \cdots & \cdots \\ 0 & 0 & \cdots & 1 & 0 & 0 & 0 & \cdots & a & b \\ n & 0 & 0 & \cdots & a & 0 & 0 & \cdots & 0 & 0 \\ 0 & n & 0 & \cdots & 0 & a & 0 & \cdots & 0 & 0 \\ 0 & 0 & n & \cdots & 0 & 0 & a & \cdots & 0 & 0 \\ \cdots & \cdots & \cdots & \cdots & \cdots & \cdots & \cdots & \cdots & \cdots & \cdots \\ 0 & 0 & \cdots & n & 0 & 0 & 0 & \cdots & a & 0 \\ 0 & 0 & 0 & \cdots & n & 0 & 0 & \cdots & 0 & a \end{vmatrix}$$

$$= n^n b^{n-1} + a^n (1 - n)^{n-1}.$$

1422

Show that $2\cos(2\pi/7)$ satisfies an irreducible cubic polynomial over $\mathbb{Q}$. Compute this cubic and show that $\mathbb{Q}(\cos(2\pi/7))$ is Galois over $\mathbb{Q}$ of degree 3.

(*Princeton*)

Solution.

Let $\{\omega_k = \cos \frac{2k\pi}{7} + i \sin \frac{2k\pi}{7} \mid k = 1, 2, \ldots, 6\}$ be the set of 7-th primitive root. Note that

$$\omega_1 + \omega_6 = \cos \frac{2\pi}{7} + i \sin \frac{2\pi}{7} + \cos \frac{12\pi}{7} + i \sin \frac{12\pi}{7} = 2 \cos \frac{2\pi}{7}.$$

Let $a = \omega_1 + \omega_6$. Then $a^2 = \omega_1^2 + 2\omega_1\omega_6 + \omega_6^2 = \omega_2 + \omega_5 + 2$, which gives $\omega_2 + \omega_5 = a^2 - 2$. Similarly, $a^3 = \omega_3 + \omega_4 + 3(\omega_1 + \omega_6) = \omega_3 + \omega_4 + 3a$ gives that $\omega_3 + \omega_4 = a^3 - 3a$. Since $\omega_1 + \omega_2 + \omega_3 + \omega_4 + \omega_5 + \omega_6 + 1 = 0$, then $1 + (\omega_1 + \omega_6) + (\omega_2 + \omega_5) + (\omega_3 + \omega_4) = 1 + a + (a^2 - 2) + (a^3 - 3a) = a^3 + a^2 - 2a - 1 = 0$. So $a = 2 \cos \frac{2\pi}{7}$ is a root of the irreducible polynomial $x^3 + x^2 - 2x - 1 \in \mathbb{Q}[x]$.

Similarly, we see that $a_1 = \omega_2 + \omega_5$ and $a_2 = \omega_3 + \omega_4$ are the other two roots of $x^3 + x^2 - 2x - 1 = 0$. Since $a_1 = a^2 - 2$ and $a_2 = a^3 - 3a$, so a_1 and a_2 are in the field $\mathbb{Q}(a)$ and $\mathbb{Q}(a)$ is the splitting field of the irreducible polynomial $x^3 + x^2 - 2x - 1$ over $\mathbb{Q}$. It follows that $\mathbb{Q}(\cos\frac{2\pi}{7}) = \mathbb{Q}(a)$ is a Galois extension over $\mathbb{Q}$ of degree 3.

1423

Explain how to construct an angle of $72°$ by ruler and compass.

(*Princeton*)

Solution.
Recall that in an isosceles triangle with angles $72°$, $72°$ and $36°$, the triangle has a ratio between the length of the long side and the short side $1 : \frac{\sqrt{5}-1}{2}$.

Hence to construct an angle of $72°$ by ruler and compass, it suffices to construct an isosceles triangle with the ratio $1 : \frac{\sqrt{5}-1}{2}$ between the length of the long side and the short side. This construction is routine.

1424

Let k be a field. Denote by $G(k)$ the set of polynomials in $k[x]$ of degree one, i.e., $G(k)$ consists of all polynomials $ax + b$ with $a \in k^*$ and $b \in k$.

(a) Show that under composition of polynomial functions, $G(k)$ forms a group.

(b) Show that if $ax + b \in G(k)$, then the map of k to itself defined by $\lambda \mapsto a\lambda + b$ is a set-theoretic automorphism of k.

(c) Show that the map $G(k) \to aut_{set}(k)$ defined by

$$ax + b \mapsto (\text{the automorphism } \lambda \mapsto a\lambda + b)$$

is an injective group homomorphism.

(d) Show that the subgroup $G_1(k)$ consisting of all elements of the form $x+b$ is a normal subgroup of $G(k)$, and that the map $x + b \mapsto b$ defines an isomorphism of $G_1(k)$ with the additive group of k.

(e) Show that the map $ax + b \mapsto a$ defines a group homomorphism of $G(k)$ onto the multiplicative group k^* whose kernel is $G_1(k)$.

(f) Let $\Gamma \subset G(k)$ be a normal subgroup. Show that if $\Gamma \neq \{e\}$, then Γ contains $G_1(k)$.

(g) Show that if $\Gamma \subseteq G(k)$ is a subgroup and if Γ contains $G_1(k)$, then Γ is a normal subgroup of $G(k)$.

(Princeton)

Solution.

(a) For any $f = f(x) = ax + b$ and $g = g(x) = cx + d \in G(k)$,

$$f \cdot g = (f \cdot g)(x) = a(cx + d) + b = acx + (ad + b).$$

Since both a and c are in the group k^*, ac is also in k^*. Likewise, since a, d and b are all in the field k, $ad+b$ is also in k. Therefore $f \cdot g$ is also in $G(k)$, hence $G(k)$ is closed under the composition of polynomial functions.

For any $f = f(x) = a_1x + b_1, g = g(x) = a_2x + b_2$ and $h = h(x) = a_3x + b_3 \in G(k)$,

$$((f \cdot g) \cdot h)(x) = (a_1a_2h(x) + a_1b_2 + b_1)$$
$$= a_1a_2a_3x + a_1a_2b_3 + a_1b_2 + b_1,$$
$$(f \cdot (g \cdot h))(x) = a_1(a_2a_3x + a_2b_3 + b_2) + b_1$$
$$= a_1a_2a_3x + a_1a_2b_3 + a_1b_2 + b_1.$$

Therefore $(f \cdot g) \cdot h = f \cdot (g \cdot h)$, and so the associativity holds in $G(k)$ for the composition of polynomial functions.

Let $e = e(x) = x \in G(k)$. Since for any $f = f(x) = ax + b \in G(k)$, $(e \cdot f)(x) = (f \cdot e)(x) = ax + b = f(x)$, i.e., $e \cdot f = f \cdot e = f$. It follows that e is the identity in $G(k)$.

For any $f = f(x) = ax + b \in G(k)$, let $f^{-1} = f^{-1}(x) = \frac{1}{a}x - \frac{b}{a} \in G(k)$. Then

$$f \cdot f^{-1} = (f \cdot f^{-1})(x) = x = e(x) = (f^{-1} \cdot f)(x),$$

i.e., $f \cdot f^{-1} = f^{-1} \cdot f = e$. Hence f^{-1} is the inverse of f in $G(k)$ under the composition of polynomial functions.

Therefore $G(k)$ forms a group under the composition of polynomial functions.

(b) Let $ax + b \in G(k)$ be fixed. Since k is a field, the following is a map

$$k \to k, \lambda \mapsto a\lambda + b.$$

Since $a \neq 0$, it is straight-forward to see that the map is bijective. Hence it is a set-theoretic automorphism of k.

(c) Let $\varphi : G(k) \to aut_{set}(k)$, $ax + b \mapsto$ (the automorphism $\lambda \mapsto a\lambda + b$) be the map. Given any $f = f(x) = a_f x + b_f$ and $g = g(x) = a_g x + b_g$ in $G(k)$, then $\varphi(f)$ is the map

$$k \to k, \lambda \mapsto a_f \lambda + b_f,$$

and $\varphi(g)$ is the map

$$k \to k, \ \lambda \mapsto a_g \lambda + b_g.$$

Since $(f \cdot g)(x) = f(a_g x + b_g) = (a_f a_g)x + (a_f b_g + b_f)$, $\varphi(f \cdot g)$ is the map

$$k \to k, \lambda \mapsto a_f a_g \lambda + a_f b_g + b_f.$$

And $\varphi(f) \cdot \varphi(g)$ is the map

$$k \to k, \lambda \mapsto \varphi(f)(a_g \lambda + b_g) = a_f a_g \lambda + a_f b_g + b_f.$$

So $\varphi(f \cdot g) = \varphi(f) \cdot \varphi(g)$. Hence φ is a group homomorphism.

Note that the identity of the automorphism group $aut_{set}(k)$ is $I : k \to k, \lambda \mapsto \lambda$. It is easy to see that the kernel of φ is $\{e = e(x) = x\}$, i.e., the kernel of φ is trivial. Hence φ is an injective group homomorphism.

(d) Pick any element $g = cx + d$ in $G(k)$. Recall that $g^{-1} = \frac{1}{c}x - \frac{d}{c}$ by part (a). Now take any element $f = x + b$ in $G_1(k)$. Then

$$gfg^{-1} = (gfg^{-1})(x) = (gf)\left(\frac{1}{c}x - \frac{d}{c}\right)$$

$$= g\left(\frac{1}{c}x - \frac{d}{c} + b\right) = x + cb,$$

which is in $G_1(k)$. It follows that $G_1(k)$ is a normal subgroup of $G(k)$.

Next we show that the map $\varphi : G_1(k) \to (k, +)$, $x+b \mapsto b$ is a group isomorphism. For any $f_1 = x+b_1$ and $f_2 = x+b_2 \in G_1(k)$,

$$\varphi(f_1 \cdot f_2) = \varphi((x + b_2) + b_1) = \varphi(x + b_1 + b_2) = b_1 + b_2$$

$$= \varphi(f_1) + \varphi(f_2),$$

so φ is a group homomorphism. Each $b \in k$ has, and can only have $x+b$ as its pre-image in the group $G_1(k)$, so φ is bijective. Therefore φ is an isomorphism.

(e) Let ψ be the map $G(k) \to k^*$, $ax + b \mapsto a$. Obviously, ψ is surjective. For any $f_1 = a_1x + b_1$ and $f_2 = a_2x + b_2 \in G(k)$,

$$\psi(f_1 \cdot f_2) = \psi(a_1(a_2x + b_2) + b_1) = \psi(a_1a_2x + (a_1b_2 + b_1))$$

$$= a_1a_2 = \psi(a_1x + b_1)\psi(a_2x + b_2)$$

$$= \psi(f_1)\psi(f_2).$$

Therefore ψ is a group homomorphism. An element of $G(k)$ is in the kernel of ψ if and only if ψ maps it to 1 in the multiplicative group k^*. Thus the kernel of ψ is precisely $G_1(k)$.

(f) Let Γ be a non-trivial normal subgroup of $G(k)$. To prove $G_1(k) \subseteq \Gamma$, we need to show that for any $x+l \in G_1(k)$, $x+l \in \Gamma$. Pick an element $e \neq f = cx+d \in \Gamma$. For any $g = ax+b \in G(k)$, $gfg^{-1} \in \Gamma$ as Γ is normal. Note that $g^{-1} = \frac{1}{a}x - \frac{b}{a}$ and

$$(gfg^{-1})(x) = gf\left(\frac{1}{a}x - \frac{b}{a}\right) = g\left(c\left(\frac{1}{a}x - \frac{b}{a}\right) + d\right)$$

$$= g\left(\frac{c}{a}x - \frac{bc}{a} + d\right) = a\left(\frac{c}{a}x - \frac{bc}{a} + d\right) + b$$

$$= cx - bc + ad + b.$$

Since Γ is a subgroup, $f^{-1} = \frac{1}{c}x - \frac{d}{c} \in \Gamma$, and consequently $gfg^{-1}f^{-1} \in \Gamma$.

$$(gfg^{-1}f^{-1})(x) = c\left(\frac{1}{c}x - \frac{d}{c}\right) - bc + ad + b$$

$$= x - d - bc + ad + b$$

$$= x + d(a - 1) - b(c - 1).$$

If $c = 1$, then $d \neq 0$ as $f \neq e$. Now given any $x + l \in G_1(k)$,

$$x + l = \begin{cases} (gfg^{-1}f^{-1})(x), & \text{if } c = 1,\ d \neq 0, -l,\ \text{taking} \\ & \qquad a = (l + d)d^{-1}, \\ f^{-1}, & \text{if } c = 1,\ d = -l, \\ (gfg^{-1}f^{-1})(x), & \text{if } c \neq 1,\ \text{taking} \\ & \qquad a = 1, b = (-l)(c - 1)^{-1}, \end{cases}$$

is in Γ. Since we have exhausted all the possibilities of c, d in the expression of $f = cx + d \in \Gamma$, we have shown that any element in $G_1(k)$ is in Γ, i.e., $G_1(k) \subseteq \Gamma$.

(g) By part (e), $G(k)/G_1(k) \cong k^*$ is an abelian group. The conclusion follows. We may also check it directly.

For any $f = ax + b \in G(k)$ and any $g = cx + d \in \Gamma$, we need to show that $fgf^{-1} \in \Gamma$. Note $(fgf^{-1})(x) = cx - cb + ad + b$. Now let $h = x - b + \frac{ad}{c} - \frac{d}{c} + \frac{b}{c}$ (we can do this because $c \neq 0$). Then $h \in \Gamma$ because $h \in G_1(k) \subseteq \Gamma$. Therefore $gh \in \Gamma$. Now we have

$$(fgf^{-1})(x) = cx - cb + ad + b$$

$$= cx - cb + ad - d + b + d$$

$$= c\left(x - b + \frac{ad}{c} - \frac{d}{c} + \frac{b}{c}\right) + d = (gh)(x) \in \Gamma.$$

It follows that Γ is a normal subgroup of $G(k)$.

1425

Following 1424, suppose now that $k = \mathbb{F}_q$ is a finite field of characteristic $p > 0$.

(a) Compute $\#G(\mathbb{F}_q)$ and $\#G_1(\mathbb{F}_q)$.

(b) Show that $G_1(\mathbb{F}_q)$ is the unique p-Sylow subgroup of $G(\mathbb{F}_q)$.

(c) Let $\Gamma \subset G(\mathbb{F}_q)$ be a non-trivial subgroup. Show that Γ is normal in $G(\mathbb{F}_q)$ if and only if $q | \#\Gamma$.

(d) Let $d \geq 1$ be a divisor of $q - 1$. Show that the unique subgroup of $G(\mathbb{F}_q)$ of order qd consists of all elements $ax + b$ with a, b in $\mathbb{F}_q$ and $a^d = 1$.

(Princeton)

Solution.

(a) Since k is of characteristic $p > 0$, $q = p^n$ for some $n > 0$. Since $G(\mathbb{F}_q)$ consists of linear polynomials $ax + b$, where $a \in \mathbb{F}_q^*$ has $q - 1$ choices and $b \in \mathbb{F}_q$ has q choices, $\#G(\mathbb{F}_q) = q(q-1) = p^n(p^n - 1)$. For $ax + b \in G_1(\mathbb{F}_q)$, there is one choice for a, i.e., $a = 1$, and q choices for b. So $\#G_1(\mathbb{F}_q) = q = p^n$.

(b) Since $p \nmid p^n - 1$, p^n has the greatest power of p among the divisors of $p^n(p^n - 1) = \#G(\mathbb{F}_q)$. Therefore a p-Sylow subgroup of $G(\mathbb{F}_q)$ is any subgroup with order p^n. Since $G_1(\mathbb{F}_q)$ has order p^n and is a subgroup, it is a p-Sylow subgroup of $G(\mathbb{F}_q)$.

Then we show its uniqueness. By the second Sylow Theorem, all p-Sylow subgroups of $G(\mathbb{F}_q)$ are conjugates of $G_1(\mathbb{F}_q)$. Since $G_1(\mathbb{F}_q)$ is a normal subgroup, any conjugate of $G_1(\mathbb{F}_q)$ is just itself. Therefore the only p-Sylow subgroup of $G(\mathbb{F}_q)$ is $G_1(\mathbb{F}_q)$.

(c) By parts (f) and (g) in 1424, the subgroup Γ is normal in $G(\mathbb{F}_q)$ if and only if it has $G_1(\mathbb{F}_q)$ as a subgroup. So to prove the statement, it is equivalent to prove that $G_1(\mathbb{F}_q) \subseteq \Gamma$ if and only if $q | \#\Gamma$.

If $q | \#\Gamma$, then there exits a p-Sylow subgroup with order p^n in Γ, and consequently in $G(\mathbb{F}_q)$. By part (b), the only p-Sylow subgroup of $G(\mathbb{F}_q)$ is $G_1(\mathbb{F}_q)$, so $G_1(\mathbb{F}_q) \subseteq \Gamma$. Conversely, if Γ contains $G_1(\mathbb{F}_q)$, then $\#G_1(\mathbb{F}_q)|\#\Gamma$, so $q|\#\Gamma$.

(d) First we show that the subset $H_{qd} = \{ax + b \in G(\mathbb{F}_q)|a^d = 1\}$ is a subgroup of $G(\mathbb{F}_q)$ of order qd. Given any $f = ax + b$ and $g = cx + d$ in H_{qd}, then $a^d = 1$, $c^d = 1$, and

$$f \cdot g^{-1} = a \left(\frac{1}{c}x - \frac{d}{c} \right) + b = \frac{a}{c}x - \frac{ad}{c} + b.$$

Since $(\frac{a}{c})^d = 1$, so $f \cdot g^{-1} \in H_{qd}$. Hence H_{qd} is a subgroup of $G(\mathbb{F}_q)$.

There are d choices for a in this subgroup because there are d distinct roots of the polynomial $x^d - 1$ in $\mathbb{F}_q$, and there are q choices for b since it can be any of the q elements in $\mathbb{F}_q$. Therefore the subgroup H_{qd} has order qd.

Next we show the uniqueness. Let H be another subgroup of $G(\mathbb{F}_q)$ with order qd. Then by part (c), H is a normal subgroup of $G(\mathbb{F}_q)$. By part (f) in 1424, $H \supseteq G_1(\mathbb{F}_q)$. By part (e) in 1424, $G(\mathbb{F}_q)/G_1(\mathbb{F}_q) \cong \mathbb{F}_q^*$, $\overline{ax + b} \mapsto a$. Therefore $H/G_1(\mathbb{F}_q)$ is isomorphic to a subgroup of k^* of order d. So, given any element $ax + b \in H$, $a^d = 1$. It follows that $H \subseteq H_{qd}$. Since both H and H_{qd} are of order qd, so $H = H_{qd}$.

1426

Let p and l be two (not-necessarily distinct) prime numbers. Consider the polynomial $x^p - l \in \mathbb{Z}[x]$, and denote by K its splitting field over $\mathbb{Q}$.

(a) Show $x^p - l$ is irreducible over $\mathbb{Q}$.

(b) Show K contains a primitive p-th root of unity ζ.

(c) Show deg $(K/\mathbb{Q}) = p(p - 1)$.

(d) Fix an element $\alpha \in K$ with $\alpha^p = l$. Show that for any $\sigma \in \mathrm{Gal}(K/\mathbb{Q})$, there exists unique elements $a \in \mathbb{F}_p^*$ and $b \in \mathbb{F}_p$

such that

$$\begin{cases} \sigma(\alpha) = \zeta^b \alpha, \\ \sigma(\zeta) = \zeta^a, \end{cases}$$

and show that the map $\sigma \mapsto ax + b$ defines an isomorphism of groups $\mathrm{Gal}(K/\mathbb{Q}) \to G(\mathbb{F}_p)$ (see notations in 1424).

(e) Show that the map of sets $\mathbb{F}_p \to \{\text{roots of } x^p - l \text{ in } K\}$ defined by $b \mapsto \zeta^b \alpha$ is bijective, and that by means of this bijection, the composite map

$$\mathrm{Gal}(K/\mathbb{Q}) \cong G(\mathbb{F}_p) \hookrightarrow aut_{set}(\mathbb{F}_p)$$
$$\cong aut_{set}(\text{roots of } x^p - l \text{ in } K)$$

is the permutation action of the Galois group on the roots.

(f) Show that if F is an intermediate field, i.e., $\mathbb{Q} \subseteq F \subseteq K$, and if $F \neq K$, then $F/\mathbb{Q}$ is Galois if and only if $F \subseteq \mathbb{Q}(\zeta)$.

(Princeton)

Solution.

(a) By Eisenstein Criterion, $x^p - l$ is irreducible over $\mathbb{Q}$ as l is prime.

(b) Let $\alpha_1, \alpha_2, \ldots, \alpha_p$ be the roots of $x^p - l$ in K. Let $\zeta_i = \alpha_i \alpha_1^{-1}$ for $1 \leq i \leq p$. Since the α_i's are distinct, so are the ζ_i's. So $\{\zeta_1, \zeta_2, \ldots, \zeta_p\} \subset K$ is the set of all p-th roots of unity, because $\zeta_i^p = (\alpha_i \alpha_1^{-1})^p = \alpha_i^p (\alpha_1^p)^{-1} = ll^{-1} = 1$ for any $1 \leq i \leq p$. Hence K contains a primitive p-th root of unity.

(c) Let $\zeta \in K$ be a primitive p-th root of unity, and $\alpha \in K$ be a root of $x^p - l$. Then over K

$$x^p - l = (x - \alpha)(x - \zeta\alpha)(x - \zeta^2\alpha) \cdots (x - \zeta^{p-1}\alpha)$$

and $K = \mathbb{Q}[\zeta, \alpha]$. Since $f(x) = x^{p-1} + x^{p-2} + \cdots + 1$ is the minimal polynomial of ζ over $\mathbb{Q}$ (by Eisenstein Criterion, $f(x + 1) = x^{p-1} + C_p^1 x^{p-2} + \cdots + C_p^2 x + p$, hence $f(x)$, is irreducible), $\deg(\mathbb{Q}(\zeta)/\mathbb{Q}) = p - 1$. Since $x^p - l$ is irreducible over $\mathbb{Q}$, $\deg(\mathbb{Q}(\alpha)/\mathbb{Q}) = p$. Since $K = \mathbb{Q}[\zeta][\alpha]$, $\deg(K/\mathbb{Q}[\zeta]) \leq p$. Hence

$$\deg(K/\mathbb{Q}) = \deg(K/\mathbb{Q}(\zeta)) \cdot \deg(\mathbb{Q}(\zeta)/\mathbb{Q}) \leq p(p - 1).$$

Since $p = \deg(\mathbb{Q}(\alpha)/\mathbb{Q})| \deg(K/\mathbb{Q})$ and $p - 1 = \deg(\mathbb{Q}(\zeta)/\mathbb{Q})|$ $\deg(K/\mathbb{Q})$,

$$\deg(K/\mathbb{Q}) = p(p-1).$$

(d) For any $\sigma \in \mathrm{Gal}(K/\mathbb{Q})$, since l is in the ground field $\mathbb{Q}$, $(\sigma(\alpha))^p = \sigma(\alpha^p) = \sigma(l) = l$. Therefore $\sigma(\alpha)$ is another root of $x^p - l$ in K. Therefore $\sigma(\alpha) = \zeta^b \alpha$ for some b. Since α is fixed, this b is uniquely determined for σ up to modulo p.

Similarly, since $(\sigma(\zeta))^p = \sigma(\zeta^p) = \sigma(1) = 1$, so $\sigma(\zeta)$ is another p-th root of unity and so it must be ζ^a for some a. Again since ζ is fixed, this a is uniquely determined up to modulo p and $a \neq 0 \bmod p$. Therefore with a fixed α and a fixed ζ, each σ corresponds uniquely to an element $(a, b) \in \mathbb{F}_p^* \times \mathbb{F}_p$. Hence there is a map:

$$\varphi : \mathrm{Gal}(K/\mathbb{Q}) \to G(\mathbb{F}_p), \sigma \mapsto ax + b$$

where $(a, b) \in \mathbb{F}_p^* \times \mathbb{F}_p$ satisfies that $\sigma(\zeta) = \zeta^a$ and $\sigma(\alpha) = \zeta^b \alpha$.

For any $\sigma_1, \sigma_2 \in \mathrm{Gal}(K/\mathbb{Q})$, if $\varphi(\sigma_1) = a_1 x + b_1$ and $\varphi(\sigma_2) = a_2 x + b_2$, i.e., $\sigma_1(\zeta) = \zeta^{a_1}$, $\sigma_1(\alpha) = \zeta^{b_1}\alpha$, and $\sigma_2(\zeta) = \zeta^{a_2}$, $\sigma_2(\alpha) = \zeta^{b_2}\alpha$, then

$$(\sigma_1\sigma_2)(\zeta) = \sigma_1(\zeta^{a_2}) = \zeta^{a_1 a_2},$$
$$(\sigma_1\sigma_2)(\alpha) = \sigma_1(\zeta^{b_2}\alpha) = \sigma_1(\zeta^{b_2})\sigma_1(\alpha) = \zeta^{a_1 b_2}\zeta^{b_1}\alpha = \zeta^{a_1 b_2 + b_1}\alpha.$$

It follows that

$$\varphi(\sigma_1\sigma_2) = (a_1 a_2)x + (a_1 b_2 + b_1)$$
$$= a_1(a_2 x + b_2) + b_1$$
$$= \varphi(\sigma_1)(a_2 x + b_2)$$
$$= \varphi(\sigma_1)\varphi(\sigma_2).$$

Therefore φ is a homomorphism. Obviously, φ is injective. Since both the order of $\mathrm{Gal}(K/\mathbb{Q})$ and $G(\mathbb{F}_p)$ are $p(p-1)$, $\varphi : \mathrm{Gal}(K/\mathbb{Q}) \to G(\mathbb{F}_p)$ is an isomorphism.

(e) The roots of $x^p - l$ in K are of the form $\zeta^b \alpha$, where $b = 0, 1, \ldots, p-1$, with the same notation as in part (d). Let ψ be the map

$$\mathbb{F}_p \to \{\text{roots of } x^p - l \text{ in } K\}, b \mapsto \zeta^b \alpha.$$

Then ψ is surjective. Obviously, ψ is also injective.

For the second part, starting with $\sigma \in \mathrm{Gal}(K/\mathbb{Q})$, we follow down the maps. By part (d), $\sigma \mapsto ax + b \in G(\mathbb{F}_q)$, where $(a, b) \in \mathbb{F}_p^* \times \mathbb{F}_p$ satisfies that $\sigma(\zeta) = \zeta^a$ and $\sigma(\alpha) = \zeta^b \alpha$. Then by part (c) in 1424, the corresponding element in $aut_{set}(\mathbb{F}_p)$ is $\mathbb{F}_p \to \mathbb{F}_p$, $\lambda \mapsto a\lambda + b$. By the bijection just established at the beginning of this question, the corresponding element in aut_{set} (roots of $x^p - l$ in K) is

$$\{\text{roots of } x^p - l \text{ in } K\} \to \{\text{roots of } x^p - l \text{ in } K\} : \zeta^\lambda \alpha \mapsto \zeta^{a\lambda + b} \alpha.$$

So the composition map is:

$$\mathrm{Gal}(K/\mathbb{Q}) \to aut_{set}(\text{roots of } x^p - l \text{ in } K), \sigma \mapsto \{\zeta^\lambda \alpha \mapsto \zeta^{a\lambda + b} \alpha\}$$

if $\sigma(\zeta) = \zeta^a$ and $\sigma(\alpha) = \zeta^b \alpha$.

Since $\sigma(\zeta^\lambda \alpha) = \sigma(\zeta)^\lambda \sigma(\alpha) = \zeta^{a\lambda} \zeta^b \alpha = \zeta^{a\lambda + b} \alpha$ for any $0 \leq \lambda \leq p - 1$, so the composition map is exactly the permutation action of the Galois group on the roots.

(f) First note that $F/\mathbb{Q}$ is Galois if and only if $\mathrm{Gal}(K/F)$ is a normal subgroup of $\mathrm{Gal}(K/\mathbb{Q})$. By the definition of Galois extension, K/F is Galois.

If $F \subseteq \mathbb{Q}(\zeta)$, then $\mathrm{Gal}(K/\mathbb{Q}(\zeta)) \subseteq \mathrm{Gal}(K/F)$. Since $\mathrm{Gal}(K/\mathbb{Q})$ is isomorphic to $G(\mathbb{F}_p)$ by part (d), we can look at $G(\mathbb{F}_p)$ instead to study $\mathrm{Gal}(K/F)$ and $\mathrm{Gal}(K/\mathbb{Q}(\zeta))$ as subgroups. $\mathrm{Gal}(K/F)$ is isomorphic to some subgroup in $G(\mathbb{F}_p)$, while $\mathrm{Gal}(K/\mathbb{Q}(\zeta))$ is isomorphic specifically to $G_1(\mathbb{F}_p)$. The latter is so because any element in $\mathrm{Gal}(K/\mathbb{Q}(\zeta))$ must fix ζ, which means that for any $\sigma \in \mathrm{Gal}(K/\mathbb{Q}(\zeta))$, $\sigma(\zeta) = \zeta^1$, so by the map provided in part (d), the corresponding a is 1, while b can be anything in $\mathbb{F}_p$. This means that the corresponding subgroup of $\mathrm{Gal}(K/\mathbb{Q}(\zeta))$ in

$G(\mathbb{F}_p)$ is $G_1(\mathbb{F}_p)$. Now by part (g) in 1424, the corresponding subgroup of $\mathrm{Gal}(K/F)$ in $G(\mathbb{F}_p)$ is normal, so $\mathrm{Gal}(K/F)$ is a normal subgroup of $\mathrm{Gal}(K/\mathbb{Q})$. Therefore $F/\mathbb{Q}$ is a Galois extension.

Conversely, if $F/\mathbb{Q}$ is Galois, then $\mathrm{Gal}(K/F)$ is a normal subgroup of $\mathrm{Gal}(K/\mathbb{Q})$. Since $F \neq K$, $\mathrm{Gal}(K/F) \neq \{e\}$. By parts (c) and (d) in 1425, any normal subgroup of $G(\mathbb{F}_p)$ consists of elements $ax + b$ with a, b in $\mathbb{F}_p$ and $a^d = 1$ for some fixed $d|p - 1$. By the map provided in part (d), back in $\mathrm{Gal}(K/\mathbb{Q})$, the normal subgroup $\mathrm{Gal}(K/F)$ must accordingly consist of map such that $\sigma(\zeta) = \zeta^a$, where $a^d = 1$ for some fixed $d|p - 1$. Now for any $\sigma \in \mathrm{Gal}(K/\mathbb{Q}(\zeta))$, since $\sigma(\zeta) = \zeta^1$ and $1^d = 1$, so $\sigma \in \mathrm{Gal}(K/F)$. It follows that $\mathrm{Gal}(K/\mathbb{Q}(\zeta)) \subseteq \mathrm{Gal}(K/F)$. Thus by the Galois correspondence, $F \subseteq \mathbb{Q}(\zeta)$.

1427

Let p be a prime number which is odd.

(a) Show that $\mathbb{F}_p^*$ has a unique subgroup of order $(p - 1)/2$, namely the subgroup consisting of squares of elements of $\mathbb{F}_p^*$.

(b) Show that the map $\chi_2 : \mathbb{F}_p^* \to \{1, -1\}$ (the multiplicative group) defined by

$$\chi_2(a) = \begin{cases} 1, & \text{if } a = b^2 \text{ for some } b \in \mathbb{F}_p^*, \\ -1, & \text{if not,} \end{cases}$$

is a homomorphism (called the "quadratic character mod p").

(c) Show that $\chi_2(a) = a^{\frac{p-1}{2}}$.

(d) Let ζ_p be the complex number $e^{2\pi i/p}$, a primitive p-th root of unity in $\mathbb{C}$. So $\mathbb{Q}(\zeta_p)$ is a subfield of $\mathbb{C}$. Denote by $c : z \mapsto \bar{z}$ the automorphism "complex conjugation" of $\mathbb{C}$, show that $\mathbb{Q}(\zeta_p)$ is mapped to itself by c, and that c induces an automorphism of $\mathbb{Q}(\zeta_p)$ which under the isomorphism $\mathrm{Gal}(\mathbb{Q}(\zeta_p)/\mathbb{Q}) \cong \mathbb{F}_p^*$ corresponds to the element $-1 \in \mathbb{F}_p^*$.

(e) Show that the number in $\mathbb{Q}(\zeta_p)$ defined by the "Gauss Sum"

$$g = \sum_{a \in \mathbb{F}_p^*} \chi_2(a)(\zeta_p)^a$$

satisfies, as complex number, the relations

$$\begin{cases} g\bar{g} = p, \\ \bar{g} = \chi_2(-1) \cdot g. \end{cases}$$

(f) Show that $\chi_2(-1) = 1$ if and only if $p \equiv 1 \bmod 4$.

(g) Deduce from (e) and (f) above that

$$\begin{cases} \sqrt{p} \in \mathbb{Q}(\zeta_p), & \text{if } p \neq 1 \bmod 4, \\ \sqrt{-p} \in \mathbb{Q}(\zeta_p), & \text{if } p \neq 3 \bmod 4. \end{cases}$$

(Princeton)

Solution.

(a) Since $\mathbb{F}_p^*$ is a cyclic group of order $p - 1$ and p is odd, $\mathbb{F}_p^*$ has a unique subgroup of order $(p-1)/2$.

Let $G = \{a \in \mathbb{F}_p^* \mid a = m^2 \text{ for some } m \in \mathbb{F}_p^*\}$. Given any $a, b \in G$, say, $a = m^2$ and $b = n^2$ for some $m, n \in \mathbb{F}_p^*$, then $ab^{-1} = m^2(n^{-1})^2 = (mn^{-1})^2 \in G$ as $mn^{-1} \in \mathbb{F}_p^*$. Hence, G, i.e., the set of squares of the elements of $\mathbb{F}_p^*$, is a subgroup of $\mathbb{F}_p^*$.

For any $m \in \{1, 2, \ldots, (p-1)/2\}$, $(p-m)^2 = p^2 - 2mp + m^2$, and so $(p-m)^2 \equiv a^2 \bmod p$. Therefore there can be at most $(p-1)/2$ different elements in $\mathbb{F}_p^*$ that are squares.

On the other hand, if $m^2 = n^2 \bmod p$ for some distinct $m, n \in \{1, 2, \ldots, (p-1)/2\}$, then $p \mid (m+n)(m-n)$. However, since $(m+n) < p - 1, -p + 2 < (m-n) < p - 2$, the prime p cannot divide $(m+n)(m-n)$. Therefore we have shown by contradiction that there is no repeat among the squares of the elements $1, 2, \ldots, (p-1)/2$.

Therefore, the order of the group G is exactly $(p-1)/2$, and G is the unique subgroup of $\mathbb{F}_p^*$ of order $(p-1)/2$.

(b) By part (a), $\mathbb{F}_p^*/G$ is of order 2, which is isomorphic to the multiplicative group $\{1, -1\}$. So the map $\chi_2 : \mathbb{F}_p^* \to \{1, -1\}$ is the composition of

$$\mathbb{F}_p^* \to \mathbb{F}_p^*/G \cong \{1, -1\},$$

which is a group homomorphism.

(c) For any $a \in \mathbb{F}_p^*$, $a^{\frac{p-1}{2}}$ is an element in $\mathbb{F}_p^*$ whose square is 1 since $a^{p-1} = 1$ in $\mathbb{F}_p^*$. In part (a) we have shown that for any distinct $m, n \in \{1, 2, \ldots, (p-1)/2\}$, $m^2 \neq n^2 \bmod p$. And none of other elements in $\mathbb{F}_p^*$ squares to give 1 except 1 and $p-1$. Thus $a^{\frac{p-1}{2}}$ must be either 1 or $p-1$ in $\mathbb{F}_p^*$, i.e., $a^{\frac{p-1}{2}} = 1$ or -1 in $\mathbb{F}_p^*$.

Next we claim that $a^{\frac{p-1}{2}} = 1$ if and only if a is a square. If $a = b^2$ for some $b \in \mathbb{F}_p^*$, then $a^{\frac{p-1}{2}} = b^{p-1} = 1$ in $\mathbb{F}_p^*$. On the other hand, all the square elements in $\mathbb{F}_p^*$ are roots of the polynomial $x^{\frac{p-1}{2}} - 1$ over $\mathbb{F}_p$, and this polynomial has exactly $\frac{p-1}{2}$ roots, which are precisely $\frac{p-1}{2}$ elements that are squares in $\mathbb{F}_p^*$ by part (a). Therefore if an element $a \in \mathbb{F}_p^*$ satisfies that $a^{\frac{p-1}{2}} = 1$, then a is a square.

We have also shown that $a^{\frac{p-1}{2}} = -1$ if and only if a is not a square in $\mathbb{F}_p^*$.

Therefore, $\chi_2(a) = a^{\frac{p-1}{2}}$ for any $a \in \mathbb{F}_p^*$ by the definition of χ.

(d) Since $c(\zeta_p) = c(e^{2\pi i/p}) = e^{-2\pi i/p} = \zeta_p^{p-1}$. Therefore $c(\mathbb{Q}(\zeta_p)) = \mathbb{Q}(\zeta_p^{p-1})$, which is exactly the same as $\mathbb{Q}(\zeta_p)$. Therefore $\mathbb{Q}(\zeta_p)$ is mapped to itself by c, and c induces an automorphism in $\mathrm{Gal}(\mathbb{Q}(\zeta_p)/\mathbb{Q})$. Note that $\mathrm{Gal}(\mathbb{Q}(\zeta_p)/\mathbb{Q})$ is a cyclic group of order $p - 1$, and the map

$$\mathrm{Gal}(\mathbb{Q}(\zeta_p)/\mathbb{Q}) \to \mathbb{F}_p^*, \ \sigma \mapsto i, \ \text{if } \sigma(\zeta_p) = \zeta_p^i \text{ for some } i \in \mathbb{F}_p^*,$$

is a group isomorphism. Since $c(\zeta_p) = \zeta_p^{p-1} = \zeta_p^{-1}$, so c corresponds to -1 under the above isomorphism.

(e) Since $\overline{\zeta_p} = \zeta_p^{p-1} = \zeta_p^{-1}$,

$$\bar{g} = \sum_{a \in \mathbb{F}_p^*} \chi_2(a) \zeta_p^{-a} = \sum_{a \in \mathbb{F}_p^*} \chi_2(-1) \chi_2(-a) \zeta_p^{-a}$$

$$= \chi_2(-1) \sum_{a \in \mathbb{F}_p^*} \chi_2(a) \zeta_p^a$$

$$= \chi_2(-1) g.$$

So

$$g\bar{g} = \chi_2(-1) g^2 = \chi_2(-1) \sum_{a,b \in \mathbb{F}_p^*} \chi_2(ab)(\zeta_p)^{a+b}$$

$$= \chi_2(-1) \sum_{a \in \mathbb{F}_p^*} \left[\sum_{b \in \mathbb{F}_p^*} \chi_2(ab)(\zeta_p)^{a+b} \right]$$

$$= \chi_2(-1) \sum_{a \in \mathbb{F}_p^*} \left[\sum_{b \in \mathbb{F}_p^*} \chi_2(a^2 b)(\zeta_p)^{a+ab} \right]$$

$$= \chi_2(-1) \sum_{b \in \mathbb{F}_p^*} \chi_2(b) \left[\sum_{a \in \mathbb{F}_p^*} (\zeta_p)^{a(b+1)} \right]$$

$$= \chi_2(-1) \chi_2(-1) \sum_{a \in \mathbb{F}_p^*} \zeta_p^{a \cdot 0}$$

$$+ \chi_2(-1) \sum_{-1 \neq b \in \mathbb{F}_p^*} \chi_2(b) \left[\sum_{a \in \mathbb{F}_p^*} (\zeta_p)^{a(b+1)} \right]$$

$$= p - 1 + \chi_2(-1) \sum_{-1 \neq b \in \mathbb{F}_p^*} \chi_2(b) \left(\sum_{a \in \mathbb{F}_p^*} \zeta_p^a \right)$$

$$= p - 1 - \chi_2(-1) \sum_{-1 \neq b \in \mathbb{F}_p^*} \chi_2(b)$$

$$= p - 1 - \sum_{1 \neq b \in \mathbb{F}_p^*} \chi_2(b)$$

$$= p - 1 + \chi_2(1)$$

$$= p.$$

(f) By part (c), $\chi_2(-1) = (-1)^{\frac{p-1}{2}} \in \mathbb{F}_p$. If $p \equiv 1 \bmod 4$, say, $p = 4k+1$, then $\chi_2(-1) = (-1)^{2k} = 1$. Since p is odd, it can only

be of the form $4k+1$ or $4k+3$. If p is of the form $4k+3$, then
$\chi_2(-1) = (-1)^{\frac{(4k+3)-1}{2}} = (-1)^{2k+1} = -1 \neq 1 \in \mathbb{F}_p$. Therefore
$\chi_2(-1) = 1$ if and only if $p \equiv 1 \mod 4$.

(g) By part (e), $p = g\bar{g} = \chi_2(-1) \cdot g^2$. By part (f),

$$\begin{cases} \chi_2(-1) = 1, & \text{if } p \equiv 1 \mod 4, \\ \chi_2(-1) = -1, & \text{if } p \equiv 3 \mod 4, \end{cases}$$

Therefore,

$$\begin{cases} p = g^2, & \text{if } p \equiv 1 \mod 4, \\ p = -g^2, & \text{if } p \equiv 3 \mod 4, \end{cases}$$

and so

$$\begin{cases} \sqrt{p} = g \in \mathbb{Q}(\zeta_p), & \text{if } p \equiv 1 \mod 4, \\ \sqrt{-p} = g \in \mathbb{Q}(\zeta_p), & \text{if } p \equiv 3 \mod 4. \end{cases}$$

1428

Prove or disprove the following:

(a) If a monic polynomial f in $\mathbb{Z}[x]$ reduces modulo a prime number p to an irreducible polynomial in $\mathbb{F}_p[x]$, then f is irreducible in $\mathbb{Q}[x]$.

(b) $x^p - x - 1$ is irreducible in $\mathbb{F}_p[x]$.

(c) Given any field F, and any finite group G, there exists a Galois extension E/F with Galois group G.

(d) If $L \supseteq E \supseteq F$ is a chain of fields with L/E Galois and E/F Galois, then L/F is Galois.

(e) If F is a field of characteristic $p > 0$, and if E/F is a finite Galois extension, then p does not divide the degree of E/F.

(Princeton)

Solution.

(a) Claim: the statement is true.

Proof of the claim. Note that a monic polynomial is irreducible in $\mathbb{Q}[x]$ if and only if it is irreducible in $\mathbb{Z}[x]$. Assume there exists a reducible monic polynomial f in $\mathbb{Z}[x]$ that reduces modulo p to an irreducible polynomial f_p in $\mathbb{F}_p[x]$. Write $f = gh$, a non-trivial factorization into two monic polynomials in $\mathbb{Z}[x]$. Since $f = gh$, $f_p = g_p h_p \in \mathbb{F}_p[x]$ is a non-trivial factorization. It follows that f_p is not irreducible, a contradiction.

(b) Claim: $x^p - x - 1$ is irreducible in $\mathbb{F}_p[x]$.

Proof of the claim. As proved in 1410, $x^p - x - 1$ is either irreducible or splits completely into p linear factors over $\mathbb{F}_p$. Since all elements in $\mathbb{F}_p$ satisfy $x^p - x = 0$. Therefore $x^p - x - 1 = 0$ has no root in $\mathbb{F}_p$. Hence $x^p - x - 1$ is irreducible in $\mathbb{F}_p[x]$.

(c) Claim: the statement is not true in general.

In the case that $F = \mathbb{Q}$, this is the Inverse Galois Problem: "Does every finite group G appear as a Galois group over $\mathbb{Q}$?" which is still open.

Proof of the claim. Let $F = \mathbb{F}_2$. Let $L = \mathbb{F}_4$ and $E = \mathbb{F}_{16}$ be the splitting field of the irreducible polynomials $x^2 + x + 1$ and $x^4 + x + 1$ over $\mathbb{F}_2$ respectively. Since both $x^2 + x + 1$ and $x^4 + x + 1$ are separable, L/F and E/F are all Galois extension. Let $\alpha \in L$ be a root of $x^2 + x + 1$. Then $L = \{0, 1, \alpha, \alpha + 1\}$. Note that $x^4 + x + 1 = (x^2 + x + \alpha)(x^2 + x + \alpha + 1)$ in $L[x]$. Let $\beta \in E$ be a root of $x^2 + x + \alpha$. Then $\beta, \beta + 1, \beta + \alpha, \beta + \alpha + 1$ are the roots of $x^4 + x + 1$ in E, $E = L[\beta]$ and E/L is also a Galois extension. It is easy to see that $\mathrm{Gal}(L/F) = \{1, \sigma\}$, where σ is the F-automorphism with $\sigma(\alpha) = \alpha + 1$, $\mathrm{Gal}(E/L) = \{1, \tau\}$, where τ is the L-automorphism with $\tau(\beta) = \beta + 1$, and $\mathrm{Gal}(E/F)$ is the Klein four group.

Since any field extension of $\mathbb{F}_2$ of degree 4 is isomorphic to the extension $\mathbb{F}_{16}/\mathbb{F}_2$, so there exists no Galois extension of $\mathbb{F}_2$ with $\mathbb{Z}/(4)$ as the Galois group.

(d) Claim: the statement is false.

Proof of the claim. Let F be $\mathbb{Q}$, and E be $\mathbb{Q}(\sqrt{2})$. Then E/F is a Galois extension because E is the splitting field of the irreducible polynomial $x^2 - 2$ in $F[x]$. Let L be $\mathbb{Q}(\sqrt[4]{2})$. Then L/E is also Galois because L is the splitting field of $x^2 - \sqrt{2}$ in

$E[x]$. The minimal polynomial of $\sqrt[4]{2}$ over F is $x^4 - 2 \in F[x]$. Note that

$$x^4 - 2 = (x + \sqrt[4]{2})(x - \sqrt[4]{2})(x + i\sqrt[4]{2})(x - i\sqrt[4]{2}).$$

For a Galois extension, if one of the roots of an irreducible polynomial over the ground field is in the extension field, then all roots of this polynomial have to be in the extension field. Now $\sqrt[4]{2}$ is in L, but it is obvious that $i\sqrt[4]{2}$ is not in L. Therefore L/F is not Galois.

(e) Claim: the statement is not true.

Proof of the claim. We give an example of Galois extension that disproves the statement. Let $F = \mathbb{F}_p$. We have shown in part (b) that $x^p - x - 1$ is irreducible in $\mathbb{F}_p[x]$. This polynomial is separable, because differentiating it gives -1. Let E be the splitting field of this polynomial. Then E/F is a finite Galois extension. Let a be a root of this polynomial in E. It is easy to see that $a+1$ is also a root. Hence $a, a+1, \ldots, a+p-1$ are all the p distinct roots of $x^p - x - 1$. Therefore $E = F(a)$ and E/F is an extension of degree p, and certainly is divisible by p. Therefore $F = \mathbb{F}_p$ is of characteristic p and E/F is a finite Galois extension with degree divisible by p, and so the statement is false.

PART 2
Topology

Section 1

Point Set Topology

2101

Let A and B be connected subspaces of a topological space X, such that $A \cup \overline{B} \neq \emptyset$. Prove that $A \cup B$ is connected. If A and B are path connected, need $A \cup B$ be path connected?

(Indiana)

Solution.

Let f be any continuous map from $A \cup B$ to $S^0 = \{-1, 1\}$. Since A is connected, $f|_A$ must be constant. Without loss of generality, we may assume that $A \subset f^{-1}(-1)$. By the same reason, $f|_B$ is also constant. Let $x_0 \in A \cap \overline{B}$. We have $f(x_0) = -1$. Since f is continuous, there exists a neighborhood of x_0, say U, such that $U \subset f^{-1}(-1)$. But since $x_0 \in \overline{B}$, there is a point of B which belongs to U. Therefore we have $B \subset f^{-1}(-1)$. Hence f is not surjective. It means that $A \cup B$ is connected.

The following example shows that if A and B are path connected then $A \cup B$ needs not to be path connected. Let

$$A = \{(0, 0)\} \subset R^2$$

and

$$B = \left\{ \left(x, \sin \frac{1}{x}\right) \,\middle|\, 0 < x \leq 1 \right\}.$$

Then A and B are path connected and $A \cap \bar{B} = A$, but $A \cup B$ is not path connected.

2102

Suppose that A and B are compact subspaces of spaces X and Y respectively, and that N is an open neighborhood of

$$A \times B \subset X \times Y.$$

Prove that there are open sets $U \subset X$ and $V \subset Y$ such that

$$A \times B \subset U \times V \subset N.$$

(Indiana)

Solution.

Let x be a point of A. For any $y \in B$, since (x, y) belongs to $A \times B$ and N is an open neighborhood of $A \times B \subset X \times Y$, there exist open sets $U_y(x) \subset X$ and $V_y(x) \subset Y$ such that

$$(x, y) \in U_y(x) \times V_y(x) \subset N.$$

Therefore the family of open sets $\{V_y(x), y \in B\}$ covers B. Since B is compact, there is a subcover $\{V_{y_i}(x), i = 1, \ldots, p\}$ such that $B \subset \cup_{i=1}^{p} V_{y_i}(x)$. Let $U(x) = \cap_{i=1}^{p} U_{y_i}(x)$ and $V(x) = \cup_{i=1}^{p} V_{y_i}(x)$. It is obvious that $U(x)$ and $V(x)$ are open sets of X and Y respectively and that

$$\{x\} \times B \subset U(x) \times V(x) \subset N.$$

On the other hand, $\{U(x), x \in A\}$ is an open cover of A. Since A is also compact, there exists a subcover $\{U(x_i), j = 1, \ldots, q\}$ such that $A \subset \cup_{j=1}^{q} U(x_j)$. Let $U = \cup_{j=1}^{q} U(x_j)$ and $V = \cap_{j=1}^{q} V(x_j)$. It is easy to see that U and V are open sets of X and Y respectively and that $A \times B \subset U \times V \subset N$.

2103

Let X be a locally compact Hausdorff space. Let A and B be disjoint subsets of X, with A compact and B closed. Does there exist a continuous function $f : X \to [0, 1]$ such that $f|_A = 0$ and $f|_B = 1$?

(Cincinnati)

Solution.

If X is compact, then X is normal and the existence of f is obvious. Hence we may assume that X is noncompact. We denote by X^* the one-point compactification of X. Since X is locally compact and Hausdorff, X^* is compact and Hausdorff, and consequently is also a normal space. Let $X^* = X \cup \{\infty\}$. It is easy to see that A and $F = B \cup \{\infty\}$ are two disjoint closed subsets of X^*. Then by the Urysohn Lemma there exists a continuous function $\tilde{f} : X^* \to [0, 1]$ such that $\tilde{f}|_A = 0$ and $\tilde{f}|_F = 1$. Therefore, the restriction of $\tilde{f}$ on X, f, satisfies the requirements that $f|_A = 0$ and $f|_B = 1$.

2104

No proofs, only the correct answers to the question asked, are required for this problem.

If X and Y are topological spaces, the join of X and Y is the quotient space

$$X * Y = (X \times Y \times [0, 1])/ \sim,$$

where

(x, y, t) is equivalent to (x', y', t') if and only if
$$\begin{cases} x = x' \quad \text{and} \quad t = t' = 0 \\ \text{or} \\ y = y' \quad \text{and} \quad t = t' = 1. \end{cases}$$

(a) $S^0 * S^0$ and $S^1 * S^0$ are homeomorphic to familiar spaces. What space are they?

(b) Describe $X * S^0$ for a general space X.

(Indiana)

Solution.

(a) $S^0 * S^0$ is homeomorphic to the unit circle S^1, and $S^1 * S^0$ is homeomorphic to the unit sphere S^2.

(b) Generally, $X * S^0$ is homeomorphic to the quotient space $X \times [0,1]/\sim$ obtained from the cylinder $X \times [0,1]$ by collapsing $X \times \{0\}$ and $X \times \{1\}$ to two points p and q respectively.

2105

(a) Define quotient map.

(b) Show that if X is compact, Y is Hausdorff and $f : X \to Y$ is continuous and onto, then f is a closed map.

(c) Show that if f satisfies the condition of (b) then f is a quotient map.

(Indiana)

Solution.

(a) Let X be a topological space and $\sim$ be an equivalence relation on X. Define by $X/\sim$ the space of equivalence classes under $\sim$. By the quotient map $\pi : X \to X/\sim$ we mean the map which assigns to $x \in X$ the equivalence class containing x. The quotient space $X/\sim$ may be topologized by defining a subset $U \subset X/\sim$ to be open if and only if $\pi^{-1}(U)$ is open in X. Under this topology, π becomes a continuous map. More generally, if $f : X \to Y$ is a continuous map, there is naturally associated an equivalence relation on X such that $x_1 \sim x_2$ if and only if $f(x_1) = f(x_2)$. If Y is homeomorphic to $X/\sim$ under the map $i : X/\sim \to Y$ and $f = i \circ \pi$ then we call f a quotient map.

(b) Let A be a closed subset of X. Since X is compact, A is compact too. It follows from the continuity of f that $f(A)$ is a compact subset of Y. Since Y is Hausdorff, $f(A)$ is closed in Y. Hence f is a closed map.

(c) Let $\sim$ be the equivalence relation on X associated to the map f. Denote by $[x]$ the equivalence class containing x. Then we define a map $i : X/\sim \to Y$ by $i([x]) = f(x)$. Since f is onto, i is obviously a $1-1$ map. By the result of (b), the quotient space $X/\sim$ is compact. Hence i is a continuous $1-1$ map from the compact space $X/\sim$ to the Hausdorff space Y, and, therefore, is a homeomorphism. It is clear that $f = i \circ \pi$. So f is a quotient map.

2106

Let $f : X \to Y$ be a continuous function from a space X to a Hausdorff space Y. Let C be a closed subspace of Y, and let U be an open neighborhood of $f^{-1}(C)$ in X.

(a) Prove that if X is compact then there is an open neighborhood V of C in Y such that $f^{-1}(V) \subset U$.

(b) Give a counterexample to show that if X is not compact, then there need not be such a neighborhood V.

(Indiana)

Solution.

(a) Let $W = X - U$, then W is closed in X. Since X is compact, W is compact too. Since f is a continuous function, $f(W)$ is a compact set of Y, and consequently is a closed set of Y because Y is a Hausdorff space. Let $V = Y - f(W)$. Then V is an open neighborhood of C in Y. Since $W \subset f^{-1}(f(W))$, we see that

$$f^{-1}(V) = X - f^{-1}(f(W)) \subset X - W = U.$$

(b) Let $X = R$ and $Y = S^1$. $f : X \to Y$ is the continuous function defined by $f(t) = e^{2\pi i t}$ for $t \in R$. Take $C = \{1\} \in S^1$. It is obvious that

$$f^{-1}(C) = \{n | n \in \mathbb{Z}n\}.$$

Let $U = \cup_{n \in \mathbb{Z}} U_n$ be the open neighborhood of $f^{-1}(C)$ in X, where

$$U_n = \left(-\frac{1}{n} + n, n + \frac{1}{n} \right).$$

Since

$$\lim_{n \to \infty} |U_n| = \frac{2}{n} = 0,$$

one can easily prove that there does not exist such a neighborhood V.

2107

Let X be a normal topological space and $A \subset X$ a closed subspace.

(a) Show that the quotient space Y obtained by collapsing A to a point is normal.

(b) Does this result hold if normality is replaced with regularity?

(*Indiana*)

Solution.

(a) Let $\pi : X \to Y$ be the identification map and $y_0 \in Y$ be the point which A collapses to. Let U and V be two nonempty closed sets in Y such that $U \cap V = \emptyset$. Then $\pi^{-1}(U)$ and $\pi^{-1}(V)$ are two nonempty disjoint closed sets in X. If $y_0 \notin U \cup V$, then, by the normality of X, it is clear that there exist two disjoint open sets W_1 and W_2 in X containing $\pi^{-1}(U)$ and $\pi^{-1}(V)$ respectively such that $W_i \cap A = \emptyset$ for i equal to 1 and 2. Thus we see that $\pi(W_1)$ and $\pi(W_2)$ are two disjoint open sets in Y and contain U and V respectively. If $y_0 \in U$ (or V), we only need to take the sets W_1 and W_2 without the restrictions that $W_i \cap A = \emptyset$.

(b) Let X be a regular space which is not a normal space. It means that there exist two disjoint closed sets A and B in X

such that they cannot be separated by disjoint open sets in X. Then the quotient space Y obtained by collapsing A to a point y_0 is not regular, because one can prove that the point y_0 and the closed set $\pi(B)$ in Y cannot be separated by disjoint open sets in Y.

2108

Let $p : E \to B$ be a covering map with E locally path connected and simply connected. Let X be a connected space, let $f : X \to B$ be a continuous map, and $f_1, f_2 : X \to E$ be two lifts of f. Prove that there is a deck transformation $g : E \to E$ such that $f_2 = gf_1$.

(*Indiana*)

Solution.

Take a point $x_0 \in X$ and let $f_1(x_0) = e_0 \in E$. Then $e_0 \in p^{-1}(b_0)$, where $b_0 = p(e_0)$. Let $f_2(x_0) = e_1$. Since $pf_1 = pf_2 = f$, we see that $e_1 \in p^{-1}(b_0)$. By the assumptions $p : E \to B$ is the universal covering map. Thus there exists a deck transformation g such that $g(e_0) = e_1$. Therefore, $gf_1(x_0) = f_2(x_0)$. Let

$$A = \{x \in X \mid gf_1(x) = f_2(x)\}.$$

It is obvious that A is not empty.

It is not difficult to prove that A is both open and closed in X. Thus, by the connectedness of X, we see that $A = X$, which means $f_2 = gf_1 : X \to E$.

2109

Let $p : \tilde{X} \to X$ be an n-sheeted covering projection, $n < \infty$. Suppose that X is compact. Prove that $\tilde{X}$ is compact.

(*Indiana*)

Solution.

Suppose that $\mathcal{U} = \{U_\lambda, \lambda \in \Lambda\}$ is an open covering of $\tilde{X}$. For any point $x \in X$, let $p^{-1}(x) = \{\tilde{x}_1, \ldots, \tilde{x}_n\}$. Let $W(x)$ be an elementary neighborhood, i.e., $W(x)$ is a path-connected open neighborhood of x such that each path component of $p^{-1}(W(x))$ is mapped topologically onto $W(x)$ by p. Let

$$p^{-1}(W(x)) = \bigcup_{i=1}^{n} \tilde{W}_i(x),$$

where $\tilde{W}_i(x)$ is a path component of $p^{-1}(W(x))$ such that $\tilde{x}_i \in \tilde{W}_i(x)$. Choose a $U_i \in \mathcal{U}$ such that $\tilde{x}_i \in U_i$. Let $\tilde{V}_i(x)$ be the path component of $\tilde{W}_i(x) \cap U_i$ containing $\tilde{x}_i$. It is obvious that each $\tilde{V}_i(x)$ is open and $\tilde{V}_i(x) \cap \tilde{V}_j(x) = \emptyset$ for $i \neq j$. Since p is an open map, $\cap_{i=1}^{n} p(\tilde{V}_i(x))$ is an open neighborhood of x in X. Choose another elementary neighborhood $V(x)$ of x such that

$$V(x) \subset \bigcap_{i=1}^{n} p(\tilde{V}_i(x)).$$

Let

$$p^{-1}(V(x)) = \bigcup_{i=1}^{n} \tilde{V}_i'(x),$$

where $\tilde{V}_i'(x)$ is the path component of $p^{-1}(V(x))$ for any i. Then it is easy to see that $\tilde{V}_i'(x) \subset \tilde{V}_i(x) \subset U_i$ for $i = 1, \ldots, n$. It follows that $p^{-1}(V(x)) \subset \cup_{i=1}^{n} U_i$. That is, we have proved that for any point $x \in X$ there exists an elementary neighborhood $V(x)$ of x such that $p^{-1}(V(x))$ can be covered by a finite number of sets in $\mathcal{U}$. Since X is compact, X can be covered by a finite number of $V(x_i)$, $i = 1, \ldots, m$, and consequently $\tilde{X}$ can be covered by a finite subcover of $\mathcal{U}$.

2110

Let $\mathcal{T}$ and $\mathcal{U}$ be two different topology on X such that X is compact and Hausdorff with respect to both. Prove that $\mathcal{T} \not\subset \mathcal{U}$.

(Recall that $\mathcal{T} \subset \mathcal{U}$ means that every set in the topology $\mathcal{T}$ is contained in $\mathcal{U}$.)

(*Indiana*)

Solution.
We use the reduction to absurdity. Suppose that $\mathcal{T} \subset \mathcal{U}$. Let $(X, \mathcal{T})$ and $(X, \mathcal{U})$ denote the topological spaces of X with respect to $\mathcal{T}$ and $\mathcal{U}$ respectively. $h : (X, \mathcal{U}) \rightarrow (X, \mathcal{T})$ is the identity map of X. Then h is a $1-1$ map from the compact space $(X, \mathcal{U})$ to the Hausdorff space $(X, \mathcal{T})$. We claim that h is a continuous map. For any point $x_0 \in X$. Let U be any open neighborhood of x_0 in $(X, \mathcal{T})$. Since $\mathcal{T} \subset \mathcal{U}$, U is also an open neighborhood of x_0 in $(X, \mathcal{U})$. It is obvious that $h(x_0) = x_0$ and $h(U) = U$. Hence h is continuous at x_0. Thus h is a homeomorphism from $(X, \mathcal{U})$ to $(X, \mathcal{T})$, which means $\mathcal{U} = \mathcal{T}$. This contradicts the assumption.

2111

Let X be a topological space and let $A \subset X$. Show that if C is a connected subset of X that intersects both A and $X - A$, then C intersects BdA. (Recall that $BdA = \bar{A} \cap \overline{(X - A)}$.)

(*Indiana*)

Solution.
We use the reduction ad absurdum. Suppose that $C \cap BdA = \emptyset$. Take $U = C \cap \bar{A}$ and $V = C \cap \overline{(X - A)}$. Since $C \cap A \subset U$ and $C \cap (X - A) \subset V$, from the assumption, we see that both U and V are nonempty subset of C. It is clear that $C = U \cup V$ and both U and V are closed subsets of C, and, consequently, that both U and V are open sets of C. But

$$U \cap V = C \cap \bar{A} \cap \overline{(X - A)} = C \cap BdA = \emptyset,$$

which is a contradiction to the connectedness of C.

2112

Let (X, d) be a metric space. For any subspace $A \subset X$ and real number $\varepsilon > 0$, let

$$O_\varepsilon(A) = \{x \in A : d(x, a) < \varepsilon \text{ for some } a \in A\},$$

$$C_\varepsilon(A) = \{x \in A : d(x, a) \leq \varepsilon \text{ for some } a \in A\}.$$

(a) Prove that $O_\varepsilon(A)$ is an open subspace of X.

(b) If A is compact, show that $C_\varepsilon(A)$ is closed in X. Must $C_\varepsilon(A)$ be closed for a general subspace A of X?

(*Indiana*)

Solution.

(a) Let x_0 be any point of $O_\varepsilon(A)$. By the definition of $O_\varepsilon(A)$, there exists a point $a \in A$ such that $d(x_0, a) < \varepsilon$, i.e., $\delta = \varepsilon - d(x_0, a) > 0$. Then, for any $x \in O_{\delta/4}(x_0)$,

$$d(x, a) \leq d(x, x_0) + d(x_0, a) < \frac{\delta}{4} + (\varepsilon - \delta) < \varepsilon,$$

which means that $O_{\delta/4}(x_0) \subset O_\varepsilon(A)$. Thus $O_\varepsilon(A)$ is an open subspace of X.

(b) Let x_0 be any cluster point of $C_\varepsilon(A)$. Then there exists a sequence $\{x_n\}$ in $C_\varepsilon(A)$ such that $x_n \neq x_0$ for any n and $x_n \to x_0$ as $n \to \infty$. By the definition of $C_\varepsilon(A)$, for each x_n there exists an $a_n \in A$ such that $d(x_n, a_n) < \varepsilon$. Since A is compact, without loss of generality, we may assume that $a_n \to a$ for some $a \in A$. Thus we have

$$d(x_0, a) = \lim_{n \to \infty} d(x_n, a_n) \leq \varepsilon,$$

which means $x_0 \in C_\varepsilon(A)$. So $C_\varepsilon(A)$ is closed in X. If A is not compact, we give a counterexample as follows. Take $X = [0, 2] \subset R$ and $A = [0, 1)$. Then $C_{\frac{1}{2}}(A) = [0, \frac{3}{2})$ is not closed in X.

2113

Suppose that X is a dense subspace of a topological space Y. Prove or give counterexamples to the following assertions:

(a) If X is Hausdorff, then Y is Hausdorff.

(b) If X is connected, then Y is connected.

(*Indiana*)

Solution.

(a) This assertion is not correct. We give a counterexample as follows. Let $Y = \{a, b, c\}$. The topology on Y is determined by the family of open sets

$$\mathcal{T} = \{\{a, b, c\}, \{a, c\}, \{b, c\}, \{c\}, \emptyset\}.$$

Since any neighborhood of the point a always contains the point c, Y is not a Hausdorff space. But it is easy to see that the subspace $\{c\}$ is dense in Y and is Hausdorff.

(b) This assertion is true. We give a proof to it as follows. If Y were not connected, then there would exist a nonempty proper subset U of Y which is both open and closed in Y. Let $A = X \cap U$. Since X is dense in Y and U is open in Y, A would be a nonempty open set of X. On the other hand, $X - A = X \cap (Y - U)$ would be an open set of X, because $Y - U$ is open in Y. Thus A would be a nonempty subset of X, which is both open and closed in X. Since X is dense in Y and $Y - U$ is open in Y, $X - A$ is nonempty. It means $A \neq X$. This contracts the connectedness of X.

2114

Let X and Y be topological spaces, $X = U \cup V$, and $f : X \to Y$ be a function so that $f|_U$ and $f|_V$ are continuous.

(a) If U and V are open in X, show that f is continuous.

(b) Give an example where U and V are not open in X and f is not continuous.

(*Indiana*)

Solution.

(a) Let N be an open set of Y. Since $f|_U$ and $f|_V$ are continuous, $f^{-1}|_U(N)$ and $f^{-1}|_V(N)$ are open in, respectively, U and V. But U and V are open in X, therefore, $f^{-1}|_U(N)$ and $f^{-1}|_V(N)$ are also open in X. Thus

$$f^{-1}(N) = f^{-1}|_U(N) \cup f^{-1}|_V(N)$$

is open in X, and consequently, f is continuous.

(b) Let $X = [0, 2]$, $U = [0, 1)$ and $V = [1, 2]$. $f : X \to R$ is defined by

$$f(x) = \begin{cases} x, & x \in U, \\ 2, & x \in V. \end{cases}$$

Then $f|_U$ and $f|_V$ are continuous, but f is not continuous.

2115

Let $q : X \to Y$ be a quotient space projection from a topological space X to a connected topological space Y. Assume that $q^{-1}(y)$ is connected for each $y \in Y$.

(a) Show that X is connected.

(b) Is X necessarily connected if the map q is not assumed to be a quotient mapping? Justify your assertion.

(Columbia)

Solution.

(a) Suppose that X is not connected. Thus, there exist two disjoint nonempty open subsets of X, U and V such that $X = U \cup V$. Then $q(U) \cap q(V) = \emptyset$. For otherwise, let $y \in q(U) \cap q(V)$. Therefore,

$$q^{-1}(y) = (q^{-1}(y) \cap U) \cup (q^{-1}(y) \cap V),$$

and it is obvious that $q^{-1}(y) \cap U$ and $q^{-1}(y) \cap V$ are both nonempty open subsets of $q^{-1}(y)$, which contradicts the connectedness of $q^{-1}(y)$. Since q is a quotient mapping, $V = q^{-1}(q(V))$

and $U = q^{-1}(q(U))$, $q(U)$ and $q(V)$ are disjoint nonempty open subsets such that $Y = q(U) \cup q(V)$, which contradicts the connectedness of Y. Thus X must be connected.

(b) The following example shows that the assumption that q is a quotient mapping is necessary for X to be connected. Let $X = U \cup V$, where

$$U = \{(x, 0) \in R^2 | 0 \leq x \leq 1\}$$

and

$$V = \{(1, y) \in R^2 | 1 \leq y \leq 2\}.$$

The topology of X is induced from the topology of R^2. Let $Y = [0, 1]$, the unit interval of R, and $q : X \to Y$ be the map defined by $q(x, 0) = x$ for $(x, 0) \in U$ and $q(1, y) = 1$ for $(1, y) \in V$. Then q is continuous but is not a quotient mapping. For each $y \in Y$, $q^{-1}(y)$ is connected. But X is not connected.

2116

Let X and Y be topological spaces, with Y compact. Let $p : X \times Y \to X$ be the usual projection onto the first factor. Show that p is a closed map.

(Cincinnati)

Solution.

Let $U \subset X \times Y$ be a closed subset and $x_0 \in X - p(U)$. Then, for any $y \in Y$, $(x_0, y) \notin U$. Since U is closed, there exist an open set $W_{x_0}(y)$ of X and open set $V_y(x_0)$ of Y such that $(x_0, y) \in W_{x_0}(y) \times V_y(x_0)$ and $(W_{x_0}(y) \times V_y(x_0)) \cap U = \emptyset$. Since Y is compact, there must exist a finite number of $V_{y_1}(x_0), \ldots, V_{y_n}(x_0)$ such that

$$Y = \bigcup_{i=1}^{n} V_{y_i}(x_0).$$

Let

$$W(x_0) = \bigcap_{i=1}^{n} W_{x_0}(y_i).$$

Then $W(x_0)$ is an open neighborhood of x_0 in X. Since $(W(x_0) \times Y) \cap U = \emptyset$, we see that $W(x_0) \cap p(U) = \emptyset$, i.e., $W(x_0) \subset X - p(U)$. Thus $X - p(U)$ is an open set of X, and consequently, $p(U)$ is closed in X, which means that p is a closed map.

2117

Let Y be a connected subset of the topological space, and let Z be a set such that Y is a subset of Z and Z is a subset of the closure of Y. Prove that Z is connected.

(Minnesota)

Solution.
According to the assumptions, we have $Y \subset Z \subset \bar{Y}$. Let $f : Z \to S^0$ be any continuous map, where $S^0 = \{-1, 1\}$ is the 0-dimensional sphere with discrete topology. Since Y is connected, without loss of generality, we may assume that $f(Y) = \{1\}$, i.e., $Y \subset f^{-1}(1)$. Taking closures of these two sets with respect to Z and noting that $f^{-1}(1)$ is a closed subset of Z, we get $(\bar{Y})_Z \subset f^{-1}(1)$. But as well-known, $(\bar{Y})_Z = \bar{Y} \cap Z = Z$. Thus we have $Z \subset f^{-1}(1)$, and it follows that f is not surjective. It means that there does not exist any continuous surjective map from Z to S^0 and therefore Z is connected.

2118

Let A be a connected subspace of a connected set X. If C is a component of $X \backslash A$, show that $X \backslash C$ is connected.

(Cincinnati)

Solution.
We first prove that if X is a connected space, U is a connected subset of X, and if V is a both open and closed subset with

respect to $X \backslash U$, then $U \cup V$ is connected. Suppose that $U \cup V$ is not connected. Then let $U \cup V = W_1 \cup W_2$ where W_1 and W_2 are two disjoint nonempty both open and closed subsets of $U \cup V$. Since U is connected, without loss of generality, we may assume that $U \subset W_2$. Thus W_1 is a both open and closed subset of V, and, consequently, a both open and closed subset of $X \backslash U$. Hence W_1 is a nonempty both open and closed subset of $(U \cup V) \cup (X \backslash U) = X$, which contradicts the assumption that X is connected. So the above statement is proved.

Now suppose that $X \backslash C$ is not connected. Let $X \backslash C = U \cup V$ where U and V are two disjoint nonempty both open and closed subsets of $X \backslash C$. Since $A \subset X \backslash C$ and A is connected, we may assume that $A \subset V$. By the above fact, $C \cup U$ is connected because C is connected. Since $C \cup U \subset X \backslash A$ and $C \cap U = \emptyset$, we see that $C \cup U$ is a connected subset of $X \backslash A$ containing C, which contradicts that C is a component of $X \backslash A$. Hence $X \backslash C$ must be connected.

2119

(1) A metric space X has property S if for every $\varepsilon > 0$ there is a cover of X by connected sets each of which has diameter $< \varepsilon$.

(a) Prove a metric space X has property S if X has a dense subset with property S.

(b) Suppose X is a subset of a metric space. Suppose the closure of X has property S. Must X have property S?

(*Cincinnati*)

Solution.

(a) Suppose that X has a dense subset A which has property S. Then, for any $\varepsilon > 0$, there is a cover $\{V_\alpha, \alpha \in \Gamma\}$ of A by connected sets such that each V_α has diameter $< \varepsilon/2$. Since each V_α is connected, $\bar{V}_\alpha$ is also connected and obviously has diameter $< \varepsilon$. Since $\bar{A} = X$ and $\bar{A} = \cup_{\alpha \in \Gamma} \bar{V}_\alpha$, we see that $\{\bar{V}_\alpha, \alpha \in \Gamma\}$ is a cover of X by connected sets each of which has diameter $< \varepsilon$. Thus X has property S.

(b) We give a counterexample as follows. Let R denote the euclidean real line, X be the set of all rational numbers. Then $\bar{X} = R$ has property S, but it is easy to see that X does not have property S.

2120

Let X be the topologist's sine curve defined by

$$X = \{(x, \sin \pi/x) | 0 < x \leq 1\} \cup \{(0, y) | -1 \leq y \leq 2\}$$
$$\cup \{(x, 2) | 0 \leq x \leq 1\} \cup \{(1, y), 0 \leq y \leq 2\} \subset R^2.$$

(i) Sketch X.

(ii) Let $f : X \to X$ be continuous.

Show that either $f(X) = X$ or else there exists $\delta > 0$ such that

$$f(X) \cap \left\{ (x, y) \,\middle|\, 0 < x < \delta, \quad -\frac{3}{2} \leq y \leq \frac{3}{2} \right\} = \emptyset$$

(*Toronto*)

Solution.

(i) X is shown in Fig. 2.1.

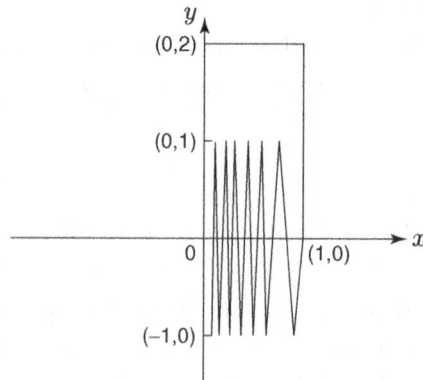

Fig. 2.1.

(ii) Let

$$A = f(X) \cap \{(x, \sin \pi/x) | 0 < x \le 1\}$$

and

$$\delta = \inf\{x | (x, \sin \pi/x) \in A\}.$$

If $\delta = 0$, we claim that $f(X) = X$. To prove it, we first note that since X is path connected, $f(X)$ is also path connected. Hence in this case there exists a δ_0, $0 < \delta_0 \le 1$, such that

$$\{(x, \sin \pi/x) | 0 < x \le \delta_0\} \subset f(X).$$

Therefore it is easy to prove that the set

$$\{(0, y) | -1 \le y \le 1\} \subset f(X).$$

Once again, using the fact that $f(X)$ is path connected, we see that $f(X) = X$. If $\delta > 0$, then it is obvious that

$$f(X) \cap \left\{(x, y) \,\middle|\, 0 < x < \delta, -\frac{3}{2} \le y \le \frac{3}{2}\right\} = \emptyset.$$

2121

Let $T = S^1 \times S^1$ denote the torus.

(i) Show that T can be covered by 3 contractible open subsets.

(ii) Show that T cannot be covered by 2 contractible open subsets.

(Toronto)

Solution.

(i) It is well-known that the torus T can be identified to the quotient space of a square X obtained by identifying opposite sides of the square X according to the directions indicated by the arrows as shown in Fig. 2.2.

Thus let $U_1 = X - \{a \cup b\}$, $U_2 = X - I$ and $U_3 = X - \Pi$. (See Fig. 2.2.) Then U_1, U_2 and U_3 are contractible open subsets and $T = U_1 \cup U_2 \cup U_3$.

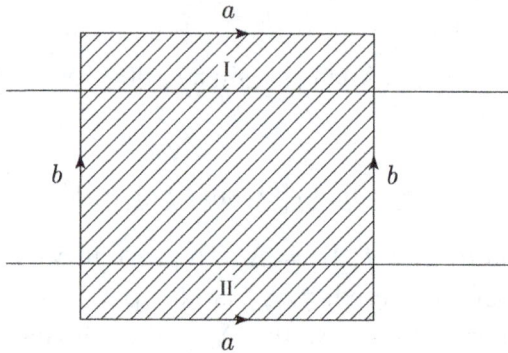

Fig. 2.2.

(ii) Suppose that T could be covered by 2 contractible open subsets. From the Van Kampen Theorem it would follows that the fundamental group $\pi_1(T) = 0$. It contradicts the known fact that $\pi_1(T) \approx \mathbb{Z} \oplus \mathbb{Z}$.

2122

Let $\mathcal{U} = \{U_\alpha\}_{\alpha \in J}$ be an open cover of the space X.

(a) Give the definition for "$\mathcal{U}$ is locally finite".

(b) If $\mathcal{U}$ is locally finite show that, for any subset $K \subset J$, $\cup_{\beta \in K} \bar{U}_\beta$ is closed.

(*Toronto*)

Solution.

(a) By definition, $\mathcal{U}$ is said to be locally finite, if each point $p \in X$ has a neighborhood which intersects only a finite number of U_α.

(b) For any point $p \notin \cup_{\beta \in K} \bar{U}_\beta$, since $\mathcal{U}$ is locally finite, there is an open neighborhood W of p which intersects only a finite number of sets in $\mathcal{U}$, particularly, only a finite number of U_β for $\beta \in K$, say, $U_{\beta_1}, \ldots, U_{\beta_n}$. Since $p \notin \bar{U}_{\beta_i}$ for each β_i, there is an open neighborhood V_i of p such that $V_i \cap \bar{U}_{\beta_i} = \emptyset$. Then let $V = W \cap V_1 \cap \cdots \cap V_n$. It is clear that V is an open neighborhood of p such that $V \cap (\cup_{\beta \in K} \bar{U}_\beta) = \emptyset$. Hence $\cup_{\beta \in K} \bar{U}_\beta$ is closed.

2123

Let S be a set and let F be a family of real valued functions on S such that $f(s_1) = f(s_2)$ for all $f \in F$ implies $s_1 = s_2$. Prove that there exists a weakest topology in S amongst all those for which all members of F are continuous. Show further that the resulting topological space satisfies the Hausdorff separation axiom.

(Harvard)

Solution.

Let

$$\mathcal{U} = \{f^{-1}((a,b)) | (a,b) \text{ is any open interval of } R \text{ and } f \in F\}.$$

Then there exists a unique topology $\mathcal{T}$ on the set S of which $\mathcal{U}$ is the topology subbase. It is easy to see that $\mathcal{T}$ is the weakest topology on S amongst all those for which all members of F are continuous. Suppose that s_1 and s_2 are two distinct points of S. By the assumption there exists at least an $f \in F$ such that $f(s_1) \neq f(s_2)$. Hence we may take two open intervals (a_1, b_1) and (a_2, b_2) such that $f(s_i) \in (a_i, b_i)$ for $i = 1, 2$ and $(a_1, b_1) \cap (a_2, b_2) = \emptyset$. Then $U_i = f^{-1}((a_1, b_1))$ and $U_2 = f^{-1}((a_2, b_2))$ are two disjoint open sets of $(S, \mathcal{T})$ such that $s_i \in U_i$ for $i = 1, 2$. So $(S, \mathcal{T})$ is Hausdorff.

2124

Let K be a compact subset of R^n and $\{B_j\}$ a sequence of open balls that covers K. Prove that there is a positive number ε such that each ε-ball centered at a point of K is contained in one of the ball B_j.

(UC, Berkeley)

Solution.

Since K is compact and $\{B_j\}$ is an open cover of K, we can choose a finite subcover $\{B_1, \ldots, B_N\}$ of K. For any $x \in K$, we define

$$f(x) = \max\{\text{dist}\,(x, R^n \backslash B_j)|x \in B_j\}.$$

It is easy to see that f is continuous and positive on K. Since K is compact, f has a positive minimum $\varepsilon > 0$. Then, by the definition of f, we see that every ε-ball centered at a point of K is contained in some B_j.

2125

Let (X_1, d_1) and (X_2, d_2) be metric spaces and $f : X_1 \to X_2$ a continuous surjective map such that $d_1(p, q) \leq d_2(f(p), f(q))$ for every pair of points p, q in X_1.

(1) If X_1 is complete, must X_2 be complete? Give a proof or a counterexample.
(2) If X_2 is complete, must X_1 be complete? Give a proof or a counterexample.

(UC, Berkeley)

Solution.

(1) Assume X_1 is complete. Suppose that $\{y_n\}$ is a Cauchy sequence in X_2. Since $d_1(p, q) \leq d_2(f(p), f(q))$ and f is surjective, each y_n can be written as $f(x_n)$ with x_n in X_1. It is obvious that $\{x_n\}$ is also a Cauchy sequence in X_1, hence convergent, say to x. By the continuity of f, we have $\lim_{n \to \infty} y_n = f(x)$. Thus X_2 is also complete.

(2) The completeness of X_2 does not imply the completeness of X_1. For example, let $X_1 = (-\frac{\pi}{2}, \frac{\pi}{2})$, $X_2 = (-\infty, +\infty)$ and $f(x) = \tan x$. Since $f'(x) = \sec^2 x \geq 1$, by Lagrange's mean value theorem, we see that the condition $|x - y| \leq |f(x) - f(y)|$ holds.

2126

Let (M, d) be a nonempty complete metric space. Let S map M into M, and write S^2 for $S \circ S$, that is, $S^2(x) = S(S(x))$. Suppose that S^2 is a strict contraction, that is, there is a constant $\lambda < 1$ such that for all points $x, y \in M$, $d(S^2(x), S^2(y)) \leq \lambda d(x, y)$. Show that S has a unique fixed point in M.

(UC, Berkeley)

Solution.

By the Contraction Mapping Theorem, S^2 has a unique fixed point x, that is, $S^2(x) = x$. Write $S(x) = y$. Then we have $S^2(y) = S^3(x) = S(S^2(x)) = S(x) = y$, which means that y is also a fixed point of S^2. By the uniqueness, we have $y = x$. Thus $S(x) = x$, so S has a fixed point. Noting that any fixed point of S is a fixed point of S^2, we see that S has a unique fixed point in M.

2127

Let K be a continuous function on the unit square $0 \leq x, y \leq 1$ satisfying $|K(x, y)| \leq 1$ for all x and y. Show that there is a continuous function $f(x)$ on $[0, 1]$ such that we have

$$f(x) + \int_0^1 K(x, y) f(y) dy = e^{x^2}.$$

Can there be more than one such function f?

(UC, Berkeley)

Solution.

Since the unit square is compact, we have $\max |K(x, y)| = M < 1$. Now we define a map $T : C[0, 1] \to C[0, 1]$ by

$$T(f)(x) = e^{x^2} - \int_0^1 K(x, y) f(y) dy.$$

Then we have

$$|T(f)(x) - T(g)(x)| = \left| \int_0^1 K(x,y)(f(y) - g(y))dy \right|$$

$$\leq M \int_0^1 |f(y) - g(y)|dy$$

$$\leq M \max_{0 \leq y \leq 1} |f(y) - g(y)| = M\|f - g\|.$$

So T is a strict contraction map, and hence T has a unique fixed point $f \in C[0,1]$ satisfying $T(f) = f$, i.e.,

$$f(x) + \int_0^1 K(x,y)f(y)dy = e^{x^2}.$$

Any such solution is a fixed point of T, so there does not exist more than one such function f.

Section 2

Homotopy Theory

2201

(a) Give generators and relations for the fundamental groups of the torus and of the oriented surface of genus 2.

(b) Compute the fundamental group of the figure 8 and draw a piece of its universal covering space.

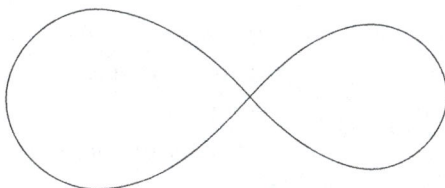

(*Harvard*)

Solution.

(a) As is well-known, we can present the torus T as the space obtained by identifying the opposite sides of a square, as shown in Fig. 2.3(b). Under the identification the sides a and b each become circles which intersect in the point x_0. Let y be the center point of the square, and let $U = T - \{y\}$. Let V be the image of the interior of the square under the identification. Since V is simply connected, by the Van Kampen Theorem,

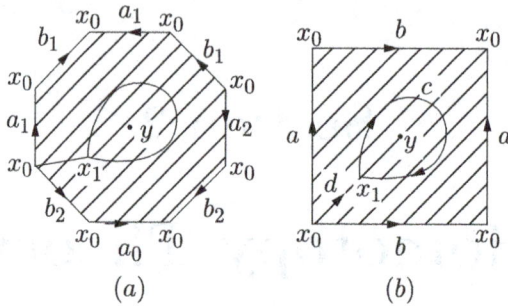

Fig. 2.3.

we conclude that $\pi_1(T, x_1)$ is isomorphic to $\pi_1(U, x_1)$ modulo the smallest normal subgroup of $\pi_1(U, x_1)$ containing the image $\phi_*(\pi_1(U \cap V, x_1))$, where ϕ_* is the homomorphism induced by the inclusion map $\phi : U \cap V \to U$. It is easily seen that $a \cup b$ is a deformation retract of U. Hence $\pi_1(U, x_0)$ is a free group on two generators α and β, where α and β are presented by circles a and b, respectively. It is also clear that $\pi_1(U, x_1)$ is a free group on two generators $\alpha' = \delta^{-1}\alpha\delta$ and $\beta' = \delta^{-1}\beta\delta$, where δ is the equivalence class of a path d from x_0 to x_1. (See Fig. 2.3(b).) On the other hand, it is easy to see that $\pi_1(U \cap V, x_1)$ is an infinite cyclic group generated by γ, the equivalence class of a closed path c which circles around the point y once, and, consequently, that $\phi_*(\gamma) = \alpha'\beta'\alpha'^{-1}\beta'^{-1}$. The smallest normal subgroup of $\pi_1(U, x_1)$ containing $\phi_*(\pi_1(U \cap V, x_1))$ is just the commutator subgroup of $\pi_1(U, x_1)$. Thus $\pi_1(T, x_1)$ is a free abelian group on two generators α' and β'. Changing to the base point x_0, we see that $\pi_1(T, x_0)$ is a free abelian group on two generators α and β, which are presented by circles a and b, respectively.

In a similar way, we can see that the fundamental group of the oriented surface of genus 2 is a free group on four generators $\alpha_1, \beta_1, \alpha_2, \beta_2$ with the single relation $[\alpha_1, \beta_1][\alpha_2, \beta_2]$, where $\alpha_1, \beta_1, \alpha_2, \beta_2$ are presented by circles a_1, b_1, a_2, b_2, respectively, and $[\alpha_i, \beta_i]$ denotes the commutor $\alpha_i\beta_i\alpha_i^{-1}\beta_i^{-1}$. (See Fig. 2.3(a).)

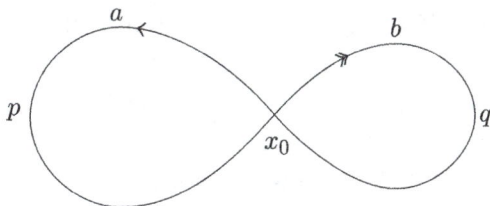

Fig. 2.4.

(b) Let X denote the figure 8 space, as shown in Fig. 2.4. Let $U = X - \{q\}$, $V = X - \{p\}$. By the Van Kampen Theorem we can prove that $\pi_1(X, x_0)$ is a free group on two generators α, β, where α, β are presented by circles a and b, respectively.

The following picture is a piece of its universal covering space.

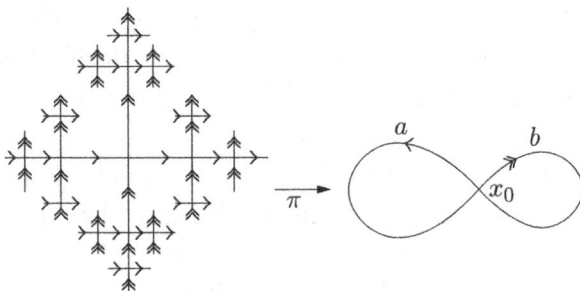

Fig. 2.5.

Under the covering map π, each level segment is mapped on the circle a according to the direction indicated by the arrow, and each vertical segment is mapped on the circle b according to the direction indicated by the double arrows.

2202

Let A be a connected, closed subspace of a compact Hausdorff space X, and suppose $f : A \to A$ is a continuous map. For each positive integer n let $f^n(A) = \underbrace{f \circ f \circ \cdots \circ f}_{n \text{ times}}(A)$.

(i) Show that $B = \cap_{n=1}^{\infty} f^n(A)$ is connected.

(ii) Suppose $\gamma : S^1 \to X - B$ is a nullhomotopic map. Show that there exists a positive integer n such that $\gamma(S^1) \subset X - f^n(A)$ and such that the induced map $\gamma' : S^1 \to X - f^n(A)$ $(\gamma'(s) = \gamma(s)$ for $s \in S^1)$ is also nullhomotopic.

(*Indiana*)

Solution.

(i) Since X is compact and A is closed, A is also compact. Thus it is easy to see that $f^n(A)$ is compact, closed and connected. Noting that $f^{n+1}(A) \subset f^n(A) \subset A$ for any n, and that A is compact, we see that $B = \cap_{n=1}^{\infty} f^n(A)$ is a non-empty closed subset of A. Suppose that B is not connected. Then $B = U \cup V$, where U and V are disjoint non-empty closed subsets of B. It is obvious that U and V are also closed subsets of A. Since A is obviously compact and Hausdorff, A is a normal space. Therefore, there exist disjoint open subsets of A, W_1 and W_2 such that $U \subset W_1$ and $V \subset W_2$. Thus it follows that $B \subset W_1 \cup W_2 = W$, which is an open subset of A. We claim that there exists a positive integer N such that $f^N(A) \subset W$. For, otherwise, there would be a squence in A, $\{x_n\}$, such that $x_n \in f^n(A)$ but $x_n \notin W$ for every n. It means that $x_n \in A - W$ for every n. Since $A - W$ is closed in A and, consequently, is a compact subset, there exists at least a limit point x_0 of the sequence $\{x_n\}$. Noting that $f^n(A)$ is compact for every n and $f^{n+1}(A) \subset f^n(A)$, we can see that $x_0 \in B \subset W$, which is a contradiction. Thus $f^N(A) \subset W$, and therefore,

$$f^N(A) = (f^N(A) \cap W_1) \cup (f^N(A) \cap W_2),$$

which contradicts the fact that $f^N(A)$ is connected.

(ii) Let $F : S^1 \times I \to X - B$ be a homotopy between γ and the constant map. Since $F(S^1 \times I)$ is compact and $X - B$ is open, $F(S^1 \times I)$ is also closed in X. Since X is normal, there exists open sets U and V such that $F(S^1 \times I) \subset U$ and $B \subset V$. From the proof for (i), it follows that there exists a positive N

such that $f^N(A) \subset V$. Therefore,

$$F(S^1 \times I) \subset U \subset X - f^N(A).$$

Particularly, $\gamma(S^1) \subset X - f^N(A)$. The remainder of (ii) is obvious.

2203

Let RP^n be the real projective n-space and T^m be the m-torus $S^1 \times \cdots \times S^1$ (m factors). Prove that any continuous map $RP^n \to T^m$ is null-homotopic.

(*Indiana*)

Solution.

It is well-known that R^m is a universal covering space of T^m. Let $P : R^m \to T^m$ be the universal covering map. It is also well-known that

$$\pi_1(T^m) \approx \mathbb{Z} \oplus \cdots \oplus \mathbb{Z} \quad (m \text{ times})$$

and that $\pi_1(RP^n) \approx \mathbb{Z}_2$. Therefore, for any continuous map $f : RP^n \to T^m$, the induced homomorphism $f_* : \pi_1(RP^n) \to \pi_1(T^m)$ is trivial. Thus there exists a lifting of f, say $\tilde{f}$, such that $p\tilde{f} = f$. Let p_0 be a fixed point of R^m. Define $H : RP^n \times I \to R^m$ by

$$H(x,t) = (1-t)\tilde{f}(x) + tp_0.$$

Then H is a homotopy between $\tilde{f}$ and the constant map p_0. So $\tilde{f}$ is null-homotopic and consequently f is also null-homotopic.

2204

Let X be the quotient space obtained by collapsing $\{pt.\} \times S^1 \subset S^1 \times S^1$ to a point

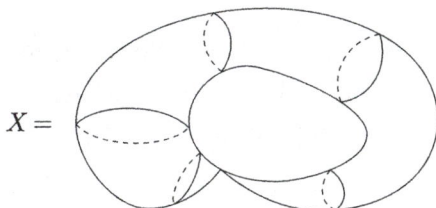

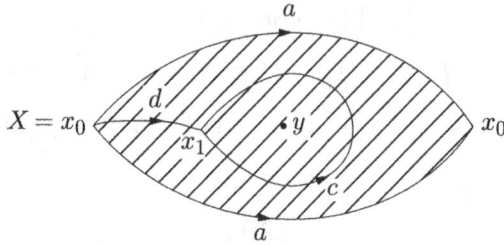

Fig. 2.6.

Compute $\pi_1(X)$ and $H_*(X)$.

(*Columbia*)

Solution.

X may be identified with the space shown in Fig. 2.6, which is obtained by identifying the sides of a 2-gon. Let y be the center of the 2-gon, $U = X - \{y\}$, and V be the interior of the 2-gon. Then, U and V are open subsets, U, V, and $U \cap V$ are path connected, and V is simply connected. Thus, by the Van Kampen Theorem, $\pi_1(X, x_1)$ is isomorphic to the quotient group of $\pi_1(U, x_1)$ modulo the smallest normal subgroup of $\pi_1(U, x_1)$ containing the image $\phi_*(\pi_1(U \cap V, x_1))$, where ϕ_* is the homomorphism induced by the inclusion map $\phi : U \cap V \to U$. It is easy to see that a is a deformation ratract of U. Therefore, $\pi_1(U, x_1)$ is an infinite cyclic group on the generator $\delta^{-1}a\delta$, where δ is the equivalence class of a path d connecting x_0 and x_1. (See Fig. 2.6.) It is also clear that $\pi_1(U \cap V, x_1)$ is an infinite cyclic group on the generator γ, where γ is the equivalence class of a closed path c which goes once round the point y. It is easy to see that $\phi_*(\gamma) = 1_{\pi_1(U, x_1)}$. Therefore we conclude that $\pi_1(X) = \mathbb{Z}$.

To compute $H_*(X)$, we may apply the Mayer–Vietoris sequence to the pair (U, V). The conclusion is that $H_i(X)$ is an infinite cyclic group for i equal to 0, 1, 2 and is zero otherwise.

2205

(i) Suppose $n \geq 2$. Does there exist a continuous map $f : S^n \to S^1$ which is not homotopic to a constant?

(ii) Suppose $n \geq 2$. Does there exist a continuous map $f : RP^n \to S^1$ which is not homotopic to a constant?

(iii) Let $T = S^1 \times S^1$ be the torus. Does there exist a continuous map $f : T \to S^1$ which is not homotopic to a constant?

(Toronto)

Solution.

(i) Let $\pi : R \to S^1$ be the universal covering map defined by $\pi(t) = e^{2\pi i t}$, and $f : S^n \to S^1$ be a continuous map. Since $\pi_1(S^n) = 0$ for $n \geq 2$, we see that there is a lifting of f, $\tilde{f} : S^n \to R$ such that $\pi \tilde{f} = f$. Since R is contractible, $\tilde{f}$ must be homotopic to a constant map $\tilde{C} : S^n \to R$, hence f is homotopic to $\pi \circ \tilde{C} = C$, which is also a constant map from S^n to S^1, i.e., there does not exist any continuous map $f : S^n \to S^1$ which is not homotopic to a constant map.

(ii) Since $\pi_1(RP^n) = \mathbb{Z}_2$, any continuous map $f : RP^n \to S^1$ induces a trivial homomorphism $f_* : \pi_1(RP^n) \to \pi_1(S^1)$, and, consequently, has a lifting $\tilde{f} : RP^n \to R$ such that $\pi \circ \tilde{f} = f$. By the same argument as in (i), f must be homotopic to a constant.

(iii) Denote $T = S^1 \times S^1$ by $T = (e^{2\pi i t_1}, e^{2\pi i t_2})$, $0 \leq t_1, t_2 \leq 1$. Define $f : T \to S^1$ by

$$f(e^{2\pi i t_1}, e^{2\pi i t_2}) = e^{2\pi i t_1} \in S^1.$$

It is easy to see that the induced homomorphism $f_* : H_1(T) \to H_1(S^1)$ maps one generator of $H_1(T)$ to the generator of $H_1(S^1)$, and another generator to zero. Hence f_* is not trivial, which means that f is not homotopic to a constant.

2206

A continuous map of topological spaces: $p : E \to B$ is called a fibration if it has the homotopy lifting property — that is, for any pair of continuous maps $I \to G : X \times I \to B$ and $h : X \times \{0\} \to E$ such that $ph = G|_{X \times \{0\}}$ there exists a continuous map $H : X \times I \to E$ such that $H|_{X \times \{0\}} = h$ and $pH = G$. Let $p : E \to B$ be a fibration, $b_0 \in B$ be a base point, $F = p^{-1}(b_0)$ (the "fiber"), and $e_0 \in F$. Let $i : F \to E$ denote the inclusion map.

(i) If F is path connected, prove that $p_\# : \pi_1(E, e_0) \to \pi_1(B, b_0)$ is surjective.

(ii) Prove that in general the 3-term sequence

$$\pi_1(F, e_0) \to \pi_1(E, e_0) \to \pi_1(B, b_0)$$

in which the homomorphisms are $i_\#$ and $p_\#$, respectively, is exact.

(*Indiana*)

Solution.

(i) Let $\alpha = [f] \in \pi_1(B, b_0)$, where $f : I \to B$ is a closed path at b_0 which represents α. Let $G : I \times I \to B$ be a continuous map defined by $G(t, s) = f(st)$ for $(s, t) \in I \times I$, and $h : I \times \{0\} \to E$ be the constant map such that $h(t, 0) \equiv e_0$. It is obvious that $ph = G|_{I \times \{0\}}$. Therefore there exists a continuous map $H : I \times I \to E$ such that $H|_{I \times \{0\}} = h$ and $pH = G$. Particularly we have $pH(t, 1) = f(t)$. Let $C : I \to E$ be a path in E defined by $c(t) = H(t, 1)$. Then $pc = f$. Let $c(0) = e_1$ and $c(1) = e_2$. It is clear that e_1 and e_2 belong to F. Since F is path-connected, we may choose two paths g_1 and g_2 in F such that $g_1(0) = e_2$, $g_1(1) = e_1$, $g_2(0) = e_1$ and $g_2(1) = e_1$. Thus $\tilde{f} = g_2 * c * g_1 * g_2^{-1}$ is a closed path at e_1, where g_2^{-1} is the inverse path of g_2. Noting that pg_1, pg_2 and pg_2^{-1} are all constant path at b_0, we have

$$p_\#[\tilde{f}] = [p\tilde{f}] = [pg_2] \cdot [pc] \cdot [pg_1] \cdot [pg_2^{-1}] = [pc] = [f] = \alpha,$$

which means that $p_\#$ is surjective.

(ii) Let $\alpha = [f] \in \pi_1(F, e_0)$. It is obvious that $p_\# \cdot i_\#(\alpha) = [pf] = [b_0]$, where b_0 is the constant path at b_0. Hence

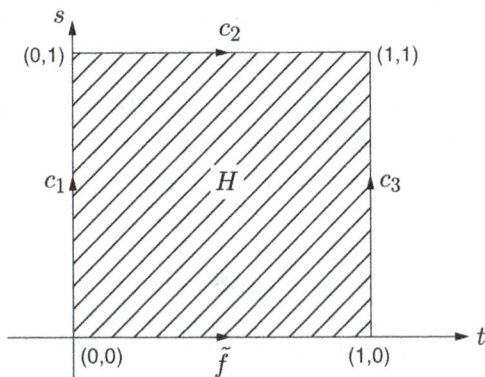

Fig. 2.7.

$\operatorname{im}_\# \subset \ker p_\#$. On the other hand, suppose that $\tilde{\alpha} = [\tilde{f}] \in \ker p_\#$. Then there exists a homotopy $G : I \times I \to B$ between $p\tilde{f}$ and the constant path b_0 such that $G(t,0) = (p\tilde{f})(t)$, $G(t,0) = b_0$, and $G(0,s) \equiv G(1,s) \equiv b_0$ for any s. By the homotopy lifting property, there exists a continuous map $H : I \times I \to E$ such that $PH = G$ and $H(t,0) = \tilde{f}(t)$. Let $c_1 = H|_{\{0\} \times I}$, $c_2 = H|_{I \times \{1\}}$ and $c_3 = H|_{\{1\} \times I}$. Then c_1, c_2 and c_3 are paths in F and $c_1(0) = e_0$, $c_1(1) = c_2(0)$, $c_2(1) = c_3^{-1}(0)$ and $c_3^{-1}(1) = e_0$. Hence $c = c_1 \times c_2 \times c_3^{-1}$ is a closed path in F with base point e_0. It is easy to see that $\tilde{f}$ is homotopic to C. (See Fig. 2.7.) Therefore, $\tilde{\alpha} = [\tilde{f}] = [c] = i_\#[c]$. That is, $\ker p_\# \subset \operatorname{im} i_\#$. Hence the sequence mentioned above is exact.

2207

Is the canonical map $q : S^2 \to RP^2$ (which identifies antipodal points of S^2) nullhomotopic? Why or why not?

(*Indiana*)

Solution.

It is well-known that q is also the universal covering map from S^2 to RP^2. Suppose that q is nullhomotopic. Then q is homotopic to a constant map $c : S^2 \to RP^2$, and therefore

q has a lifting $\tilde{q} : S^2 \to S^2$ which is also homotopic to the lifting of c, a constant map $\tilde{c} : S^2 \to S^2$. But it is obvious that $\tilde{q}$ is the identity map or the antipodal map from S^2 to S^2. Hence $\deg \tilde{g} = \pm 1$. On the other hand, we have $\deg \tilde{g} = \deg \tilde{c} = 0$. This is a contradiction. Thus we conclude that q is not nullhomotopic.

2208

Let $p : E \to B$ be a universal cover with E and B path-connected and locally path-connected. Let $T : B \to B$ be a map so that $T^n = Id$ and so that $T(b) = b$ for some $b \in B$. (Here $T^n = T \circ T \circ \cdots \circ T$, n times.) Show that there is a map $\tilde{T} : E \to E$ so that $p \circ \tilde{T} = T \circ p$ and $\tilde{T}^n = Id$.

(*Indiana*)

Solution.
Choose $e_0 \in p^{-1}(b)$ and consider the map $T \circ p : E \to B$. We have $T \circ p(e_0) = T(b) = b$. Since $\pi_1(E)$ is trivial, there is a lifting of $T \circ p$, $\tilde{T} : (E, e_0) \to (E, e_0)$ such that $p \circ \tilde{T} = T \circ p$. Since

$$p \circ \tilde{T}^n = (p \circ \tilde{T}) \circ \tilde{T}^{n-1} = T \circ (p\tilde{T}^{n-1}) = \cdots = T^n \circ p$$

and

$$\tilde{T}^n(e_0) = \tilde{T}^{n-1}(\tilde{T}(e_0)) = \tilde{T}^{n-1}(e_0) = \cdots = \tilde{T}(e_0) = e_0,$$

we see that $\tilde{T}^n$ is a lifting of $T^n \circ p$ at the base point e_0. On the other hand, since $T^n = Id$, it follows that identity map $Id : (E, e_0) \to (E, e_0)$ is obviously a lifting of $T^n p$ at the base point e_0. Therefore, by the uniqueness of lifting, we conclude that $\tilde{T}^n = Id$.

2209

Let $T^2 = S^1 \times S^1$, and let $X \subset T^2$ be the subset $S^1 \times \{1\} \cup \{1\} \times S^1$. Prove that there is no retraction of T^2 to X.

(*Indiana*)

Solution.

We use the reduction to absurdity. Suppose that there is a retraction of T^2 to X, denoted by r. Let $i : X \to T^2$ be the inclusion map. Then $r \circ i : X \to X$ is the identity map. Hence $r_* \circ i_* : \pi_1(X) \to \pi_1(X)$ is the identity homomorphism. It is well-known that $\pi_1(T^2)$ is abelian and that $\pi_1(X)$ is a non-abelian free group on two generators denoted by a and b. Hence we have $ab \neq ba$ and $i_*(a)i_*(b) = i_*(b)i_*(a)$. Therefore we have

$$ab = r_* i_*(ab) = r_*(i_*(a)i_*(b)) = r_*(i_*(b)i_*(a))$$
$$= r_* i_*(b) r_* i_*(a) = ba,$$

which is a contradiction.

2210

Let $A \subset R^3$ be the union of the x and y-axis,

$$A = \{(x, y, z) | (y^2 + z^2)(x^2 + z^2) = 0\},$$

and let $p = (0, 0, 1)$.
(a) Compute $H_1(R^3 - A)$.
(b) Prove that $\pi_1(R^3 - A, p)$ is not abelian.

(*Indiana*)

Solution.

Let $X = \{(x, y, z) | x^2 + z^2 = 1\}$ be a circular cylindrical surface. p_1 and p_2 denote the points $(1, 0, 0)$ and $(-1, 0, 0)$ respectively. Then it is easy to see that $X - \{p_1, p_2\}$ is a deformation retract of $R^3 - A$. It is also clear that $X - \{p_1, p_2\}$ is homotopically equivalent to the space Y as shown in Fig. 2.8.

(a) Y has a structure of a graph with 4 vertices and 6 edges. The Euler Characteristic $K(Y) = 4 - 6 = -2$. Hence the rank of $H_1(Y)$ is equal to $1 - (-2) = 3$. Therefore we conclude that $H_1(R^3 - A) \approx H_1(Y) \approx Z \oplus Z \oplus Z$.

(b) Let e be a point on the arc ab different from a and b. Take $U = Y - \{c\}$ and $V = Y - \{e\}$. Then we have $Y = U \cup V$. It is

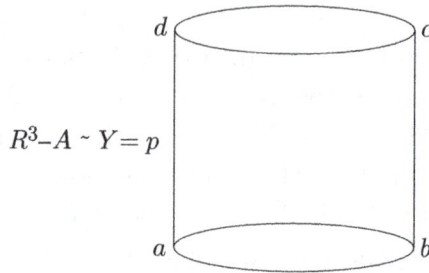

$$R^3 - A \sim Y = p$$

Fig. 2.8.

easy to see that U has the same homotopy type as S^1 and that V has the same homotopy type as the "eight figure space". Since $U \cap V$ is contractible, by the Van Kampen Theorem we see that $\pi_1(Y)$ is a free group generated by $\pi_1(U)$ and $\pi_1(V)$. Therefore we conclude that $\pi_1(R^3 - A, p) \approx \pi_1(Y)$ is a free group on three generators and, consequently, that $\pi_1(R^3 - A, p)$ is not abelian.

2211

Let $p : (\tilde{Y}, \tilde{y}) \rightarrow (Y, y)$ be a regular covering space; that is, $p_*(\pi_1(\tilde{Y}, \tilde{y}))$ is a normal subgroup of $\pi_1(Y, y)$. Suppose that $f : X \rightarrow Y$ is a continuous function from the path-connected space X to Y with $f(x_0) = f(x_1) = y$, and that there is a lifting of f, $\tilde{f}_0 : (X, x_0) \rightarrow (\tilde{Y}, \tilde{y})$. Show that there is a second lifting, $\tilde{f}_1 : (X, x_1) \rightarrow (\tilde{Y}, \tilde{y})$. ($\tilde{f}_1$ need not be distinct from $\tilde{f}_0$.)

(*Indiana*)

Solution.
    Let $\tilde{f}_0(x_1) = \tilde{y}'$. Since

$$p(\tilde{y}') = p\tilde{f}_0(x_1) = f(x_1) = y, \quad \tilde{y}' \in p^{-1}(y).$$

Since p is a regular covering space, there is a deck transformation γ such that $\gamma(\tilde{y}') = \tilde{y}$. Then let

$$\tilde{f}_1 = \gamma \circ \tilde{f}_0 : X \rightarrow \tilde{Y}.$$

Therefore

$$\tilde{f}_1(x_1) = \gamma(\tilde{f}_0(x_1)) = \gamma(\tilde{y}') = \tilde{y}.$$

Hence $\tilde{f}_1$ is the lifting of f we want.

2212

Let X be the identification space obtained from a unit 2-disk by identifying points on its boundary if the arc distance between them on the boundary circle is $\frac{2\pi}{3}$. Compute the fundamental group of X.

(*Indiana*)

Solution.

Take a point y in the open disk. (See Fig. 2.9.) Let $U = X - \{y\}$ and let V be the open disk. Then both U and V are path connected subsets of X and $X = U \cup V$. Since V is simply connected, by the Van Kampen Theorem, $\pi_1(X)$ is isomorphic to the quotient group of $\pi_1(U)$ with respect to the least normal subgroup containing $\phi_*(\pi_1(U \cap V))$, where $\phi : (U \cap V) \to \pi_1(U)$ is the homomorphism induced by the inclusion $\phi : U \cap V \to U$. Take a point $x_1 \in U \cap V$ as the base point. (See Fig. 2.9.)

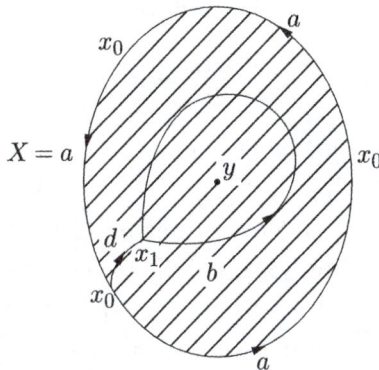

Fig. 2.9.

It is clear that $\pi_1(U \cap V, x_1)$ is an infinite cyclic group generated by γ_0 the closed path class of a closed path c which circles around the point y once. Since the circle a is a deformation retract of U, it is clear that $\pi_1(U, x_0)$ is an infinite cyclic group generated by a' the closed path class of a. Therefore $\pi_1(U, x_1)$ is an infinite cyclic group on generator $\tilde{a} = \gamma^{-1} a' \gamma$, where γ is the path class of a path d from x_0 to x_1. It is also clear that $\phi_*(\gamma_c) = 3\tilde{a}$. Hence the least normal subgroup containing $\phi_*(\pi_1(U \cap V))$ is isomorphic to $3\mathbb{Z}$. It follows that $\pi_1(X) \approx \mathbb{Z}/3\mathbb{Z} = \mathbb{Z}_3$.

2213

Sketch a proof of the Fundamental Theorem of Algebra (every nonconstant polynomial with complex coefficients has a complex zero) using techniques of algebraic topology.

(Indiana)

Solution.

Let $\mathbb{C}$ denote the complex plane and $f(z)$ be a polynomial of positive degree with complex coefficients. We may consider f to be a continuous nonconstant map $f : \mathbb{C} \to \mathbb{C}$. Note that $|f(z)| \to \infty$ as $|z| \to \infty$; hence, we may extend f to a map of the one-point compactification of $\mathbb{C}$

$$f : S^2 \to S^2$$

by setting $f(\infty) = \infty$, where ∞ denotes the north pole. Then we may first prove that if $f(z) = z^k$, $k > 0$, then the degree of the extension $f : S^2 \to S^2$ is equal to k. Furthermore, we may prove that if f is any polynomial of degree $k > 0$ then the degree of the extension $f : S^2 \to S^2$ is still equal to k. Noting the fact that if a continuous map $f : S^2 \to S^2$ is not surjective then the degree of f is zero, we may prove the Fundamental Theorem of Algebra by means of the reduction to absurdity.

2214

Let X denote the subspace of R^3 that is the union of the unit sphere S^2, the unit disk D^2 in the $x - y$ plane, and the portion, call it A, of the z axis lying within S^2.

(a) Compute the fundamental group of X.
(b) Compute the integral homology groups of X.

(Indiana)

Solution.

(a) It is clear that X has the homotopy type of the one point union $X_1 \vee X_2$ where X_1 and X_2 are each homeomorphic to the union of the unit sphere and the portion of the z axis lying within S^2. (See Fig. 2.10.) To compute $\pi_1(X_1 \vee X_2)$, we take $U = X_1 \vee X_2 - \{p\}$ and $V = X_1 \vee X_2 - \{q\}$. Then we have $U \cup V = X_1 \vee X_2$. Since $U \cap V$ is contractible, by the Van Kampen, $\pi_1(X_1 \vee X_2)$ is a free product of the groups $\pi_1(U)$ and $\pi_1(V)$ with respect to the homomorphisms induced by the inclusion maps. It is obvious that

$$\pi_1(U) \approx \pi_1(V) \approx \pi_1(X_1) \approx \pi_1(X_2) \approx \pi_1(S^1).$$

Therefore, $\pi_1(X)$ is a free product generated by two generators. We can take as generators the closed path classes which are determined by the closed paths omp and omq respectively. (See Fig. 2.10.)

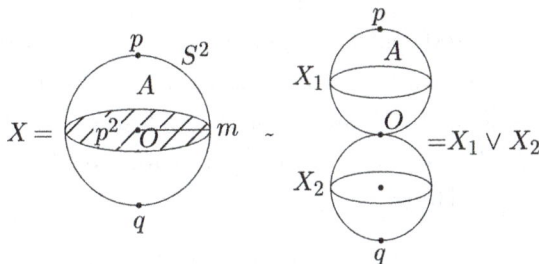

Fig. 2.10.

(b) Applying the Mayer–Vietoris sequence to the pair (U, V), we see that

$$H_i(X) \approx H_i(U) \oplus H_i(V) \approx H_i(X_1) \oplus H_i(X_2).$$

Noting that $H_i(X_1) \approx H_i(X_2)$ which are infinite cyclic for i equal to 0, 1, 2 and are zero otherwise, we conclude that

$$H_i(X) = \begin{cases} \mathbb{Z} \oplus \mathbb{Z}, & i = 1, 2, \\ \mathbb{Z}, & i = 0, \\ 0, & \text{otherwise.} \end{cases}$$

2215

Let X be a path-connected space, $f : X \to Y$ a continuous function, and $x_0, x_1 \in X$. Suppose that the induced homomorphism

$$f_* : \pi_1(X, x_0) \to \pi_1(Y, f(x_0))$$

is surjective. Show that

$$f_* : \pi_1(X, x_1) \to \pi_1(Y, f(x_1))$$

is also surjective.

(Indiana)

Solution.
Since X is path-connected, there exists a path $C : [0, 1] \to X$ such that $c(0) = x_0$ and $c(1) = x_1$. Then $\tilde{c} = f \circ c$ is a path connecting $f(x_0)$ and $f(x_1)$ in Y. For any $[a] \in \pi_1(Y, f(x_1))$, where a is a closed path at $f(x_1)$, $\check{c}a\tilde{c}^{-1}$ is a closed path at $f(x_0)$. By the assumption, there is a closed path h at x_0 such that $f_*([h]) = [\check{c}a\tilde{c}^{-1}]$. It means that $f \circ h$ and $\check{c}a\tilde{c}^{-1}$ are homotopic. Thus $f \circ (c^{-1}hc)$ and a are homotopic, and, consequently, $f_*([c^{-1}hc]) = [a]$. Hence

$$f_* : \pi_1(X, x_1) \to \pi_1(Y, f(x_1))$$

is surjective.

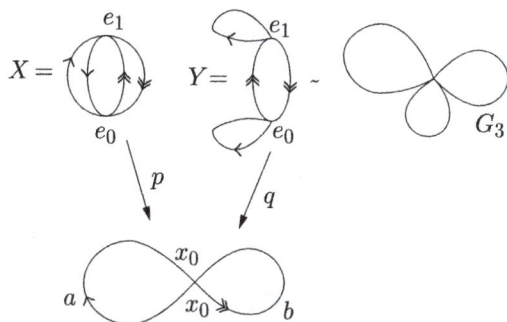

Fig. 2.11.

2216

Let B denote the "figure eight space". Let $p : X - B$ and $q : Y \to B$ be 2-fold covering maps, where both X and Y are connected. Prove that X and Y are homotopy equivalent, but not necessarily homeomorphic.

(*Indiana*)

Solution.

For any 2-fold covering map $p : X \to B$, let $p^{-1}(x_0) = \{e_0, e_1\}$. (See Fig. 2.11.) Since the automorphism group $A(X, p) \approx \mathbb{Z}_2$ and X is connected, it is not difficult to see, by considering the liftings of the circles a and b in X, that, in substance, X has only two different types as shown in Fig. 2.11. Then it is easy to see that they are homotopy equivalent to a 3-leaved rose G_3. But the spaces X and Y shown in Fig. 2.11 are not homeomorphic. Otherwise, suppose that $f : X \to Y$ is a homeomorphism. Then $X - \{e_0\}$ is homeomorphic to $Y - \{f(e_0)\}$. Since $X - \{e_0\}$ is contractible and $Y - \{f(e_0)\}$ is obviously not contractible, we come to a contradiction.

2217

Calculate the fundamental group of the space $RP^2 \times S^2$.

(*Indiana*)

Solution.

By the formula

$$\pi_1(X \times Y) \approx \pi_1(X) \oplus \pi_1(Y),$$

we see that

$$\pi_1(RP^1 \times S^2) \approx \pi_1(RP^2) \oplus \pi_1(S^2) \approx \mathbb{Z}_2 \oplus \{0\} \approx \mathbb{Z}_2.$$

2218

Let Z denote the figure 8 space, $Z = X \vee Y$, X and Y circles. Let $x, y \in \pi_1(Z, *)$ be the elements in Z defined by X, Y, where $*$ denotes the vertex

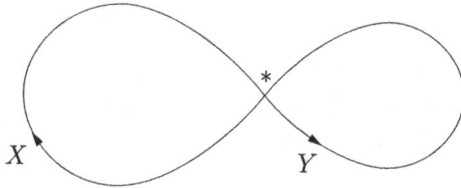

Fig. 2.12.

(a) Let $h : \pi_1(Z, *) \rightarrow \mathbb{Z}/6\mathbb{Z}$ be the homomorphism satisfying $h(x) = 2$ and $h(y) = 3$, and let $p : \tilde{Z} \rightarrow Z$ denote the covering space corresponding to the kernel of h. $(p_*(\pi_1(\tilde{Z}, *)) = \ker(h).)$ If A is a path component of $p^{-1}(X)$ and B is a path component of $p^{-1}(Y)$, how many intersection points of A and B are there? (That is, what is the cardinality of the sat $A \cap B$?)

(b) If G is a finite group, $h : \pi_1(Z, *) \rightarrow G$ a surjection, and $p : \tilde{Z} \rightarrow Z$ the corresponding cover, prove that the number of intersection points of a path component A of $p^{-1}(X)$ with a path component B of $p^{-1}(Y)$ divides the order of G.

(Indiana)

Solution.

(a) Since $h(x) = 2$ and $h(y) = 3$ and $\pi_1(Z, *)$ is generated by x and y, it follows that the homomorphism h is surjective. Thus

$\pi_1(Z, *)/\ker h$ is isomorphic to the group $\mathbb{Z}/6\mathbb{Z} = \mathbb{Z}_6$. Hence the covering space $p : \tilde{Z} \to Z$ is a 6-fold cover. We denote $p^{-1}(*)$ by

$$p^{-1}(*) = \{e_0, e_1, e_2, e_3, e_4, e_5\}.$$

Then, from $h(x) = 2$, we see that the path component of $p^{-1}(X)$ contains exactly the points $\{e_0, e_2, e_4\}$ of $p^{-1}(*)$ which correspond to the elements $\{0, 2, 4\}$ of $\mathbb{Z}_6$ respectively. By the same reason, we see that the path component of $p^{-1}(Y)$ contains exactly the points $\{e_0, e_3\}$ of $p^{-1}(*)$. Since

$$p^{-1}(X) \cap p^{-1}(Y) = p^{-1}(*),$$

we conclude that $A \cap B = \{e_0\}$.

(b) Since h is a surjection and G is a finite group, the corresponding cover $p : \tilde{Z} \to Z$ is a finite fold cover. Suppose that $h(x) = r$ and $h(y) = s$. Then r generates a subgroup H_1 of G and s generates a subgroup of H_2 of G. In a similar way as in (a), we see that the number of intersection points of A and B is equal to the number of elements in $H_1 \cap H_2$. By Lagrange's Theorem, it divides the order of G.

2219

Let X be the result of attaching a 2-cell D^2 to the circle S^1 by the map $f : S^1 \to S^1$ given in terms of complex numbers by $z \to z^6$.

(a) Compute, with proof, the fundamental group of X.
(b) Compute, with proof, the homology of the universal cover of X.

(*Indiana*)

Solution.

(a) Represent X as the space obtained by identifying the edges of a hexagon, as shown in Fig. 2.13. Under the identification the edges a become a circle through the point x_0. Let y

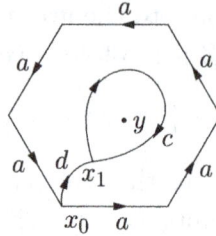

Fig. 2.13.

be the center point of the hexagon, and let $U = X - \{y\}$. Let V be the image of the interior of the hexagon under the identification. Then, U and V are open subsets, U, V, and $U \cap V$ are arcwise connected, and V is simply connected. Let x_1 be a point in $U \cap V$. It is clear that $U \cap V$ has the same homotopy type with S^1, and that $\pi_1(U \cap V, x_1)$ is an infinite cyclic group generated by γ, the homotopy class of a closed path c which circles around the point y once. (See Fig. 2.13.)

Applying the Van Kampen Theorem, we conclude that

$$\psi_1 : \pi_1(U, x_1) \to \pi_1(X, x_1)$$

is an epimorphism and its kernel is the smallest normal subgroup containing the image of the homomorphism

$$\phi_1 : \pi_1(U \cap V, x_1) \to \pi_1(U, x_1),$$

where ψ_1 and ϕ_1 are homomorphism induced by inclusion maps.

It is obvious that the circle a is a deformation retract of U. Thus $\pi_1(U, x_0)$ is an infinite cyclic group generated by a and, consequently, $\pi_1(U, x_1)$ is an infinite cyclic group generated by $a' = \tilde{\gamma}^{-1} a \tilde{\gamma}$, where $\tilde{\gamma}$ is the homotopy class of a path d from x_0 to x_1.

It is obvious that $\phi_1(\gamma) = a'^6$. Hence the smallest normal subgroup containing $\phi_1(\pi_1(U \cap V, x_1))$ is isomorphic to $6\mathbb{Z}$. Thus we conclude that

$$\pi_1(X) \approx \mathbb{Z}/6\mathbb{Z} = \mathbb{Z}_6.$$

(b) Since X is a finite 2-dimensional CW complex and $\pi_1(X) = \mathbb{Z}_6$, its universal covering space $\tilde{X}$ is a 6-fold covering space and is also a 2-dimensional CW complex. It is well-known that the Euler characteristic $\mathcal{X}(\tilde{X}) = 6\mathcal{X}(X)$. But it is easy to see that $\mathcal{X}(X) = 1$, hence $\mathcal{X}(\tilde{X}) = 6$. From $H_0(\tilde{X}) \approx \mathbb{Z}$, $H_1(\tilde{X}) = 0$ and $H_2(\tilde{X}) \approx H_2(\tilde{X}, \tilde{X}^1)$, where $\tilde{X}^1$ is the 1-skeleton of $\tilde{X}$, we see that $H_2(\tilde{X})$ is a free Abelian group of rank 5. So we conclude that

$$H_i(\tilde{X}) = \begin{cases} 0, & i \geq 3, \\ \mathbb{Z} \oplus \mathbb{Z} \oplus \mathbb{Z} \oplus \mathbb{Z} \oplus \mathbb{Z}, & i = 2, \\ 0, & i = 1, \\ \mathbb{Z}, & i = \emptyset. \end{cases}$$

2220

If X is any topological space and S^1 denotes the unit circle in the complex plane with its usual topology as a topological group with multiplication given by the multiplication of complex numbers, then it is known that the set $[X, S^1]$ of homotopy classes of maps from X to S^1 inherits a natural group structure.

(a) Define this group operation explicitly and indicate the group identity and how inverses are formed. You do not need to prove your assertions.

(b) Compute this group explicitly for $X = $ point, S^1, S^2, and $T^2 = S^1 \times S^1$.

(*Indiana*)

Solution.

(a) We denote the homotopy class of a map $f : X \to S^1$ by $[f]$ and write S^1 as

$$S^1 = \{e^{2\pi i\theta} \in \mathbb{C} | 0 \leq \theta \leq 1\}.$$

Then the multiplication of $[X, S^1]$ is defined by $[f] \cdot [g] = [f \cdot g]$, where the map $f \cdot g : X \to S^1$ is defined by $(f \cdot g)(x) = f(x) \cdot g(x)$

for any $x \in X$. Here $f(x) \cdot g(x)$ is defined by the multiplication of complex numbers. The identity of this multiplication is $[e]$, where the map $e : X \to S^1$ is defined by $e(x) \equiv e^{2\pi i}$ for any $x \in X$. The inverse of $[f]$ is the homotopy class of a map $\tilde{f}$, which is defined by $\tilde{f}(x) = 1/f(x)$ for any $x \in X$.

(b) If $X = $ point, then $[X, S^1]$ obviously has only one element. So the group $[X, S^1]$ is trivial. For the case of $X = S^1$, one can easily prove that

$$[S^1, S^1] \approx \pi_1(S^1) \approx \mathbb{Z}.$$

For the case of $X = S^2$, it is easy to see that each homotopy class $[f]$ can be represented by a map $f : S^2 \to S^1$ which maps the northpole N of S^2 to the point $p_0 = e^{2\pi i}$ of S^1. Then by the facts that S^2 is simply connected and the universal covering space of S^1, R is contractible, one can easily prove that $[S^2, S^1] \approx \pi_2(S^1) = 0$. Now we discuss the case of $X = T^2 = S^1 \times S^1$. For any map f from $S^1 \times S^1 \to S^1$ we define two maps f_1 and f_2 from $S^1 \to S^1$ by $f_1(\theta) = f(e^{2\pi i\theta}, p_0)$ and $f_2(\theta) = f(p_0, e^{2\pi i\theta})$ for any $e^{2\pi i\theta} \in S^1$, respectively. Then let $\phi : [X, S^1] \to \mathbb{Z} \oplus \mathbb{Z}$ is defined by $\phi([f]) = (\deg f_1, \deg f_2)$. We have

$$
\begin{aligned}
\phi([f] \cdot [g]) = \phi([f \cdot g]) &= (\deg(f \cdot g)_1, \deg(f \cdot g)_2) \\
&\equiv (\deg(f_1 \cdot g_1), \deg(f_2 \cdot g_2)) \\
&= (\deg f_1 + \deg g_1, \deg f_2 + \deg g_2) \\
&= (\deg f_1, \deg f_2) + (\deg g_1, \deg g_2) \\
&= \phi([f]) + \phi([g]).
\end{aligned}
$$

Therefore, ϕ is a homomorphism from $[X, S^1]$ to $\mathbb{Z} \oplus \mathbb{Z}$. Note that f_1 can be extended to a map from X to S^1, still denoted by f_1, by

$$f_1(e^{2\pi i\theta}, e^{2\pi i\psi}) \equiv f_1(e^{2\pi i\theta}, p_0),$$

which is homotopic to f under the homotopy map $F : X \times I \to S^1$ defined by

$$F(e^{2\pi i\theta}, e^{2\pi i\psi}, t) = f(e^{2\pi i\theta}, e^{2\pi it + 2\pi i(1-t)\psi}).$$

Thus one can easily prove that ϕ is a monomorphism. It is clear that ϕ is an epimorphism, and consequently ϕ is an isomorphism. We conclude that $[X, S^1] \approx \mathbb{Z} \oplus \mathbb{Z}$.

2221

If X is a path-connected space whose universal cover is compact, show that $\pi_1(X, x_0)$ is finite.

(*Indiana*)

Solution.

Let $\pi : \tilde{X} \to X$ be the universal cover of X. If $\pi_1(X, x_0)$ were not finite, then $\pi^{-1}(x_0)$ would be a closed set of infinite points in $\tilde{X}$. Since $\tilde{X}$ is compact, $\pi^{-1}(x_0)$ must have at least a limit point, say $\tilde{x}$, such that $\pi(\tilde{x}) = x_0$. Thus it is easy to see that π is not a local homeomorphism at $\tilde{x}$, which is a contradiction.

2222

Prove that if X is locally path connected and simply connected then every map $X \to S^1$ is homotopic to a constant. What can you say if we just assume that X is path connected, locally path connected and the fundamental group of X is finite?

(*Indiana*)

Solution.

Let $\exp : R \to S^1$ denote the exponential covering map, i.e., the universal covering space of S^1. Since X is locally path connected and simply connected, $\pi_1(X) = 0$, and $f_*(\pi_1(X)) = 0$ for any map $f : X \to S^1$. Hence there exists a lifting of f, $\tilde{f} : X \to R$ such that $\exp(\tilde{f}) = f$. Since R is simply connected, $\tilde{f}$ is homotopic to a constant map $\tilde{c}$. Denote the homotopy between

$\tilde{f}$ and $\tilde{c}$ by $\tilde{H}$. Then $\exp(\tilde{H})$ is the homotopy between f and c, i.e., f is homotopic to a constant map.

Suppose that $\pi_1(X)$ is finite. Since $\pi_1(S^1) \approx \mathbb{Z}$ and $f_*(\pi_1(X))$ is a finite subgroup of $\pi_1(S^1)$, we see that $f_*(\pi_1(X))$ must be trivial. So the above argument still works in this case, and the same conclusion holds.

2223

Assume that a 6-element group Γ, isomorphic to the group of permutation on three letters S_3, acts on $X = S^3$ freely. Compute $\pi_1(Y)$ and $H_1(Y, \mathbb{Z})$, where $Y = X/\Gamma$.

(*Indiana*)

Solution.

It is obvious that we have a covering map $p : X \to X/\Gamma$. Since X is simply connected and Hausdorff, X is a universal cover and so

$$\pi_1(Y)/p_*(\pi_1(X)) \cong S_3.$$

But

$$p_*(\pi_1(X)) = p_*(0) = 0.$$

Thus we have $\pi_1(Y) \cong S_3$.

By Hurewicz Theorem, $H_1(Y, \mathbb{Z})$ is the abelianization of S_3, that is,

$$H_1(Y, \mathbb{Z}) \cong S_3/A_3 \cong \mathbb{Z}/2\mathbb{Z},$$

since $[S_3 : A_3] = 2$.

2224

Let T^2 be the two-dimensional torus and let $\varphi : S^2 \to T^2$ be smooth map. Show that for any top de Rham cohomology class $[v] \in H^2(T^2)$, we have $\varphi^*[v] = 0$.

(*Indiana*)

Solution.

Let $p : R^2 \to T^2$ be the well-known universal cover mapping from the two-dimensional Euclidean space R^2 to T^2. Then there exists a unique lift $\tilde{\varphi}$ of φ from S^2 to R^2 such that the following diagram commutes, i.e., $\varphi = p \circ \tilde{\varphi}$:

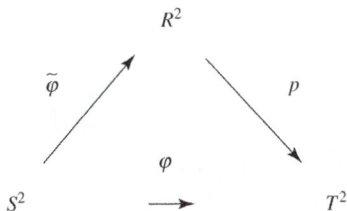

$$
\begin{array}{ccc}
 & R^2 & \\
\tilde{\varphi} \nearrow & & \searrow p \\
 & \varphi & \\
S^2 & \longrightarrow & T^2
\end{array}
$$

Then the induced maps on cohomology give the following commutative diagram

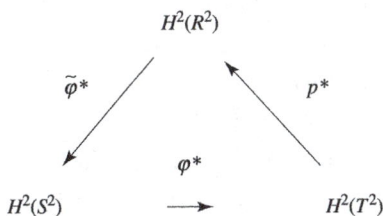

$$
\begin{array}{ccc}
 & H^2(R^2) & \\
\tilde{\varphi}^* \swarrow & & \searrow p^* \\
 & \varphi^* & \\
H^2(S^2) & \longrightarrow & H^2(T^2)
\end{array}
$$

But R^2 has trivial second cohomology, so $p^*([v]) = 0$ for any top de Rham cohomology class $[v] \in H^2(T^2)$, and so $\varphi^*([v]) = \tilde{\varphi}^*(p^*([v])) = \tilde{\varphi}^*(0) = 0$.

2225

A closed surface is a compact, connected metrizable 2-manifold without boundary.

(a) State the Classification Theorem for closed surfaces. Include a definition of "connected sum".

(b) Three surfaces with boundary are shown below. If 2-disks are attached to each component of their boundaries, closed surfaces result. Identify these 3 closed surfaces among the model surfaces given in your statement of the Classification Theorem.

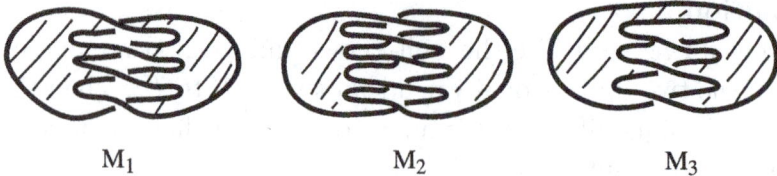

$$M_1 \qquad M_2 \qquad M_3$$

(*Wisconsin*)

Solution.

(a) The connected sum of two closed surfaces M_1 and M_2, denoted by $M_1 \# M_2$, is obtained by choosing a closed 2-disk $D_i \subset M_i$ and identifying $M_1 - \operatorname{int} D_1$ with $M_2 - \operatorname{int} D_2$ via a homeomorphism of ∂D_1 with ∂D_2. The resulted close surface does not depend on the choices made in this construction.

The Classification Theorem says that if M is an orientable closed surface then M is homeomorphic to the connected sum of S^2 and a finite number $k \geq 0$ of tori $T^2 = S^1 \times S^1$ and the Euler characteristic of M is $2 - k$; if M is a non-orientable closed surface then M is homeomorphic to the connected sum of S^2 and a finite number $k \geq 0$ of real projective planes P^2 and the Euler characteristic of M is $2 - k$.

(b) We see that the resulted closed surface M_1 is a torus T^2, M_2 is the connected sum $T^2 \# T^2$, and M_3 is the connected sum $T^2 \# P^2 = P^2 \# P^2 \# P^2$.

2226

(a) Define the meaning of "$p : E \to B$ is a covering map".

(b) Let X be the familiar "figure-8" subspace of R^2 : X is the union of two circles with a single point x_0 in common. Give a covering space argument to show that $\pi_1(X, x_0)$ contains elements a and b such that $(ab)^2 \neq a^2 b^2$. Give a careful statement of any theorems from covering space theory that you use. Do not use results from the theory of free groups.

(*Wisconsin*)

Solution.

(a) A continuous map $p : E \to B$ is called a covering map if it is surjective and each point $x \in B$ has a neighborhood U such that $p^{-1}(U)$ can be expressed as the union of disjoint open sets V_j in E such that for each j, p restricts to a homeomorphism of V_j onto U.

(b) We first give a 3 to 1 covering of X as shown in the following figure:

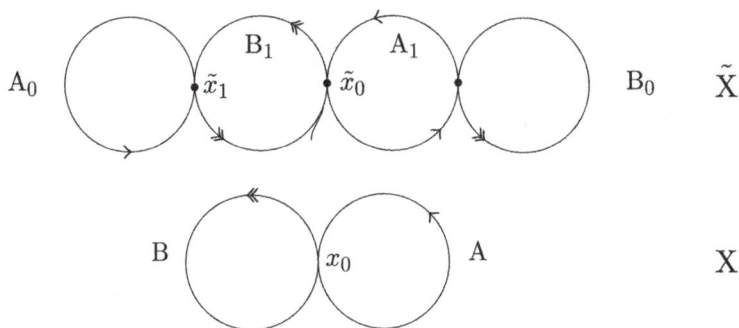

Let $I = [0,1]$. Define a map $f : I \to X$ with $f(I) = A$, $f(0) = x_0 = f(1)$, and $f|_{(0,1)}$ being a homeomorphism into $A \backslash \{x_0\}$. Similarly, define a map $g : I \to X$ with $g(I) = B$, $g(0) = x_0 = g(1)$ being a homeomorphism onto $B \backslash \{x_0\}$. Then f and g have liftings $\tilde{f}, \tilde{g} : I \to X$ with $\tilde{f}(0) = x_0 = \tilde{g}(0)$ (see the figure). Denote the path homotopy class of the loop f and g at x_0 by a and b respectively. Then it is easy to see that $(ab)^2 \neq a^2 b^2$, since the final points of the liftings of $(ab)^2$ and $a^2 b^2$ are x_1 and x_0 respectively.

Section 3

Homology Theory

2301

Prove the 3×3 Lemma.

Consider the following commutative diagram of abelian groups

$$
\begin{array}{ccccccccc}
& & 0 & & 0 & & 0 & & \\
& & \downarrow & & \downarrow & & \downarrow & & \\
0 & \to & A_3 & \xrightarrow{\alpha_2} & A_2 & \xrightarrow{\alpha_1} & A_1 & \to & 0 \\
& & \delta_2 \downarrow & & \varepsilon_2 \downarrow & & \zeta_2 \downarrow & & \\
0 & \to & B_3 & \xrightarrow{\beta_2} & B_2 & \xrightarrow{\beta_1} & B_1 & \to & 0 \\
& & \delta_1 \downarrow & & \varepsilon_1 \downarrow & & \zeta_1 \downarrow & & \\
0 & \to & C_3 & \xrightarrow{\gamma_2} & C_2 & \xrightarrow{\gamma_1} & C_1 & \to & 0 \\
& & \downarrow & & \downarrow & & \downarrow & & \\
& & 0 & & 0 & & 0 & &
\end{array}
$$

If all 3 columns and the first two rows are short exact, then the last row is also short exact.

(Harvard)

Solution.

To prove the exactness at C_3, we show that γ_2 is injective. Let $c \in C_3$ and $\gamma_2(c) = 0$. Since δ_1 is surjective, there is a $b \in B_3$

such that $c = \delta_1(b)$. By the commutativity we see $\beta_2(b) \in \ker \varepsilon_1$. Hence there is an $a_2 \in A_2$ such that $\varepsilon_2(a_2) = \beta_2(b)$. Then since

$$\zeta_2(\alpha_1(a_2)) = \beta_1(\varepsilon_2(a_2)) = \beta_1(\beta_2(b)) = 0$$

and ζ_2 is injective, we have $\alpha_1(a_2) = 0$, i.e., $a_2 \in \ker \alpha_1 = \operatorname{im} \alpha_2$. Thus there is an $a \in A_3$ such that $\alpha_2(a) = a_2$. Since

$$\beta_2(\delta_2(a) - b) = \beta_2 \delta_2(a) - \beta_2(b) = \varepsilon_2 \alpha_2(a) - \beta_2(b)$$

$$= \varepsilon_2(a_2) - \beta_2(b) = 0$$

and β_2 is injective, we see $\delta_2(a) = b$. Therefore

$$c = \delta_1(b) = \delta_1 \delta_2(a) = 0.$$

Hence γ_2 is injective.

Now we prove that $\ker \gamma_1 \subset \operatorname{im} \gamma_2$. For any $c \in \ker \gamma_1$, since ε_1 is surjective, there is a $b \in B_2$ such that $c = \varepsilon_1(b)$. Thus it is easy to see that $\beta_1(b) \in \ker \zeta_1$, and consequently that there is an $a_1 \in A_1$ such that $\zeta_2(a_1) = \beta_1(b)$. Since α_1 is surjective, there is an $a_2 \in A_2$ such that $\alpha_1(a_2) = a_1$. Thus by the commutativity, we have $b - \varepsilon_2(a_2) \in \ker \beta_1$. Therefore there is a $b_3 \in B_3$ such that $\beta_2(b_3) = b - \varepsilon_2(a_2)$. Then

$$\gamma_2(\delta_1(b_3)) = \varepsilon_1 \beta_2(b_3) = \varepsilon_1(b) - \varepsilon_1 \varepsilon_2(a_2) = c.$$

Hence $c \in \operatorname{im} \gamma_2$. In a similar way we may prove that $\operatorname{im} \gamma_2 \subset \ker \gamma_1$. Thus the exactness at C_2 is proved.

We leave the proof of the exactness at C_1 to the reader.

2302

Prove that if M is a compact manifold of odd dimension, then $\mathcal{X}(M) = 0$. Show examples of compact 4-manifolds with $\mathcal{X} = 0, 1, 2, 3, 4$.

(Columbia)

Solution.

Since M is compact, M is $\mathbb{Z}_2$-orientable. Suppose that $\dim M = 2m + 1$. Therefore,

$$\mathcal{X}(M) = \sum_{i=0}^{2m+1} (-1)^i \dim H_i(M, \mathbb{Z}_2).$$

By the Poincaré duality Theorem,

$$\dim H_i(M, \mathbb{Z}_2) = \dim H_{2m+1-i}(M, \mathbb{Z}_2)$$

for any i. Thus, since i and $2m+1-i$ have different parity, they appear in the sum with opposite signs. Therefore $\mathcal{X}(M) = 0$.

Let $X_0 = T^2 \times T^2$, $X_1 = RP^2 \times RP^2$. Denote by U_h the connected sum of h projective planes. Then it is well-known that $\mathcal{X}(U_h) = 2 - h$. Let $X_2 = U_3 \times U_4$, $X_3 = U_3 \times U_5$, and $X_4 = S^2 \times S^2$. Using the fact that

$$\mathcal{X}(M_1 \times M_2) = \mathcal{X}(M_1) \times \mathcal{X}(M_2),$$

we see that $\mathcal{X}(X_i) = i$ for i equal to 0, 1, 2, 3 and 4.

2303

Let X be a topological space. The suspension ΣX of X is defined to be the identification space obtained from $X \times [-1, 1]$ by identifying $X \times \{-1\}$ to a point and $X \times \{1\}$ to another point. For example the sphere S^n is the suspension of S^{n-1} with the north and south poles corresponding to the two identification points.

Compute the homology of ΣX in terms of the homology of X.

(Illinois)

Solution.

Let p_1 and p_2 be the identification point respectively. Set $U = \Sigma X - \{p_1\}$ and $V = \Sigma X - \{p_2\}$. Then U and V are open sets of ΣX, and $\Sigma X = U \cup V$. It is obviously to see that U and

V are contractible spaces and X is a deformation retractor of $U \cap V$. By Mayer–Vietoris sequence, we have

$$H_q(\Sigma X) \approx \begin{cases} \tilde{H}_{q-1}(X), & q > 0, \\ \mathbb{Z}, & q = 0, \end{cases}$$

where $\tilde{H}_{q-1}(X)$ denotes the reduced homology of X.

2304

Consider the chain complex

$$C = \{C_n | n \geq 0\} \ C_n = \mathbb{Z}^3$$
$$\partial_n : C_n \to C_{n-1} \text{ defined by}$$
$$(r, s, t) \to (s - t, 0, 0) \ n \text{ even}$$
$$(r, s, t) \to (0, s + t, s + t) \ n \text{ odd}.$$

Compute $H_n(C)$ for all n.

(Illinois)

Solution.

When n is even, $(r, s, t) \in Z_n(c)$ if and only $s = t$. So $Z_n(c) = \{(r, s, s) \in C_n\}$, i.e., $Z_n(c)$ is isomorphic to $\mathbb{Z} \oplus \mathbb{Z}$.

Noting that $\text{im} \, \partial_{n+1} = \{(0, t, t) \in C_n\}$ for even n, we have $H_n(c) = \mathbb{Z}$ for even n.

When n is odd, $(r, s, t) \in Z_n(c)$ if and only if $s + t = 0$. So $Z_n(c)$ is isomorphic to $\mathbb{Z} \oplus \mathbb{Z}$, and noting $\text{im} \, \partial_{n+1} = \{(r, 0, 0) \in C_n\} \approx \mathbb{Z}$, we have $H_n(c) = \mathbb{Z}$.

Therefore, $H_n(c) \approx \mathbb{Z}$ for any n.

2305

Construct a *CW* complex which has the following $\mathbb{Z}$-homology groups:

$$H_0(X) = \mathbb{Z},$$
$$H_1(X) = \mathbb{Z} \oplus \mathbb{Z}/2\mathbb{Z},$$

$$H_2(X) = \mathbb{Z}/3\mathbb{Z},$$

$$H_3(X) = \mathbb{Z},$$

$$H_n(X) = 0, \quad \text{if } n \geq 4.$$

(Columbia)

Solution.

Denote by X^1 the "figure eight space" as shown below,

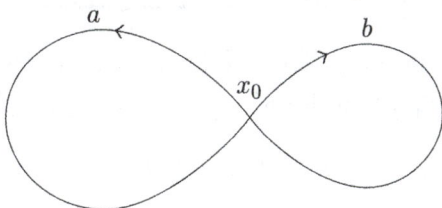

Let X^2 be the space obtained by attaching two 2-cells to X^1, one by the map $f_1 : S^1 \to x_0$ and the other by the map $f_2 : S^1 \to a$ such that $f_2(z) = z^2$. Therefore, the image of one 2-cell under f_1 is homeomorphic to S^2, the 2-sphere. It is well-known that $H_1(X^1) = \mathbb{Z} \oplus \mathbb{Z}$, which has two generators a and b, consider the following diagram

$$0 \to H_2(X^2) \xrightarrow{j_*} H_2(X^2, X^1) \xrightarrow{\partial_*} H_1(X^1) \xrightarrow{i_*} H_1(X^2) - H_1(X^2, X^1)$$
$$\uparrow f_{i*} \qquad\qquad\qquad \uparrow f_i|_{S^1*}$$
$$0 \to H_2(D^2, S^1) \xrightarrow{\partial'_*} H_1(S^1) \to 0$$

The square is commutative and the level rows are exact. Since

$$H_2(X^2, X^1) = \operatorname{im} f_{1*} \oplus \operatorname{im} f_{2*} \approx \mathbb{Z} \oplus \mathbb{Z},$$

$\operatorname{im} f_1|_{S^1*} = 0$ and $\operatorname{im} f_2|_{S^1*} = 2\mathbb{Z}$, we may see that $\operatorname{im} \partial_* \approx 2\mathbb{Z}$ and $\ker \partial_* \approx \mathbb{Z}$. Since $H_1(X^2, X^1) = 0$, we have $H_1(X^2) \approx \mathbb{Z} \oplus \mathbb{Z}_2$. It is also easy to see that $H_2(X^2) = \mathbb{Z}$.

Now let X be the space obtained by attaching two 3-cells to X^2, one by the map $g_1 : S^2 \to x_0$ and the other by a map $g_2 : S^2 \to S^2$ such that $\deg g_2 = 3$. Then in a similar way as above, we may conclude that the space X satisfies the requirements in the problem.

2306

Compute $H_p(S^n \vee S^n \vee \cdots \vee S^n)$, $n \geq 0$, for all p.

(Columbia)

Solution.
Denote by $S^n(q)$ the space $\underbrace{S^n \vee S^n \vee \cdots \vee S^n}_{q \text{ times}}$. When $n = 0$, $S^0(q)$ has $q + 1$ points. Then $H_p(S^0(q))$ is a free abelian group of rank q for p equal to 0 and is zero otherwise. In fact

$$S^n(q+1) = S^n(q) \vee S^n.$$

Let $a \in S^n(q)$ and $b \in S^n$. It is easy to see that $U = S^n(q+1) - \{a\}$ has the homotopy type of $S^n(q)$ and $V = S^n(q+1) - \{b\}$ also has the homotopy type of $S^n(q)$. It is also clear that $S^n(q+1) = U \cup V$ and $U \cap V$ has the homotopy type of $S^n(q-1)$. Thus, when $n > 0$, by induction on q and the Mayer–Vietoris sequence of the pair (U, V), we may prove that $\tilde{H}_p(S^n(q))$ is a free abelian group of rank q for p equal to n and is zero otherwise, where $\tilde{H}_p$ is the reduced homology group.

2307

Build a CW complex X by adding two 2-cells to S^1, one by the map $z \to z^4$ and the other by the map $z \to z^6$. What is the homology of this space?

(Indiana)

Solution.
X is a 2-dimensional CW complex with 2-skeleton $K^2 = X$, 1-skeleton $K^1 = S^1$ and 0-skeleton $K^0 = \{p\}$. K^2 is obtained from K^1 by attaching two 2-cells via the maps f and g as indicated in Fig. 2.14. K^1 is obtained from K^0 by attaching one 1-cell. Let $K^n = K^2$ for $n \geq 3$. Then we have a chain

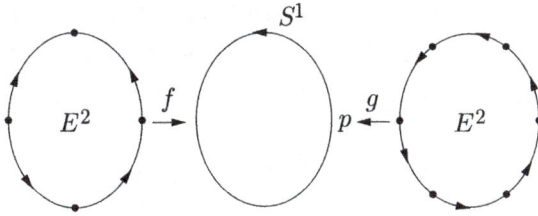

Fig. 2.14.

complex $K = \{C_n(K), d_n\}$, where $C_n(K) = H_n(K^n, K^{n-1})$ and $d_n : C_n(K) \to C_{n-1}(K)$ is defined to be the composition of homomorphisms,

$$H_n(K^n, K^{n-1}) \xrightarrow{\partial_*} H_{n-1}(K^{n-1}) \xrightarrow{j_{n-1}} H_{n-1}(K^{n-1}, K^{n-2}),$$

where δ_* is the boundary operator of the pair (K^n, K^{n-1}) and j_{n-1} is the homomorphism induced by the inclusion map. It is well-known that $H_n(X) \approx H_n(K)$. It is obvious that $H_n(X) = H_n(K) = 0$ for $n \geq 3$. To compute $H_2(k)$, consider the following diagram:

$$
\begin{array}{ccccccc}
H_2(K^1) \to H_2(K^2) & \xrightarrow{j_2} & H_2(K^2, K^1) & \xrightarrow{\partial_*} & H_1(K^1) & \xrightarrow{j_1} & H_1(K^1, K^0) \\
& & \uparrow f_* & & \uparrow (f|_{S^1})_* & & \\
& & H_2(E^2, S^1) & \xrightarrow{\partial'_*} & H_1(S^1) & &
\end{array}
$$

The square is commutative and it is well known that f_* is a monomorphism and ∂'_* is an isomorphism. Noting $f|_{S^1}$ is the map $z \to z^4$, we see that the image of $(f|_{S^1})_*$ is $4\mathbb{Z}$. In the same way we see that the image of $(g|_{S^1})_*$ is $6\mathbb{Z}$. Since

$$H_2(K^2, K^1) \approx \operatorname{im} f_* \oplus \operatorname{im} g_*,$$

it follows that $\operatorname{im} \partial_*$ is isomorphic to the subgroup $2\mathbb{Z}$ of $H_1(K^1)$, and that $\ker \partial_*$ is isomorphic to $\mathbb{Z} \oplus \mathbb{Z}_2$. Thus, since j_2 is injective, we see that

$$H_2(K^2) \approx \operatorname{im} j_2 \approx \ker \partial_* \approx \mathbb{Z} \oplus \mathbb{Z}_2.$$

It is clear that j_1 is an isomorphism. Therefore, $\operatorname{im} d_2 \approx 2\mathbb{Z}$. Noting that $\ker d_1 \approx H_1(K^1, K^0) \approx \mathbb{Z}$, we have $H_1(K) \approx \mathbb{Z}/2\mathbb{Z} \approx \mathbb{Z}_2$. Thus we conclude that

$$H_i(X) = \begin{cases} \mathbb{Z} \oplus \mathbb{Z}_2, & i = 2, \\ \mathbb{Z}_2, & i = 1, \\ \mathbb{Z}, & i = 0, \\ 0, & \text{otherwise.} \end{cases}$$

2308

Let

$$X = \{(x, y, z) \in R^3 | xyz = 0\}.$$

(a) Compute $H_*(X, X - \{0\})$.

(b) Prove that any homeomorphism $h : X \to X$ must leave the origin fixed.

(*Indiana*)

Solution.

(a) Let $U = D^3 \cap X$, where

$$D^3 = \{(x, y, z) \in R^3 | x^2 + y^2 + z^2 < 2\}.$$

It is easy to see that U is contractible. Thus

$$H_q(X, X - \{0\}) \approx H_q(U, U - \{0\}) \approx \tilde{H}_{q-1}(U - \{0\})$$

for any q. Let $S = S^2 \cap X$, where S^2 is the unit 2-sphere. Then S is a deformation retract of $U - \{0\}$. So $\tilde{H}_*(U - \{0\}) \approx \tilde{H}_*(S)$. To compute $\tilde{H}_*(S)$, let

$$W_1 = S - \{(0, 0, 1), (0, 0, -1)\}$$

and

$$W_2 = S - \{(-1, 0, 0), (1, 0, 0)\}.$$

Applying the Mayer–Vietoris sequence of the pair (W_1, W_2), we see that

$$H_q(X, X - \{0\}) = \begin{cases} \mathbb{Z} \oplus \mathbb{Z} \oplus \mathbb{Z} \oplus \mathbb{Z} \oplus \mathbb{Z} \oplus \mathbb{Z} \oplus \mathbb{Z}, & q = 2, \\ 0, & \text{otherwise.} \end{cases}$$

(b) Let x be any point of X different from the origin. Similarly we obtain

$$H_2(X, X - \{x\}) = \begin{cases} \mathbb{Z} \oplus \mathbb{Z} \oplus \mathbb{Z}, & \text{if } x \text{ is in a coordinate axis,} \\ \mathbb{Z}, & \text{otherwise.} \end{cases}$$

Let $h : X \to X$ be any homeomorphism. Since the local homology groups are invariant under homeomorphism,

$$H_2(X, X - \{0\}) \approx H_2(X, X - \{f(0)\}).$$

From the above results, it follows that $f(0) = 0$.

2309

(a) Write down the Mayer–Vietoris sequence in reduced homology which relates the spaces $S^1 \times R^1$, $S^1 \times R^1 - \{p\}$, and a disk D in $S^1 \times R^1$ about the point p. (S^1 = 1-sphere.)

(b) Use (a) to calculate the homology of $S^1 \times R^1 - \{p\}$.

(Indiana)

Solution.

(a) The Mayer–Vietoris sequence is

$$\cdots \to \tilde{H}_q(D - \{p\}) \xrightarrow{\phi} \tilde{H}_q(S^1 \times R^1 - \{p\}) \oplus \tilde{H}_q(D)$$
$$\xrightarrow{\psi} \tilde{H}_q(S^1 \times R^1) \xrightarrow{\Delta} \tilde{H}_{q-1}(D - \{p\}) \to \cdots.$$

(b) Since $S^1 \times R^1$ and $D - \{p\}$ have the same homotopy type with S^1, it follows that

$$\tilde{H}_q(S^1 \times R^1) \approx \tilde{H}_q(D - \{p\}) \approx \tilde{H}_q(S^1).$$

Since D is contractible, the nontrivial part of the Mayer–Vietoris sequence is

$$0 \to \tilde{H}_1\,(D - \{p\}) \xrightarrow{\phi} \tilde{H}_1(S^1 \times R^1 - \{p\}) \xrightarrow{\psi} \tilde{H}_1\,(S^1 \times R^1) \to 0,$$
$$\underset{\approx \mathbb{Z}}{} \qquad\qquad\qquad\qquad \underset{\approx \mathbb{Z}}{}$$

which is split exact. Thus we have

$$H_i(S^1 \times R^1 - \{p\}) = \begin{cases} \mathbb{Z} \oplus \mathbb{Z}, & i = 1, \\ \mathbb{Z}, & i = 0, \\ 0, & \text{otherwise.} \end{cases}$$

2310

Let the real projective plane RP^2 be embedded in the standard way in the real projective 5-space RP^5. Compute $H_q(RP^5/RP^2)$, where RP^5/RP^2 is the space obtained from RP^5 by identifying RP^2 to a point.

(Indiana)

Solution.
Since RP^5 is compact Hausdorff and RP^2 is a strong deformation retract of a compact neighborhood of RP^2 in RP^5, we see that

$$\tilde{H}_q(RP^5/RP^2) \approx H_q(RP^5, RP^2) \text{ for any } q.$$

It is well-known that

$$H_q(RP^5) = \begin{cases} \mathbb{Z}, & q = 0, 5, \\ \mathbb{Z}_2, & q = 1, 3, \\ 0, & \text{otherwise,} \end{cases}$$

and

$$H_q(RP^2) = \begin{cases} \mathbb{Z}, & q = 0, \\ \mathbb{Z}_2, & q = 1, \\ 0, & \text{otherwise.} \end{cases}$$

Thus, from the exact sequence of the pair (RP^5, RP^2), it follows that

$$H_q(RP^5/RP^2) = \begin{cases} \mathbb{Z}, & q = 0, 5, \\ \mathbb{Z}_2, & q = 3, \\ 0, & \text{otherwise.} \end{cases}$$

2311

Let

$$Y = \{(x_1, x_2) \in R^2 | x_1 x_2 = 0 \text{ and } x_2 \geq 0\}.$$

Prove that $Y \times R$ is not homeomorphic to R^2.

(*Indiana*)

Solution.
Let

$$Y \times R = \{(x_1, x_2, x_3) \in R^3 | x_1 x_2 = 0, \ x_2 \geq 0\}.$$

Then the point $O = (0,0,0) \in Y \times R$. To compute the local homology groups

$$H_i(Y \times R, \ Y \times R - \{0\}), \quad i = 0, 1, 2, \ldots,$$

we take an open neighborhood of the point $O, U = D^3 \cap (Y \times R)$, where D^3 is an open ball centered at O. Therefore

$$H_i(Y \times R, \ Y \times R - \{0\}) \approx H_i(U, U - \{0\}) \approx \tilde{H}_{i-1}(U - \{0\}),$$

because U is contractible. Let X be the space as shown in the following figure

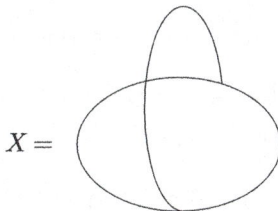

$X =$

It is easy to see that X is a deformation retract of $U - \{0\}$ and that $H_1(X) \approx \mathbb{Z} \oplus \mathbb{Z}$. Thus

$$H_2(Y \times R, \ Y \times R - \{0\}) \approx \mathbb{Z} \oplus \mathbb{Z}.$$

It is well-known that for any point $p \in R^2$ the local homology group

$$H_2(R^2, R^2 - \{p\}) \approx \mathbb{Z}.$$

Since the local homology groups are isomorphic under homeomorphism, it follows that $Y \times R$ is not homeomorphic to R^2.

2312

Let A be a non-empty subset of X and $X \cup CA$ be the union of X with the cone of A; that is, $X \cup CA$ is obtained from the subset $X \times \{0\} \cup A \times [0,1]$ of $X \times [0,1]$ by identifying $A \times \{1\}$ to a point.

Prove that the reduced homology group $\tilde{H}_n(X \cup CA)$ is isomorphic to $H_n(X, A)$, for every n.

(Indiana)

Solution.

Let $Y = X \times \{0\} \cup A \times [0,1]$ and $B = A \times \{1\}$. Then $X \cup CA = Y/B$. Let $\pi : Y \to Y/B$ be the identification map and y_0 denote the point $\pi(B)$ in Y/B. Let $U = A \times [\frac{1}{2}, 1] \subset Y$. Then y_0 is a strong deformation retract of $\pi(U)$. Thus, in the exact sequence of the triple $(Y/B, \pi(U), y_0)$,

$$\cdots \to H_n(\pi(U), y_0) \to H_n(Y/B, y_0)$$
$$\to H_n(y/B, \pi(U)) \to H_{n-1}(\pi(U), y_0) \to \cdots$$

it follows that $H_*(\pi(U), y_0) = 0$. Hence, the inclusion map of pairs induces an isomorphism

$$H_*(Y/B, y_0) \approx H_*(Y/B, \pi(U)).$$

Let $V = A \times [\frac{3}{4}, 1]$. By the excision property, we have an isomorphism

$$H_*(Y, U) \approx H_*(Y - V, U - V).$$

Since B is a strong deformation retract of U, it follows from the exact sequence that

$$H_*(Y, B) \approx H_*(Y, U).$$

Thus, we have

$$H_*(Y, B) \approx H_*(Y - V, U - V).$$

In a similar way the set $\pi(V)$ may be excised from the pair $(Y/B, \pi(U))$ to give an isomorphism

$$H_*(Y/B, y_0) \approx H_*(Y/B, \pi(U))$$
$$\approx H_*(Y/B - \pi(V), \pi(U) - \pi(V)).$$

Now the restriction of the map π gives a homeomorphism of pairs

$$\pi : (Y - V, U - V) \to (Y/B - \pi(V), \pi(U) - \pi(V)),$$

and so an isomorphism of their homology groups. All of these combine to give an isomorphism

$$H_*(Y/B, y_0) \approx H_*(Y, B).$$

Since Y admits obviously a strong deformation retraction onto $X \times \{0\}$ which maps B onto $A \times \{0\}$, we have

$$H_*(Y, B) \approx H_*(X, A).$$

On the other hand, from the exact sequence of the pair $(Y/B, y_0)$, it follows that

$$H_*(Y/B, y_0) \approx \tilde{H}_*(Y/B) = \tilde{H}_*(X \cup CA).$$

Therefore, $\tilde{H}_n(X \cup CA)$ is isomorphic to $H_n(X, A)$ for every n.

2313

For any topological space X let ΣX denote its (unreduced) suspension. (ΣX is the quotient space $(X \times [0,1] / \sim$, where $\sim$ denotes the equivalence relation generated by requiring that $(x,t) \sim (y,s)$ if $s = t = 0$ or $s = t = 1$.) If $f : X \to Y$ is a map, let $\Sigma f : \Sigma X \to \Sigma Y$ be the map of suspensions induced by the map $(x,t) \to (f(x),t)$ of $X \times [0,1]$.

(i) Prove that if $f : X \to Y$ is a homotopy equivalence then so is Σf.

(ii) Using only the Eilenberg–Steenrod axioms (Homotopy, Exactness, Excision, Dimension) for a homology theory prove that $\tilde{H}_i(\Sigma X)$ is naturally isomorphic to $\tilde{H}_{i-1}(X)$. (Here $\tilde{H}_i$ denotes reduced singular homology.)

(*Indiana*)

Solution.

(i) Denote by $\Pi_X : X \times I \to \Sigma X$ the quotient map, and by p_0, p_1 the equivalence classes of $(x,0)$ and $(x,1)$, $x \in X$, respectively. Let $f : X \to Y$ be a continuous map. By the definition we see that

$$\Sigma f = \Pi_Y \circ (f \times id) \circ \Pi_X^{-1},$$

where $id : I \to I$ is the identity map. Suppose that $g : X \to Y$ is another continuous map which is homotopic to f by a homotopy $G : X \times I \to Y$ such that $G(x,0) = f(x)$ and $G(x,1) = g(x)$ for any $x \in X$. Then define $\tilde{G} : (X \times I) \times I \to \Sigma Y$ by

$$\tilde{G}((x,t),s) = \Pi_Y(G(X,s),t).$$

Let $H : \Sigma X \times I \to \Sigma Y$ be a map defined by $H(\tilde{x}, s) = \tilde{G}(\Pi_X^{-1}(\tilde{x}), s)$ for any $(\tilde{x}, s) \in \Sigma X \times I$, where, if $\tilde{x} = p_0$ or p_1, $\Pi_X^{-1}(\tilde{x})$ means any point $(x,0)$ or $(x,1)$. It is easy to check that H is well-defined and continuous. It is also easy to see that $H(\tilde{x}, 0) = \Sigma f(\tilde{x})$ and $H(\tilde{x}, 1) = \Sigma g(\tilde{x})$. Therefore Σf is homotopic to Σg, and consequently it follows that if $f : X \to Y$ is a homotopy equivalence then so is Σf.

(ii) Let

$$U = \left\{ \Pi_X(x, t) \;\middle|\; t > \frac{1}{3} \right\}$$

and

$$V = \left\{ \Pi_X(x, t) \;\middle|\; t < \frac{2}{3} \right\}.$$

It is obvious that U and V are both open sets of ΣX and that U and V are both contractible. Hence $\tilde{H}_i(U) = \tilde{H}_i(V) = 0$ for all i. It follows from the homology sequence of the pair $(\Sigma X, U)$ that $\tilde{H}_i(\Sigma X) \approx H_i(\Sigma X, U)$. Let $W = \{\Pi_X(x, t) | t \geq 2/3\}$. Then $\bar{W} \subset U$. By the excision property we have

$$H_i(\Sigma X, U) \approx H_i(\Sigma X - W, U - W) = H_i(V, U - W).$$

It is easy to see that $U - W$ has the same homotopy type of X. On the other hand, from the homolopy sequence of the pair $(V, U - W)$ we see that

$$H_i(V, U - W) \approx \tilde{H}_{i-1}(U - W).$$

Thus we conclude that $\tilde{H}_i(\Sigma X) \approx \tilde{H}_{i-1}(X)$.

2314

Consider the following commutative diagram of abelian groups in which the row and column are exact sequences.

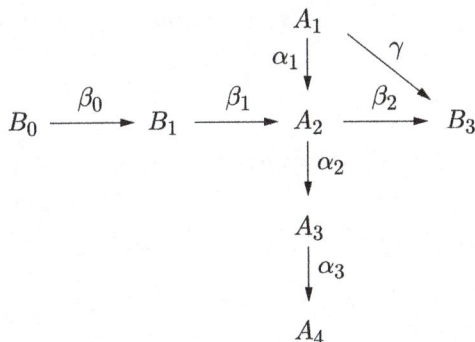

Suppose that γ and β_0 are surjective. Prove that α_3 is injective.

(*Indiana*)

Solution.

Let $a_3 \in A_3$ such that $\alpha_3(a_3) = 0$. Then from the exactness there is an $a_2 \in A_2$ such that $\alpha_2(a_2) = a_3$. Since γ is surjective, there is an $a_1 \in A_1$ such that $\gamma(a_1) = \beta_2(a_2)$. Hence by the commutativity we have

$$\gamma(a_1) = \beta_2(\alpha_1(a_1)) = \beta_2(a_2),$$

so $\alpha_1(a_1) - a_2 \in \ker \beta_2$, and there is a $b_1 \in B_1$ such that $\beta_1(b_1) = \alpha(a_1) - a_2$. Since β_0 is surjective, there is a $b_0 \in B_0$ such that $\beta_0(b_0) = b_1$. Once again by the exactness we have $0 = \beta_1\beta_0(b_0) = \beta_1(b_1)$, which means that $\alpha_1(a_1) - a_2 = 0$. Hence $a_3 = \alpha_2(a_2) = \alpha_2\alpha_1(a_1) = 0$. It means that $\ker \alpha_3 = 0$, i.e., α_3 is injective.

2315

Suppose that a topological space X is expressed as the union $U \cup V$ of two open path-connected subspaces such that $U \cap V$ is path connected, $H_1(U) = 0$, and the inclusion $U \cap V \to V$ induces a surjective homomorphism $H_1(U \cap V) \to H_1(V)$. (Here H denotes singular homology.)

(a) Prove that $H_1(X) = 0$.

(b) Give an example to show that the analogous assertion becomes false in general if H_1 is replaced everywhere by H_2.

(*Indiana*)

Solution.

(a) Consider the Mayer–Vietoris sequence of the pair (U, V)

$$\cdots \to H_2(X) \xrightarrow{\Delta} H_1(U \cap V) \xrightarrow{\phi_*} H_1(U) \oplus H_1(V)$$

$$\xrightarrow{\psi_*} H_1(X) \xrightarrow{\Delta} \tilde{H}_0(U \cap V).$$

Since $U \cap V$ is path-connected, $\tilde{H}_0(U \cap V) = 0$. It means $H_1(X) = \operatorname{im} \psi_*$. By the hypothesis, ϕ_* is surjective so that

$$\ker \psi_* = \operatorname{im} \phi_* = H_1(U) \oplus H_1(V).$$

Therefore $\operatorname{im} \psi_* = 0$, i.e., $H_1(X) = 0$.

(b) Let $X = S^2$, the unit 2-sphere,

$$U = \left\{ (x, y, z) \in S^2 \,\Big|\, z > -\frac{1}{2} \right\}$$

and

$$V = \left\{ (x, y, z) \in S^2 \,\Big|\, z < \frac{1}{2} \right\}.$$

Then $X = U \cup V$. It is obvious that U, V and $U \cap V$ are path-connected and that $H_2(U) = H_2(V) = 0$. So the homomorphism $H_2(U \cap V) \to H_2(V)$ induced by the inclusion map is surjective. But $H_2(X) \approx \mathbb{Z}$.

2316

Let

$$D^n = \{ x \in R^n \,|\, |x| \leq 1 \}$$

denote the standard unit ball in Euclidean space R^n, and

$$S^{n-1} = \{ x \in R^n \,|\, |x| = 1 \}$$

denote the standard $(n-1)$-sphere. Suppose that $f : D^n \to D^n$ is a continuous map such that the restriction of f to S^{n-1} is a homeomorphism from S^{n-1} to S^{n-1}. Prove that f is surjective.

(*Indiana*)

Solution.

We use the reduction to absurdity. Suppose that there is a point $x_0 \in D^n$ such that $x_0 \notin f(D^n)$. By the assumption, we see that $x_0 \in D^n - S^{n-1}$.

Therefore S^{n-1} is a deformation retract of $D^n - \{x_0\}$. Let $r : D^n - \{x_0\} \to S^{n-1}$ be a retraction. Then $g = r \circ f$ is a continuous map from D^n to S^{n-1} and $g|_{S^{n-1}} : S^{n-1} \to S^{n-1}$ is just the restriction of f to S^{n-1}, which is a homeomorphism. Let $i : S^{n-1} \to D^n$ be the inclusion map. Therefore $g \circ i : S^{n-1} \to S^{n-1}$ is a homeomorphism. Hence

$$(g \circ i)_* : H_{n-1}(S^{n-1}) \to H_{n-1}(S^{n-1})$$

is an isomorphism. But $(g \circ i)_* = g_* \circ i_*$ and $i_* : H_{n-1}(S^{n-1}) \to H_{n-1}(D^n)$ is obviously a trivial homomorphism because of $H_{n-1}(D^n) = 0$, and consequently $(g \circ i)_*$ is trivial. This is a contradiction.

2317

Let X be a connected CW complex with two 0-cells, three 1-cells, three 2-cells, and no higher-dimensional cells. Assume $H_1(X) \approx \mathbb{Z} \oplus \mathbb{Z}/3$. Compute the Euler characteristic of X and determine (with proof) all possibilities for $H_2(X)$.

(Indiana)

Solution.

The Euler characteristic of X, $\mathcal{X}(X) = 2 - 3 + 3 = 2$. On the other hand,

$$\mathcal{X}(X) = \operatorname{rank} H_0(X) - \operatorname{rank} H_1(X) + \operatorname{rank} H_2(X).$$

Therefore rank $H_2(X) = 2$. Let X^k denote the k-skeleton of X. By the assumption, we see that $H_3(X^3, X^2) = 0$, and $H_2(X^2, X^1) = \mathbb{Z} \oplus \mathbb{Z} \oplus \mathbb{Z}$. Therefore

$$H_2(X) \approx \ker \{d_* : H_2(X^2, X^1) \to H_1(X^1, X^0)\},$$

where d_* is the composition of homomorphisms

$$H_2(X^2, X^1) \xrightarrow{\partial_*} H_1(X^1) \xrightarrow{j_*} H_1(X^1, X^0).$$

Thus because $H_2(X^2, X^1)$ is free abelian, $H_2(X)$ is also a free abelian group of rank 2, i.e., $H_2(X) \approx \mathbb{Z} \oplus \mathbb{Z}$.

2318

Let X be a CW complex with exactly one $k+1$-cell. Prove that if $H_k(X;\mathbb{Z})$ is a non-trivial finite group then $H_{k+1}(X;\mathbb{Z}) = 0$.

(*Indiana*)

Solution.

Denote by X^k the k-skeleton of the CW complex X, and by d^k the composition of homomorphisms of pairs

$$H_k(X^k, X^{k-1}) \xrightarrow{\partial_*} H_{k-1}(X^{k-1}) \xrightarrow{j_*} H_{k-1}(X^{k-1}, X^{k-2}).$$

It is well-known that

$$H_k(X;\mathbb{Z}) \approx \ker d^k / \operatorname{im} d^{k+1}$$

and

$$H_{k+1}(X;\mathbb{Z}) \approx \ker d^{k+1} / \operatorname{im} d^{k+2}.$$

By the hypothesis,

$$H_{k+1}(X^{k+1}, X^k) \approx \mathbb{Z}.$$

We choose a generator a in $H_{k+1}(X^{k+1}, X^k)$. We claim that $d^{k+1}(a) \neq 0$. Otherwise we would have $\operatorname{im} d^{k+1} = 0$. Therefore

$$H_k(X;\mathbb{Z}) \approx \ker d^k \subset H_k(X^k, X^{k-1}).$$

But it is well-known that $H_k(X^k, X^{k-1})$ is a free abelian group, and, consequently, $\ker d^k$ is trivial or free abelian. It means that $H_k(X;\mathbb{Z})$ is not a non-trivial finite group and contradicts the hypothesis. Thus we have $d^{k+1}(a) \neq 0$. It follows that $\ker d^{k+1} = 0$ and, therefore, $H_{k+1}(X;\mathbb{Z}) = 0$.

2319

Let

$$A = \{(x_1, x_2, x_3, x_4) \in R^4 | x_1 = 0\}$$

and

$$B = \{(x_1, x_2, x_3, x_4) \in R^4 | x_4 = 0\}.$$

Let $X = A \cup B$. Compute the relative homology groups $H_i(X, X - (0, 0, 0, 0))$ for all i.

(*Indiana*)

Solution.

Let $\tilde{U}$ be the unit open ball centered at the point $(0, 0, 0, 0)$ in R^4. Then $U = \tilde{U} \cap X$ is an open neighborhood of $(0, 0, 0, 0)$ in X. By the excision property we have

$$H_i(X, X - (0,0,0,0)) \approx H_i(U, U - (0,0,0,0))$$

for all i. It is obvious that U is contractible and that $U - (0, 0, 0, 0)$ has a deformation retract $C \cup D$, where

$$C = \{(x_1, x_2, x_3, x_4) \in R^4 | x_1 = 0, x_2^2 + x_3^2 + x_4^2 = 1\}$$

and

$$D = \{(x_1, x_2, x_3, x_4) \in R^4 | x_4 = 0, x_1^2 + x_2^2 + x_3^2 = 1\}.$$

(See Fig. 2.15.)

Let $W_1 = C \cup D - \{p_1, p_2\}$ and $W_2 = C \cup D - \{p_3, p_4\}$. (See Fig. 2.15.) Then W_1 and W_2 are open subsets of $C \cup D$ and $W_1 \cup W_2 = C \cup D$. It is clear that both C and D are homeomorphic to the unit sphere S^2 and that D and C are deformation retracts of W_1 and W_2 respectively. It is also clear that

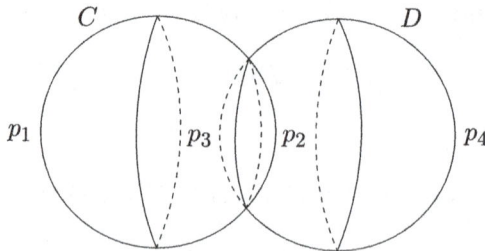

Fig. 2.15.

S^1 is a deformation retract of $W_1 \cap W_2$. Therefore, applying the Mayer–Vietoris sequence to the pair (W_1, W_2), we see that

$$H_i(C \cup D) = \begin{cases} \mathbb{Z} \oplus \mathbb{Z} \oplus \mathbb{Z}, & i = 2, \\ \mathbb{Z}, & i = 0, \\ 0, & \text{otherwise.} \end{cases}$$

From the homology sequence of the pair $(U, U - (0,0,0,0))$, we see that

$$H_{i+1}(U, U - (0,0,0,0)) \approx H_i(U - (0,0,0,0))$$

for all $i \geq 0$. Hence we conclude that

$$H_i(X, X - (0,0,0,0)) = \begin{cases} \mathbb{Z} \oplus \mathbb{Z} \oplus \mathbb{Z}, & i = 3, \\ \mathbb{Z}, & i = 0, 1 \\ 0, & \text{otherwise.} \end{cases}$$

2320

Let X be a path connected and locally path connected space and let $x \in X$. Let $Y = S^1 \times S^1 \times S^2$. Show that if $\pi_1(X, x)$ is finite, then any continuous map $f : X \to Y$ induces the trivial map $f_* : H_i(X, \mathbb{Z}) \to H_i(Y, \mathbb{Z})$ for all i different from 0 and 2.

(*Indiana*)

Solution.

It is easy to see that $R^2 \times S^2$ is the universal covering space of Y. By the Künneth Theorem, we have

$$H_i(R^2 \times S^2) = \sum_{j+k=i} H_j(R^2) \otimes H_k(S^2) = \begin{cases} \mathbb{Z}, & i = 0, 2, \\ 0, & \text{otherwise.} \end{cases}$$

It is clear that

$$\pi_1(S^1 \times S^1 \times S^1) = \pi_1(T^2) \oplus \pi_1(S^2) = \mathbb{Z} \oplus \mathbb{Z}.$$

Since $\pi_1(X, x)$ is finite, the homomorphism $f_* : \pi_1(X, x) \to \pi_1(Y, f(x))$ must be trivial. Therefore, we may lift f to a map $\tilde{f} : X \to R^2 \times S^2$ such that $\pi \circ \tilde{f} = f$, where $\pi : R^2 \times S^2 \to Y$ is the covering map. Hence we have

$$f_* = \pi_* \circ \tilde{f}_* : H_i(X, \mathbb{Z}) \to H_i(Y, \mathbb{Z})$$

for any i. But it is obvious that

$$\tilde{f}_* : H_i(X, \mathbb{Z}) \to H_i(R^2 \times S^2, \mathbb{Z})$$

is trivial for i different from 0 and 2, so is

$$f_* : H_i(X, \mathbb{Z}) \to H_i(Y, \mathbb{Z}).$$

2321

Call a commutative diagram of abelian groups

$$
\begin{array}{ccc}
A & \xrightarrow{\alpha} & B \\
\beta \downarrow & & \downarrow \gamma \\
C & \xrightarrow{\delta} & D
\end{array}
$$

"exact" if the sequence of groups and homomorphisms

$$0 \to A \xrightarrow{(\alpha, \beta)} B \oplus C \xrightarrow{\gamma - \delta} D \to 0 \text{ is exact.}$$

It is an interesting fact that if

$$
\begin{array}{ccc}
A & \xrightarrow{\alpha} & B \\
\beta \downarrow & & \downarrow \gamma \\
C & \xrightarrow{\delta} & D
\end{array}
$$

and

$$
\begin{array}{ccc}
B & \xrightarrow{\alpha'} & E \\
\gamma \downarrow & & \downarrow \varepsilon \\
D & \xrightarrow{\delta'} & F
\end{array}
$$

are exact, then so is

$$
\begin{array}{ccc}
A & \xrightarrow{\alpha'\alpha} & E \\
\beta \downarrow & & \downarrow \varepsilon \\
C & \xrightarrow[\delta'\delta]{} & F
\end{array}
$$

Prove exactness of

$$0 \to A \xrightarrow{(\beta,\alpha'\alpha)} C \oplus E \xrightarrow{\delta'\delta-\varepsilon} F \to 0 \text{ at } C \oplus E.$$

$$(Indiana)$$

Solution.

By the assumptions, we have $\delta\beta = \gamma\alpha$ and $\varepsilon\alpha' = \delta'\gamma$. Therefore, we have

$$(\delta'\delta - \varepsilon)(\beta, \alpha'\alpha) = \delta'\delta\beta - \varepsilon\alpha'\alpha = \delta'\gamma\alpha - \delta'\gamma\alpha = 0.$$

It follows that im $(\beta, \alpha'\alpha) \subset \ker(\delta'\delta - \varepsilon)$. Let $(c, e) \in \ker(\delta'\delta - \varepsilon)$, i.e., $\delta'\delta(c) - \varepsilon(e) = 0$. By the assumptions, we have im $(\alpha, \beta) = \ker(\gamma - \delta)$ and im $(\alpha', \gamma) = \ker(\varepsilon - \delta')$. Thus, noting that

$$(e, \delta(c)) \in \ker(\varepsilon - \delta'),$$

we see that there exists a $b \in B$ such that $\alpha'(b) = e$ and $\gamma(b) = \delta(c)$. Hence we have $(b, c) \in \ker(\gamma - \delta)$. Therefore, there exists an $a \in A$ such that $\alpha(a) = b$ and $\beta(a) = c$, and, consequently,

$$(c, e) = (\beta, \alpha'\alpha)(a) \in \text{im}(\beta, \alpha'\alpha).$$

It means that

$$\ker(\delta'\delta - \varepsilon) \subset \text{im}(\beta, \alpha'\alpha).$$

The exactness at $C \oplus E$ is proved.

2322

Let X be a non-empty compact Hausdorff space and $f : X \to X$ be a continuous map. Prove that there exists a non-empty closed subset A of X such that $f(A) = A$. Give an example to show that compactness is essential for this assertion.

(*Indiana*)

Solution.

Let $F_1 = f(X)$. Then F_1 is a non-empty closed subset of X. We define $F_{n+1} = f(F_n)$ inductively for $n \in \mathbb{Z}^+$. It is clear that $\{F_n, n \in \mathbb{Z}^+\}$ is a sequence of non-empty closed subsets in X and that

$$F_1 \supset F_2 \supset \cdots \supset F_n \supset F_{n+1} \supset \cdots .$$

Since X is compact, we see that the subset $A = \cap_{n=1}^{\infty} F_n$ is a non-empty closed subset of X. We claim that $f(A) = A$ holds.

We give an example to show that compactness is essential for this assertion. Let $X = (0, 1]$ and $f : X \to X$ is defined by $f(x) = \frac{1}{2}x$ for $x \in X$. Suppose that a non-empty closed A of X satisfies $f(A) = A$. Then there would exist an $x_0 \in A$ such that $x \leq x_0$ for any $x \in A$. In fact $x_0 = \sup_{x \in A} x$. Since $f(A) = A$ and $x_0 \in A$, there would exists an $x_1 \in A$ such that $f(x_1) = x_0$, i.e., $x_0 = \frac{1}{2}x_1$. Therefore $x_0 \geq x_1 = 2x_0$, which is a contradiction.

2323

Recall that

$$H_3(S^3; \mathbb{Z}) \approx H_3(RP^3; \mathbb{Z}) \approx \mathbb{Z}.$$

(RP^n is the n-dimensional real projective space.) Prove that there is no function $f : S^3 \to RP^3$ inducing an isomorphism on the third homology.

(*Indiana*)

Solution.

It is well-known that S^3 is the 2-fold universal covering space of RP^3. Let $\pi : S^3 \to RP^3$ denote the universal covering map. Then it is clear that the degree of π is equal to 2. Let f be a function from S^3 to RP^3. Since $\pi_1(S^3) = \{0\}$, there is a lifting of f, $\tilde{f} : S^3 \to S^3$ such that $\pi\tilde{f} = f$. Denote by a and b the generators of $H_3(S^3)$ and $H_3(RP^3)$ respectively. Then we have

$$f_*(a) = \pi_*(\tilde{f}_*(a)) = \deg \tilde{f} \cdot \pi_*(a) = 2 \cdot \deg \tilde{f} \cdot b,$$

because $\deg \tilde{f} \in \mathbb{Z}$, it is obvious that f_* is not an isomorphism.

2324

Let

$$X = \{(x, y, z) | xy = 0\}.$$

(a) Compute $H_1(X - (0,0,0))$.
(b) Using part a, show that X is not homeomorphic to R^2.
(c) Prove or disprove: X is homotopy equivalent to R^2.

(*Indiana*)

Solution.

(a) In fact, $X = \pi_1 \cup \pi_2$ where π_1 is the plane $y = 0$ and π_2 is the plane $x = 0$. Denote by A and B the unit circle in π_1 and π_2 respectively. Then it is easy to see that $A \cup B$ is a deformation retract of $X - (0,0,0)$. Take $U = A \cup B - (0,0,1)$ and $V = A \cup B - (0,0,-1)$. Therefore, U and V are both contractible, $U \cup V = A \cup B$ and $U \cap V$ has four path components which are all contractible. Applying the Mayer–Vietoris sequence to the pair (U, V), we see that

$$H_1(X - (0,0,0)) \approx H_1(A \cup B) = \mathbb{Z} \oplus \mathbb{Z} \oplus \mathbb{Z}.$$

(b) Suppose that there exists a homeomorphism $f : X \to R^2$. Then $f|_{X-(0,0,0)} : X - (0,0,0) \to R^2 - f(0,0,0)$ is also a homeomorphism, which would induce an isomorphism from $H_1(X - (0,0,0))$ to $H_1(R^2 - f(0,0,0))$. But it is clear that

$$H_1(R^2 - f(0,0,0)) \approx H_1(S^1) = \mathbb{Z}.$$

It is a contradiction.

(c) Let $F : X \times I \to X$ be a map defined as follows.

$$F((x, y, z), t) = \begin{cases} (x, y, z), & \text{if } x = 0, \\ (tx, y, z), & \text{if } x \neq 0. \end{cases}$$

Then F is a deformation retraction of X onto the plane π_2. Hence X is homotopy equivalent to R^2.

2325

Let (B^n, S^{n-1}) be the standard ball and sphere pair in R^n, $n > 1$. Suppose that $f : (B^n, S^{n-1}) \to (X, A)$ is a continuous map and $f|_{S^{n-1}} : S^{n-1} \to A$ is a homeomorphism. Show that if $H_n(X) = 0$ then $H_n(X, A) = \mathbb{Z}$.

(*Indiana*)

Solution.

Consider the following diagram

$$\begin{array}{ccccccc} H_n(X) & \xrightarrow{j_*} & H_n(X, A) & \xrightarrow{\partial_*} & H_{n-1}(A) & \xrightarrow{i_*} & H_{n-1}(X) \\ & & \uparrow f_* & & \uparrow f_{1*} & & \uparrow f_* \\ & & H_n(B^n, S^{n-1}) & \xrightarrow{\partial'_*} & H_{n-1}(S^{n-1}) & \xrightarrow{i'_*} & H_{n-1}(B^n) \end{array}$$

in which the level rows are exact and the squares are commutative. By the assumption, f_{1*} is an isomorphism, and therefore

$$H_{n-1}(A) \approx H_{n-1}(S^{n-1}) = \mathbb{Z}.$$

It is well-known that ∂'_* is an isomorphism and $H_{n-1}(B^n) = 0$. Thus we have

$$\ker i_* = H_{n-1}(A) = \mathbb{Z}.$$

Since $H_n(X) = 0$, ∂_* is a monomorphism. Hence we have

$$H_n(X, A) \approx \operatorname{im} \partial_* \approx \ker i_* = \mathbb{Z}.$$

2326

(a) Describe a CW structure on $S^2 \times S^5$ and use it to compute the homology of $S^2 \times S^5$.

(b) Compute the homology of $S^2 \times S^5$ with 2 points removed.

(*Indiana*)

Solution.

(a) Let $x_0 \in S^2$, $x_1 \in S^5$. Then S^2 is obtained by attaching a 2-cell to x_0, and S^5 is obtained by attaching a 5-cell to x_1. Denote by $S^2 \vee S^5$ the one point union of S^2 and S^5, which can be considered as the space obtained by attaching a 2-cell and a 5-cell to the point $(x_0, x_1) \in S^2 \times S^5$. It is easy to see that $S^2 \times S^5$ is homeomorphic to the space obtained by attaching a 7-cell to $S^2 \vee S^5$. Therefore the CW structure on $S^2 \times S^5$ has a 0-cell, a 2-cell, a 5-cell and a 7-cell. Hence it is obvious that $H_i(S^2 \times S^5)$ is an infinite cyclic group for i equal to 0, 2, 5, and 7, and is zero otherwise.

(b) Let $X = S^2 \times S^5 - \{p_1, p_2\}$, $U = S^2 \times S^5 - \{p_1\}$ and $V = S^2 \times S^5 - \{p_2\}$. Then U, V are both open sets of $S^2 \times S^5$ and $U \cap V = X$, and $U \cup V = S^2 \times S^5$. It is easy to see that $S^2 \vee S^5$ is a deformation retract of both U and V. Applying the Mayer–Vietoris sequence to the pair (U, V) and using the fact that

$$\tilde{H}_i(S^2 \vee S^5) \approx \tilde{H}_i(S^2) \oplus \tilde{H}_i(S^5),$$

we conclude that

$$H_i(X) = \begin{cases} \mathbb{Z}, & i = 0, 2, 5, 6, \\ \mathbb{Z} \oplus \mathbb{Z} \oplus \mathbb{Z} \oplus \mathbb{Z}, & i = 1, \\ 0, & \text{otherwise.} \end{cases}$$

2327

Let A be a subspace of $S^2 \times S^2$ homeomorphic to the 2-sphere S^2. State what the homology groups $H_q(S^2 \times S^2)$ are (no proof is required). What can you say about the possibilities for the relative homology groups $H_q(S^2 \times S^2, A)$?

(Indiana)

Solution.

By the Künneth formula we conclude that

$$H_q(S^2 \times S^2) = \begin{cases} \mathbb{Z}, & q = 0, 4, \\ \mathbb{Z} \oplus \mathbb{Z}, & q = 2, \\ 0, & \text{otherwise.} \end{cases}$$

From the homology sequence of the pair $(S^2 \times S^2, A)$, using the above result and the fact that $H_i(A) \approx H_i(S^2)$, we see that

$$H_i(S^2 \times S^2, A) \approx H_i(S^2 \times S^2)$$

for $i \geq 4$ and that the non-trivial parts of the sequence are

$$0 \to H_3(S^2 \times S^2, A) \xrightarrow{\partial_*} H_2(A) \xrightarrow{i_*} H_2(S^2 \times S^2)$$
$$\xrightarrow{j_*} H_2(S^2 \times S^2, A) \to 0$$

and

$$0 \to H_1(S^2 \times S^2, A) \xrightarrow{\partial_*} \tilde{H}_0(A) \to \tilde{H}_0(S^2 \times S^2)$$
$$\to H_0(S^2 \times S^2, A) \to 0.$$

Since $\tilde{H}_0(A) = 0$, $H_i(S^2 \times S^2, A) = 0$ for i equal to 0 and 1. By the exactness, we have $H_3(S^2 \times S^2, A) \approx \ker i_*$ and $H_2(S^2 \times S^2, A) \approx \mathbb{Z} \oplus \mathbb{Z}/\operatorname{im} i_*$, where i_* is the homomorphism induced by the inclusion map $i : A \to S^2 \times S^2$.

2328

Compute the homology groups of the space X obtained as the union of the 2-sphere S^2 and the x-axis in R^3.

(Indiana)

Solution.

Let

$$X^* = X - \{\text{the open interval } (-1,1) \text{ in the } x\text{-axis}\}.$$

Then the space X may be viewed as a space obtained from X^* by attaching a 1-cell. It is clear that $H_i(X, X^*)$ is infinite cyclic for i equal to 1 and is zero otherwise. It is easy to see that X^* has the homotopy type of S^2. Thus from the homology sequence of the pair (X, X^*) we conclude that $H_i(X)$ is infinite cyclic for i equal to 0, 1, 2 and is zero otherwise.

2329

Define the "unreduced suspension" ΣX of a space X to be the quotient space of $I \times X$ obtained by identifying $\{0\} \times X$ to one point and $\{1\} \times X$ to one point. (This is the union of two "cones" on X.) Show that there is a natural isomorphism $\sigma_X : \tilde{H}_i(X) \to \tilde{H}_{i+1}(\Sigma X)$, for all $i \geq 0$.

(Indiana)

Solution.

Let $\{0\} \times X$ and $\{1\} \times X$ be identified to the points x_0 and x_1 respectively. Take $U = \Sigma X - \{x_0\}$ and $V = \Sigma X - \{x_1\}$. Then U and V are open sets of ΣX, and $U \cup V = \Sigma X$. Consider the Mayer–Vietories sequence of the pair (U, V):

$$\cdots \to \tilde{H}_{i+1}(U \cap V) \xrightarrow{\Phi_*} \tilde{H}_{i+1}(U) \oplus \tilde{H}_{i+1}(V) \xrightarrow{\Psi_*} \tilde{H}_{i+1}(\Sigma X)$$

$$\xrightarrow{\Delta} \tilde{H}_i(U \cap V) \to \cdots .$$

It is clear that U and V are contractible and that $U \cap V$ has the homotopy type of X. Thus the homomorphism $\Delta : \tilde{H}_{i+1}(\Sigma X) \to \tilde{H}_i(U \cap V)$ is an isomorphism and we may obtain an isomorphism $\Delta^* : \tilde{H}_{i+1}(\Sigma X) \to \tilde{H}_i(X)$. The inverse of Δ^*, $\sigma_X : \tilde{H}_i(X) \to \tilde{H}_{i+1}(\Sigma X)$ is just what we are looking for. The naturality of σ_X can be derived from the naturality of the Mayer–Vietories sequence.

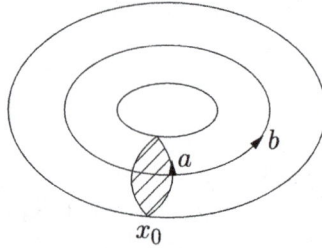

Fig. 2.16.

2330

Denote by X the union of the torus $S^1 \times S^1$ with the disc D^2, where D^2 is attached to T^2 by identifying ∂D^2 with a meridian curve $S^1 \times \{x_0\}$ in the torus, where $x_0 \in S^1$. (See Fig. 2.16.)
(a) Calculate $H_n(X)$ for all $n \geq 0$.
(b) Is T^2 a retract of X? Why or why not?

(*Indiana*)

Solution.
(a) Consider the homology sequence of the pair (X, T^2). It is well-known that $H_i(X, T^2) = 0$ for $i \neq 2$ and $H_2(X, T^2) \approx f_*(H_2(D^2, \partial D^2)) \approx \mathbb{Z}$, where f_* is the homomorphism induced by the adjunction map $f : D^2 \to X$. Thus the nontrivial part of the sequence is as follows.

$$0 \to H_2(T^2) \xrightarrow{i_*} H_2(X) \xrightarrow{j_*} H_2(X, T^2) \xrightarrow{\partial'_*} H_1(T^2) \xrightarrow{i_*} H_1(X) \to 0$$
$$\approx\uparrow f_* \qquad\qquad \uparrow f_{1*}$$
$$0 \to H_2(D^2, \partial D^2) \xrightarrow{\partial_*}_{\approx} H_1(\partial D^2) \to 0$$

It is easy to see that $\operatorname{im} \partial'_* = \operatorname{im} f_{1*}$ and $\ker \partial'_* = \ker f_{1*}$. We know that $H_1(T^2) = \mathbb{Z} \oplus \mathbb{Z}$ is generated by a and b. (See Fig. 2.16.) It is clear that $f_{1*}(H_1(\partial D^2)) = \mathbb{Z} \oplus \{0\}$. Thus we see that $H_1(X) \approx \mathbb{Z} \oplus \mathbb{Z}/\mathbb{Z} \approx \mathbb{Z}$ and $H_2(X) \approx H_2(T^2) \approx \mathbb{Z}$. Hence we conclude that

$$H_n(X) = \begin{cases} 0, & n \geq 3, \\ \mathbb{Z}, & n = 0, 1, 2. \end{cases}$$

(b) We claim that T^2 is not a retract of X. Otherwise, there would exist a retraction map $r : X \to T^2$ such that $r|_{T^2} = id_{T^2}$. Let $i : T^2 \to X$ be the inclusion map. Then $r \circ i = id_{T^2} : T^2 \to T^2$, and, consequently, we have $r_* \circ i_* = id : H_1(T^2) \to H_1(T^2)$. In other words, we have the following commutative diagram

$$\mathbf{Z} \oplus \mathbf{Z} \xrightarrow{\ i_*\ } \mathbf{Z} \xrightarrow{\ r_*\ } \mathbf{Z} \oplus \mathbf{Z}$$
$$\underset{id}{\underbrace{\qquad\qquad\qquad}}$$

because of $H_1(T^2) = \mathbb{Z} \oplus \mathbb{Z}$ and $H_1(X) = \mathbb{Z}$. This is a contradiction.

2331

Given homomorphism $h : A \to B$ and $g : C \to B$, the pull back of h via g is the group

$$g^*(A) = \{(c, a) \in C \times A | g(c) = h(a)\}.$$

(a) Let $g^*(h) : g^*(A) \to C$ be the homomorphism obtained by restricting the projection onto the first factor $C \times A \to C$ to $g^*(A)$. Prove that the kernel of $g^*(A)$ is isomorphic to the kernel of h.

$$\begin{array}{ccc} g^*(A) & & A \\ \downarrow g^*(h) & & \downarrow h \\ C & \xrightarrow{\ g\ } & B \end{array}$$

(b) Let $A = \mathbb{Z}/4\mathbb{Z}$, $B = \mathbb{Z}/2\mathbb{Z}$, and let $h : A \to B$ be the surjection defined by $h(1) = 1$. Let $C = \mathbb{Z}/8\mathbb{Z}$ and let $g : C \to B$ be the surjection defined by $g(1) = 1$. Identify the group $g^*(A)$, with explanation.

(Indiana)

Solution.

(a) Suppose that $(c, a) \in \ker g^*(h)$. Then $g^*(h)(c, a) = 1_c$, and, therefore, $C = 1_c$, the identity of C. By the definition of $g^*(A)$, we see that

$$\ker g^*(h) = \{(1_c, a) \,|\, a \in \ker h\}.$$

Let $F : \ker g^*(h) \to \ker h$ be defined by $F(1_c, a) = a$. It is easy to see that F is an isomorphism.

(b) In this case, we have $h^{-1}(0) = \{0, 2\}$, $h^{-1}(1) = \{1, 3\}$, $g^{-1}(0) = \{0, 2, 4, 6\}$ and $g^{-1}(1) = \{1, 3, 5, 7\}$. Thus it is clear that the group $g^*(A)$ consists of the following 16 elements: (0, 0), (0, 2), (2, 0), (2, 2), (4, 0), (4, 2), (6, 0), (6, 2), (1, 1), (1, 3), (3, 1), (3, 3), (5, 1), (5, 3), (7, 1), (7, 3).

2332

Recall that if C is a homeomorphic copy of the circle in S^3, then $H_i(S^3 - C)$ is infinite cyclic for i equal to 0 or 1 and is zero otherwise. Assuming this fact compute

(a) The homology of $R^3 - C$, when C is a homeomorphic copy of the circle in R^3.

(b) The homology of $Y = R^3 - X$, where $X \subset R^3$ is a homeomorphic copy of the "figure-eight space" (i.e., the one-point union of two circles.)

(Indiana)

Solution.

(a) Denote $S^3 = R^3 \cup \{\infty\}$. Let $A = S^3 - C$, $B = S^3 - \{\infty\}$. Then A and B are open subsets in S^3, and $A \cup B = S^3$ and $A \cap B = R^3 - C$. We have the following Mayer–Vietoris sequence.

$$\cdots \to \tilde{H}_{i+1}(S^3) \xrightarrow{\Delta} \tilde{H}_i(R^3 - C) \xrightarrow{\phi_*} \tilde{H}_i(S^3 - C)$$
$$\oplus \tilde{H}_i(S^3 - \{\infty\}) \xrightarrow{\Psi_*} \tilde{H}_i(S^3) \xrightarrow{\Delta} \tilde{H}_{i-1}(R^3 - C) \to \cdots$$

Noting that

$$\tilde{H}_i(S^3 - \{\infty\}) = \tilde{H}_i(R^3) = 0$$

for any i and that

$$\tilde{H}_i(S^3) = \begin{cases} \mathbb{Z}, & i = 3, \\ 0, & \text{otherwise}, \end{cases}$$

and

$$\tilde{H}_i(S^3 - C) = \begin{cases} \mathbb{Z}, & i = 1, \\ 0, & \text{otherwise}, \end{cases}$$

we can see that

$$H_i(R^3 - C) = \begin{cases} \mathbb{Z}, & i = 0, 1, 2, \\ 0, & \text{otherwise}. \end{cases}$$

(b) Represent X as shown in Fig. 2.17. Let $U = X - \{p_1\}$ and $V = X - \{p_2\}$. Then we see that U and V have homotopy type of S^1 and that $U \cap V$ is contractible. Since $U \cup V = X$, we have

$$Y = R^3 - X = (R^3 - U) \cap (R^3 - V)$$

and

$$(R^3 - U) \cup (R^3 - V) = R^3 - (U \cap V).$$

From the result of (a), we see that $\tilde{H}_i(R^3 - U)$ and $\tilde{H}_i(R^3 - V)$ is infinite cyclic for i equal to 1 or 2 and is zero otherwise. It is

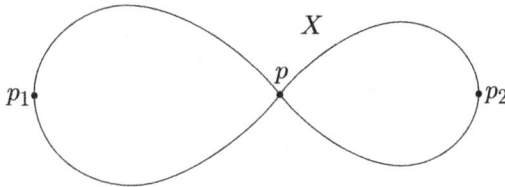

Fig. 2.17.

easy to see that

$$\tilde{H}_i(R^3 - (U \cap V)) \approx \tilde{H}_i(S^2).$$

Due to the above facts, the non-trivial part of the Mayer–Vietoris sequence of the pair $(R^3 - U, R^3 - V)$ is

$$0 \to \tilde{H}_2(Y) \xrightarrow{\phi_*} \tilde{H}_2(R^3 - U) \oplus \tilde{H}_2(R^3 - V) \xrightarrow{\psi_*} \tilde{H}_2(R^3 - U \cap V)$$
$$\approx \mathbb{Z} \oplus \mathbb{Z} \qquad\qquad\qquad \approx \mathbb{Z}$$
$$\xrightarrow{\Delta} \tilde{H}_1(Y) \xrightarrow{\phi_*} \tilde{H}_1(R^3 - U) \oplus \tilde{H}_1(R^3 - V) \to 0.$$
$$\approx \mathbb{Z} \oplus \mathbb{Z}$$

We claim that the homomorphism ψ_* is an epimorphism. Take a sufficiently large $r > 0$ such that the sphere $S^2(r)$ belongs to $R^3 - X$. Then $S^2(r)$ is a deformation retract of $R^3 - U \cap V$. Hence we can consider the generator of $\tilde{H}_2(S^2(r))$, $[c]$, as the generator of $\tilde{H}_2(R^3 - U \cap V)$. Since the representative chain c of $[c]$ can also be considered as a chain of $(R^3 - U) \cap (R^3 - V)$, by the definition of ψ_* we have $[c] = \psi_*([c], 0)$. The claim is proved, which means $\operatorname{im} \Delta = 0$ too. Thus we get $\tilde{H}_1(Y) = \mathbb{Z} \oplus \mathbb{Z}$ and a split exact sequence

$$0 \to \tilde{H}_2(Y) \xrightarrow{\phi_*} \mathbb{Z} \oplus \mathbb{Z} \xrightarrow{\psi_*} \mathbb{Z} \to 0.$$

It follows that $\tilde{H}_2(Y) = \mathbb{Z}$. So we conclude that

$$\tilde{H}_i(Y) = \begin{cases} 0, & i \geq 3, \\ \mathbb{Z}, & i = 2, \\ \mathbb{Z} \oplus \mathbb{Z}, & i = 1, \\ \mathbb{Z}, & i = 0. \end{cases}$$

2333

It is known that if $X \subset S^3$ is homeomorphic to S^1 then $H_*(S^3 - X) \approx H_*(S^1)$. Use this fact to compute the homology of $S^3 - Y$ where Y is a subspace of S^3 homeomorphic to the disjoint union of two copies of S^1.

(*Indiana*)

Solution.

Denote Y by $Y = A \cup B$, where $A \cap B = \emptyset$ and both A and B are homeomorphic to S^1. Therefore,

$$S^3 - Y = (S^3 - A) \cap (S^3 - B).$$

Noting $S^3 - A$ and $S^3 - B$ are open sets of S^3 and

$$(S^3 - A) \cup (S^3 - B) = S^3,$$

in the Mayer–Vietoris sequence we have

$$\cdots \to \tilde{H}_{q+1}(S^3) \xrightarrow{\Delta} \tilde{H}_q(S^3 - Y) \xrightarrow{\phi_*} \tilde{H}_q(S^3 - A) \oplus \tilde{H}_q(S^3 - B)$$
$$\xrightarrow{\psi_*} \tilde{H}_q(S^3) \xrightarrow{\Delta} \tilde{H}_{q-1}(S^3 - Y) \xrightarrow{\phi_*} \cdots$$

Using the fact that

$$\tilde{H}_q(S^3 - A) \approx \tilde{H}_q(S^3 - B) \approx \tilde{H}_q(S^1) = \begin{cases} \mathbb{Z}, & q = 1, \\ 0, & q \neq 1, \end{cases}$$

and that

$$\tilde{H}_q(S^3) = \begin{cases} \mathbb{Z}, & q = 3, \\ 0, & q \neq 3, \end{cases}$$

we can easily see that

$$\tilde{H}_q(S^3 - Y) = \begin{cases} 0, & q \geq 3, \\ \mathbb{Z}, & q = 2, \\ \mathbb{Z} \oplus \mathbb{Z}, & q = 1, \\ \mathbb{Z}, & q = 0. \end{cases}$$

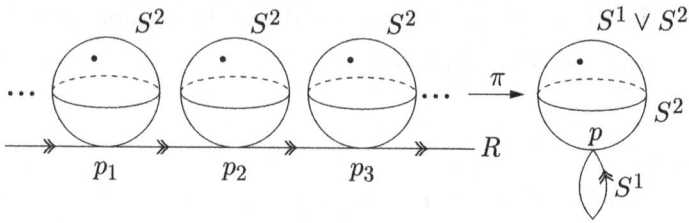

Fig. 2.18.

2334

(a) Sketch pictures of the universal covering of the one point union $S^1 \vee S^2$ and of the connected 2-fold covering (no proofs required).

(b) Compute the homology of the connected 2-fold covering space of $S^1 \vee S^2$.

(*Indiana*)

Solution.

(a) The universal covering of the one point union $S^1 \vee S^2$ is shown as in Fig. 2.18. The covering map π, restricted on each S^2, is the identity map, and, restricted on R, is the exponential map: $R \to S^1$. The connected 2-fold covering of $S^1 \vee S^2$ is shown as follows.

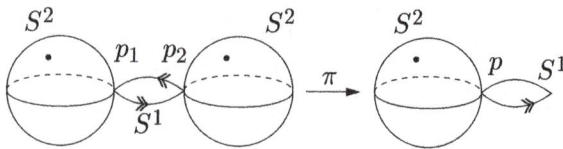

Fig. 2.19.

The covering map π, restricted on each S^2, is the identity map, and, restricted on S^1, is the 2-fold covering map: $z \to z^2$ from S^1 to S^1.

(b) To compute the homology of the connected 2-fold covering space of $S^1 \vee S^2$, i.e., the homology of $S^2 \vee S^1 \vee S^2$, we take

$$U = S^2 \vee S^1 \vee S^2 - \{\text{the antipodal point of } p_1\}$$

and

$$V = S^2 \vee S^1 \vee S^2 - \{\text{the antipodal point of } p_2\}$$

(see the above figure). Then it is easy to see that S^1 is a deformation retract of $U \cap V$, and that $S^1 \vee S^2$ is a deformation retract of U and V.

Thus, it is clear that

$$\tilde{H}_q(U \cap V) = \begin{cases} \mathbb{Z}, & q = 1, \\ 0, & q \neq 1, \end{cases}$$

and

$$\tilde{H}_q(U) \approx \tilde{H}_q(V) \approx \tilde{H}_q(S^2) \oplus \tilde{H}_q(S^1) = \begin{cases} \mathbb{Z}, & q = 1, 2, \\ 0, & q \neq 1, 2. \end{cases}$$

Therefore, the non-trivial part of the Mayer–Vietoris sequence is

$$0 \to \tilde{H}_2(U) \oplus \tilde{H}_2(V) \xrightarrow{\psi_*} \tilde{H}_2(S^2 \vee S^1 \vee S^2) \xrightarrow{\Delta} \tilde{H}_1(U \cap V)$$

$$\xrightarrow{\phi_*} \tilde{H}_1(U) \oplus \tilde{H}_1(V) \xrightarrow{\psi_*} \tilde{H}_1(S^2 \vee S^1 \vee S^2) \to 0.$$

It is easy to see that the inclusion maps $k : U \cap V \to U$ and $l : U \cap V \to V$ induce injective homomorphisms $k_* : \tilde{H}_1(U \cap V) \to \tilde{H}_1(U)$ and $l_* : \tilde{H}_1(U \cap V) \to \tilde{H}_1(V)$, respectively. Hence the homomorphism

$$\phi_* : \tilde{H}_1(U \cap V) \to \tilde{H}_1(U) \oplus \tilde{H}_1(V)$$

is injective, i.e., $\ker \phi_* = 0$. Thus $\operatorname{im} \Delta = 0$. It means that

$$\tilde{H}_2(S^2 \vee S^1 \vee S^2) \approx \ker \Delta \approx \operatorname{im} \psi_* \approx \tilde{H}_2(U) \oplus \tilde{H}_2(V) = \mathbb{Z} \oplus \mathbb{Z}.$$

It is clear that

$$\tilde{H}_1(S^2 \vee S^1 \vee S^2) \approx \tilde{H}_1(U) \oplus \tilde{H}_1(V)/\operatorname{im} \phi_* \approx \mathbb{Z}.$$

Hence, we have

$$\tilde{H}_q(S^2 \vee S^1 \vee S^2) = \begin{cases} \mathbb{Z} \oplus \mathbb{Z}, & q = 2, \\ \mathbb{Z}, & q = 1, \\ 0, & \text{otherwise.} \end{cases}$$

2335

Compute the homology of $S^1 \times S^1$-point.

(Indiana)

Solution.

It is easy to see that the space $X = A \cup B$ is a deformation retract of the space $H = S^1 \times S^1$-point, where A and B are each homeomorphic to S^1 and $A \cap B = \{x_0\}$ as shown in Fig. 2.20. Choose points $a \in A$ and $b \in B$ such that $a \neq x_0$ and $b \neq x_0$. Let $U = X - \{b\}$, and let $V = X - \{a\}$. It is clear that A and B are deformation retracts of U and V, respectively, and that $U \cap V = X - \{a, b\}$ is contractible. Applying Mayer–Vietoris sequence to U and V, we have

$$H_n(S^1 \times S^1 - \text{point}) = H_n(X) = H_n(A) \oplus H_n(B)$$

$$= \begin{cases} 0, & n \geq 2, \\ \mathbb{Z} \oplus \mathbb{Z}, & n = 1, \\ \mathbb{Z}, & n = 0. \end{cases}$$

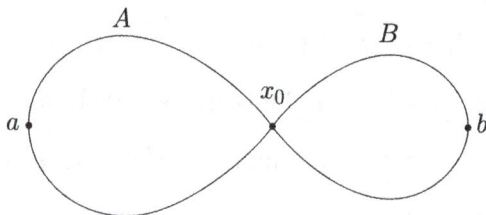

Fig. 2.20.

2336

Suppose the following diagram is commutative, the rows are exact, and γ_n is an isomorphism for all n.

$$
\begin{array}{ccccccccc}
\cdots \to & A_n & \xrightarrow{i_n} & B_n & \xrightarrow{j_n} & C_n & \xrightarrow{\delta_n} & A_{n-1} & \to \cdots \\
& \downarrow \alpha_n & & \downarrow \beta_n & & \downarrow \gamma_n & & \downarrow \alpha_{n-1} & \\
\cdots \to & A'_n & \xrightarrow{i'_n} & B'_n & \xrightarrow{j'_n} & C'_n & \xrightarrow{\delta'_n} & A'_{n-1} & \to \cdots
\end{array}
$$

Construct an exact sequence

$$\cdots \to A_n \to A'_n \oplus B_n \to B'_n \to A_{n-1} \to \cdots .$$

Write out the proof of exactness at B'_n.

(Indiana)

Solution.

The following sequence is exact:

$$\cdots \to A_n \xrightarrow{(\alpha_n, i_n)} A'_n \oplus B_n \xrightarrow{\Delta} B'_n \xrightarrow{\delta_n \circ \gamma_n^{-1} \circ j'_n} A_{n-1} \to \cdots$$

where Δ is defined by

$$\Delta(a', b) = i'_n(a') - \beta_n(b)$$

for any $(a', b) \in A'_n \oplus B_n$. We give the proof of exactness at B'_n as follows.

Let $u \in \ker(\delta_n \circ \gamma_n^{-1} \circ j'_n)$. Then $\gamma_n^{-1} \circ j'_n(u) \in \ker \delta_n$. Due to the exactness at C_n, there exists a $b \in B_n$ such that $j_n(b) = \gamma_n^{-1} \circ j'_n(u)$, and consequently, $\gamma_n \cdot j_n(b) = j'_n(b)$. From the commutativity, we have $j'_n \circ \beta_n(b) = j'_n(u)$. Thus $\beta_n(b) - u \in \ker j'_n$ and there exists an $a' \in A'_n$ such that $i'_n(a') = \beta_n(b) - u$. It means that

$$u = \beta_n(b) - i'_n(a') = \Delta(-a', -b).$$

So $\ker(\delta_n \circ \gamma_n^{-1} \circ j'_n) \subset \operatorname{im} \Delta$.

Now we prove

$$\operatorname{im} \Delta \subset \ker(\delta_n \circ \gamma_n^{-1} \circ j'_n).$$

Suppose that $u \in \mathrm{im}\, \Delta$, i.e., $u = i'_n(a') - \beta_n(b)$ for some $a' \in A'_n$ and $b \in B_n$. Since $j'_n \circ i'_n = 0$ and $j'_n \circ \beta_n = \gamma_n \circ j_n$, it is easy to see that

$$\delta_n \circ \gamma_n^{-1} \circ j'_n(u) = -\delta_n \circ j_n(b) = 0.$$

Thus the exactness at B'_n is proved.

2337

Suppose that X is a space and $f : X \to Y$, $g : X \to Z$ are two maps of X into contractible spaces Y and Z. Let M be the mapping cylinder of f and g, that is, M is the identification space obtained from the disjoint union of Y, $X \times I$ and Z by identifying each $(x, 0)$ with $f(x)$, and each $(x, 1)$ with $g(x)$. Prove that $\tilde{H}_q M \cong \tilde{H}_{q-1} X$.

(*Indiana*)

Solution.

Let $U = X \times [0, 3/4] \overset{\|}{=} Y/ \sim$, where $X \times [0, 3/4] \overset{\|}{=} Y$ denotes the disjoint union of $X \times [0, 3/4]$ and Y, the equivalence relation $\sim$ is determined by $(x, 0) \sim f(x)$. In the same way, let $V = X \times (\frac{1}{2}, 1] \overset{\|}{=} Z/ \sim'$, where the equivalence relation $\sim'$ is determined by $(x, 1) \sim g(x)$. Then we have $M = U \cup V$ and $U \cap V = X \times (1/2, 3/4)$. Noting U, V and $U \cap V$ are open sets of M, by Mayer–Vietoris sequence, we have the following exact sequence:

$$\cdots \to \tilde{H}_q(U \cap V) \to \tilde{H}_q(U) \oplus \tilde{H}_q(V) \to \tilde{H}_q(M)$$
$$\overset{\Delta}{\to} \tilde{H}_{q-1}(U \cap V) \to \tilde{H}_{q-1}(U) \oplus \tilde{H}_{q-1}(V) \to \cdots \qquad (1)$$

Since $X \times \{0\} \overset{\|}{=} Y/ \sim$ is homeomorphic to Y and is a deformation retract of U, from the assumption that Y is contractible, we have $\tilde{H}_q(U) \cong \tilde{H}_q(Y) = 0$. In the same way, $\tilde{H}_q(V) \cong \tilde{H}_q(Z) = 0$.

It is obvious that $U \cap V$ has the same homotopy type with X. So $\tilde{H}_q(U \cap V) \cong \tilde{H}_q(X)$. Thus, from (1), $\tilde{H}_q(M) \cong \tilde{H}_{q-1}(X)$.

2338

Let $\{X_i, i = 1, 2, 3, \ldots\}$ be an infinite sequence of topological spaces and $f_i : X_i \rightarrow X_{i+1}$ be maps. The mapping telescope of this sequence is defined to be the identification space $T = (\prod_{i=1}^{\infty} X_i \times [0,1])/ \sim$, where $(x,1) \sim (y,0)$ if $x \in X_i$, $y \in X_{i+1}$ and $y = f_i(x)$ describes the identification. Consider the case where all $X_i = S^m$ $(m > 0)$ and $f_i : X_i \rightarrow X_{i+1}$ is a map of degree d.

(a) Compute $H_*(T)$ when $d = 0$.

(b) Compute $H_*(T; \mathbb{Q})$ when $d \neq 0$.

(Wisconsin)

Solution.

Let

$$T_n = \left(\left(\prod_{i=1}^{n-1} X_i \times [0,1] \right) \prod X_n \times \left[0, \frac{1}{2} \right] \right) / \sim$$

with a similar identification. Then we see that $\{T_n, n = 2, 3, \ldots\}$ is an increasing sequence of open subspaces of T and $T = \bigcup_{n=2}^{\infty} T_n$. For the case that all $X_i = S^m$ $(m > 0)$, consider the following commutative diagram:

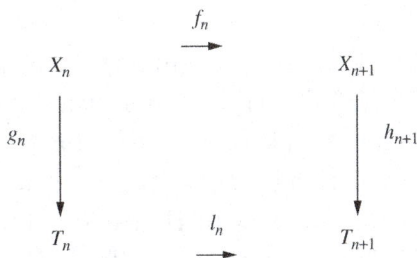

$$
\begin{array}{ccc}
X_n & \xrightarrow{f_n} & X_{n+1} \\
\downarrow{g_n} & & \downarrow{h_{n+1}} \\
T_n & \xrightarrow{l_n} & T_{n+1}
\end{array}
$$

where g_n, h_{n+1} and l_n are inclusions. Noting that X_n is a deformation retract of T_n, we see that g_n and h_{n+1} are homotopy equivalences. Since degree of f_n is d, l_n induces a multiplication by d in $H_*(T_n) \rightarrow H_*(T_{n+1})$.

(a) If $d = 0$, then f is null-homotopic. Thus passing to the increasing union: $\bigcup_n T_n \to \bigcup_n T_{n+1}$ shows that the identity map: $T \to T$ is null-homotopic. Hence T is contractible and $H_i(T) = 0$ for $i > 0$ and $H_0(T) = \mathbb{Z}$.

(b) If $d \neq 0$, then l_n induces an isomorphism $H_*(T_n; \mathbb{Q}) \to H_*(T_{n+1}; \mathbb{Q})$. Hence, the non-zero homology groups are obtained as direct limits of isomorphisms $\mathbb{Q} \to \mathbb{Q}$. Thus we see that $H_0(T; \mathbb{Q}) = H_m(T; \mathbb{Q}) = \mathbb{Q}$, and all other homology groups vanish.

2339

Let R be a commutative ring with 1. Define an R-homology sphere to be a Hausdorff topological space X such that $H_i(X; R) \cong H_i(S^n; R)$ for some sphere S^n.

(a) Let W be a compact connected $(n + 1)$-manifold and let $X = \partial W$. Suppose that $H_i(W, \mathbb{Z}_2) = 0$ for all $i > 0$. Prove that X is a $\mathbb{Z}_2$-homology sphere.

(b) Give an example of R and an R-homology sphere which is a non-orientable connected compact manifold. Justify your answer by quoting appropriate theorems. You may quote standard computations without proof.

(Wisconsin)

Solution.

(a) By Poincaré–Lefschetz duality Theorem, we have $H_i(W, \partial W, \mathbb{Z}_2) \cong H^{n+1-i}(W, \mathbb{Z}_2)$. On the other hand, from Universal Coefficients Theorem, we have $H^r(W, \mathbb{Z}_2) = \mathrm{Hom}(H_r(W; \mathbb{Z}_2), \mathbb{Z}_2)$. Since $H_i(W, \mathbb{Z}_2) = 0$ for all $i > 0$, we obtain $H^r(W, \mathbb{Z}_2) = 0$ for $r > 0$. Hence $H_i(W, \partial W, \mathbb{Z}_2) = 0$ for $i < n+1$. Since W is a connected manifold, $H_{n+1}(W, \partial W, \mathbb{Z}_2) \cong \mathbb{Z}_2$. Using the homology exact sequence for $(W, \partial W)$ with $\mathbb{Z}_2$-coefficient:

$$\cdots \to H_{i+1}(W, \partial W, \mathbb{Z}_2) \to H_i(\partial W, \mathbb{Z}_2) \to H_i(W, \mathbb{Z}_2)$$
$$\to H_i(W, \partial W, \mathbb{Z}_2) \to \cdots,$$

we have $H_i(\partial W, \mathbb{Z}_2) = 0$ unless $i = 0$ or n and $H_0(\partial W, \mathbb{Z}_2) = H_n(\partial W, \mathbb{Z}_2) = \mathbb{Z}_2$.

(b) Take $R = \mathbb{Z}_2$ for a prime number $p \neq 2$. Consider $X = RP^{2k} \otimes S^m$ for any $m > 0$. By Künneth formula, we have

$$H_{2k+m}(X) = H_{2k}(RP^{2k}) \otimes H_m(S^m)$$
$$\oplus \operatorname{Tor}(H_1(RP^{2k}), H_{2k+m-i-1}(S^m)) = 0.$$

So X is a non-orientable connected compact manifold without boundary. From the Universal Coefficient Theorem, one can get $H_1(RP^{2k}; \mathbb{Z}_p) = \mathbb{Z}_2 \otimes \mathbb{Z}_p = 0$ for $i > 0$. On the other hand, we have

$$H_1(X; \mathbb{Z}_p) = \oplus_r H_r(RP^{2k}; \mathbb{Z}_p) \otimes H_{i-r}(S^m; \mathbb{Z}_p) \oplus$$
$$\oplus_s \operatorname{Tor}(H_s(RP^{2k}; \mathbb{Z}_p), H_{i-1-s}(S^m; \mathbb{Z}_p)) = 0$$

for $i \neq 0, m$, and $H_0(X; \mathbb{Z}_p) = H_m(X; \mathbb{Z}_p) = \mathbb{Z}_p$. Hence X is a $\mathbb{Z}_p$-homology sphere.

PART 3
Differential Geometry

Section 1

Differential Geometry of Curves

3101

Let $\alpha(s)$ be a closed plane curve. Define the diameter d_α of $\alpha(s)$ to be

$$d_\alpha = \sup_{t,s \in \mathbb{R}} \|\alpha(s) - \alpha(t)\|.$$

Now assume that the curvature $k(s) \geq 1$ for all s.

(i) For any $N \in \mathbb{Z}^+$ sketch an example of such an $\alpha(s)$ with $d_\alpha > N$.

(ii) Assume further that α is a simple closed curve. Prove that $d_\alpha \leq 2$ (or some other constant independent of α; 2 is the best possible such constant).

(*Indiana*)

Solution.

(i) The following is an example of such an $\alpha(s)$ with $k(s) \geq 1$ and $d_\alpha > N$.

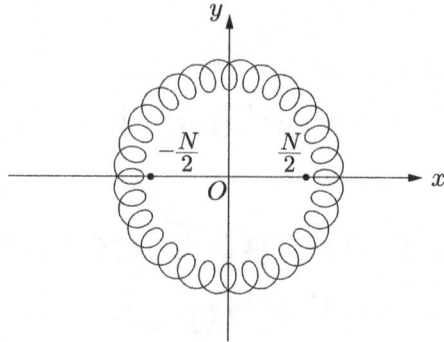

Fig. 3.1.

(ii) From the hypothesis that $k(s) \geq 1$ for all s, we know that the simple closed curve α is an oval.

For every oval, by Blaschke, if we take the origin O as shown in the figure and denote by $p(\theta)$ the distance from O to the tangent l at the point (x, y) of the oval, where the oval is counterclockwise orientated and θ denotes the oriented angle from the x-axis to l, then the oval can be parameterized by θ as follows

$$\begin{cases} x(\theta) = p(\theta) \sin \theta + p'(\theta) \cos \theta, \\ y(\theta) = -p(\theta) \cos \theta + p'(\theta) \sin \theta. \end{cases}$$

$p(\theta)$ is called the support function of the oval. From this we can conclude that, by direct computation, the relative curvature of the oval is $k_r(\theta) = (p(\theta) + p''(\theta))^{-1}$.

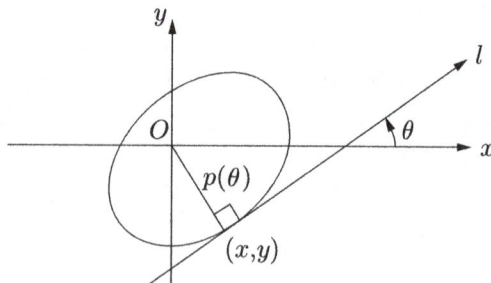

Fig. 3.2.

Now we can prove a more general result of Blaschke:

Let two ovals C and C_1 in a plane be internally tangent at a point O. Suppose that, at every pair of points P and P_1 where C and C_1 have the same tangent orientation, the curvatures of C and C_1 satisfy the inequality $k_1(P_1) \leq k(P)$. Then the domain encircled by C_1 must contain the domain encircled by C.

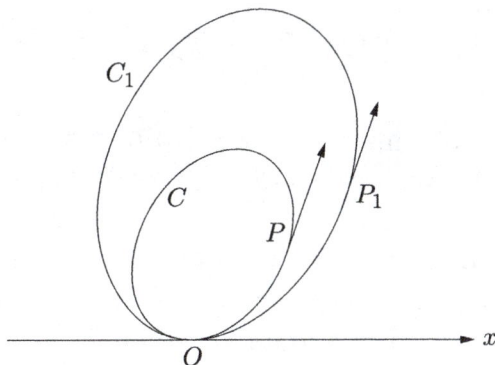

Fig. 3.3.

In fact, take the tangent point O of C and C_1 as the origin, and their common tangent line as the x-axis. Let $p(\theta)$ and $p_1(\theta)$ be the support functions of C and C_1, respectively. From the above, we can say that the support function $p(\theta)$ must be the solution to the following initial value problem of ODE

$$\begin{cases} p''(\theta) + p(\theta) = \frac{1}{k(\theta)}, \\ p(0) = p'(0) = 0, \end{cases}$$

because $p'(\theta)$ is exactly the distance from O to the normal line of C at point $(x(\theta), y(\theta))$. Hence, $p(\theta)$ can be uniquely determined by $k(\theta) = k_r(\theta)$ of C,

$$p(\theta) = \int_0^\theta \frac{\sin(\theta - \phi)}{k(\phi)} d\phi.$$

Analogously, for the curve C_1, we also have the similar expression of $p_1(\theta)$. Thus,

$$p_1(\theta) - p(\theta) = \int_0^\theta \frac{k(\phi) - k_1(\phi)}{k(\phi)k_1(\phi)} \sin(\theta - \phi)d\phi.$$

By the hypothesis, we know that

$$\frac{k(\phi) - k_1(\phi)}{k(\phi)k_1(\phi)} \geq 0.$$

As for the sign of $\sin(\theta - \phi)$, firstly, if $0 \leq \theta \leq \pi$, then owing to $0 \leq \phi \leq \theta$, we have $\sin(\theta - \phi) \geq 0$. Therefore, $p_1(\theta) - p(\theta) \geq 0$. Secondly, if $\pi \leq \theta \leq 2\pi$, we make the reflections of the ovals C and C_1 with respect to their common tangent line at O and reverse the orientation of the x-axis. Then we can get $p_1(\tilde{\theta}) - p(\tilde{\theta}) \geq 0$, where $\tilde{\theta}$ and its corresponding original θ satisfy $\tilde{\theta} + \theta = 2\pi$. Hence, we always have $p_1(\theta) \geq p(\theta)$. Noticing that every oval is the envelope of all its tangent lines, we see that the domain encircled by C must be contained in the domain encircled by C_1. The assertion of Blaschke is proved.

If we take a circle with radius 1 and centered at $\alpha(s_0) + N(s_0)$ as C_1, and take α as C, then (ii) follows immediately from the above assertion.

3102

Let α be a regular C^∞ curve in $\mathbb{R}^3$ with nonvanishing curvature. Suppose the normal vector $N(t)$ is proportional to the position vector; that is, $N(t) = c(t)\alpha(t)$ for all t, where c is a smooth function. Determine all such curves.

(Indiana)

Solution.
For convenience, we assume that the parameter t is the arc length of the curve α. Differentiating both sides of the equality

$N(t) = c(t)\alpha(t)$ with respect to t, we have, by the Frenet formula,

$$-k(t)T(t) - \tau(t)B(t) = c'(t)\alpha(t) + c(t)T(t).$$

Thus we can immediately obtain $k(t) = $ constant, $\tau(t) = 0$ and $c(t) = -k$. Therefore, by the fundamental theorem of the theory of curves, we know that the curve α must be a circle (or a part of it).

3103

A surface $S \subset \mathbb{R}^3$ is called triply ruled if at every $p \in S$ we can find three open line segments L_1, L_2, L_3 lying in S such that $L_1 \cap L_2 \cap L_3 = \{p\}$. Determine all triply ruled surfaces.

(*Indiana*)

Solution.
If S is a triply ruled surface, then, by the hypothesis, there are three different asymptotic directions at every point of S. Observe that every asymptotic direction (du, dv) satisfies

$$L(u,v)du^2 + 2M(u,v)dudv + N(u,v)dv^2 = 0,$$

where $L(u,v), M(u,v), N(u,v)$ are the coefficients of the second fundamental form of S at point $p(u,v)$. Noticing that the above equation is of 2nd order with respect to $du : dv$ and it has two roots, we obtain $L(u,v) = M(u,v) = N(u,v) = 0$ for all (u,v). In other words, every point of S is a planar point. Therefore, S must be a plane (or a part of it).

3104

Let $\gamma : (a,b) \to \mathbb{R}^3$ be smooth with $|\gamma'| = 1$ and curvature k and torsion τ, both nonvanishing. Denote the Frenet frame by $\{T, N, B\}$. Assume that there exists a unit vector $a \in \mathbb{R}^3$ with

$$T \cdot a = \text{constant} = \cos\alpha.$$

(a) Show that a circular helix is an example of such a curve.

(b) Show that $N \cdot a = 0$.

(c) Show that $k/\tau = \text{constant} = \pm \tan \alpha$.

(Indiana)

Solution.

(a) Let a circular helix be parameterized as follows

$$\gamma(s) = (r \cos \omega s, r \sin \omega s, h \omega s),$$

where $r, h, \omega = (r^2 + h^2)^{-\frac{1}{2}}$ are all constants. Then it is easy to verify that s is the arc length parameter. Hence

$$T(s) = (-r\omega \sin \omega s, r\omega \cos \omega s, h\omega).$$

If we take $a = (0, 0, 1)$, then $T(s) \cdot a = h\omega = \text{constant}$.

(b) Differentiating $T \cdot a = \text{constant}$ with respect to the arc length parameter s, we obtain $k(s)N(s) \cdot a = 0$, from which follows $N(s) \cdot a = 0$ for all $s \in (a, b)$.

(c) By the property of (b), we may assume that the constant vector

$$a = \cos \alpha \cdot T(s) \pm \sin \alpha \cdot B(s).$$

Differentiating the above equality with respect to s, we have

$$0 = (\cos \alpha \cdot k(s) \pm \sin \alpha \cdot \tau(s))N(s).$$

Thus, for all $s \in (a, b)$, $k(s)/\tau(s) = \pm \tan \alpha$.

3105

Show that if γ is a geodesic on the cone $z = \sqrt{x^2 + y^2}$, $(x, y) \in \mathbb{R}^2 \backslash \{0, 0\}$, then γ intersects itself at most a finite number of times.

(Indiana)

Solution.

If we cut the cone along a generator l and develop it into a plane, then the cone becomes an infinite sector without the vertex, and the geodesic γ becomes a straight line on the developed infinite sector. Noticing that the central angle of the sector is $\sqrt{2}\pi$, an obtuse angle, and the image of γ on the sector must be one of the following three cases:

a generator,

a straight line never intersecting the generator l,

two rays which start from the generator l,

then we can conclude that γ never intersects itself.

3106

Let $\gamma : (a, b) \rightarrow \mathbb{R}^3$ be a C^∞ curve parameterized by arc length, with curvature and torsion $k(s)$ and $\tau(s)$. Assume $k(s) \neq 0$, $\tau(s) \neq 0$ for all $s \in (a, b)$, and let T and N denote the unit tangent and normal vectors to γ. The curve γ is called a Bertrand curve if there exists a regular curve $\bar{\gamma} : (a, b) \rightarrow \mathbb{R}^3$ such that for each $s \in (a, b)$ the normal lines of γ and $\bar{\gamma}$ at s are equal. In this case, $\bar{\gamma}$ is called the Bertrand mate of γ and we can write $\bar{\gamma}(s) = \gamma(s) + rN(s)$ for some $r = r(s) \in \mathbb{R}$. (Note that s might not be an arc length parameter for $\bar{\gamma}$.)

(a) Prove that r is constant.

(b) Prove that if γ is a Bertrand curve (with r as above), then there exists a constant C such that $rk(s) + C\tau(s) = 1$ for all $s \in (a, b)$.

(c) Prove that if γ has more than one Bertrand mate, then γ is a circular helix.

(*Indiana*)

Solution.

(a) From $\bar{\gamma}(s) = \gamma(s) + r(s)N(s)$ it follows by differentiation that

$$\bar{T}(s)\frac{d\bar{s}}{ds} = (1 - r(s)k(s))T(s) + r'(s)N(s) - r(s)\tau(s)B(s).$$

Taking inner products at both sides with $N(s) = \pm\bar{N}(s)$, we have that $r'(s) = 0$, namely $r = $ const.

(b) From (a), now we have

$$\bar{T}(s) = \frac{ds}{d\bar{s}}(1 - rk(s))T(s) - \frac{ds}{d\bar{s}}r\tau(s)B(s).$$

If we denote

$$\bar{T}(s) = a(s)T(s) + b(s)B(s),$$

then by differentiating with respect to s, we obtain

$$\bar{k}(s)\bar{N}(s)\frac{d\bar{s}}{ds} = a'(s)T(s)$$
$$+ (a(s)k(s) + b(s)\tau(s))N(s) + b'(s)B(s).$$

Hence, from $N(s) = \pm\bar{N}(s)$ we know that $a'(s) = b'(s) = 0$, namely

$$\frac{ds}{d\bar{s}}(1 - rk(s)) = \text{const}, \quad \frac{ds}{d\bar{s}}r\tau(s) = \text{const}.$$

Therefore, there exists a constant C such that $rk(s) + C\tau(s) = 1$ for all $s \in (a, b)$.

(c) Suppose that γ_1 and γ_2 are the Bertrand mates of γ. Then, by (b), there exist constants r_1, r_2, C_1, C_2 such that for all $s \in (a, b)$,

$$\begin{cases} r_1k(s) + C_1\tau(s) = 1, \\ r_2k(s) + C_2\tau(s) = 1. \end{cases}$$

Because the non-zero constants r_1, r_2 are not equal, the above system of linear algebraic equations has solution $k(s) = \text{const}$, $\tau(s) = \text{const}$. Hence γ must be a circular helix.

3107

Let $x(s)$ be a curve in $\mathbb{R}^3$ parameterized by arc-length. Assume that $\tau(s) \neq 0$ and $k'(s) \neq 0$ for all s. Show that a necessary and sufficient condition for $x(s)$ to lie on a sphere is that

$$\frac{1}{k^2(s)} + \frac{1}{\tau^2(s)} \cdot \frac{k'^2(s)}{k^4(s)} = \text{constant.}$$

(Indiana)

Solution.

Suppose that $x(s)$ lies on a sphere centered at the origin. Then we may assume that $x(s) = a(s)T(s) + b(s)N(s) + c(s)B(s)$, where $\{T(s), N(s), B(s)\}$ is the Frenet frame field along $x(s)$, and $a(s)$, $b(s)$, $c(s)$ are suitable functions to be ascertained later. Differentiating $\langle x(s), x(s) \rangle = R^2$ with respect to s, we have $\langle x(s), T(s) \rangle = 0$, from which it follows that $a(s) = 0$, $\forall s$. Differentiating $\langle x(s), T(s) \rangle = 0$ with respect to s again, we obtain

$$1 + k(s)\langle x(s), N(s) \rangle = 0,$$

which means that $b(s) = -\frac{1}{k(s)}$. At last, still differentiating $\langle x(s), N(s) \rangle = -\frac{1}{k(s)}$ with respect to s, we obtain

$$\langle x(s), B(s) \rangle = -\frac{k'(s)}{k^2(s)\tau(s)},$$

namely,

$$c(s) = -\frac{k'(s)}{k^2(s)\tau(s)}.$$

Therefore,

$$R^2 = \langle x(s), x(s) \rangle = \frac{1}{k^2(s)} + \frac{1}{\tau^2(s)} \cdot \frac{k'^2(s)}{k^4(s)}.$$

Conversely, differentiating

$$x(s) + \frac{1}{k(s)}N(s) + \frac{k'(s)}{k^2(s)\tau(s)}B(s)$$

with respect to s, we have

$$\left(x(s) + \frac{1}{k(s)}N(s) + \frac{k'(s)}{k^2(s)\tau(s)}B(s)\right)'$$

$$= \left[-\frac{\tau(s)}{k(s)} + \left(\frac{k'(s)}{k^2(s)\tau(s)}\right)'\right]N(s)$$

that vanishes identically because $k'(s) \neq 0$ and

$$0 = \frac{d}{ds}\left[\frac{1}{k^2(s)} + \frac{1}{\tau^2(s)}\frac{k'^2(s)}{k^4(s)}\right]$$

$$= -2\cdot\frac{k'(s)}{k^3(s)} + 2\cdot\frac{k'(s)}{k^2(s)\tau(s)}\cdot\left(\frac{k'(s)}{k^2(s)\tau(s)}\right)'$$

$$= \frac{2k'(s)}{k^2(s)\tau(s)}\left[-\frac{\tau(s)}{k(s)} + \left(\frac{k'(s)}{k^2(s)\tau(s)}\right)'\right].$$

Then

$$x(s) + \frac{1}{k(s)}N(s) + \frac{k'(s)}{k^2(s)\tau(s)}B(s)$$

is a constant vector, denoted by m. Hence $\langle x(s)-m, x(s)-m\rangle =$ constant, namely, $x(s)$ lies on a sphere centered at m.

3108

Let $M \subset \mathbb{R}^3$ be the torus obtained by rotating the circle $\{(0, y, z) : (y-2)^2 + z^2 = 1\}$ around the z-axis, and let $c(t) = (2\cos t, 2\sin t, 1)$ ("top circle"). Is this curve a geodesic on M? Explain without long computations.

(Indiana)

Solution.
Observe that the geodesic curvature of a curve on M can be computed as follows

$$k_g = \pm k(t)\sin\theta(t),$$

where $k(t)$ is the curvature of the curve, and $\theta(t)$ is the angle between the normal of M and the principal normal of the curve at the point corresponding to the parameter t.

Then, the curve $c(t)$ is not a geodesic on M, because neither the top circle is a straight line, nor its principal normal, which is orthogonal to the z-axis, is parallel to the normal of M along $c(t)$, which is parallel to the z-axis.

3109

Let $\gamma : [0,1] \to \mathbb{R}^3$ be a C^∞ curve with $|\dot\gamma| = 1$ and non-vanishing curvature. Assume the torsion $\tau = 0$.

(a) Show that γ lies in a plane.

(b) What happens if the curvature is allowed to vanish at a point?

(Indiana)

Solution.

(a) $|\dot\gamma| = 1$ implies that the curve is parameterized by arclength s. From the Frenet formula we know that the binormal vector field of γ, denoted by B, is constant. Thus, $\frac{d}{ds}(\gamma(s)B) = 0$, which means that $\gamma(s)B = $ constant, that is, γ lies in a plane.

(b) The vanishing curvature at point s_0 means $\ddot\gamma(s_0) = 0$, i.e., s_0 is a stationary point of γ's tangent vector field. If the point is the strict extreme value point of the tangent vector field, then it is an inflection point of the curve γ.

3110

Let $f : \mathbb{R} \to \mathbb{R}$ be positive and smooth. Let M be the surface in $\mathbb{R}^3$ obtained by rotating the graph $\{(x, f(x)) : x \in \mathbb{R}\}$ of f in the xz plane about the x axis. Characterize in terms of f the set of x such that $\pm\frac{1}{f(x)}$ is a principal curvature of M at $(x, 0, f(x))$.

(Hint. Local coordinate computations are not necessary.)

(Indiana)

Solution.

At $P = (x_p, 0, f(x_p))$, in the direction of the circle of latitude, the corresponding normal curvature

$$k_n = k\cos\theta = \frac{1}{f(x_p)}\cos\theta(x_p),$$

where $\theta(x_p)$ is the angle between the normal of M and the principal normal of the circle of latitude at P. Since every circle of latitude on M is a line of curvature, then the corresponding normal curvature k_n is a principal curvature. Thus, $k_n = \pm\frac{1}{f(x_p)}$ implies that $\cos\theta(x_p) = \pm 1$, that is, the curve $\{(x, 0, f(x)) : x \in \mathbb{R}\}$ has a tangent parallel to the x axis at P. Namely, $f'(x_p) = 0$.

Besides, the meridian passing P is also a line of curvature. Its curvature k at P is just the other principal curvature of M. Thus, the second possible case is

$$k = \frac{|f''(x_p)|}{(1 + f'^2(x_p))^{3/2}} = \frac{1}{f(x_p)}.$$

Therefore, we conclude that the set of x such that $\pm\frac{1}{f(x)}$ is a principal curvature of M at $(x, 0, f(x))$ is

$$\left\{ x \in \mathbb{R} : f'(x) = 0 \text{ or } \frac{f''(x)}{(1 + f'^2(x))^{3/2}} = \pm\frac{1}{f(x)} \right\}.$$

3111

Let $\alpha(s) \subset \mathbb{R}^3$ be a smooth curve parameterized by arclength. Assume that the position vector $\alpha(s)$ is always a linear combination of the binormal and normal vector $B(s)$, $N(s)$ of $\alpha(s)$. Show that $\alpha(s)$ does not pass through $O \in \mathbb{R}^3$.

(Indiana)

Solution.

If the position vector $\alpha(s)$ is always a linear combination of the binormal and normal vectors, then $\langle T(s), \alpha(s) \rangle \equiv 0$.

Integrating the obtained equality, we have $\langle \alpha(s), \alpha(s) \rangle = \text{const.}$ Therefore, if $\alpha(s)$ passes through the origin, we will get $\alpha(s) \equiv 0$, the trivial case.

3112

Let $M^2 \subset \mathbb{R}^3$ be the cylinder $x^2 + y^2 = 1$. Suppose the curve $\alpha(s) \in M^2$ is parameterized by arclength.

(i) If $k(s) > 0$ and $\tau(s) \equiv 0$, show that $\alpha(s)$ is a closed curve. (Here k, τ are curvature and torsion of α in $\mathbb{R}^3$.)

(ii) If $k_g(s) \equiv 1$, show that $\alpha(s)$ is a closed curve (k_g is the geodesic curvature in M).

(Indiana)

Solution.

(i) The hypothesis of torsion $\tau(s) \equiv 0$ implies that the curve α is a plane curve, whereas the hypothesis of curvature $k(s) > 0$ implies that the plane π where the curve α lies does not parallel the generating line of the cylinder M. Therefore, the curve $\alpha \subset M \cap \pi$ must be closed.

(ii) If one develops M into a plane π, then the corresponding plane curve of α, denoted by $\tilde{\alpha}$, has the same geodesic curvature $\tilde{k}_g(s) \equiv 1$. Thus the curvature of $\tilde{\alpha}$ is $\tilde{k} = \sqrt{\tilde{k}_g^2 + \tilde{k}_n^2} \equiv 1$ which means that it is a circle with radius 1 in π. So, α must be closed.

3113

Let T be a two-dimensional distribution in $\mathbb{R}^3$ defined by

$$T_{(x,y,z)} = \text{span}\left\{ \frac{\partial}{\partial x}, \frac{\partial}{\partial y} + x\frac{\partial}{\partial z} \right\}.$$

(i) Show that T is not involutive.

(ii) Given a C^∞ curve $\alpha(s) \in \mathbb{R}^2 = \{(x, y, 0) : x, y \in \mathbb{R}\} \subset \mathbb{R}^3$ and $\alpha(s_0)$, show that there exists a unique C^∞ curve $\beta(s) \in \mathbb{R}^3$ such that $\beta'(s) \in T_{\beta(s)}$, $\beta(s_0) = \alpha(s_0)$ and $\pi(\beta(s)) = \alpha(s)$, where $\pi(x, y, z) = (x, y, 0)$.

(iii) Show that if $\alpha(s)$ is a simple closed curve of length L bounding the region Ω in $\mathbb{R}^2$ then $\beta(s_0+L) - \beta(s_0) = (0,0,\pm A)$ where $A = $ area of Ω.

(*Indiana*)

Solution.

(i) That $\left[\frac{\partial}{\partial x}, \frac{\partial}{\partial y} + x\frac{\partial}{\partial z}\right] = \frac{\partial}{\partial z}$ and $\left\{\frac{\partial}{\partial x}, \frac{\partial}{\partial y} + x\frac{\partial}{\partial z}, \frac{\partial}{\partial z}\right\}$ are linearly independent shows that T is not involutive.

(ii) Let $\alpha(s) = (x(s), y(s), 0)$. By $\pi(\beta(s)) = \alpha(s)$, we may assume that $\beta(s) = (x(s), y(s), z(s))$ where $z(s)$ is unknown. Then $\beta'(s) \in T_{\beta(s)}$ implies that

$$\beta'(s) = a(s)\frac{\partial}{\partial x} + b(s)\left(\frac{\partial}{\partial y} + x(s)\frac{\partial}{\partial z}\right)$$

for suitable functions $a(s)$ and $b(s)$, i.e.,

$$\left(\frac{dx(s)}{ds}, \frac{dy(s)}{ds}, \frac{dz(s)}{ds}\right) = (a(s), b(s), b(s)x(s)).$$

Hence

$$a(s) = \frac{dx(s)}{ds}, \quad b(s) = \frac{dy(s)}{ds}.$$

Thus the problem of finding $\beta(s)$ reduces to solving the following initial value problem

$$\begin{cases} \dfrac{dz}{ds} = x(s)\dfrac{dy(s)}{ds}, \\ z|_{s=s_0} = 0. \end{cases}$$

When the given plane curve $\alpha(s)$ is C^∞, the unknown function $z(s)$ can be uniquely determined by

$$z(s) = \int_{s_0}^{s} x(s)\frac{dy(s)}{ds}ds$$

and so is the space curve $\beta(s)$, i.e.,

$$\beta(s) = \left(x(s), y(s), \int_{s_0}^{s} x(s)\frac{dy(s)}{ds}ds\right).$$

(iii) If $\alpha(s)$ is simple and closed, then we easily have

$$\beta(s_0 + L) - \beta(s_0)$$

$$= \left(x(s_0 + L) - x(s_0), y(s_0 + L) - y(s_0), \oint_\alpha x\,dy \right)$$

$$= (0, 0, \pm A),$$

where the sign is determined by the orientation of the curve α.

3114

Call a normal vector field ν along a space curve γ parallel if $\dot{\nu}$ is always tangent to γ.

(i) Show that the angle through which a parallel normal vector field ν turns relative to the principal normal N along γ is given by the total torsion of γ, i.e.,

$$\int_0^L \tau(s)ds.$$

Here s and L denote arclength and the length of γ, respectively.

(ii) Show that the total torsion of any closed curve γ which lies on a sphere in $\mathbb{R}^3$ must vanish.

(Indiana)

Solution.

(i) The hypothesis that $\dot{\nu}$ is always tangent to γ means that $\langle \nu, \dot{\nu} \rangle = 0$, and hence, the normal vector field ν has constant norm. Therefore, without loss of generality, we may assume that $|\nu| \equiv 1$, and $\nu(s) = \cos\theta(s) \cdot N(s) + \sin\theta(s) \cdot B(s)$, where $\theta(s)$ is a smooth function globally defined on γ, which measures the angle between $\nu(s)$ and $N(s)$. Thus we have

$$\dot{\nu} = (-\sin\theta \cdot N + \cos\theta \cdot B)\dot{\theta} + \cos\theta \cdot (-kT - \tau B) + \sin\theta \cdot \tau N$$

$$= -\cos\theta \cdot kT - (\dot{\theta} - \tau)\sin\theta \cdot N + (\dot{\theta} - \tau)\cos\theta \cdot B.$$

Noting that $\dot{\nu}$ is parallel to T, we see that the above equality implies that $\dot{\theta}(s) = \tau(s)$, and hence

$$\theta(L) - \theta(0) = \int_0^L \tau(s)ds.$$

(ii) For any closed curve γ which lies on a sphere in $\mathbb{R}^3$, the unit normal vector field ν of the sphere along γ satisfies the above mentioned hypothesis. Therefore, we have the relation

$$\theta(s) = \int_0^s \tau(s)ds + \theta(0).$$

On the other hand, the smooth function $\theta(s)$, $s \in [0, L]$ can be regarded as a lift of a certain differentiable map $f : [0, L] \to S^1$ into $\mathbb{R}^1$. The fact that γ is closed means that

$$\int_0^L \tau(s)ds = \theta(L) - \theta(0) = 2n\pi,$$

where the integer n is just the degree of the map f, i.e., $n = \deg f$. Let $p_0 \in \gamma$ and γ contract to p_0 smoothly on S^2. Thus we get a family of curves $\{\gamma_t\}$, $t \in [0, 1]$. Furthermore, we may assume that for every $t \in [0, 1)$, γ_t overlaps with $\gamma = \gamma_0$ about $p_0 = \gamma_1$. Also, for every γ_t, we have the corresponding map f_t such that $f_0 = f$, $f_1 = $ the constant map into p_0. Because the degree of a map is homotopically invariant, finally we have

$$\int_0^L \tau(s)ds = 2\pi \cdot \deg f_1 = 0.$$

3115

Let M be a surface in $\mathbb{R}^3$ and let P be a plane. Suppose M and P intersect orthogonally. Show that the intersection curve (parameterized by arclength) is a geodesic on M.

(Indiana)

Solution.
Let $k(s)$ be the curvature of the intersection curve $C = M \cap P$. If $k \equiv 0$, then C is naturally a geodesic on M. Otherwise, along the segment where $k(s) \neq 0$, the normal of M is parallel to the principal normal of C, hence the segment is also a geodesic one.

3116

Let $\alpha(s)$ be a C^2 curve in $\mathbb{R}^3$ parameterized by arclength. Suppose that for some function $f(s)$, $\alpha''(s) = f(s)\alpha(s)$. What can you deduce about $f(s)$, $\alpha(s)$?

(Indiana)

Solution.
From

$$f(s)\alpha(s)\alpha'(s) = \frac{1}{2}d(\alpha'(s))^2 = 0$$

we have

$$f(s)d(\alpha(s))^2 = 0.$$

If $f(s) \equiv 0$, then $\alpha''(s) \equiv 0$, i.e., $\alpha(s)$ is a straight line.
If $f(s) \neq 0$, then $(\alpha(s))^2 = $ const, i.e., $\alpha(s)$ is a spherical curve. Furthermore, differentiating

$$\alpha''(s) = k(s)N(s) = f(s)\alpha(s)$$

with respect to s, we have

$$k'(s)N(s) + k(s)(-k(s)T(s) - \tau(s)B(s)) = f'(s)\alpha(s) + f(s)T(s)$$

which implies that

$$f(s) = -k^2(s), \quad \tau(s) = 0, \quad k'(s) = 0.$$

Hence $k(s) = $ const, and $\alpha(s)$ is a circle or part of it with radius $\frac{1}{\sqrt{-f}}$.

3117

Suppose $\gamma(t)$ parameterizes a space curve with curvature function $\kappa(t)$. Define a new curve $\tilde{\gamma}$ by setting, for each $t \in \mathbb{R}$, $\tilde{\gamma}(t) := c\gamma(t/c)$, where $c \in \mathbb{R}$ is an arbitrary fixed constant. Derive the curvature function for this new curve.

(Indiana)

Solution.

The hypothesis is

$$\kappa(t) = \frac{|\gamma'(t) \times \gamma''(t)|}{|\gamma'(t)|^3}.$$

Therefore, from $\tilde{\gamma}'(t) = \gamma'(t/c)$ and $\tilde{\gamma}''(t) = \gamma''(t/c)/c$ we have

$$\tilde{\kappa}(t) = \frac{|\tilde{\gamma}'(t) \times \tilde{\gamma}''(t)|}{|\tilde{\gamma}'(t)|^3} = \frac{\kappa(\frac{t}{c})}{|c|}.$$

3118

Let $\alpha : (a, b) \rightarrow \mathbb{R}^3$ be a smooth curve with non-vanishing curvature.

(i) Show that if the torsion of α vanishes identically then there is a plane $\pi \subset \mathbb{R}^3$ containing α, that is, $\alpha(t) \in \pi$ for all $t \in (a, b)$. Is π unique?

(ii) Is the conclusion of (i) still true if the curvature is allowed to vanish at a single point $c \in (a, b)$? [Of course, the torsion is not defined at c, but is assumed to be zero at all other points.]

(Indiana)

Solution.

(i) Let α be reparameterized by arclength s; namely, suppose $\alpha = \alpha(t(s))$, $s \in (s_1, s_2)$ with $t(s_1) = a$, $t(s_2) = b$. The condition $k \neq 0$ means that we can define the Frenet frame field $\{T, N, B\}$ along α. Then the hypothesis $\tau \equiv 0$ implies that B is a constant vector field. Hence, $\frac{d}{ds}\langle\alpha, B\rangle = 0$, from which follows

$$\langle\alpha(t(s)), B\rangle = \text{const} = \langle\alpha(t(s_0)), B\rangle.$$

Therefore, the plane $\pi : \langle\rho - \alpha(t(s_0)), B\rangle = 0$ contains the curve α. Obviously, the connectedness of α and the smoothness of the Frenet frame field imply that π is unique.

(ii) If the curvature k is allowed to vanish at a single point $c \in (a, b)$, then, according to the above discussion, $\alpha(t), t \in (a, c)$ must be on a plane π_1; and $\alpha(t)$, $t \in (c, b)$ must be on another plane π_2. Of course, maybe, $\pi_1 \neq \pi_2$. A counterexample is as follows. Let

$$\alpha(t) = \begin{cases} (t, f(t), 0), & \text{if } t < 0, \\ (0, 0, 0), & \text{if } t = 0, \\ (t, 0, f(t)), & \text{if } t > 0, \end{cases}$$

where the function f is defined by

$$f(t) = \begin{cases} e^{-\frac{1}{t^2}}, & \text{if } t \neq 0, \\ 0, & \text{if } t = 0. \end{cases}$$

3119

Let α and β be two regular curves in $\mathbb{R}^3$. The curve β is called an involute of α if for all t, $\beta(t)$ lies on the line tangent to α at $\alpha(t)$ and $\langle\alpha'(t), \beta'(t)\rangle = 0$. Show that every involute of a generalized helix α is a plane curve. (Recall that α is a generalized helix if for some constant vector $u \neq 0$, $\langle u, \alpha'(s)\rangle \equiv$ const, where s is arclength for α.)

(*Indiana*)

Solution.

For convenience, let α and β be parameterized by their arclengths s and s_1 respectively, and let s, s_1 represent their corresponding points. Then we may assume $\beta(s_1) = \alpha(s) + \lambda(s)T(s)$. Differentiate the equation with respect to s. Using the Frenet formulas and the hypothesis $\langle T(s), T_1(s_1) \rangle = 0$, we can ascertain that $\beta(s_1) = \alpha(s) + (s_0 - s)T(s)$. Differentiate the obtained expression for β successively. Noting that α being a generalized helix implies

$$\begin{vmatrix} k & \tau \\ k' & \tau' \end{vmatrix} \equiv 0,$$

we obtain by straightforward calculation

$$\left(\frac{d}{ds}\beta(s_1), \frac{d^2}{ds^2}\beta(s_1), \frac{d^3}{ds^3}\beta(s_1) \right) \equiv 0,$$

i.e., the torsion of β vanishes everywhere. Therefore, β is a plane curve.

3120

Suppose that the unit normal vector to a surface $M \subset \mathbb{R}^3$ is constant along a regular curve $\alpha \subset M$. Deduce that in this case, α is an asymptotic (Note. A curve in a surface is called asymptotic if its acceleration is everywhere tangent to that surface.) curve, that a plane contains it, and that the Gauss curvature of M vanishes at each point of α.

(*Indiana*)

Solution.
Let α be parameterized by arclength s, and M be locally parameterized by $X(u^1, u^2)$. Firstly, by the Weingarten formula, we have, along α,

$$0 = \frac{dn}{ds} = -\sum_{i,j=1}^{2} \omega_j^i \frac{du^j}{ds} \frac{\partial X}{\partial u^i} = -W\left(\frac{dX}{ds}\right) = -W(\alpha').$$

Further, by the Gauss formula, we immediately deduce that, along α

$$\alpha'' = \sum_{i=1}^{2}\left(\frac{d^2u^i}{ds^2} + \sum_{j,k=1}^{2}\Gamma^i_{jk}\frac{du^j}{ds}\frac{du^k}{ds}\right)\frac{\partial X}{\partial u^i} + \sum_{j,k=1}^{2}\Omega_{jk}\frac{du^j}{ds}\frac{du^k}{ds}\cdot n$$

$$= k_g n \times \alpha' + \langle W(\alpha'), \alpha'\rangle n = k_g n \times \alpha',$$

where k_g is geodesic curvature. Hence α'' is everywhere tangent to the surface.

Secondly, along α, noting that n is constant, we have $d\langle\alpha(s), n\rangle = 0$. Therefore,

$$\langle\alpha(s), n\rangle = \text{const} = \langle\alpha(s_0), n\rangle,$$

namely, $\langle\alpha(s) - \alpha(s_0), n\rangle = 0$, which means that the curve α is contained by a plane $\langle\rho - \alpha(s_0), n\rangle = 0$.

Thirdly, from $\text{III} - 2H\text{II} + K\text{I} = 0$ it follows that along α

$$K = -\left(\frac{dn}{ds}\right)^2 + 2H\langle W(\alpha'), \alpha'\rangle = 0.$$

3121

Sketch the closed regular plane curve $\beta : [-\pi, \pi] \to \mathbb{R}^2$ having $\beta(0) = (0,0)$ and $\beta'(0) = (1,0)$, if β's curvature function $k(s)$ ($s = $ arclength) is odd, satisfies

$$\int_0^\pi k(s)ds \approx 3\pi/2,$$

and has the following graph. (Include an explanation with your sketch.)

(Indiana)

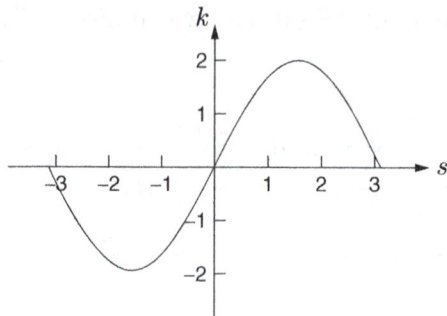

Fig. 3.4.

Solution.

The image of the plane curve β is a "figure eight", as shown in Fig. 3.5, which has x-axis as its tangent line at $(0, 0)$. And y-axis is almost its "tangent" line at $(0, 0)$, too. This assertion follows from

$$\int_0^\pi k(s)ds = \int_0^\pi \frac{d\theta}{ds}ds = \theta(\pi) - \theta(0) \approx \frac{3\pi}{2},$$

where $\theta(s)$ is the oriented angle formed by $\beta'(s)$ and $\beta'(0)$.

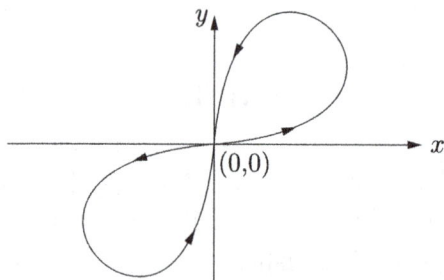

Fig. 3.5.

3122

Let $X \subset \mathbb{R}^2$ be a connected one-dimensional real analytic submanifold, not contained in a line. Prove that not every tangent line to X is bitangent — that is, it is not the case that for all $p \in X$ there exists $q \neq p \in X$ such that the tangent line to X at p equals the tangent line to X at q as lines in $\mathbb{R}^2$.

(Harvard)

Solution.

Let X be parametrized by arclength s as $r = r(s)$. If every tangent line to X is bitangent, then, for every s, there is a function $s_1 = s_1(s)$ such that the tangent lines at $r(s)$ and $r(s_1(s))$ coincide. Hence we can assume that $r(s_1(s)) = r(s) + a(s)T(s)$, where $a(s)$ is an appropriate function. By the Frenet formulas on the plane, we have that

$$T(s_1(s))\frac{ds_1}{ds} = T(s) + a'(s)T(s) + a(s)k(s)N(s).$$

Because $T(s_1(s)) = \pm T(s)$, The above equality yields $a(s)k(s)N(s) = 0$. Since for every s the points $r(s_1(s))$ and $r(s)$ are distinct, $r(s_1(s)) - r(s) = a(s)T(s) \neq 0$. Thus we get $k(s) = 0$ for every s. By a characterization of straight lines using curvature, we know that $r(s)$, $s \in [s, s_1]$ (if we suppose that $s < s_1$) is a straight line segment. Because X is connected and analytic, X must be a straight line which contradicts the assumption that X is not contained in a line. Therefore, not every tangent line to X is bitangent.

Section 2

Differential Geometry of Surfaces

3201

(i) Show that there exists a metric on the plane so that some geodesics are simple closed curves.

(ii) Let $\alpha(s)$ be a simple closed geodesic as described above and $K(x)$ be the Gaussian curvature of the above metric at $x \in \mathbb{R}^2$. Compute

$$\int_{\text{interior of } \alpha} K.$$

Here the integral is with respect to the Riemannian volume induced by the metric.

(Indiana)

Solution.

(i) Let $S^2 = \{(x_1, x_2, x_3) \in \mathbb{R}^3 \colon x_1^2 + x_2^2 + x_3^2 = 1\}$, and $\mathbb{R}^2$ be the $x_1 x_2$ plane. For every $p \in \mathbb{R}^2$, its coordinates with respect to the x_1 and x_2 axes are denoted by (u, v). Suppose that $\pi : S^2 \backslash \{N\} \to \mathbb{R}^2$ is the stereo-graphic projection from the north pole of S^2 into the plane $\mathbb{R}^2$, which maps $(x_1, x_2, x_3) \in S^2 \backslash \{N\}$ to $(u, v) \in \mathbb{R}^2$. By direct calculation, we obtain

$$x_1 = \frac{2u}{u^2 + v^2 + 1}, \quad x_2 = \frac{2v}{u^2 + v^2 + 1}, \quad x_3 = \frac{u^2 + v^2 - 1}{u^2 + v^2 + 1}.$$

Then the metric of $S^2\backslash\{N\}$ can be expressed by

$$ds^2 = \frac{4(du^2 + dv^2)}{(u^2 + v^2 + 1)^2}.$$

Now, using the pull back of ds^2 by $(\pi^{-1})^*$, we can obtain a 2-dimensional Riemannian manifold $(\mathbb{R}^2, (\pi^{-1})^* ds^2)$. Thus, the map $\pi : (S^2\backslash\{N\}, ds^2) \rightarrow (\mathbb{R}^2, (\pi^{-1})^* ds^2)$ is an isometry. Since all great circles which do not pass through the north pole are closed geodesics on $S^2\backslash\{N\}$, then their images are simple closed geodesics on $(\mathbb{R}^2, (\pi^{-1})^* ds^2)$.

(ii) Denote by D the simply connected domain encircled by the simple closed geodesic α. Then, using the famous Gauss–Bonnet formula, we immediately have

$$\int_{\text{interior of } \alpha} K(x) = 2\pi\chi(D) = 2\pi.$$

3202

Let $M^2 \subset \mathbb{R}^3$ be a surface containing $x = 0$. Assume that $\frac{\partial}{\partial x_1}$ and $\frac{\partial}{\partial x_2}$ are tangent to M at $x = 0$, and that in that basis the Weingarten map

$$L = \begin{pmatrix} 4 & 3 \\ 3 & -4 \end{pmatrix}.$$

Let α be the curve (near $x = 0$) obtained by intersecting M with the $x_1 x_3$ plane.

(i) What is the normal curvature of α at $x = 0$?
(ii) What is the geodesic curvature of α at $x = 0$?
(iii) What is the Gaussian curvature of M^2 at $x = 0$?
(iv) Sketch the surface near $x = 0$. You may assume that the normal to the surface at $x = 0$ is $(0, 0, 1)$.

(*Indiana*)

Solution.

(i) By the Meusnier Theorem, the normal curvature of α at $x = 0$ is

$$k_n\left(0, \frac{\partial}{\partial x_1}\right) = \frac{\left\langle L(\frac{\partial}{\partial x_1}), \frac{\partial}{\partial x_1} \right\rangle}{\left\langle \frac{\partial}{\partial x_1}, \frac{\partial}{\partial x_1} \right\rangle} = \left\langle 4\frac{\partial}{\partial x_1} + 3\frac{\partial}{\partial x_2}, \frac{\partial}{\partial x_1} \right\rangle = 4.$$

(ii) Since α is obtained by intersecting M with $x_1 x_3$ plane, i.e., α is a normal section, then its curvature at $x = 0$ is $k(0) = |k_n(0, \frac{\partial}{\partial x_1})| = 4$. By the relation among $k(0)$, $k_n(0, \frac{\partial}{\partial x_1})$ and the geodesic curvature $k_g(0, \frac{\partial}{\partial x_1})$

$$k^2(0) = k_n^2\left(0, \frac{\partial}{\partial x_1}\right) + k_g^2\left(0, \frac{\partial}{\partial x_1}\right),$$

we immediately obtain $k_g(0, \frac{\partial}{\partial x_1}) = 0$.

(iii) Because the eigenvalues of the Weingarten map L are ± 5, we know that the Gaussian curvature of M^2 at $x = 0$ is $K(0) = -25$; or more directly,

$$K(0) = \det \begin{pmatrix} 4 & 3 \\ 3 & -4 \end{pmatrix} = -25.$$

(iv) From (iii), we know that $x = 0$ is a hyperbolic point of M^2. Noting that the normal to the surface at $x = 0$ is $(0, 0, 1)$ and $k_n(0, \frac{\partial}{\partial x_1}) = 4 > 0$, we sketch the surface near $x = 0$ as follows.

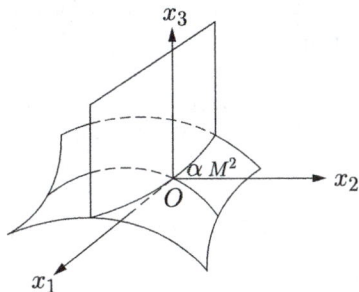

Fig. 3.6.

3203

Let $M = \{(x, y, z) \in \mathbb{R}^3 : z = 6 - (x^2 + y^2)\}$ (a paraboloid of revolution). $D = \{(x, y, z) \in M : z > 2\}$, and $\omega = yz^2 dx + xzdy + x^2 y^2 dz$. Orient M and evaluate

$$\int_D zdx \wedge dy + d\omega.$$

(Indiana)

Solution.

Choose the unit outward pointing normal as the orientation of M. Let $P = \{(x, y, z) \in \mathbb{R}^3 : x^2 + y^2 \leq 4, z = 2\}$ and take $(0, 0, -1)$ as its normal vector. Then, by the Stokes' formula, we have that

$$\int_{D \cup P} d\omega = 0.$$

Hence,

$$\int_D d\omega = -\int_P d\omega.$$

Therefore,

$$\int_D zdx \wedge dy + d\omega = \int_D [6 - (x^2 + y^2)]dx \wedge dy - \int_P d\omega.$$

Firstly, we have

$$\int_D [6 - (x^2 + y^2)]dx \wedge dy = \int_0^{2\pi} d\theta \int_0^2 (6 - r^2)rdr = 16\pi.$$

Secondly, noticing that

$$d\omega = z^2 dy \wedge dx + 2yzdz \wedge dx + zdx \wedge dy$$
$$+ xdz \wedge dy + 2xy^2 dx \wedge dz + 2x^2 ydy \wedge dz,$$

we have

$$-\int_P d\omega = -\int_P 4dy \wedge dx + 2dx \wedge dy = -2\int_P dy \wedge dx = -8\pi.$$

Hence,

$$\int_D z\,dx \wedge dy + d\omega = 8\pi.$$

3204

Let S be a surface diffeomorphic to the ordinary 2-sphere. Suppose there is a C^∞ metric of positive curvature on S for which there exist two simple closed geodesics γ_1 and γ_2. Show that γ_1 and γ_2 must intersect.

(*Indiana*)

Solution.

Here we assume the Jordan curve Theorem.

If γ_1 and γ_2 do not intersect, then γ_1 and γ_2 encircle a domain D such that the Euler characteristic $\chi(D) = 0$ and $\partial D = \gamma_1 \cup \gamma_2$. Applying the global Gauss–Bonnet formula to the domain $D \subset S$ and noting that γ_1 and γ_2 are geodesics of S, we have

$$\int_D K\,d\sigma = 2\pi\chi(D) = 0,$$

which contradicts the assumption of positive curvature.

3205

Let D be the disk $\{(x, y) \in \mathbb{R}^2 : x^2 + y^2 < \frac{1}{4}\}$ and let $O = (0,0) \in D$. Let $(x_1, x_2) = (r, \theta)$ be the usual polar coordinates and define a metric on $D \backslash \{O\}$ by

$$(g_{ij}) = \begin{pmatrix} 1 & 0 \\ 0 & h(r, \theta) \end{pmatrix},$$

where $h(r, \theta) = r^2[1 - 2r^2 + r^4 \sin^2 \theta]^2$.

(a) Prove that this metric extends to a smooth metric on D. (Hint. Use a smooth coordinate system.)

(b) Show that line segments in D which pass through the origin (suitably parameterized) are geodesics.

(Indiana)

Solution.

(a) Set

$$\begin{cases} x = r\cos\theta, \\ y = r\sin\theta. \end{cases}$$

Then $\frac{\partial(x,y)}{\partial(r,\theta)} = r$. So, on $D\backslash\{O\}$, we can choose (x,y) as the coordinates. From

$$\begin{cases} dx = \cos\theta dr - r\sin\theta d\theta \\ dy = \sin\theta dr + r\cos\theta d\theta \end{cases}$$

follows

$$\begin{cases} dr = \cos\theta dx + \sin\theta dy \\ d\theta = \frac{1}{r}(\cos\theta dy - r\sin\theta dx). \end{cases}$$

Hence the metric of $D\backslash\{O\}$ may be rewritten as

$$\begin{aligned} ds^2 &= dr^2 + h(r,\theta)d\theta^2 \\ &= [\cos^2\theta + (1 - 2r^2 + 4r^4\sin^2\theta)^2\sin^2\theta]dx^2 \\ &\quad + 2\cos\theta\sin\theta[1 - (1 - 2r^2 + 4r^4\sin^2\theta)^2]dxdy \\ &\quad + [\sin^2\theta + \cos^2\theta(1 - 2r^2 + 4r^4\sin^2\theta)^2]dy^2 \\ &= \tilde{g}_{11}dx^2 + 2\tilde{g}_{12}dxdy + \tilde{g}_{22}dy^2. \end{aligned}$$

When $r \to 0$, we see that $\tilde{g}_{11} = 1 + o(r)$, $\tilde{g}_{12} = o(r)$, $\tilde{g}_{22} = 1 + o(r)$, and $\frac{\partial \tilde{g}_{ij}}{\partial x} = o(1)$, $\frac{\partial \tilde{g}_{ij}}{\partial y} = o(1)$. Therefore, this metric can extend to a smooth metric on D.

(b) For any line segment l in D, we can express $l\backslash\{O\}$ as the union of $l_1 = \{(r,\theta) : 0 < r < \frac{1}{4}, \theta = \theta_0\}$ and $l_2 = \{(r,\theta) : 0 < r < \frac{1}{4}, \theta = \theta_0 + \pi\}$. Using $ds^2 = dr^2 + h(r,\theta)d\theta^2$, we know that both l_1 and l_2 are geodesics in $D\backslash\{O\}$. If we use the extended metric, the geodesic curvature k_g of l must be a continuous function of the coordinates. Because l_1 and l_2 are geodesics by either metric, we know that $k_g|_O = 0$. Hence, as a whole line segment, l is a geodesic, too.

3206

(i) Let M be a surface (without boundary) in $\mathbb{R}^3$. Suppose M is inside a ball of radius R and suppose a point $p \in M$ is on the boundary sphere. Show that the Gauss curvature of M at p is $\geq \frac{1}{R^2}$.

(ii) Show that there is no closed minimal surface in $\mathbb{R}^3$.

(*Indiana*)

Solution.

(i) The surface M and the sphere S^2, the boundary of the ball, have the same tangent plane at p. Let π be a normal plane of M and S^2 at p. Then, using the local canonical forms of the normal section curves $M \cap \pi$ and $S^2 \cap \pi$ at p, one can easily show that, at p, the curvature of $M \cap \pi$ is not less than that of $S^2 \cap \pi = S^1$, and p is an elliptic point of M. Hence the Gauss curvature of M at p is greater than or equal to $\frac{1}{R^2}$.

(ii) Suppose that M is a closed minimal surface in $\mathbb{R}^3$. Then there is a family of spheres that contain the surface M inside and have a fixed center. Let R be the infimum of their radii. Then, there must be at least one common point of M and the sphere with radius R and centered at the fixed point. Using the result of (i), one concludes that the Gauss curvature of M at p, $K(p) \geq \frac{1}{R^2}$. But it contradicts the fact that, for minimal surfaces,

$$K = k_1 k_2 = -k_1^2 \leq 0.$$

3207

Let $H^2 = \{(x, y) \in \mathbb{R}^2, y > 0\}$ be the upper half-plane in $\mathbb{R}^2$ and let H^2 have the Riemannian metric g such that

$$g(x, y) = \frac{1}{y^2}(x, y) \quad \text{for } (x, y) \in H^2,$$

where (x, y) is the usual inner product on $\mathbb{R}^2$.

(a) Compute the components of the Levi–Civita connection (i.e., the Christoffel symbols).

(b) Let $V(0) = (0, 1)$ be a tangent vector at the point $(0, 1) \in H^2$. Let $V(t) = (a(t), b(t))$ be the parallel transport of $V(0)$ along the curve $x = t$, $y = 1$. Show that $V(t)$ makes an angle t with the direction of the y-axis in the clockwise direction.

(Hint. Write $a(t) = \cos\theta(t)$, $b(t) = \sin\theta(t)$ where $\theta(t)$ is the angle $V(t)$ makes with the x-axis.)

(Indiana)

Solution.

(a) From the hypothesis we have $E = G = \frac{1}{y^2}$, $F = 0$ and hence

$$\Gamma^1_{11} = \frac{E_1}{2E} = 0, \quad \Gamma^1_{12} = \frac{E_2}{2E} = -\frac{1}{y}, \quad \Gamma^1_{22} = -\frac{G_1}{2E} = 0,$$

$$\Gamma^2_{11} = -\frac{E_2}{2G} = \frac{1}{y}, \quad \Gamma^2_{12} = \frac{G_1}{2G} = 0, \quad \Gamma^2_{22} = \frac{G_2}{2G} = -\frac{1}{y}.$$

(b) For convenience, denote the curve $(x, y) = (t, 1) = (u^1(t), u^2(t))$ and the parallel transport vector field $V(t) = (a(t), b(t)) = (v^1(t), v^2(t))$. Then $V(t)$ must be the solution to the following initial value problem

$$\begin{cases} \dfrac{dv^i(t)}{dt} + \displaystyle\sum_{j,k=1}^{2} \Gamma^i_{jk} v^j(t) \dfrac{du^k(t)}{dt} = 0, \quad i = 1, 2, \\ (v^1(0), v^2(0)) = (0, 1). \end{cases}$$

Noticing the above expression of Γ^i_{jk}, along the curve, we see that the problem is equivalent to

$$\begin{cases} \dfrac{d\alpha(t)}{dt} = b(t), & \dfrac{db(t)}{dt} = -a(t), \\ (a(0), b(0)) = (0, 1). \end{cases}$$

Therefore, writing $a(t) = \cos\theta(t)$, $b(t) = \sin\theta(t)$, we have immediately $\theta = -t$.

3208

Consider the hyperboloid S of one sheet $x^2 + y^2 - z^2 = 4$.
(a) Define the Gauss curvature K of S.
(b) Use the definition from part (a) to compute K at $(0, 2, 0)$.
(*Indiana*)

Solution.
(a) The Gauss curvature K of S at P is defined by $K = k_1 k_2$, where k_1, k_2 are the two principal curvatures of S at P.

(b) At $P = (0, 2, 0)$, in the direction of the circle of latitude, the normal section is a circle $x^2 + y^2 = 4$. Thus, by taking the outer unit normal vector field as the orientation of S, the corresponding principal curvature $k_1 = -\frac{1}{2}$; whereas in the direction of the meridian, the normal section is a component of the hyperbola $y^2 - z^2 = 4$, thus the corresponding principal curvature $k_2 = \frac{1}{2}$. Therefore, the Gauss curvature $K = k_1 k_2 = -\frac{1}{4}$.

3209

Calculate the total geodesic curvature of a circle of radius r on a sphere of radius $R > r$.

(*Indiana*)

Solution.

By the Gauss–Bonnet formula, the total geodesic curvature is given by

$$\int_C k_g ds = -\int\int_\Omega K d\sigma + 2\pi\chi(\Omega)$$
$$= -\frac{1}{R^2}\cdot 2\pi R(R-\sqrt{R^2-r^2}) + 2\pi$$
$$= 2\pi\frac{\sqrt{R^2-r^2}}{R},$$

where C is the given circle, and Ω is the spherical cap encircled by C.

3210

Let S be an embedded closed surface in $\mathbb{R}^3$ with the position vector $X(p)$ and the unit outward normal vector $N(p)$ for $p \in S$. For a fixed (small) t, define a surface S_t to be the set $S_t = \{X(p)+tN(p)\in\mathbb{R}^3 | p\in S\}$. Let κ_1, κ_2 be the principal curvatures of S at the point p with respect to the outward normal vector. Let H_t be the mean curvature of S_t at the point $X(p)+tN(p)$ with respect to the outward normal vector (mean curvature is defined to be the sum of the two principal curvatures). Show that

$$H_1 = \frac{\kappa_1}{1-t\kappa_1} + \frac{\kappa_2}{1-t\kappa_2}.$$

(Harvard)

Solution.

For an arbitrary point $p \in S$, firstly, we assume that p is not an umbilical point of S. Then, in a neighborhood of p, there exists a curvature line net which we can take as a coordinate net formed by u-curves and v-curves. Define the position vector of the point p_t of S_t associated with $p(u,v)$ by $Y_t(u,v) = X(u,v)+tN(u,v)$. Noting that all coordinate curves are curvature lines of S, by the Rodriques equation, we have that

$$Y_{tu} = X_u + tN_u = (1 - t\kappa_1)X_u,$$
$$Y_{tv} = X_v + tN_v = (1 - t\kappa_2)X_v,$$
$$Y_{tu} \times Y_{tv} = (1 - t\kappa_1)(1 - t\kappa_2)X_u \times X_v.$$

The last equation means that for a fixed small t, there is a neighborhood of p_t on S_t, which is also regular and has the unit normal vector $N_t(u, \nu) = N(u, \nu)$ associated with the point $p(u, \nu)$ on S. If we rewrite the Rodriques equations for the surface S as

$$N_u = -\kappa_1 X_u = -\frac{\kappa_1}{1 - t\kappa_1}Y_{tu}, \quad N_\nu = -\kappa_2 X_\nu = -\frac{\kappa_2}{1 - t\kappa_2}Y_{tv},$$

and notice that $N_{tu}(u, \nu) = N_u(u, \nu)$, $N_{tv}(u, \nu) = N_\nu(u, \nu)$, the above two equations become the Rodriques equations for the surface S_t. Hence, the principal curvatures at p_t of S_t are $\kappa_1/(1 - t\kappa_1)$ and $\kappa_2/(1 - t\kappa_2)$ which gives the conclusion

$$H_t = \frac{\kappa_1}{1 - t\kappa_1} + \frac{\kappa_2}{1 - t\kappa_2}.$$

Next, if p is an umbilical point of S, then the normal curvatures at p of S are independent of the directions at the tangent plane at p. Hence the coefficients of the first and second fundamental forms of S satisfy the relation $\Omega_{ij}(p) = \kappa(p)g_{ij}(p)$. Then, by the fundamental formulas for the surface S, if denoting $( )_1 = \frac{\partial}{\partial u}( ), ( )_2 = \frac{\partial}{\partial \nu}( )$,

$$N_i(p) = -\sum_{k,l} g^{kl}(p)\Omega_{li}(p)X_k(p) = -\kappa(p)X_i(p) \qquad i, j = 1, 2.$$

From $Y_{ti}(p_t) = (1 - t\kappa(p))X_t(p)$, $i = 1, 2$, we can sill get $N_t(p_t) = N(p)$ and the coefficients of the first and second fundamental forms of S_t at point p_t are as follows:

$$\bar{g}_{ij}(p_t) = Y_{ti}(p_t)Y_{tj}(p_t) = (1 - t\kappa(p))^2 g_{ij}(p),$$
$$\bar{\Omega}_{ij}(p_t) = -Y_{ti}(p_t)N_{tj}(p_t) = \kappa(p)(1 - t\kappa(p))g_{ij}(p)$$
$$= \frac{\kappa(p)}{(1 - t\kappa(p))}\bar{g}_{ij}(p_t), \quad i, j = 1, 2.$$

Thus p_t is also an umbilical point of S_t, and the normal curvature at p_t along every direction is $\kappa(p)/(1 - t\kappa(p))$, which means the final conclusion is still valid.

3211

Let $\lambda : \mathbb{R} \to \mathbb{R}$ be a smooth function with compact support, and consider the Riemannian surface obtained by equipping $\mathbb{R}^2$ with a metric g of the form

$$g(\mathbf{v}, \mathbf{w}) := e^{\lambda(y)} \langle \mathbf{v}, \mathbf{w} \rangle$$

for $\mathbf{v}$, $\mathbf{w} \in T_{(x,y)}\mathbb{R}^2$. Assuming the Gauss curvature K of this metric is everywhere non-negative, use the Gauss–Bonnet Theorem to deduce that in fact, the surface is flat.

(*Indiana*)

Solution.

Using the Descartes coordinates (x, y) in $\mathbb{R}^2$, we can express the metric of the Riemannian surface $\tilde{\mathbb{R}}^2$ by $ds^2 = e^{\lambda(y)}(dx^2 + dy^2)$, which is obviously conformal to that of the Euclidean plane $\mathbb{R}^2$. Take a positive number a and consider a square domain $\mathcal{D}$ in $\mathbb{R}^2$, whose vertices are $A(-a, -a), B(a, -a), C(a, a), D(-a, a)$ respectively. Suppose that, in $\tilde{\mathbb{R}}^2$, the corresponding part of $\mathcal{D}$ is $\tilde{\mathcal{D}}$, and the corresponding points of A, B, C, D are $\tilde{A}, \tilde{B}, \tilde{C}, \tilde{D}$, respectively.

Then, by the Liouville formula, we can see that the geodesic curvatures of the segments of coordinate curves $\tilde{A}\tilde{B}$ and $\tilde{C}\tilde{D}$ are

$$k_g|_{\tilde{A}\tilde{B}} = \left(\frac{d\theta}{ds} - \frac{1}{2\sqrt{G}} \frac{\partial \ln E}{\partial y} \cos\theta + \frac{1}{2\sqrt{E}} \frac{\partial \ln G}{\partial x} \sin\theta \right)_{\theta \equiv 0, y \equiv -a}$$

$$= \left(-\frac{1}{2\sqrt{e^{\lambda}}} \frac{d\lambda}{dy} \right)_{y \equiv -a},$$

$$k_g|_{\tilde{C}\tilde{D}} = \left(\frac{d\theta}{ds} - \frac{1}{2\sqrt{G}} \frac{\partial \ln E}{\partial y} \cos\theta + \frac{1}{2\sqrt{E}} \frac{\partial \ln G}{\partial x} \sin\theta \right)_{\theta \equiv \pi, y \equiv a}$$

$$= \left(\frac{1}{2\sqrt{e^\lambda}} \frac{d\lambda}{dy} \right)_{y \equiv a} = 0,$$

whereas both $\tilde{B}\tilde{C}$ and $\tilde{D}\tilde{A}$ are geodesics of $\tilde{\mathbb{R}}^2$. Applying the Gauss–Bonnet formula to $\tilde{\mathcal{D}} \subset \tilde{\mathbb{R}}^2$, we have

$$\int_{\partial \tilde{\mathcal{D}}} k_g ds + \int\int_{\mathcal{D}} K d\sigma = 0,$$

where the Gauss curvature

$$K = -\frac{1}{2e^\lambda} \Delta \ln e^\lambda = -\frac{1}{2e^\lambda} \frac{d^2\lambda}{dy^2}.$$

Noting the area element $d\sigma = e^\lambda dxdy$ and the line element $ds^2 = e^{\lambda(y)}dx^2$ along $\tilde{A}\tilde{B}$ and $\tilde{C}\tilde{D}$, from the above integral equality we can obtain

$$\left(\frac{d\lambda}{dy} \right)_{y=-a} = \left(\frac{d\lambda}{dy} \right)_{y=a}.$$

Because the number a is arbitrarily chosen, letting $a \to 0$, it leads to

$$\left(\frac{d\lambda}{dy} \right)_{y=0} = 0.$$

Besides, by the hypothesis $K \geq 0$, we have

$$\frac{d^2\lambda}{dy^2} \leq 0;$$

namely, $\frac{d\lambda}{dy}$ is a decreasing function. Thus,

$$\left(\frac{d\lambda}{dy} \right)_{y=-a} \geq 0 \geq \left(\frac{d\lambda}{dy} \right)_{y=a},$$

which combining the above equality shows that $\frac{d\lambda}{dy} = 0$ in $\mathbb{R}^1$. Therefore, $K \equiv 0$, i.e., $\tilde{\mathbb{R}}^2$ is flat.

3212

Let $M^2 \subset \mathbb{R}^3$ be a smooth compact surface such that $M^2 \subset \{(x, y, z) : z \geq 0\}$. Assume that $M^2 \cap \{(x, y, z) : z = 0\}$ is a smooth curve $\alpha(s)$, parameterized by arclength.

(i) Show that $\alpha(s)$ is an asymptotic curve on M^2.

(ii) Show that $\alpha'(s)$ is always a principal direction.

(iii) Let us now drop the assumption that $M^2 \cap \{(x, y, z) : z = 0\}$ is a curve. What kind of a set could $M^2 \cap \{(x, y, z) : z = 0\}$ be?

(*Indiana*)

Solution.

(i) The hypotheses imply that along the curve $\alpha(s)$, the plane $\{(x, y, z) : z = 0\}$ is a tangent plane of the surface M^2. Thus, the unit normal vector field $n(s)$ of M^2 is constant along $\alpha(s)$. Therefore, along $\alpha(s)$, $\frac{dn(s)}{ds} = 0$, which means that the tangent vector of $\alpha(s)$ for every s is an asymptotic direction. Hence, $\alpha(s)$ is an asymptotic curve on M^2.

(ii) Noticing the above fact, we see that along $\alpha(s)$, $\frac{dn(s)}{ds} = -0 \cdot \alpha'(s)$, from which the Rodriques equation says that $\alpha'(s)$ is always a principal direction with the principal curvature 0.

(iii) If we drop the assumption that $M^2 \cap \{(x, y, z) : z = 0\}$ is a curve, then the set $M^2 \cap \{(x, y, z) : z = 0\}$ consists of elliptic or parabolic points of M^2, at which the plane $\{(x, y, z) : z = 0\}$ is tangent to M^2.

3213

(a) Construct an example of a non-compact C^∞ surface in $\mathbb{R}^3$ with a sequence of closed geodesics $\{\sigma_i\}$ such that length $(\sigma_i) \to 0$.

(b) Show that this is not possible if the surface is compact.

(*Indiana*)

Solution.

(a) Consider the following surface S of revolution generated by a C^∞ vibrating curve C, illustrated in Fig. 3.7, rotating around the x-axis which is the asymptote of the curve. Let P_i denote the points where C has horizontal tangents. Then, on S, P_i draw closed geodesics σ_i, and length $(\sigma_i) \to 0$.

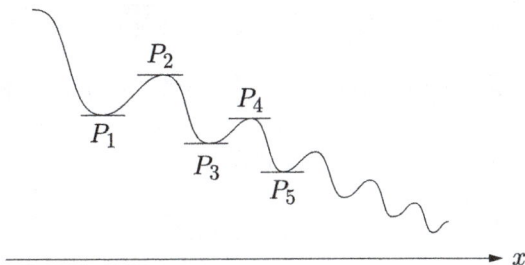

Fig. 3.7.

(b) Suppose S is a compact surface. If there exist closed geodesics σ_i such that length $(\sigma_i) \to 0$, then by the Gauss–Bonnet formula we have

$$\int\int_{\Omega_i} K d\sigma = 2\pi\chi(\Omega_i),$$

where Ω_i denotes the domain encircled by σ_i. Because S is compact, when i is sufficiently large, we may suppose Ω_i is simply connected. Thus, setting $i \to \infty$ in the above equality, we come to a contradictory result $0 = 2\pi$.

3214

Let $(r(s), 0, z(s))$ be a unit speed curve in $\mathbb{R}^3$ with $r(s) > 0$. Consider the surface of revolution $(s, \theta) \to (r(s)\cos\theta, r(s)\sin\theta, z(s))$. On this surface compute the covariant derivatives $\nabla_{\frac{\partial}{\partial\theta}}\frac{\partial}{\partial s}$ and $\nabla_{\frac{\partial}{\partial\theta}}\frac{\partial}{\partial\theta}$ in terms of $\frac{\partial}{\partial\theta}$ and $\frac{\partial}{\partial s}$.

(*Indiana*)

Solution.

The direct computation gives

$$I = ds^2 + r^2(s)d\theta^2 = Eds^2 + Gd\theta^2.$$

Denoting $(s, \theta) = (u^1, u^2)$, we have

$$\nabla_{\frac{\partial}{\partial\theta}} \frac{\partial}{\partial\theta} = \Gamma_{22}^1 \frac{\partial}{\partial s} + \Gamma_{22}^2 \frac{\partial}{\partial\theta} = -\frac{G_1}{2E}\frac{\partial}{\partial s} + \frac{G_2}{2G}\frac{\partial}{\partial\theta} = -r(s)r'(s)\frac{\partial}{\partial s},$$

$$\nabla_{\frac{\partial}{\partial\theta}} \frac{\partial}{\partial s} = \Gamma_{21}^1 \frac{\partial}{\partial s} + \Gamma_{21}^2 \frac{\partial}{\partial\theta} = \frac{E_2}{2E}\frac{\partial}{\partial s} + \frac{G_1}{2G}\frac{\partial}{\partial\theta} = \frac{r'(s)}{r(s)}\frac{\partial}{\partial\theta},$$

where Γ_{jk}^i denotes the Christoffel symbol $(i, j, k = 1, 2)$.

3215

Let $M^2 \subset \mathbb{R}^3$ be an embedded compact surface of genus ≥ 1. Show that the Gauss curvature of M must vanish somewhere on M.

(*Indiana*)

Solution.

By the Gauss–Bonnet formula, we have

$$\int\int_{M^2} K d\sigma = 2\pi\chi(M^2) = 4\pi(1 - g) \leq 0.$$

Firstly, we claim that because of the compactness of M^2, there must be a point P with $K(P) > 0$; hence, by the continuity, there exists a domain U of P such that $K|_U > 0$. Secondly, we show that there exists another point Q with $K(Q) < 0$. Otherwise, we would have

$$0 \geq \int\int_{M^2} K d\sigma \geq \int\int_U K d\sigma > 0,$$

a contradiction. Finally, by the connectedness, the continuous function K must vanish somewhere on M^2.

3216

Consider the torus-of-revolution T obtained by rotating the circle $(x - a)^2 + z^2 = r^2$ around the z-axis:

$$T = \{(x, y, z) : (x^2 + y^2 + z^2 + a^2 - r^2)^2 - 4a^2(x^2 + y^2) = 0\}.$$

Parameterize this torus, compute its Gauss curvature function K, and verify that $\int_T K dA = 0$ by explicit calculation.

(*Indiana*)

Solution.

Thus obtained torus-of-revolution T can be parameterized by

$$X(u, v) = ((a + r \cos u) \cos v, (a + r \cos u) \sin v, r \sin u),$$
$$0 < u, v < 2\pi.$$

A straightforward computation gives the coefficients of its first and second fundamental forms

$$E = r^2, \quad F = 0, \quad G = (a + r \cos u)^2;$$
$$L = r, \quad M = 0, \quad N = \cos u \cdot (a + r \cos u).$$

Therefore, its Gauss curvature function

$$K = \frac{LN - M^2}{EG - F^2} = \frac{\cos u}{r(a + r \cos u)}.$$

Noting that the area element on T is

$$dA = \sqrt{EG - F^2} du dv = r(a + r \cos u) du dv,$$

we have immediately

$$\int_T K dA = \int_0^{2\pi} dv \int_0^{2\pi} \cos u du = 0.$$

3217

Let

$$x(t) = \left(\frac{1}{2}\cos(t), \frac{1}{2}\sin(t), \frac{\sqrt{3}}{2} \right), \quad 0 \le t < 2\pi$$

be a curve on $S^2 \subset \mathbb{R}^3$. Let $X_0 = \frac{\partial}{\partial x_2} \in T_{\left(\frac{1}{2},0,\frac{\sqrt{3}}{2}\right)}S^2$. Compute the parallel translation of X_0 along $x(t)$.

(*Indiana*)

Solution.
Consider the cone that is tangent to the unit sphere S^2 along the curve x. This cone minus one generator is isometric to an open set $D \subset \mathbb{R}^2$ given in polar coordinates by $0 < \rho < +\infty$, $0 < \theta < \sqrt{3}\pi$. Because the parallel translation in the plane coincides with the normal Euclidean one, we obtain the result that, for a displacement t of a moving point p along x starting from $\left(\frac{1}{2},0,\frac{\sqrt{3}}{2}\right)$ (corresponding to the central angle $\theta = \frac{\sqrt{3}}{2}t$ in the domain D), the oriented angle formed by the tangent vector $x'(t)$ with the parallel translation vector $X(t)$ of X_0 is given by $2\pi - \theta = 2\pi - \frac{\sqrt{3}}{2}t$.

3218

Show that for any Riemannian metric on S^2 with $|K| \le 1$, where K is the Gauss curvature, the area of S^2 is not less than 4π.

(*Indiana*)

Solution.
By the Gauss–Bonnet formula, for any Riemannian metric on S^2, we have

$$\int_{S^2} K \, dv = 2\pi\chi(S^2) = 4\pi.$$

Thus, if $|K| \leq 1$, then

$$4\pi \leq \int_{S^2} |K| dv \leq \text{Area } (S^2).$$

3219

Define "geodesics", and characterize (with proof):

(i) All geodesics on the unit sphere $S^2 \subset \mathbb{R}^3$.
(ii) All geodesics on the surface $\{(x, y, z) \in \mathbb{R}^3 : x^2 + y^2 \equiv 1\}$.

(*Indiana*)

Solution.

Let $\alpha(s)$ be a curve on a surface M parameterized by arclength. Along α, we have

$$\alpha''(s) = k(s)N(s)$$

$$= \sum_i \left(\frac{d^2 u^i}{ds^2} + \sum_{j,k} \Gamma^i_{jk} \frac{du^j}{ds} \frac{du^k}{ds} \right) \frac{\partial X}{\partial u^i} + \sum_{j,k} \Omega_{jk} \frac{du^j}{ds} \frac{du^k}{ds} n$$

$$= k_g n \times \alpha' + k_n n,$$

where u^1, u^2 are local coordinates on M, n is the unit normal vector field of M along α, and k_g is the geodesic curvature of α. Then, α is a geodesic of M if and only if along α, $k_g \equiv 0$ or

$$\frac{d^2 u^i}{ds^2} + \sum_{j,k} \Gamma^i_{jk} \frac{du^j}{ds} \frac{du^k}{ds} = 0 \quad i = 1, 2.$$

(i) On the unit sphere, every great circle is a geodesic, because along α, the principal normal vector N is parallel to n. On the contrary, owing to the uniqueness of the initial value problem

$$\begin{cases} \dfrac{d^2 u^i}{ds^2} + \displaystyle\sum_{j,k} \Gamma^i_{jk} \dfrac{du^j}{ds} \dfrac{du^k}{ds} = 0, \quad i = 1, 2, \\[4mm] u^i \big|_{s_0} = u^i_0, \quad \dfrac{du^i}{ds} \Big|_{s_0} = v^i_0, \end{cases}$$

every geodesic on M, that starts from a given point and is tangent to a given direction, must be a great circle.

(ii) Since geodesics are intrinsic objects, then if we develop the cylinder to a plane, every geodesic must become a straight line. Therefore, every geodesic on the cylinder must be a helix, or a circle of latitude, or a straight generating line.

3220

Give an example (e.g., draw a picture) to show that a connected surface can have two points which are not jointed by any geodesic. What is the usual topological hypothesis that prevents this problem?

(Indiana)

Solution.

Let π be a plane and p, q two points in π. Let r be an interior point of the line segment pq. Then the surface $\pi \backslash \{r\}$ is a case in point. "Completeness" is the usual topological hypothesis that prevents this problem.

3221

Define "minimal surface in $\mathbb{R}^3$", and prove that the catenoid, obtained by rotating the graph of $y = \cosh(x)$ around the x-axis in $\mathbb{R}^3$, is minimal.

(Indiana)

Solution.

A minimal surface is a surface with mean curvature

$$H = \frac{1}{2} \frac{EN - 2FM + GL}{EG - F^2} = 0.$$

The straightforward calculation shows that the catenoid

$$X(x, \theta) = (x, \cosh(t)\cos(\theta), \cosh(t)\sin(\theta))$$

is a minimal surface in $\mathbb{R}^3$.

3222

The "Clifford" torus in S^3 can be parameterized using charts
of the form

$$X(u,v) = \frac{1}{\sqrt{2}}(\cos u, \sin u, \cos v, \sin v),$$

where u and v are constrained to lie within intervals length $< 2\pi$.

(i) Compute the metric $[g_{ij}]$ on the Clifford torus for a coordinate chart of the indicated type.

(ii) Figure out the Clifford torus' Gauss curvature function.
(Hint. Calculation is not necessary here.)

(iii) Deduce from the result of (ii) that the Euler characteristic of a torus is zero.

(*Indiana*)

Solution.

(i) From

$$X_u = \frac{1}{\sqrt{2}}(-\sin u, \cos u, 0, 0), \quad X_v = \frac{1}{\sqrt{2}}(0, 0, -\sin v, \cos v)$$

we easily have

$$g_{11} = g_{22} = \frac{1}{2}, \quad g_{12} = g_{21} = 0.$$

(ii) The Gauss equation implies that the Clifford torus Gauss
curvature function $K \equiv 0$.

(iii) Using the Gauss–Bonnet formula, we see that the Euler
characteristic of a torus T

$$\chi(T) = \frac{1}{2\pi} \int \int_T K d\sigma = 0.$$

3223

Let $f : \mathbb{R}^2 \to \mathbb{R}$ be a smooth function with a critical point
(e.g., a minimum or maximum) at the origin $x_1 = x_2 = 0$.

(i) Show that the principal curvatures of the graph

$$z = f(x_1, x_2)$$

at $(0, 0, f(0, 0)) \in \mathbb{R}^3$ are the same as the eigenvalues of the Hessian matrix $[\partial^2 f/\partial x_i \partial x_j]$ at $(0, 0)$.

(ii) Show that $\mathbb{R}^3$ contains no compact embedded surface with strictly negative Gauss curvature at all its points.

(Hint. Look at the "lowest" point on the surface.)

(*Indiana*)

Solution.

(i) The hypothesis means that $\frac{\partial f}{\partial x_1}$ and $\frac{\partial f}{\partial x_2}$ vanish at $(0, 0)$. Then, by straightforward calculation, we have

$$[g_{ij}]_{0,0} = \begin{pmatrix} 1 & 0 \\ 0 & 1 \end{pmatrix}, \quad [\Omega_{ij}]_{(0,0)} = \begin{pmatrix} \dfrac{\partial^2 f}{\partial x_1^2} & \dfrac{\partial^2 f}{\partial x_1 \partial x_2} \\ \dfrac{\partial^2 f}{\partial x_1 \partial x_2} & \dfrac{\partial^2 f}{\partial x_2^2} \end{pmatrix}_{(0,0)}.$$

Therefore, the principal curvatures of the graph

$$z = f(x_1, x_2)$$

at $(0, 0, f(0, 0)) \in \mathbb{R}^3$ are the roots of the following equation

$$\det[\Omega_{ij} - \lambda g_{ij}]_{(0,0)} = \begin{vmatrix} \dfrac{\partial^2 f}{\partial x_1^2} - \lambda & \dfrac{\partial^2 f}{\partial x_1 \partial x_2} \\ \dfrac{\partial^2 f}{\partial x_1 \partial x_2} & \dfrac{\partial^2 f}{\partial x_2^2} - \lambda \end{vmatrix} = 0.$$

(ii) Let $p(x_0, y_0, z_0)$ be the "lowest" point on the surface. (The existence of such a point follows from the compactness of the surface.) Since the surface is above the plane $\pi : z = z_0$ and p is the common point of π and the surface, then π is the tangent plane of the surface at p. By observing the normal section at p, it is easy to conclude that any normal curvature of the surface at p is greater than or equal to zero, if we take $(0, 0, 1)$ as the unit normal of the surface at p. Thus the Gauss curvature of the surface at p is not less than zero.

3224

Let M be a 2-dimensional manifold smoothly embedded in $\mathbb{R}^3$ with unit normal n.

(a) Prove that for each $p \in M$ there exist an open neighborhood U_p of p in $\mathbb{R}^3$ and a smooth function $F : U_p \to \mathbb{R}$ such that $F^{-1}(0) = U_p \cap M$.

(b) Find F if M is the graph of a smooth function $f : \mathbb{R}^2 \to \mathbb{R}$.

(*Indiana*)

Solution.

Since the inclusion map $M \to \mathbb{R}^3$ is an embedding, for each coordinate neighborhood $V \subset M$ of $p \in M$, there exists a neighborhood U of p in $\mathbb{R}^3$, such that $V = U \cap M$. Using the local coordinates (u, v) for V and (x, y, z) for U, we can express this embedding as $(u, v) \to (x(u, v), y(u, v), z(u, v))$. Noticing that

$$\text{rank} \left(\frac{\partial(x, y, z)}{\partial(u, v)} \right) = 2,$$

we know from the implicit function theorem that there exists a neighborhood $V_p \subset V$ of p, such that on V_p,

$$\begin{cases} x = x(u, v) \\ y = y(u, v) \end{cases}$$

has smooth inverse

$$\begin{cases} u = u(x, y) \\ v = v(x, y). \end{cases}$$

Then, we can find a neighborhood $U_p \subset U$ of p in $\mathbb{R}^3$, such that $V_p = U_p \cap M$, and in terms of the local coordinates (x, y), the embedding can be expressed by $(x, y) \to (x, y, f(x, y))$, where $f(x, y) = z(u(x, y), v(x, y))$.

Therefore, setting $F : U_p \to \mathbb{R}$ by $F(x, y, z) = z - f(x, y)$, we have $F^{-1}(0) = U_p \cap M$.

(b) If M is the graph of a smooth function $f : \mathbb{R}^2 \to \mathbb{R}$, then for each $p \in M$, we can take $U_p = \mathbb{R}^3$, and $F : \mathbb{R}^3 \to \mathbb{R}$ defined by $F(x, y, z) = z - f(x, y)$.

3225

Let M be a 2-dimensional manifold smoothly embedded in $\mathbb{R}^3$ with unit normal n. Assume that M is the boundary of a bounded convex open set. Assume that n is the exterior normal and that the Gauss curvature K of M is everywhere positive. [Recall that $S \subseteq \mathbb{R}^3$ is defined to be convex if for each two points of S the line segment joining these points lies in S. You may use without proof the fact that M lies on one side of each of its tangent planes.]

(a) Define the Gauss (or sphere) map

$$\eta : M \to S^2 = \{(x, y, z) : x^2 + y^2 + z^2 = 1\}.$$

(b) Prove that η is one-to-one.

(c) Assuming that η is one-to-one, prove that η is a diffeomorphism onto S^2.

(d) Show that

$$\int_M K(p)dp = 4\pi.$$

(You may assume (b) and (c) if you wish.)

(Indiana)

Solution.

(a) For each $p \in M$, define $\eta(p)$ as the end point of the unit exterior normal $n(p)$ after parallel translating it to the origin of $\mathbb{R}^3$.

(b) and (c). We first prove that $\eta : M \to S^2$ is a local diffeomorphism. For each $p \in M$, there exist a coordinate neighborhood $\Sigma \subset M$ of p and a coordinate map $h : \Sigma \to U \subset \mathbb{R}^2$, which is a diffeomorphism, such that on U the Gauss map has the following expression

$$\eta \circ h^{-1}(u, v) = (\alpha(u, v), \beta(u, v), \gamma(u, v)) = n(u, v),$$

where $n(u, v)$ is the unit normal at $X(u, v) \in \Sigma \subset M$.

Since $n_u \times n_v = K X_u \times X_v$ and $K > 0$, $X_u \times X_v \neq 0$, the rank of the Jacobi matrix

$$\mathrm{rank} \begin{pmatrix} \alpha_u & \alpha_v \\ \beta_u & \beta_v \\ \gamma_u & \gamma_v \end{pmatrix} = 2.$$

Thus, the implicit function theorem says that $\eta \circ h^{-1}$ has a smooth inverse on a neighborhood of $h(p)$ (which may be smaller than U); hence $\eta \circ h^{-1}$ is a local diffeomorphism. Therefore, in the neighborhood of $h(p)$, $\eta = (\eta \circ h^{-1}) \circ h$ is a diffeomorphism. Thus, on M, η is a local diffeomorphism.

Next, we show that $\eta : M \to S^2$ is surjective. Since $\eta : M \to S^2$ is a local diffeomorphism, η is an open map. Besides, because M is compact and S^2 is a Hausdorff space, η is also a closed map. Thus $\eta(M)$ is an open, closed and non-empty set of S^2. The connectedness of S^2 implies that $\eta(M) = S^2$.

Now we prove that $\eta : M \to S^2$ is globally one-to-one by leading to a contradiction. Suppose that there are two distinct points $P, Q \in M$ such that $\eta(P) = \eta(Q)$. From the above argument we know that there are neighborhoods Σ_P, Σ_Q of P, Q respectively, such that $\eta|_{\Sigma_P} : \Sigma_P \to \eta(\Sigma_P), \eta|_{\Sigma_Q} : \Sigma_Q \to \eta(\Sigma_Q)$ are diffeomorphisms. Because $P \neq Q$, in M we can choose

Σ_P, Σ_Q so small that $\Sigma_P \cap \Sigma_Q = \emptyset$. Now take the inverse images of $\eta(\Sigma_P) \cap \eta(\Sigma_Q)$ under $\eta|_{\Sigma_P}$ and $\eta|_{\Sigma_Q}$, respectively. Namely, set

$$U = (\eta|_{\Sigma_P})^{-1}(\eta(\Sigma_P) \cap \eta(\Sigma_Q)),$$
$$V = (\eta|_{\Sigma_Q})^{-1}(\eta(\Sigma_P) \cap \eta(\Sigma_Q)).$$

Thus, $U \cap V = \emptyset$ and $\eta(U) = \eta(V)$, which implies $\eta(M \backslash U) = S^2$. On the other hand, it is easy to show that M is compact, connected and oriented. Then the Gauss–Bonnet formula gives

$$\int_{S^2} d\bar{\sigma} = \int_M K d\sigma = 4\pi, \quad \text{(by noting } K > 0\text{)},$$

where $d\sigma$, $d\bar{\sigma}$ have local expressions

$$d\sigma = |X_u \times X_v| du dv, \quad d\bar{\sigma} = |n_u \times n_v| du dv.$$

Hence

$$4\pi = \int_M K d\sigma = \int_{M \backslash U} K d\sigma + \int_U K d\sigma$$
$$\geq \int_{S^2} d\bar{\sigma} + \int_U K d\sigma = 4\pi + \int_U K d\sigma.$$

Since $\int_U K d\sigma > 0$, we arrive at a contradiction.

In the end, noticing that differentiability is a local property, we see that the globally one-to-one, surjective local diffeomorphism is naturally a global diffeomorphism.

Remark. In fact, this problem is the famous Hadamard Theorem. Using the theory of covering map, we can simplify its proof as follows.

Firstly, as the above, show that $\eta : M \to S^2$ is a local diffeomorphism.

Since M is compact and S^2 is connected, then the local diffeomorphism $\eta : M \to S^2$ is also a covering map.

Further, because S^2 is simply connected, and $M \subset \mathbb{R}^3$ is connected and hence path connected, we know that the covering map η must be a homeomorphism and hence a global diffeomorphism.

3226

Let M be a 2-dimensional manifold smoothly embedded in $\mathbb{R}^3$ with unit normal n.

(a) Explain what is meant by intrinsic and extrinsic quantities on M.

(b) Are the principal curvatures intrinsic?

(c) Discuss why the covariant derivative on M, defined using the covariant derivative on $\mathbb{R}^3$, is intrinsic.

(d) Assuming (c), discuss why the Gauss curvature of M is intrinsic.

(*Indiana*)

Solution.

(a) The terminology "intrinsic quantities" means those geometrical quantities that are definable only by the first fundamental form of M and its derivatives. Otherwise, they are called "extrinsic quantities". In other words, intrinsic quantities are those that are invariant under isometric correspondence, but extrinsic ones are not.

(b) The principal curvatures are not intrinsic; they are extrinsic. For example, consider a plane and a cylinder.

(c) Although the covariant derivative on M is defined by using the covariant derivative on $\mathbb{R}^3$, the last local expression of the covariant derivative of M involves the tangent vector field and the Christoffel symbols of M. Therefore, it is intrinsic.

(d) For each $p \in M$, let C be a simple closed curve encircling a simply connected domain D where p lies. Let $\Delta\omega$ denote the angle variance caused by a tangent vector after parallel translating it around C once. Using the Hopf's rotation index theorem and the Gauss–Bonnet formula, we can show that the Gauss curvature of M at p can be expressed by

$$K(p) = \lim_{D \to p} \frac{\Delta \omega}{\int \int_D d\sigma}.$$

Since parallel translation is intrinsic, then the Gauss curvature is intrinsic, too.

3227

By revolving the curve γ sketched below around the x-axis, we get a surface of revolution $M^2 \subset \mathbb{R}^3$. Compute $\int_M K dA$, where K is the Gauss curvature and dA the area form, on M. (Make sure to justify your answer.)

(Indiana)

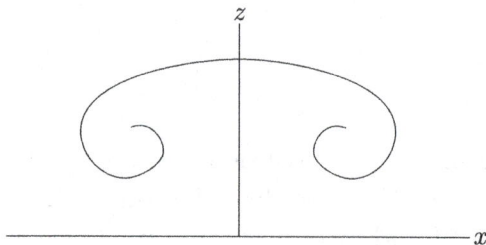

Fig. 3.8.

Solution.

The surface M^2 of revolution generated by the curve γ is homeomorphic to a section of a cylindrical surface. Thus the Euler characteristic number $\chi(M^2) = 0$. Besides, the boundary of M^2 consists of two circles which are just geodesics of M^2, because along these circles the normal vector of M^2 is parallel to the principal normal vector of the two circles respectively. Therefore, by the famous Gauss–Bonnet formula, we have immediately

$$\int_M K dA = 0.$$

3228

A surface Σ^2 immersed in a Riemannian manifold N is said to be ruled if it can be parameterized near any point by a mapping $X : (0,1)^2 \to N$ such that for each fixed $v_0 \in (0,1)$, the u-parameter curve $X(u, v_0)$ is a geodesic in N.

(a) When $n = 3$, show that the Gauss curvature of a ruled surface in $\mathbb{R}^n$ is nowhere positive.

(b) Show it for arbitrary n.

(*Indiana*)

Solution.

(a) Since every geodesic in $N = \mathbb{R}^3$ is a straight line, then Σ^2 can be characterized locally by $X(u, v) = \alpha(v) + \left(u - \frac{1}{2}\right) l(v)$, where $l(v)$ is the direction of the geodesic corresponding to $v \in (0,1)$. Thus, through computation, we can easily deduce that the first coefficient of the second fundamental form of Σ^2

$$L = \left\langle n, \frac{\partial^2 X}{\partial u^2} \right\rangle \equiv 0,$$

which shows that the Gauss curvature of the surface

$$K = \frac{LN - M^2}{EG - F^2} = -\frac{M^2}{EG - F^2} \leq 0.$$

(b) Similar to the above, suppose Σ^2 can be characterized locally by

$$X(u, v) = \alpha(v) + ul(v),$$

where, for convenience, we may assume that $v \in (0,1)$ is the arclength of the curve $\alpha(v), |l(v)| \equiv 1$, and $\langle \alpha'(u), l(v) \rangle \equiv 0$. Then, by a routine work, we see that the first fundamental form of Σ^2 is

$$ds^2 = du^2 + (1 + 2u\langle \alpha'(v), l'(v) \rangle + u^2 |l'(v)|^2) dv^2.$$

Since the Gauss equation says

$$K = -\frac{\frac{\partial^2 \sqrt{G}}{\partial u^2}}{\sqrt{G}},$$

it suffices to show that $\frac{\partial^2 \sqrt{G}}{\partial u^2} \geq 0$. However, we have

$$\frac{\partial \sqrt{G}}{\partial u} = \frac{\langle \alpha'(v), l'(v) \rangle + u|l'(v)|^2}{\sqrt{1 + 2u\langle \alpha'(v), l'(v) \rangle + u^2|l'(v)|^2}},$$

$$\frac{\partial^2 \sqrt{G}}{\partial u^2} = \frac{|l'(v)|^2 - \langle \alpha'(v), l'(v) \rangle^2}{(1 + 2u\langle \alpha'(v), l'(v) \rangle + u^2|l'(v)|^2)^{3/2}}.$$

Noting that

$$|l'(v)|^2 - \langle \alpha'(v), l'(v) \rangle^2 = |\alpha'(v) \times l'(v)|^2 \geq 0,$$

we obtain the desired result.

3229

Let $\alpha : (0,1) \to \mathbb{R}^3$ be any regular arc (that is, α is differentiable and α' is nowhere zero). Let $T(u)$, $N(u)$ and $B(u)$ be the unit tangent, normal and binormal vectors to α at $\alpha(u)$. Consider the *normal tube of radius* ε around α, that is, the surface given parametrically by $\phi(u, v) = \alpha(u) + \varepsilon \cos(v)N(u) + \varepsilon \sin(v)B(u)$.

(a) For what values of ε is this an immersion?

(b) Assuming that α itself has finite length, find the surface area of the normal tube of radius ε around α.

(*Harvard*)

Solution.

(a) we assume that $s(u)$ represents the arclength of the curve α defined in the interval $(0, u)$. By the Frenet formulas, we have that

$$\phi_u = \{[1 - \varepsilon k(u)\cos(v)]T(u) + \varepsilon \tau(u)[-\sin(v)N(u) + \cos(v)B(u)]\}\frac{ds}{du},$$

$$\phi_v = \varepsilon[-\sin(v)N(u) + \cos(v)B(u)];$$

$$\phi_u \times \phi_v = \varepsilon[1 - \varepsilon k(u)\cos(v)][-\cos(v)N(u) - \sin(v)B(u)]\frac{ds}{du}.$$

Assuming $\frac{ds}{du} > 0$, from the last equality, we know that ϕ is an immersion when $\varepsilon \neq 0$ and $1 - \varepsilon k(u) \cos(v) \neq 0$. For example, when $\varepsilon > 0$ is sufficiently small, then ϕ is an immersion.

(b) The surface area of the normal tube of radius ε around α is

$$
\begin{aligned}
\text{Area}(\phi) &= \int \int_D |\phi_u \times \phi_v| du dv \\
&= \int_0^1 du \int_0^{2\pi} \varepsilon[1 - \varepsilon k(u) \cos(v)] \frac{ds}{du} dv \\
&= \int_0^1 \left(2\pi\varepsilon \cdot \frac{ds}{du} \right) du \\
&= \int_{s(0)}^{s(1)} 2\pi\varepsilon \cdot ds = 2\pi\varepsilon[s(1) - s(0)] = 2\pi\varepsilon L,
\end{aligned}
$$

where the domain $D = (0, 1) \times (0, 2\pi)$ and L is the total length of α in (0,1).

3230

Show that the ellipsoid $x^2 + 2y^2 + 3z^2 = 1$ is not isometric to any sphere $x^2 + y^2 + z^2 = r^2$.

(Harvard)

Solution.

Firstly, we use a classical differential geometry method to prove this proposition. In the parametric domain $D = \{(u, v) | (u, v) \in (-\pi/2, \pi/2) \times (0, 2\pi)\}$, the ellipsoid defined by $x^2 + 2y^2 + 3z^2 = 1$ can be parametrized as

$$
r(u, v) = \left(\cos(u) \cos(v), \frac{1}{\sqrt{2}} \cos(u) \sin(v), \frac{1}{\sqrt{3}} \sin(u) \right).
$$

By direct calculation, we have that

$$
r_u = \left(-\sin(u) \cos(v), -\frac{1}{\sqrt{2}} \sin(u) \sin(v), \frac{1}{\sqrt{3}} \cos(u) \right),
$$

$$
r_v = \left(-\cos(u) \sin(v), \frac{1}{\sqrt{2}} \cos(u) \cos(v), 0 \right),
$$

$$r_u \times r_v = \left(-\frac{1}{\sqrt{6}} \cos^2(u) \cos(v), -\frac{1}{\sqrt{3}} \cos^2(u) \sin(v), \right.$$
$$\left. -\frac{1}{\sqrt{2}} \sin(u) \cos(u) \right);$$

$$r_{uu} = \left(-\cos(u) \cos(v), -\frac{1}{\sqrt{2}} \cos(u) \sin(v), -\frac{1}{\sqrt{3}} \sin(u) \right),$$

$$r_{uv} = \left(\sin(u) \sin(v), -\frac{1}{\sqrt{2}} \sin(u) \cos(v), 0 \right),$$

$$r_{vv} = \left(-\cos(u) \cos(v), -\frac{1}{\sqrt{2}} \cos(u) \sin(v), 0 \right).$$

By denoting $\sigma = \sqrt{\frac{1}{6} \cos^2(u) \cos^2(v) + \frac{1}{3} \cos^2(u) \sin^2(v) + \frac{1}{2} \sin^2(u)}$
and noting that the unit normal vector is

$$n(u, v) = \left(-\frac{1}{\sqrt{6}} \cos(u) \cos(v), -\frac{1}{\sqrt{3}} \cos(u) \sin(v), \right.$$
$$\left. -\frac{1}{\sqrt{2}} \sin(u) \right) \bigg/ \sigma$$

then we can obtain that

$$EG - F^2 = (r_u \times r_v) \cdot (r_u \times r_v) = \sigma^2 \cos^2(u),$$
$$L = r_{uu} \cdot n = 1/\sqrt{6}\sigma, \quad M = r_{uv} \cdot n = 0,$$
$$N = r_{vv} \cdot n = \cos^2(u)/\sqrt{6}\sigma.$$

Thus the Gaussian curvature of the ellipsoid is

$$K_e = (LN - M^2)/(EG - F^2) = 1/6\sigma^4.$$

On the other hand, the Gaussian curvature of the sphere $x^2 + y^2 + z^2 = r^2$ is $K_s = 1/r^2$. Obviously, $K_e \neq K_s$, which means that the ellipsoid cannot be isometric to any sphere, because the Gaussian curvature is an isometric invariant.

Remark. If we use Liebman's Theorem (which says that if a compact connected regular surface in $\mathbb{R}^3$ has the constant Gaussian curvature, it must be a sphere) or more general Cohn–Vossen's Theorem (which says that two isomeric ellipsoids are identical up to a orthogonal linear transformation in $\mathbb{R}^3$), then this problem is trivial.

3231

Let $(f(v)\cos(u), f(v)\sin(u), g(v))$ be a parametrization of a surface of revolution $S \subset \mathbb{R}^3$ where $(u,v) \in (0, 2\pi) \times (a,b)$. If S is given the induced metric from $\mathbb{R}^3$, prove that the following map from S to $\mathbb{R}^2$ is locally conformal where $\mathbb{R}^2$ is given the standard Euclidean metric:

$$(u,v) \mapsto \left(u, \int \frac{\sqrt{(f'(v))^2 + (g'(v))^2}}{f(v)} dv \right).$$

(Harvard)

Solution.

The surface S is obtained by rotating the curve $x = f(v)$, $z = g(v)$ around the z-axis. Without loss of generality, we assume that $f(u) > 0$. By direct computation, the first fundamental form of S with the induced metric from $\mathbb{R}^3$ is

$$I_S = f^2(v)du^2 + [(f'(v))^2 + (g'(v))^2]dv^2, \quad (u,v) \in (0, 2\pi) \times (a,b).$$

By the definition of the given map ψ from S to $\mathbb{R}^2$ defined by

$$\bar{u} = u, \quad \bar{v} = \int \frac{\sqrt{(f'(v))^2 + (g'(v))^2}}{f(v)} dv,$$

the pull-back metric by ψ^* on the surface S is

$$\psi^* I_{R^2} = (d\bar{u}^2 + d\bar{v}^2)|_\psi = du^2 + \frac{(f'(v))^2 + (g'(v))^2}{f^2(v)} dv^2 = \frac{1}{f^2(v)} I_S.$$

Therefore ψ is locally conformal.

3232

Let S be a smooth surface in $\mathbb{R}^3$ defined by $r(u, v)$, where r is the radius vector of $\mathbb{R}^3$ and (u, v) are curvilinear coordinates on S. Let H and K be respectively the mean curvature and the Gaussian curvature of S. Let A and B be respectively the supremum of the absolute value of H and K on S. Let a be a positive number and n be the unit normal vector of S. Consider the surface $\tilde{S}$ defined by $\rho(u, v) = r(u, v) + an(u, v)$. Let C be a curve in S defined by $u = u(t)$ and $v = v(t)$. Let $\tilde{C}$ be the curve in $\tilde{S}$ defined by $t \to \rho(u(t), v(t))$. Show that the length of $\tilde{C}$ is no less than the length of C multiplied by $1 - a(A + \sqrt{A^2 + 4B})$.

(Hint. Compare the first fundamental form of $\tilde{S}$ with the difference of the first fundamental form of S and $2a$ times the second fundamental form of S.)

(Harvard)

Solution.
Firstly, we know that in order to guarantee the regularity of the surface $\tilde{S}$, the positive number a should be small. Then, starting from

$$\rho_u = r_u + an_u \quad \text{and} \quad \rho_v = r_v + an_v,$$

we have, by routine calculation, that

$$\tilde{E} = \rho_u \rho_u = E - 2aL + a^2 e,$$
$$\tilde{F} = \rho_u \rho_v = F - 2aM + a^2 f,$$
$$\tilde{G} = \rho_v \rho_v = G - 2aN + a^2 g;$$

namely, $\tilde{I} = I - 2aII + a^2 III$, where I, II, III are the three fundamental forms of S and $\tilde{I}$ is the first fundamental form of $\tilde{S}$. Because the third fundamental form III of the surface S is always non-negative, the inequality $\tilde{I} \geq I - 2aII$ always holds.

Secondly, by the definition of the normal curvature $\kappa_n = \kappa_n(p, dr)$ at point p of S and in the direction of $dr = r_u du + r_v dv$, we have that

$$\kappa_n = (L du^2 + 2M dudv + N dv^2)/(E du^2 + 2F dudv + G dv^2) = II/I.$$

Thus, using Euler's formula for normal curvatures, we can express

$$I - 2aII = (1 - 2a\kappa_n)I = \{1 - 2a[\kappa_1 \cos^2(\theta) + \kappa_2 \sin^2(\theta)]\} \cdot I.$$

The principal curvatures κ_1, κ_2 can be obtained by using the relations $\kappa_1 + \kappa_2 = H$ and $\kappa_1 \kappa_2 = K$. In other words, κ_1, κ_2 are two real roots of the following quadratic equation $\kappa^2 - H\kappa + K = 0$, because $H^2 - 4K \geq 0$. Therefore,

$$\kappa_1 = \frac{H + \sqrt{H^2 - 4K}}{2}, \quad \kappa_2 = \frac{H - \sqrt{H^2 - 4K}}{2}.$$

Thus

$$|\kappa_i| \leq \frac{A + \sqrt{A^2 + 4B}}{2}, \quad i = 1, 2.$$

Hence

$$\tilde{I} \geq I - 2aII = (1 - 2a\kappa_n)I$$
$$= \{1 - 2a[\kappa_1 \cos^2(\theta) + \kappa_2 \sin^2(\theta)]\} \cdot I$$
$$\geq [1 - 2a(A + \sqrt{A^2 + 4B}] \cdot I$$

Noticing that a is a small positive constant, the length of the curve $\tilde{C}$ from t_0 to t_1 is

$$L_{\tilde{C}} = \int_{t_0}^{t_1} \frac{\sqrt{\tilde{I}}}{dt} dt \geq \int_{t_0}^{t_1} \sqrt{1 - 2a(A + \sqrt{A^2 + 4B}} \frac{\sqrt{I}}{dt} dt$$
$$\approx L_C[1 - a(A + \sqrt{A^2 + 4B})],$$

where L_C is the length of the curve C.

3233

Let $u \mapsto \rho(u)$ be a smooth curve in $\mathbb{R}^3$. Let S be the surface in $\mathbb{R}^3$ defined by $(u, v) \mapsto \rho(u) + v\rho'(u)$, where $\rho'(u)$ means the first-order derivative of $\rho(u)$ with respect to u. Assume that the two vectors $\rho'(u)$ and $\rho'(u) + v\rho''(u)$ are $\mathbb{R}$-linearly independent at the point $(u, v) = (u_0, v_0)$, where $\rho''(u)$ means the second-order derivative of $\rho(u)$ with respect to u. Verify directly from the definition of Gaussian curvature that the Gaussian curvature of S is zero at the point $(u, v) = (u_0, v_0)$.

(Harvard)

Solution.

We discuss the problem in a neighborhood of (u_0, v_0), where $\rho'(u)$ and $\rho'(u) + v\rho''(u)$ are $\mathbb{R}$-linearly independent. Let us derive some relations related to the curve $\rho(u)$. If we denote by $s = s(u)$ the arclength of $\rho(u)$, and let $T(u), N(u), B(u)$ be the unit tangent, normal and binormal vectors of the curve $\rho(u)$ respectively, then the Frenet formulas give that

$$\rho'(u) = T(u)s'(u), \quad \rho''(u) = k(u)N(u)[s'(u)]^2 + T(u)s''(u).$$

Therefore,

$$\rho'(u) \times \rho''(u) = k(u)[s'(u)]^3 B(u).$$

Then, for the surface $S : r(u, v) = \rho(u) + v\rho'(u)$, we have

$$r_u(u, v) = \rho'(u) + v\rho''(u), r_v(u, v) = \rho'(u).$$

Hence

$$r_u(u, v) \times r_v(u, v) = -vk(u)[s'(u)]^3 B(u),$$

which shows that we can choose the unit normal vector $n(u, v) = B(u)$ for the surface S, and $n_u(u, v) = B'(u) = -\tau(u)s'(u)N(u)$, $n_v(u, v) = 0$. Now we can obtain some vanishing coefficients of the second fundamental form of S as follows.

$$M(u, v) = -r_v(u, v) \cdot n_u(u, v) = 0,$$
$$N(u, v) = -r_v(u, v) \cdot n_v(u, v) = 0.$$

Thus, in the neighborhood of (u_0, v_0),

$$K(u, v) = \frac{L(u, v)N(u, v) - M^2(u, v)}{E(u, v)G(u, v) - F^2(u, v)} = 0.$$

3234

Let $M^2 \in \mathbb{R}^3$ be an embedded oriented surface and let S^2 be the unit sphere. The Gauss map $G : M \to S^2$ is defined to be $G(x) = N(x)$ for any $x \in M$, where $N(x)$ is the unit normal vector of M at x. Let h and g denote the induced Riemannian metrics on M and S^2 from $\mathbb{R}^3$ respectively. Prove that if the mean curvature of M is zero everywhere, then the Gauss map G is a conformal map from (M, h) to (S^2, g). (Recall: If (Σ_1, g_1), (Σ_2, g_2) are two Riemannian manifolds, a map $\phi : (\Sigma_1, g_1) \to (\Sigma_2, g_2)$ is called *conformal* if $g_1 = \lambda \phi^* g_2$ for some scalar function λ on Σ_1.)

(Harvard)

Solution.

For an arbitrary point $p \in M^2$, if p is not an umbilical point of the surface M, then in a domain D of p, there exists a curvature line net, which we take as the (u, v) coordinate curves net. We assume that κ_1 and κ_2 are the principal curvatures associated with u-curves and v-curves, respectively. Thus, in the domain D, M can be parametrized as $X = X(u, v)$. Let $N = N(u, v)$ denote the unit normal vector of M at the point $p(u, v)$.

Then, by the Rodriques equation, we have that $N_u = -\kappa_1 X_u$, $N_v = -\kappa_2 X_v$. If we denote the metric of M by $h_M = dX \cdot dX = h_{11} du^2 + h_{22} dv^2$, then the third fundamental form of M, i.e., the pull-back of the metric of S^2 to M by the Gauss map G, is

$$G^* g_{s^2} = dN \cdot dN = (N_u du + N_v dv)^2 = (-\kappa_1 X_u du - \kappa_2 X_v dv)^2$$
$$= \kappa_1^2 h_{11} du^2 + \kappa_2^2 h_{22} dv^2.$$

By the assumption that the mean curvature $H = (\kappa_1+\kappa_2)/2 = 0$ for M, we see that $\kappa_1^2 = \kappa_2^2$. Hence $G^*g_{s^2} = \kappa_1^2(h_{11}du^2+h_{22}dv^2) = \kappa_1^2 h_M$. Because $\kappa_1 \neq 0$ follows from the fact that p is not an umbilical point, we get $h_M = \kappa_1^{-2}G^*g_{s^2}$.

If p is an umbilical point, then the normal curvature κ_n at p is independent of directions, which means that, at the point p, $\Omega_{ij} = \kappa_n h_{ij}$. Then, by using the fundamental formulas for surfaces, we have

$$dN = N_u du + N_v dv = -\kappa_n X_u du - \kappa_n X_v dv.$$

Therefore, we can still obtain that $G^*g_{s^2} = \kappa_n^2(h_{11}du^2 + h_{22}dv^2) = \kappa_n^2 h_M$ is valid at the point p, and we can also claim that $\kappa_n \neq 0$, otherwise, the image $G(M)$ is not regular at point $G(p)$. Hence $h_M = \kappa_n^{-2}G^*g_{s^2}$. Thus we complete the proof for that G is a conformal map from (M,h) to (S^2, g).

Remark. Note that there is another simpler way to solve this problem. Using the following relation $III - 2HII + KI = 0$ among the three fundamental forms of the surface M, now we have $G^*g_{s^2} - 2H \cdot II + K \cdot h_M = 0$. Hence, if M is minimal, then we have $K \cdot h_M = G^*g_{s^2}$.

3235

Let $f : [a, b] \to \mathbb{R}_+$ be smooth, and let $X \subset \mathbb{R}^3$ be the surface of revolution formed by revolving $z = f(x)$ about x-axis.

(a) For which function f is the Gaussian curvature of X

- always positive?
- always negative?
- identically zero?

(b) Characterize by a differential equation those functions f such that X is a minimal surface.

(*Harvard*)

Solution.

(a) By the assumption, the surface of rotation X can be parametrized as

$$r(x, \theta) = (x, f(x)\cos(\theta), f(x)\sin(\theta)). \quad (f(x) > 0)$$

Thus, by taking derivatives, we get that

$$r_x = (1, f'(x)\cos(\theta), f'(x)\sin(\theta)),$$
$$r_\theta = (0, -f(x)\sin(\theta), f(x)\cos(\theta)).$$

From $r_x \times r_\theta = (f(x)f'(x), -f(x)\cos(\theta), -f(x)\sin(\theta))$, the unit normal vector of X is

$$n(x, \theta) = \frac{r_x \times r_\theta}{|r_x \times r_\theta|} = \frac{(f'(x), -\cos(\theta), -\sin(\theta))}{\sqrt{1 + (f'(x))^2}}.$$

By taking the second-order derivatives, we have that

$$r_{xx} = (0, f''(x)\cos(\theta), f''(x)\sin(\theta)),$$
$$r_{x\theta} = (0, -f'(x)\sin(\theta), f'(x)\cos(\theta)),$$
$$r_{\theta\theta} = (0, -f(x)\cos(\theta), -f(x)\sin(\theta)).$$

Then, the coefficients of the first and second fundamental forms of X are

$$E = 1 + (f'(x))^2, \quad F = 0, \quad G = f^2(x);$$
$$L = -\frac{f''(x)}{\sqrt{1 + (f'(x))^2}}, \quad M = 0, \quad N = \frac{f(x)}{\sqrt{1 + (f'(x))^2}}.$$

Therefore, the Gaussian curvature of the surface X is

$$K = \frac{LN - M^2}{EG - F^2} = -\frac{f''(x)}{f(x)[1 + (f'(x))^2]^2}.$$

- In order that $K > 0$, we need $f''(x) < 0$, which means that $z = f(x)$ is a convex function of x.
- In order that $K < 0$, we need $f''(x) > 0$, which means that $z = f(x)$ is a concave function of x.

• In order that $K = 0$, we need $f''(x) = 0$, which means that $z = ax + b$. Hence, the surface of rotation is either a cone, or a cylinder with x-axis as its symmetric axis.

(b) In order that X is a minimal surface, noticing that $F = M = 0$, we need

$$\frac{L}{E} + \frac{N}{G} = -\frac{f''(x)}{[1 + (f'(x))^2]^{3/2}} + \frac{1}{[1 + (f'(x))^2]^{1/2} f(x)} = 0.$$

It is equivalent to $f''(x)f(x) = 1 + (f'(x))^2$ the solution of which is $f(x) = a \cosh(\frac{x}{a} + b)$. Hence the surface of revolution x is a catenoid.

3236

Let γ be a geodesic curve on a regular surface of revolution $S \subset \mathbb{R}^3$. Let $\theta(p)$ denote the angle the curve forms with the parallel at a point $p \in \gamma$ and $r(p)$ be the distance to the axis of revolution. Prove Clairaut's relation: $r \cos \theta = \text{const.}$

(*Harvard*)

Solution.

Assume that the distance $r = r(p)$ is parametrized by the arclength s of γ, denoted by $r(s)$ for simplicity, and the surface of rotation S is obtained by rotating the curve $x = r(s), z = s$ about the z-axis. Then the surface can be parametrized via

$$X(u, s) = (r(s) \cos(u), r(s) \sin(u), s),$$

where without loss of generality we assume that $0 < u < \pi/2$. By direct calculation, we have that

$$I_S = r^2(s) du^2 + [1 + (r'(s))^2] ds^2, \text{ or}$$
$$E = r^2(s), \ F = 0, \ G = 1 + (r'(s))^2.$$

Because $\gamma \subset S$, γ can be parametrized with some $u = u(s)$ by

$$X(u(s), s) = (r(s) \cos(u(s)), \ r(s) \sin(u(s)), s).$$

Noting the definition of $\theta(p)$, along γ, we have

$$\frac{dX(u(s),s)}{ds} = X_u\frac{du}{ds} + X_s\frac{ds}{ds} = \sqrt{E}\frac{du}{ds}e_1 + \sqrt{G}e_2$$
$$= \cos(\theta)e_1 + \sin(\theta)e_2,$$

where e_1, e_2 are the unit vectors at p along the directions of X_u, X_s respectively. Thus we get

$$\sqrt{E}\frac{du(s)}{ds} = \cos(\theta), \quad \sqrt{G} = \sin(\theta).$$

Now the Liouville formula for geodesic curvatures

$$\kappa_g = \frac{d\theta}{ds} - \frac{1}{2\sqrt{G}}\frac{\partial \ln E}{\partial s}\cos(\theta) + \frac{1}{2\sqrt{E}}\frac{\partial \ln G}{\partial u}\sin(\theta)$$

gives that, for the geodesic γ,

$$\frac{d\theta}{ds} = \frac{1}{2\sqrt{G}}\frac{d\ln E}{ds}\cos(\theta),$$

which, by $\sqrt{G} = \sin(\theta)$ derived above, is equivalent to

$$\frac{d\theta}{ds} = \frac{1}{2\sin(\theta)}\cdot\frac{2r}{r^2}\frac{dr}{ds}\cos(\theta), \quad \text{namely,} \quad -d\ln\cos(\theta) = d\ln r.$$

Therefore, Clairaut's relation $r(s)\cos(\theta) = \text{const}$ holds along the geodesic γ.

3237

Let $F(x, y, z)$ be a smooth homogenous function of degree k, i.e., $F(\lambda x, \lambda y, \lambda z) = \lambda^k F(x, y, z)$. Prove that away from the origin, the induced metric on the conical surface $\Sigma = \{(x, y, z)|F(x, y, z) = 0\}$ has Gaussian curvature equal to 0.

(Harvard)

Solution.

In order to guarantee the regularity of the surface Σ, we always assume that the gradient $\nabla F = (F_x, F_y, F_z) \neq 0$. From the fact that, on the surface Σ,

$$dF = F_x dx + F_y dy + F_z dz = \nabla F \cdot dr = 0,$$

we know that the unit normal vector n at every point of Σ is parallel to ∇F. By using the Euler Theorem related to homogeneous functions of order k, for $F(x, y, z)$, we have that $r \cdot \nabla F = kF = 0$ holds on Σ, i.e., the radius vector r of Σ is orthogonal to the unit normal vector n everywhere.

For $p \in \Sigma$, we assume that $r(u, v) = (x(u, v), y(u, v), z(u, v))$ is a locally parametrized equation of Σ near the point p. Differentiating the obtained equality $r(u, v) \cdot n(u, v) = 0$ with respect to u and v, respectively, we get that

$$r_u \cdot n + r \cdot n_u = 0, \quad r_v \cdot n + r \cdot n_v = 0.$$

Thus, we get $r \cdot n_u = 0$, $r \cdot n_v = 0$ as well as $r \cdot n = 0$. Because $r \neq 0$, it implies that

$$(n, n_u, n_v) = n \cdot (n_u \times n_v) = 0.$$

Noticing that n is parallel to $r_u \times r_v$, we have $(r_u \times r_v, n_u, n_v) = 0$. By the formulas $(r_u \times r_v, n_u, n_v) = (r_u \times r_v) \cdot (n_u \times n_v) = (r_u \cdot n_u)(r_v \cdot n_v) - (r_u \cdot n_v)(r_v \cdot n_u) = LN - M^2$, we eventually obtain the Gaussian curvature

$$K = \frac{LN - M^2}{EG - F^2} = 0.$$

3238

Let $X \subset \mathbb{R}^3$ be the cone $x^2 = y^2 + z^2$, and let Y be the torus $(\sqrt{x^2 + y^2} - 2)^2 + z^2 = 1$, that is, the torus obtained by rotating the circle $(x - 2)^2 + z^2 - 1 = y = 0$ around the z-axis.

(a) Show that for any point $p \in X$ other that the vertex $(0, 0, 0)$, there is a neighborhood of p in X isometric to an open subset of the Euclidean plane $\mathbb{R}^2$.

(b) Show that no open subset of Y is isometric to any open subset of the Euclidean plane.

(*Harvard*)

Solution.

(a) Let $F(x, y, z) = x^2 - y^2 - z^2$. Hence $F(\lambda x, \lambda y, \lambda z) = \lambda^2 F(x, y, z)$. Then F is a smooth homogeneous function of degree 2 and $X \subset \mathbb{R}^3$ is defined by $F(x, y, z) = 0$. Therefore, by the solution to the previous problem, we know that away from the origin, the Gaussian curvature $K = 0$. For any $p \in S$, except the origin, we take a geodesic polar coordinate system (ρ, θ) around p, in which

$$I = d\rho^2 + G d\theta^2 \quad \text{and} \quad \lim_{\rho \to 0} \sqrt{G} = 0, \ \lim_{\rho \to 0} (\sqrt{G})_\rho = 1.$$

By the Gauss equation, now we have that

$$K = -\frac{1}{\sqrt{EG}} \left\{ \left[\frac{(\sqrt{E})_\theta}{\sqrt{G}} \right]_\theta + \left[\frac{(\sqrt{G})_\rho}{\sqrt{E}} \right]_\rho \right\} = -\frac{(\sqrt{G})_{\rho\rho}}{\sqrt{G}} = 0.$$

From the last equation, it follows that $(\sqrt{G})_{\rho\rho} = 0$. Integrating with respect to ρ, we get $(\sqrt{G})_\rho = f(\theta)$. The property $\lim_{\rho \to 0} (\sqrt{G})_\rho = 1$ shows that $f(\theta) = 1$. Integrating $(\sqrt{G})_\rho = 1$ again with respect to ρ gives $\sqrt{G} = \rho + g(\theta)$. Then $\lim_{\rho \to 0} \sqrt{G} = 0$ implies that $g(\theta) = 0$. Therefore, near the point p, $I = d\rho^2 + \rho^2 d\theta^2$, which shows that there is a neighborhood of p isometric to an open set including the origin on the Euclidean plane $\mathbb{R}^2$, furnished with the polar coordinates.

(b) By the definition, Y is obtained by rotating the circle $(x - 2)^2 + z^2 = 1$, $y = 0$ around the z-axis. Thus Y can be parametrized via

$$r(u, v) = ((2 + \cos(u)) \cos(v), (2 + \cos(u)) \sin(v), \sin(u)).$$

By taking derivatives, we get that

$$r_u = (-\sin(u) \cos(v), -\sin(u) \sin(v), \cos(u),$$
$$r_v = (-(2 + \cos(u)) \sin(v), (2 + \cos(u)) \cos(v), 0);$$

$$r_{uu} = (-\cos(u)\cos(v), -\cos(u)\sin(v), -\sin(u)),$$

$$r_{uv} = (\sin(u)\sin(v), -\sin(u)\cos(v), 0),$$

$$r_{vv} = (-(2+\cos(u))\cos(v), -(2+\cos(u))\sin(v), 0).$$

Therefore, we have that $E = 1$, $F = 0$, $G = (2+\cos(u))^2$; and also

$$L = (r_u, r_v, r_{uu})/\sqrt{EG - F^2} = 1,$$

$$M = (r_u, r_v, r_{uv})/\sqrt{EG - F^2} = 0,$$

$$N = (r_u, r_v, r_{vv})/\sqrt{EG - F^2} = \cos(u)(2+\cos(u)).$$

Hence we obtain the Gaussian curvature

$$K = (LN - M^2)/(EG - F^2) = \cos(u)/(2+\cos(u)).$$

Therefore, when $u = \pi/2$ or $3\pi/2, K = 0$; when $\pi/2 < u < 3\pi/2, K < 0$; and when $3\pi/2 < u < 2\pi$ or $0 \le u < \pi/2, K > 0$. We know that two surfaces are locally isometric if and only if they have the same Gaussian curvature at the corresponding points. Now that, on the torus Y, only at highest and lowest meridians the Gaussian curvature vanishes, hence there is no open subset of the torus Y which is isometric to any open subset of the Euclidean plane $\mathbb{R}^2$.

3239

Let $\Sigma \subset \mathbb{R}^3$ be a smooth 2-dimensional submanifold, and $n : \Sigma \to \mathbb{R}^3$ a smooth map such that $n(p)$ is a unit length normal to Σ at p. Identify the tangent bundle $T\Sigma$ as the subspace of pairs $(p, v) \in \Sigma \times \mathbb{R}^3$ such that $v * n(p) = 0$, where $*$ designates the Euclidean inner product. Suppose now the $t \to p(t)$ is a smoothly parametrized curve in $\mathbb{R}^3$ that lies on Σ. Prove that this curve is a geodesic if and only if

$$p''(t) * (n(p(t)) \times p'(t)) = 0, \forall t.$$

Here, p' is the derivative of the map $t \to p(t)$ and p'' is the second derivative.

(*Harvard*)

Solution.

We recall some relations related to the curve $\gamma : t \to p(t)$. Let $s = s(t)$ be the arclength of γ. By the Frenet formulas, we have that

$$p'(t) = T(s(t))\frac{ds}{dt}, \quad p''(t) = k(s(t))N(s(t))\left(\frac{ds}{dt}\right)^2 + T(s(t))\frac{d^2s}{dt^2},$$

$$p'(t) \times p''(t) = k(s(t))\left(\frac{ds}{dt}\right)^3 B(s(t)).$$

Hence, the equation $p''(t) * (n(p(t)) \times p'(t)) = 0, \forall t$ can be rewritten as

$$\begin{aligned} p''(t) * (n(p(t)) \times p'(t)) &= (p''(t), n(p(t)), p'(t)) \\ &= (n(p(t)), p'(t), p''(t)) \\ &= k(s(t))\left(\frac{ds}{dt}\right)^3 [n(p(t)) * B(s(t))] = 0, \quad \forall t. \end{aligned} \tag{3.1}$$

If γ is a geodesic, then $n((p(t)) = \pm N(s(t))$, hence Eqs. (3.1) hold.

Conversely, if (3.1) hold, then under the condition $k(s(t))\left(\frac{ds}{dt}\right) \neq 0$, we have that

$$n(p(t)) * B(s(t)) = 0.$$

In addition, we also have $n(p(t)) * T(s(t)) = 0$. Thus we can see that the vector

$$n(p(t)) = \pm B(s(t)) \times T(s(t)) = \pm N(s(t))$$

which shows that γ is a geodesic. The remaining problem is what happens if $k(s(t))\left(\frac{ds}{dt}\right) = 0$. Generally speaking, we always assume that $\frac{ds}{dt} \neq 0$, then $k(s(t)) = 0$. If $k(s(t)) = 0, t \in (t_1, t_2)$, then this portion of γ is a line segment, hence is still geodesic; if $k(s(t_0)) = 0$ only for a fixed $t = t_0$, then by the relation $k^2 = \kappa_g^2 + \kappa_n^2$, we also have that $\kappa_g(s(t_0)) = 0$, where κ_g and κ_n are the geodesic and normal curvatures of γ at $s(t_0)$, respectively.

Section 3

Differential Geometry
of Manifolds

3301

Let M^n be a Riemannian manifold and π a 2-plane in T_*M.

(i) Let $\{X, Y\}$ be an orthonormal basis of π. Use this basis to define the sectional curvature $K(\pi)$ and show that it depends only on π and not the particular basis chosen.

(ii) Define the Riemann curvature tensor in terms of covariant differentiation. Explain why it is a tensor.

(iii) Recall that if $M^n \subset \mathbb{R}^{n+1}$ is a hypersurface, then

$$R(X, Y)Z = \langle LY, Z \rangle LX - \langle LX, Z \rangle LY,$$

where L is the Weingarten map. Using this or any other valid method, compute all sectional curvatures of the sphere $\{x \in \mathbb{R}^4 : |x| = 3\}$.

(Indiana)

Solution.

(i) The sectional curvature is defined by $K(\pi) = -R(X, Y, X, Y)$, where $R(\ )$ is the Riemannian curvature tensor field of M^n. If $\{\tilde{X}, \tilde{Y}\}$ is another orthonormal basis of π, then $\tilde{X} = aX + bY$, $\tilde{Y} = cX + dY$. Noticing that $\begin{pmatrix} a & b \\ c & d \end{pmatrix}$ is an orthogonal matrix, and $R(\ )$ is a 4th order covariant tensor field, we can easily have $R(\tilde{X}, \tilde{Y}, \tilde{X}, \tilde{Y}) = R(X, Y, X, Y)$, which proves that $K(\pi)$ depends only on π.

(ii) For any vector fields X, Y, Z, W on M^n, define the Riemannian curvature operator by $R(X, Y)Z = \nabla_X \nabla_Y Z - \nabla_Y \nabla_X Z - \nabla_{[X,Y]} Z$ and the Riemannian curvature tensor field by $R(X, Y, Z, W) = \langle R(X, Y)Z, W \rangle$, respectively. Then for any C^∞ function f on M^n, using the properties of covariant differentiation and of the inner product, we can conclude from straightforward computation that

$$R(fX, Y, Z, W) = R(X, fY, Z, W) = R(X, Y, fZ, W)$$
$$= R(X, Y, Z, fW) = fR(X, Y, Z, W).$$

Thus, in terms of local coordinate frame field, one can show that $\forall p \in M^n$, $R(X, Y, Z, W)|_p$ is dependent only on $X_p, Y_p, Z_p, W_p \in T_p(M^n)$. Therefore, $R(\ )|_p$ is a well-defined 4th order multilinear function. Besides, the inner product and covariant differentiation are all C^∞. Hence, $R(\ )$ thus defined is a C^∞ tensor field.

(iii) For the sphere $M^n = \{x \in \mathbb{R}^4, |x| = 3\}$, we can regard the position vector field x as the normal vector field of M^n. Let $\{e_i\}$ be a local orthonormal frame field about x (as a point) of M^n. Then by the original definition of covariant differentiation, one can easily show that $\tilde{\nabla}_{e_i} x = e_i$, where $\tilde{\nabla}$ denotes the covariant differentiation in $\mathbb{R}^4$. Hence, the Weingarten map is $L : X \mapsto L(X) = -\tilde{\nabla}_X \frac{x}{3} = -X/3$. Therefore, if $\{X, Y\}$ is any orthonormal basis of an arbitrary 2-plane π in $T_p(M^n)$,

$$K(\pi) = -\langle R(X, Y)X, Y \rangle = \langle LX, X \rangle \langle LY, Y \rangle$$
$$-\langle LY, X \rangle \langle LX, Y \rangle = 1/9.$$

3302

Let $C = \{x \in \mathbb{R}^3 : 0 \leq x_i \leq 1\}$ be the unit cube in $\mathbb{R}^3$. Suppose $F : \mathbb{R}^3 \to \mathbb{R}^4$ is a $1 - 1\ C^\infty$ immersion in some neighborhood of C. The image $F(C)$ is then a compact Riemannian submanifold of $\mathbb{R}^4$ with boundary and therefore has a volume. Justify the following formula:

$$\text{vol}(F(C)) = \int_0^1 \int_0^1 \int_0^1 \left\| F_* \left(\frac{\partial}{\partial x_1} \right) \wedge F_* \left(\frac{\partial}{\partial x_2} \right) \right.$$

$$\left. \wedge F_* \left(\frac{\partial}{\partial x_3} \right) \right\| dx_1 dx_2 dx_3.$$

Also, evaluate the integrand if

$$F(x) = ((1 + x_1)^2, (1 + x_2)^2, (1 + x_3)^2, (2 + x_3)^3).$$

(*Indiana*)

Solution.

Consider the following four points in $F(C)$

$$P_1 = F(x_1, x_2, x_3), \qquad P_2 = F(x_1 + \Delta x_1, x_2, x_3),$$
$$P_3 = F(x_1, x_2 + \Delta x_2, x_3), \qquad P_4 = F(x_1, x_2, x_3 + \Delta x_3),$$

where $|\Delta x_1|, |\Delta x_2|, |\Delta x_3|$ are sufficiently small. Construct three vectors as follows

$$\overrightarrow{P_2 P_1} = F(x_1 + \Delta x_1, x_2, x_3) - F(x_1, x_2, x_3) = \frac{\partial F}{\partial x_1} \Delta x_1 + \cdots,$$

$$\overrightarrow{P_3 P_1} = F(x_1, x_2 + \Delta x_2, x_3) - F(x_1, x_2, x_3) = \frac{\partial F}{\partial x_2} \Delta x_2 + \cdots,$$

$$\overrightarrow{P_4 P_1} = F(x_1, x_2, x_3 + \Delta x_3) - F(x_1, x_2, x_3) = \frac{\partial F}{\partial x_3} \Delta x_3 + \cdots.$$

Let $\Delta \sigma$ be the volume of the parallelopiped spanned by the vectors $\overrightarrow{P_2 P_1}$, $\overrightarrow{P_3 P_1}$ and $\overrightarrow{P_4 P_1}$ in $\mathbb{R}^4$. Then

$$\Delta \sigma = \| \overrightarrow{P_2 P_1} \wedge \overrightarrow{P_3 P_1} \wedge \overrightarrow{P_4 P_1} \|$$

$$= \left\| F_* \left(\frac{\partial}{\partial x_1} \right) \wedge F_* \left(\frac{\partial}{\partial x_2} \right) \wedge F_* \left(\frac{\partial}{\partial x_3} \right) \Delta_{x_1} \Delta_{x_2} \Delta_{x_3} + \cdots \right\|$$

is an infinitesimal when $\Delta x_1 \to 0$, $\Delta x_2 \to 0$, $\Delta x_3 \to 0$. We take the principal part of the infinitesimal $\Delta \sigma$ as the volume element, namely, the volume element of $F(C)$ is

$$d\sigma = \left\| F_* \left(\frac{\partial}{\partial x_1} \right) \wedge F_* \left(\frac{\partial}{\partial x_2} \right) \wedge F_* \left(\frac{\partial}{\partial x_3} \right) \right\| dx_1 dx_2 dx_3.$$

Hence the desired formula follows immediately.

Furthermore, if $F(x) = ((1+x_1)^2, (1+x_2)^2, (1+x_3)^2, (2+x_3)^3)$, then, by denoting $F(x) = F(x_1, x_2, x_3) = (y_1, y_2, y_3, y_4)$, we have that

$$F_* \left(\frac{\partial}{\partial x_1} \right) = 2(1 + x_1) \frac{\partial}{\partial y_1},$$

$$F_* \left(\frac{\partial}{\partial x_2} \right) = 2(1 + x_2) \frac{\partial}{\partial y_2},$$

$$F_* \left(\frac{\partial}{\partial x_3} \right) = 2(1 + x_3) \frac{\partial}{\partial y_3} + 3(2 + x_3)^2 \frac{\partial}{\partial y_4}.$$

Since $\frac{\partial}{\partial y_1}, \frac{\partial}{\partial y_2}$ and

$$\left[2(1 + x_3) \frac{\partial}{\partial y_3} + 3(2 + x_3)^2 \frac{\partial}{\partial y_4} \right] \Big/ \sqrt{4(1 + x_3)^2 + 9(2 + x_3)^4}$$

form an orthonormal frame, we obtain that

$$\left\| F_* \left(\frac{\partial}{\partial x_1} \right) \wedge F_* \left(\frac{\partial}{\partial x_2} \right) \wedge F_* \left(\frac{\partial}{\partial x_3} \right) \right\|$$

$$= 4(1 + x_1)(1 + x_2) \sqrt{4(1 + x_3)^2 + 9(2 + x_3)^4}.$$

3303

There is no submersion from S^3 into $\mathbb{R}^2$.

(*Indiana*)

Solution.

Let $S^3 \subset \mathbb{R}^4$ be defined by $\{(x^1, x^2, x^3, x^4) \in \mathbb{R}^4; (x^1)^2 + (x^2)^2 + (x^3)^2 + (x^4)^2 = 1\}$. Observe the following three vector fields

$$X_1 = x^2 \frac{\partial}{\partial x^1} - x^1 \frac{\partial}{\partial x^2} + x^4 \frac{\partial}{\partial x^3} - x^3 \frac{\partial}{\partial x^4},$$

$$X_2 = x^4 \frac{\partial}{\partial x^1} + x^3 \frac{\partial}{\partial x^2} - x^2 \frac{\partial}{\partial x^3} - x^1 \frac{\partial}{\partial x^4},$$

$$X_3 = -x^3 \frac{\partial}{\partial x^1} + x^4 \frac{\partial}{\partial x^2} + x^1 \frac{\partial}{\partial x^3} - x^2 \frac{\partial}{\partial x^4},$$

where $(x^1, x^2, x^3, x^4) \in S^3$. It is easy to see that they are nowhere-vanishing tangent vector fields on S^3. Since $(x^1)^2 + (x^2)^2 + (x^3)^2 + (x^4)^2 = 1$, then, without loss of generality, we may assume $x^4 \neq 0$. Hence, from the fact that

$$\det \begin{pmatrix} x^2 & -x^1 & x^4 \\ x^4 & x^3 & -x^2 \\ -x^3 & x^4 & x^1 \end{pmatrix} = (x^4)^3 + x^4(x^3)^2 + x^4(x^2)^2 + x^4(x^1)^2$$

$$= x^4 \neq 0,$$

we know that X_1, X_2, X_3 are three linearly independent vector fields. Furthermore, by direct calculation, we obtain

$$[X_1, X_2] = 2X_3, \quad [X_2, X_3] = 2X_1, \quad [X_3, X_1] = 2X_2.$$

Now we suppose that there is such a submersion $\pi : S^3 \to \mathbb{R}^2$. Then, $\pi_* : T_p(S^3) \to T_{\pi(p)}(\mathbb{R}^2)$ is a surjection for every point $p \in S^3$. Thus, we may assume that, for example, $\pi_* X_2, \pi_* X_3$ are linearly independent, and $\pi_* X_1 = a\pi_* X_2 + b\pi_* X_3$ at p. Then, on the one hand,

$$[\pi_* X_1, \pi_* X_2] = b[\pi_* X_3, \pi_* X_2] = b\pi_*[X_3, X_2]$$

$$= -2b\pi_* X_1 = -2ab\pi_* X_2 - 2b^2 \pi_* X_3;$$

on the other hand,

$$[\pi_* X_1, \pi_* X_2] = \pi_*[X_1, X_2] = 2\pi_* X_3.$$

Therefore,

$$\pi_* X_3 = -ab\pi_* X_2 - b^2 \pi_* X_3$$

from which we get $ab = 0$ and $-b^2 = 1$. Similarly, from $[X_3, X_1] = 2X_2$ we can deduce that $-a^2 = 1$ and $ab = 0$. They are all contradictory. So, there is no submersion from S^3 into $\mathbb{R}^2$.

3304

Let M^n and N^k be Riemannian manifolds. Then $M \times N$ is naturally a Riemannian manifold with the product metric. If $x_1, \ldots, x_n$ are local coordinates on M and $y_1, \ldots, y_k$ are local coordinates on N, the product metric in local coordinates $(x_1, \ldots, x_n, y_1, \ldots, y_k)$ looks like

$$g = \left(\begin{array}{c|c} g_{ij}(x) & 0 \\ \hline 0 & g_{pq}(y) \end{array} \right).$$

(i) Let X be a vector field on $M \times N$ "along" M (i.e., in local coordinates no y_p and $\frac{\partial}{\partial y_q}$ are present) and Y be a vector field along N. Show that $D_X Y \equiv 0$.

(ii) Show that at $z \in M \times N$, some sectional curvatures always vanish. (For example, product manifolds in the product metric never have strictly positive curvature.)

(*Indiana*)

Solution.

(i) Using the properties of the Riemannian connection on $M \times N$, we only need to verify $D_{\frac{\partial}{\partial x_i}} \frac{\partial}{\partial y_p} = 0$. In fact, observing that the inverse matrix of g is of the form

$$g^{-1} = \left(\begin{array}{c|c} g^{ij}(x) & 0 \\ \hline 0 & g^{pq}(y) \end{array} \right),$$

and the first class of Christoffel symbols of $M \times N$ satisfy

$$\Gamma_{ip,h} = \frac{1}{2} \left(\frac{\partial g_{ih}}{\partial y_p} + \frac{\partial g_{ph}}{\partial x_i} - \frac{\partial g_{ip}}{\partial x_h} \right) = 0,$$

$$\Gamma_{ip,r} = \frac{1}{2} \left(\frac{\partial g_{ir}}{\partial y_p} + \frac{\partial g_{pr}}{\partial x_i} - \frac{\partial g_{ip}}{\partial y_r} \right) = 0,$$

we easily conclude that

$$D_{\frac{\partial}{\partial x_i}} \frac{\partial}{\partial y_p} = \sum_j \Gamma^j_{ip} \frac{\partial}{\partial x_j} + \sum_q \Gamma^q_{ip} \frac{\partial}{\partial y_q}$$

$$= \sum_{j,h} g^{jh} \Gamma_{ip,h} \frac{\partial}{\partial x_j} + \sum_{q,r} g^{qr} \Gamma_{ip,r} \frac{\partial}{\partial y_q} = 0.$$

Obviously, we also have $D_{\frac{\partial}{\partial y_p}} \frac{\partial}{\partial x_i} = 0$.

(ii) By the definition of the curvature operator, using the result of (i), we have

$$R\left(\frac{\partial}{\partial x_i}, \frac{\partial}{\partial y_p}\right) \frac{\partial}{\partial x_h} = 0, \quad R\left(\frac{\partial}{\partial x_i}, \frac{\partial}{\partial y_p}\right) \frac{\partial}{\partial y_q} = 0.$$

This means that $R(\frac{\partial}{\partial x_i}, \frac{\partial}{\partial y_p}) = 0$, Hence, for arbitrary X along M and Y along N, using the $C^\infty(M \times N)$-linearity of the curvature operator, we have

$$R(X, Y, X, Y) = 0.$$

Therefore, the sectional curvature determined by X and Y always vanishes. Obviously, here X may involve y_p, and Y may involve x_i.

3305

Let $F : M \to N$ be smooth, X and Y be smooth vector fields on M and N, respectively, and assume $F_* X = Y$; that is, that

$$F_{*p}(X(p)) = Y(F(p)) \quad \text{for } p \in M.$$

(a) Let ω be a smooth 1-form on N. Define the Lie derivative $L_Y \omega$ of ω with respect to Y. (If you use local coordinates, you must verify independence of choice of coordinates.)

(b) Prove that

$$F^*(L_Y \omega) = L_X(F^* \omega).$$

(c) Let Z be a smooth vector field on M such that $F_*Z = W$, where W is a smooth vector field on N. Show that $L_Y W = F_*(L_X Z)$.

(Indiana)

Solution.

(a) $L_Y \omega$ is defined by $L_Y \omega = d(Y \lfloor \omega) + Y \lfloor d\omega$, where the symbol "$\lfloor$" denotes the interior product of a vector field with a form, for example, if Ω is a p-form, then $Y \lfloor \Omega$ is a $(p-1)$-form defined by

$$(Y \lfloor \Omega)(Y_1, \ldots, Y_{p-1}) = \Omega(Y, Y_1, \ldots, Y_{p-1}).$$

Thus, $L_Y \omega$ is a well-defined 1-form on N.

(b) Let Z be a smooth vector field on M such that $F_*Z = W$, where W is a smooth vector field on N. Then we have

$$
\begin{aligned}
F^*(L_Y \omega)(Z) &= (F^*(d(Y \lfloor \omega) + Y \lfloor d\omega))(Z) \\
&= (dF^*(Y \lfloor \omega))(Z) + (F^*(Y \lfloor d\omega))(Z) \\
&= (dF^*(\omega(Y)))(Z) + (Y \lfloor d\omega)(F_*Z) \\
&= (d\omega(F_*X))(Z) + d\omega(F_*X, F_*Z) \\
&= (d(F^*\omega)(X))(Z) + (F^*d\omega)(X, Z) \\
&= (d(X \lfloor F^*\omega))(Z) + dF^*\omega(X, Z) \\
&= (d(X \lfloor F^*\omega) + X \lfloor dF^*\omega)(Z) \\
&= (L_X F^*\omega)(Z).
\end{aligned}
$$

(c) Noticing that $L_Y W = [Y, W]$ and $[F_*X, F_*Z] = F_*[X, Z]$, we immediately obtain $L_Y W = F_*(L_X Z)$.

3306

Let r, θ, z be the usual cylindrical coordinates in $\mathbb{R}^3$. Let $\omega = [2rz \sin\theta + 3z^2 r \cos\theta + 5r^2 \sin^2\theta]d\theta + [-2r\cos\theta + 6zr\sin\theta]dz + [4zr^2 \sin\theta]dr$. Let the curve γ be given in rectangular coordinates by

$$\gamma(t) = (x(t), y(t), z(t)) = (\cos t, \sin t, 4\sin^5 t + \sin^2 t \cos^8 t),$$
$$0 \leq t \leq 2\pi.$$

Evaluate $\int_r \omega$.

(*Indiana*)

Solution.

Let a function f be defined by

$$f(r, \theta, z) = -2rz\cos\theta + 3z^2 r\sin\theta, \quad (r, \theta, z) \in \mathbb{R}^3.$$

Then f is a C^∞ function in $\mathbb{R}^3$ and we have

$$df = [2rz\sin\theta + 3z^2 r\cos\theta]d\theta + [-2r\cos\theta + 6zr\sin\theta]dz$$
$$+ [-2z\cos\theta + 3z^2\sin\theta]dr.$$

It is easy to see that $f|_\gamma$ is also a C^∞ function, and by the invariance of the form of first order differentiation, the above expression of df is an exact 1-form on the closed curve γ. Noticing that γ is a closed curve, we have, by using the Stokes' Theorem,

$$\int_\gamma \omega = \int_\gamma (\omega - df) = \int_\gamma 5r^2 \sin^2\theta d\theta$$
$$+ (4zr^2\sin\theta + 2z\cos\theta - 3z^2\sin\theta)dr.$$

Without loss of generality, we may assume $0 \leq \theta < 2\pi$ and $r = 1$. Then

$$\int_\gamma \omega = \int_0^{2\pi} 5\sin^2\theta d\theta = 5\int_0^{2\pi} \frac{1 - \cos 2\theta}{2} d\theta = 5\pi.$$

3307

Let M be a C^∞ Riemannian manifold. Assume the theorem that there is a unique C^∞ mapping $\nabla : \mathcal{X}(M) \times \mathcal{X}(M) \to \mathcal{X}(M)$ denoted by $\nabla : (X, Y) \to \nabla_X Y$ which has the following linearity properties: For all $f, g \in C^\infty(M)$ and $X, X', Y, Y' \in \mathcal{X}(M)$, we have

$$\nabla_{fX+gX'}Y = f(\nabla_X Y) + g(\nabla_{X'}Y),$$
$$\nabla_X(fY + gY') = f\nabla_X Y + g\nabla_X Y' + (Xf)Y + (Xg)Y',$$
$$[X,Y] = \nabla_X Y - \nabla_Y X,$$
$$X\langle Y, Y'\rangle = \langle \nabla_X Y, Y'\rangle + \langle Y, \nabla_X Y'\rangle.$$

(a) Suppose $\alpha : [a, b] \to M$ is a smooth curve. Define what it means for a vector field Y on α to be parallel along α. Derive the differential equations that must be satisfied if Y is parallel along α.

(b) Let $X, Y \in \mathcal{X}(M)$. Let $p \in M$ and let $\alpha : [0, b] \to M$ be an integral curve of X such that $\alpha(0) = p$, $d\alpha/dt = X(\alpha(t))$. Show that

$$(\nabla_X Y)(p) = \frac{d}{dt}[P_{\alpha;0,t}^{-1}Y(\alpha(t))]|_{t=0},$$

where $P_{\alpha;0,t} : T_{\alpha(0)}M \to T_{\alpha(t)}M$ is the parallel transport along α.

(*Indiana*)

Solution.

(a) As known, thus defined mapping ∇ is the Riemannian connection. Using the properties of ∇, we can prove that $\nabla_{\frac{d\alpha}{dt}}Y$ is completely determined by $\alpha(t)$ and $Y(\alpha(t))$. So $\nabla_{\frac{d\alpha}{dt}}Y$ is well-defined for every vector field along α.

Now we give a definition that a vector field Y on α is parallel along α, if and only if $\nabla_{\frac{d\alpha}{dt}}Y = 0$, $\forall t \in [a, b]$. In order to derive the differential equations, we choose a coordinate neighborhood with local coordinates $\{x^i\}$. Then

$$\frac{d\alpha}{dt} = \sum_i \frac{d\alpha^i(t)}{dt}\frac{\partial}{\partial x^i}(\alpha(t)), \quad Y(t) = \sum_i Y^i(t)\frac{\partial}{\partial x^i}(\alpha(t)).$$

Denoting

$$\nabla_{\frac{\partial}{\partial x^i}}\frac{\partial}{\partial x^j} = \sum_k \Gamma_{ij}^k \frac{\partial}{\partial x^k},$$

we have, by the properties of ∇,

$$\nabla_{\frac{d\alpha}{dt}} Y = \sum_j \nabla_{\frac{d\alpha}{dt}} \left(Y^j(t) \frac{\partial}{\partial x^j}(\alpha(t)) \right)$$

$$= \sum_j \frac{dY^j(t)}{dt} \frac{\partial}{\partial x^j}(\alpha(t)) + \sum_j Y^j(t) \nabla_{\frac{d\alpha}{dt}} \frac{\partial}{\partial x^j}(\alpha(t))$$

$$= \sum_k \left[\frac{dY^k(t)}{dt} + \sum_{ij} (\Gamma_{ij}^k \circ \alpha) \frac{d\alpha^i(t)}{dt} Y^j(t) \right] \frac{\partial}{\partial x^k}(\alpha(t)).$$

Therefore, the desired differential equations are

$$\frac{dY^k(t)}{dt} + \sum_{i,j} (\Gamma_{ij}^k \circ \alpha) \frac{d\alpha^i(t)}{dt} Y^j(t) = 0, \quad \forall k.$$

(b) Choose a basis $\{e_1, \ldots, e_n\}$ of $T_p(M)$, where $n = \dim M$. Let

$$e_i(t) = P_{\alpha;0,t}(e_i).$$

Since $P_{\alpha;0,t}$ is a parallel isomorphism for every t, then $\{e_1(t), \ldots, e_n(t)\}$ is a basis of $T_{\alpha(t)}(M)$. Hence we can denote

$$Y(\alpha(t)) = \sum_i Y^i(t) e_i(t).$$

Then, by the properties of the connection ∇ and by the fact that $e_i(t)$ is parallel along $\alpha(t)$, $i = 1, \ldots, n$, we immediately have

$$(\nabla_X Y)(p) = \nabla_{\frac{d\alpha}{dt}(0)} Y = \sum_i \frac{dY^i}{dt}(0) e_i.$$

On the other hand, because

$$P_{\alpha;0,t}(e^i) = e^i(t)$$

is equivalent to

$$e_i = P_{\alpha;0,t}^{-1}(e_i(t)),$$

we can write

$$P_{\alpha;0,t}^{-1}(Y(\alpha(t))) = \sum_i Y^i(t)e_i.$$

Therefore,

$$\frac{d}{dt}[P_{\alpha;0,t}^{-1}(Y(\alpha(t)))]_{t=0} = \sum_i \frac{dY^i}{dt}(0)e_i = (\nabla_X Y)(p).$$

3308

Let M be a Riemannian manifold with the property that given any two points $p, q \in M$ the parallel transport of a vector from $T_p M$ to $T_q M$ is independent of the curve joining p and q. Prove that the curvature of M is identically zero, i.e., $R(X, Y)Z = 0$ for all $X, Y, Z \in \mathcal{X}(M)$.

(*Indiana*)

Solution.

For an arbitrary point $p \in M$, take a coordinate neighborhood D of p with local coordinates $x^1, \ldots, x^n$. Let

$$V_{pi} = \sum_j v_{i|}^j \frac{\partial}{\partial x^j}(x^p)$$

be n linearly independent vectors in $T_p M$. Using the hypothesis that the parallel transport of a vector from $T_p M$ to $T_q M$ is independent of the curve joining p and q, then for $q \in D$, we can transport every V_{pi} from $T_p M$ to $T_q M$. Thus, we can define n linearly independent vector fields $V_1, \ldots, V_n$ in D. Obviously, all V_i's are well-defined, and if we transport along special coordinate curves, then

$$0 = \nabla_{\frac{\partial}{\partial x^j}} V_j = \sum_k \left(\frac{\partial v_{i|}^k(x)}{\partial x^j} + \sum_l \Gamma_{jl}^k v_{i|}^l(x) \right) \frac{\partial}{\partial x^k}(x)$$

$$\stackrel{\text{def.}}{=} \sum_k v_{i|,j}^k(x) \frac{\partial}{\partial x^k}(x),$$

namely

$$v^k_{i|,j}(x) = 0 \quad i, j, k = 1, \ldots, n.$$

Now, by the Ricci identity, we have

$$v^k_{i|,jm}(x) - v^k_{i|,mj}(x) = -\sum_l v^l_{i|}(x) R^k_{ljm}(x) = 0.$$

Noting the linear independence of V_i's, we immediately have, in D,

$$R^k_{ljm} = 0 \quad k, l, j, m = 1, \ldots, n,$$

i.e., the curvature of M is identically zero.

3309

Let M be an n-dimensional Riemannian manifold. Suppose there are n orthonormal vector fields $X_1, \ldots, X_n$ that commute with each other (i.e., $[X_i, X_j] = 0$; $i, j = 1, \ldots, n$), show that the sectional curvature of M is identically zero.

(*Indiana*)

Solution.
Set

$$\nabla_{X_i} X_j = \sum_k \Gamma^k_{ij} X_k.$$

Then, from $\langle X_i, X_j \rangle = \delta_{ij}$ it follows that

$$\langle \nabla_{X_k} X_i, X_j \rangle + \langle X_i, \nabla_{X_k} X_j \rangle = 0,$$

i.e., $\Gamma^j_{ki} + \Gamma^i_{kj} = 0$; $i, j, k = 1, \ldots, n$. On the other hand, from $[X_i, X_j] = 0$ it follows that

$$\nabla_{X_i} X_j - \nabla_{X_j} X_i = [X_i, X_j] = 0,$$

i.e., $\Gamma^k_{ij} - \Gamma^k_{ji} = 0$; $i, j, k = 1, \ldots, n$. Namely, the Christoffel symbols Γ^k_{ij}'s are antisymmetric with respect to k, j, and symmetric with respect to i, j. Therefore,

$$\Gamma^k_{ij} = -\Gamma^j_{ik} = -\Gamma^j_{ki} = \Gamma^i_{kj} = \Gamma^i_{jk} = -\Gamma^k_{ji} = -\Gamma^k_{ij},$$

which means $\Gamma_{ij}^k = 0$; $i, j, k = 1, \ldots, n$. Hence the sectional curvature of M is identically zero.

3310

Let X be a smooth vector field on a Riemannian manifold M. The divergence of X, denoted by $\text{div}(X)$, is defined by the function trace (∇X).

(i) If M is closed (i.e., compact without boundary), show that

$$\int_M \text{div}(X) dv = 0.$$

(ii) If M is compact with boundary ∂M, show that

$$\int_M \text{div}(X) dv = \int_{\partial M} \langle X, N \rangle ds,$$

where N is the outer normal vector field of ∂M.

(Hint. Consider $\omega \in \Lambda^1(M)$ defined by $\omega(Y) = \langle X, Y \rangle$, and try to use the Stokes' Theorem.)

(Indiana)

Solution.

In fact, this problem is the famous Green Theorem. The outline of the proof is as follows.

Firstly, one can show that, for example, by means of the normal coordinate system about $p \in M$, thus defined $\text{div}(X)$ satisfies $\text{div}(X)\Omega_M = d(i(X)\Omega_M)$, where Ω_M is the volume form of M and $i(\ )$ is the interior product operator.

Next, for $p \in \partial M$, choose an oriented orthonomal frame field about p $\{e_1, \ldots, e_n\}$ such that, at p, $e_1 = N_p$. Let $\{\omega^1, \ldots, \omega^n\}$ be the dual frame field of $\{e_1, \ldots, e_n\}$. Then the volume elements of M and ∂M are respectively

$$dv = \Omega_M(p) = \omega^1 \wedge \cdots \wedge \omega^n, \quad ds = \Omega_{\partial M}(p) = \omega^2 \wedge \cdots \wedge \omega^n.$$

Observing that, at p,

$$i(X)\Omega_m = i(X)(\omega^1 \wedge \cdots \wedge \omega^n)$$
$$= \omega^1(X)\omega^2 \wedge \cdots \wedge \omega^n + (\text{terms involve } \omega^1)$$
$$= \langle X, N \rangle \Omega_{\partial M} + (\text{terms involve } \omega^1),$$

and along $\partial M \, \omega^1 \equiv 0$, one can obtain, by Stokes' Theorem,

$$\int_M \operatorname{div}(X) dv = \int_M d(i(X)\Omega_M)$$
$$= \int_{\partial M} \langle X, N \rangle \Omega_{\partial M} = \int_{\partial M} \langle X, N \rangle ds.$$

If M is compact without boundary, the right-hand side of the above formula vanishes naturally.

3311

Let $w^1, \ldots, w^k$ be one-forms. Show that $\{w^i\}_{i=1}^k$ are linearly independent if and only if $w^1 \wedge w^2 \wedge \cdots \wedge w^k \neq 0$.

(Indiana)

Solution.

Let $w^1, \ldots, w^k$ be defined on an n-dimensional manifold with $n \geq k$.

Suppose $w^1 \wedge \cdots \wedge w^k \neq 0$. If $w^1, \ldots, w^k$ are linearly dependent, then without loss of generality, we may suppose that

$$w^k = a_1 w^1 + \cdots + a_{k-1} w^{k-1}$$

with suitable functions $a_1, \ldots, a_{k-1}$. Thus we have

$$w^1 \wedge \cdots \wedge w^k = w^1 \wedge \cdots \wedge w^{k-1} \wedge (a_1 w^1 + \cdots + a_{k-1} w^{k-1}) = 0,$$

which contradicts the above hypothesis.

On the contrary, if $w^1, \ldots, w^k$ are linearly independent, then we can extend them to a basis $\{w^1, \ldots, w^k, w^{k+1}, \ldots, w^n\}$. Thus

$$w^1 \wedge \cdots \wedge w^k \wedge w^{k+1} \wedge \cdots \wedge w^n \neq 0$$

implies that

$$w^1 \wedge \cdots \wedge w^k \neq 0.$$

3312

Let M be a Riemannian manifold. Let $p \in M$.

(a) Show that there exists $\delta > 0$ such that

$$\exp_p : B_\delta(0) \subset T_pM \to M$$

is a diffeomorphism onto its image.

(b) Show that there exists $\varepsilon > 0$ such that $\exp_p(B_\varepsilon(0))$ is a convex set.

(Hint. Let $d(x) =$ distance from x to p. Show that d^2 is convex in a neighborhood of p.)

(*Indiana*)

Solution.

(a) This is just the existence of the normal neighborhood of p. Use the fact that $d\exp_p$ is non-singular at p, and then the implicit function theorem.

(b) The existence of convex neighborhoods is a classic result due to J. H. C. Whitehead. Refer to every standard textbook on differential geometry.

3313

Let

$$A = \left\{ \begin{pmatrix} a & b \\ c & d \end{pmatrix} : a, b, c, d \in \mathbb{R}, ad - bc = 1 \right\}.$$

Show that

(i) A is a differentiable manifold.

(ii) A is a Lie group with the standard matrix multiplication as a product.

(*Indiana*)

Solution.

Define a map $F : Gl(2,\mathbb{R}) \to Gl(1,\mathbb{R})$ by $F(X) = \det X$. Then F is a smooth homomorphism between Lie groups, and the rank of F is constant. Therefore, the kernel of F

$$\ker F = F^{-1}(1) = A$$

is a closed regular submanifold of $Gl(2,\mathbb{R})$ and thus a Lie group.

3314

(a) Let f be a smooth function on a Riemannian manifold M. Let grad f be the vector field defined by the equation

$$\langle \text{grad } f, v \rangle_p = d_p f v, \quad v \in T_p M.$$

Let $(x^1, \ldots, x^n)$ be local coordinates around p. Find the expression for grad f in terms of $x^1, \ldots, x^n$.

(b) For a vector field X define the divergence of X, $\text{div}(X)$ as the trace of the operator $Y \to D_Y X$ where D is the Levi–Civita connection. Find the expression for the divergence of X in a local coordinate system $(x^1, \ldots, x^n)$.

(c) Use (a) and (b) to find the expression for the Laplacian Δ in local coordinates, where Δ acting on a smooth function f is defined by $\Delta f = \text{div}(\nabla f)$.

(*Indiana*)

Solution.

For convenience, we omit the suffix p.

(a) Let

$$\text{grad } f = \sum_l a^l \frac{\partial}{\partial x^l}.$$

Then, from

$$\left\langle \text{grad } f, \frac{\partial}{\partial x^k} \right\rangle = \sum_l a^l g_{lk} = df\left(\frac{\partial}{\partial x^k}\right) = \frac{\partial f}{\partial x^k}$$

it follows that

$$a^l = \sum_k g^{lk} \frac{\partial f}{\partial x^k}.$$

Thus,

$$\operatorname{grad} f = \sum_{k,l} g^{lk} \frac{\partial f}{\partial x^k} \frac{\partial}{\partial x^l}.$$

(b) Denote

$$X = \sum_i X^i \frac{\partial}{\partial x^i}.$$

Then, by the definition of divergence, we have

$$\operatorname{div}(X) = \sum_{k,l} \left\langle D_{\frac{\partial}{\partial x^k}} X, \frac{\partial}{\partial x^l} \right\rangle g^{kl} = \sum_i \left(\frac{\partial X^i}{\partial x^i} + \sum_k \Gamma_{ki}^i X^k \right).$$

(c)

$$\Delta f = \operatorname{div}(\operatorname{grad} f) = \sum_i \left[\frac{\partial}{\partial x^i} \left(\sum_k g^{ik} \frac{\partial f}{\partial x^k} \right) + \sum_{k,l} \Gamma_{ki}^i g^{kl} \frac{\partial f}{\partial x^l} \right]$$

$$= \sum_{i,k} g^{ik} \left(\frac{\partial^2 f}{\partial x^i \partial x^k} - \sum_m \Gamma_{ik}^m \frac{\partial f}{\partial x^m} \right).$$

3315

Let M be a compact connected Riemannian manifold without boundary. Let f be a smooth function satisfying $\Delta f = 0$. Show that $f = \operatorname{const}$.

(Hint. Use the definitions in the previous problem to show that $\operatorname{div}(f \nabla f) = |\nabla f|^2 + f \Delta f$.)

(*Indiana*)

Solution.

For any function f, by straight calculation, we have

$$\operatorname{div}(f \nabla f) = |\nabla f|^2 + f \Delta f.$$

Now, noticing the hypothesis of this problem, by Green's Theorem we have

$$\int_M |\nabla f|^2 dv = 0,$$

which implies $df = 0$ everywhere, that is, $f = \text{const}$.

3316

Suppose $F : M \to N$ is a smooth map between differentiable manifolds, and is homotopically trivial. Show that in this case, $F^*\omega$ will be exact whenever ω is a closed 1-form on N. (Note. F is homotopically trivial if it extends to a smooth mapping $\bar{F} : M \times [0, 1] \to N$ such that $\bar{F}(x, 0) = F(x)$, for all $x \in M$, while $\bar{F}(x, 1) \equiv q \in N$ (q constant) for all x.)

(*Indiana*)

Solution.
Consider the map $G : M \to \{q\}$ and the inclusion $i : \{q\} \to N$. Then the map $F : M \to N$ is homotopic to the composition $i \circ G$. Furthermore, we know that F^* and $(i \circ G)^* = G^* \circ i^*$ induce the same homomorphism

$$F^{**} = (i \circ G)^{**} : H^1(N, d) \to H^1(M, d).$$

Since $i^*(Z^1(N, d)) \subset Z^1(\{q\}, d) = 0$, the induced homomorphism $F^{**} = (i \circ G)^{**}$ is a zero homomorphism, i.e., $F^*(H^1(N, d)) = 0$. In other words, for every $\omega \in Z^1(N, d)$, $F^*\omega \in B^1(M, d)$, namely, $F^*\omega$ is exact.

3317

Regard $\mathbb{R}^9$ as the space of all 3×3 matrices with real entries. Does the subset

$$\Sigma = \{A \in \mathbb{R}^9 : \det(A) = 0\}$$

form a smooth submanifold of $\mathbb{R}^9$?

(*Indiana*)

Solution.

Observe that $\{A \in \mathbb{R}^9 : \text{rank } A \leq 1\}$, the union of all axis in $\mathbb{R}^9$, is closed. Then $M := \mathbb{R}^9 \backslash \{A \in \mathbb{R}^9 : \text{rank } A \leq 1\}$ is an open submanifold of $\mathbb{R}^9$. Define a map $F : M \to \mathbb{R}^1$ by $F(A) = \det(A)$, $A \in M$. Noting that $dF = (A_{11}, A_{12}, A_{13}, A_{21}, A_{22}, A_{23}, A_{31}, A_{32}, A_{33})$ where A_{ij} is the algebraic complement of the corresponding entry a_{ij} of A, we see that $\text{rank } dF = 1$ on M. Therefore, $F^{-1}(0) = \{A \in \mathbb{R}^9 : \det(A) = 0\}$ is a closed regular submanifold of M. Hence, it is also a submanifold of $\mathbb{R}^9$.

3318

Let

$$\omega = x_1 dx_2 \wedge dx_3 \wedge dx_4 + x_2 dx_1 \wedge dx_3 \wedge dx_4$$
$$+ x_3 dx_1 \wedge dx_2 \wedge dx_4.$$

Compute $\int_{S^3} \omega$, where $S^3 = \{x \in \mathbb{R}^4 : |x| = 1\}$, oriented as the boundary of the unit ball (assume standard orientation on $\mathbb{R}^4$).

(*Indiana*)

Solution.

Denote $D^4 = \{x \in \mathbb{R}^4 : |x| \leq 1\}$. Then, by the Stokes Theorem we have

$$\int_{S^3} \omega = \int_{\partial D^4} \omega = \int_{D^4} d\omega$$

$$= \int_{D^4} (dx_1 \wedge dx_2 \wedge dx_3 \wedge dx_4$$
$$+ dx_2 \wedge dx_1 \wedge dx_3 \wedge dx_4$$
$$+ dx_3 \wedge dx_1 \wedge dx_2 \wedge dx_4)$$

$$= \int_{D^4} dx_1 \wedge dx_2 \wedge dx_3 \wedge dx_4 = \text{vol}(D^4) = \frac{\pi^2}{2}.$$

3319

Prove that

$$\left\{ (x_1, x_2, x_3, x_4) : \text{rank}\left(\begin{pmatrix} x_1 & x_2 \\ x_3 & x_4 \end{pmatrix} \right) = 1 \right\}$$

is a three-dimensional submanifold of $\mathbb{R}^4$.

(*Indiana*)

Solution.
Define a map F from $\mathbb{R}^4 \backslash \{(0,0,0,0)\}$ into $\mathbb{R}^1$ by

$$F(x_1, x_2, x_3, x_4) = \begin{vmatrix} x_1 & x_2 \\ x_3 & x_4 \end{vmatrix} = x_1 x_4 - x_2 x_3.$$

Then F is a C^∞ map with $(x_4 - x_3 - x_2 \ x_1)$ as its Jacobi matrix which has constant rank 1 on $\mathbb{R}^4 \backslash \{(0,0,0,0)\}$. Thus

$$F^{-1}(0) = \left\{ (x_1, x_2, x_3, x_4) : \text{rank}\left(\begin{pmatrix} x_1 & x_2 \\ x_3 & x_4 \end{pmatrix} \right) = 1 \right\}$$

is a regular submanifold of $\mathbb{R}^4 \backslash \{(0,0,0,0)\}$ with dimension 3. Noting that $\mathbb{R}^4 \backslash \{(0,0,0,0)\}$ is an open submanifold of $\mathbb{R}^4$, we see that

$$\left\{ (x_1, x_2, x_3, x_4) : \text{rank}\left(\begin{pmatrix} x_1 & x_2 \\ x_3 & x_4 \end{pmatrix} \right) = 1 \right\}$$

is also a three-dimensional submanifold of $\mathbb{R}^4$.

3320

Let vector fields X_1, X_2 on $\mathbb{R}^4$ be defined by

$$X_1 = \frac{\partial}{\partial x_2} + x_1 \frac{\partial}{\partial x_3}, \quad X_2 = \frac{\partial}{\partial x_1} + x_2 \frac{\partial}{\partial x_4}.$$

(i) Is there a 2-dimensional submanifold M^2 of $\mathbb{R}^4$ such that for each $p \in M^2$, $X_1(p), X_2(p) \in T_p M^2$?

(ii) Is there a non-constant function f in the neighborhood of $0 \in \mathbb{R}^4$ such that $X_1 f \equiv 0$ and $X_2 f \equiv 0$?

(Indiana)

Solution.

(i) No, there is not. The reason is that the bracket $[X_1, X_2] = \frac{\partial}{\partial x_4} - \frac{\partial}{\partial x_3}$ does not satisfy the Frobenius condition.

(ii) Yes, for example, we can set $f = x_1 x_2 - (x_3 + x_4)$.

3321

Let $M = \{(x, y) : x, y \in \mathbb{R}^3, \|x\| = 1, \|y\| = 1, \langle x, y \rangle = 0\}$.

(i) Show that M is a smooth compact embedded submanifold of $\mathbb{R}^6$ and explain how M can be identified with the unit tangent bundle of S^2.

(ii) Show that M is orientable.

(Indiana)

Solution.

(i) Identify $\mathbb{R}^6$ with $\{(x, y) : x, y \in \mathbb{R}^3\}$ and define a map $F : \mathbb{R}^6 \to \mathbb{R}^3$ by

$$F(x, y) = (f_1, f_2, f_3) = (\|x\|^2, \|y\|^2, \langle x, y \rangle).$$

It is easy to verify that the Jacobi matrix of the C^∞ map F

$$\left(\frac{\partial(f_1, f_2, f_3)}{\partial(x_1, x_2, x_3, y_1, y_2, y_3)} \right) = \begin{pmatrix} 2x_1 & 2x_2 & 2x_3 & 0 & 0 & 0 \\ 0 & 0 & 0 & 2y_1 & 2y_2 & 2y_3 \\ y_1 & y_2 & y_3 & x_1 & x_2 & x_3 \end{pmatrix}$$

has constant rank three when $f_1 = \|x\|^2 = 1$, $f_2 = \|y\|^2 = 1$ and $f_3 = \langle x, y \rangle = 0$. Therefore, $F^{-1}(1, 1, 0) = M$ is a closed embedded submanifold of $\mathbb{R}^6$. Besides, since $\|(x, y)\| = (\|x\|^2 + \|y\|^2)^{\frac{1}{2}} = \sqrt{2}$ which means M is bounded in $\mathbb{R}^6$, M must be compact.

Naturally, for every $(x, y) \in M$, if we regard $x \in S^2$ and $y \in T_x(S^2)$ with $\|y\| = 1$, then we can identify M with the unit tangent bundle of S^2.

(ii) Because S^2 is orientable, S^2 has a covering $\{(U_\alpha, \phi_\alpha)\}$ of coherently oriented coordinate neighborhoods. By using identification of M with the unit tangent bundle of S^2, for every $(x, y) \in M$, x has the local coordinates $(u_\alpha^1, u_\alpha^2) \in U_\alpha$, and y is uniquely determined by the oriented angle θ_α at x from $\frac{\partial}{\partial u_\alpha^1}$ to the unit tangent vector $y \in T_x(S^2)$. Thus, $(U_\alpha \times I_\alpha, \phi_\alpha \times \psi_\alpha)$ is a coordinate neighborhood of $(x, y) \in M$, where $I_\alpha = (\theta_\alpha - \varepsilon, \theta_\alpha + \varepsilon)$ with ε being a suitable positive real number and ψ_α is the map from (x, y) to θ_α.

If $(U_\alpha \times I_\alpha, \phi_\alpha \times \psi_\alpha) \cap (U_\beta \times I_\beta, \phi_\beta \times \psi_\beta) \neq \emptyset$, then the transition function has the following Jacobian

$$\frac{\partial(u_\alpha^1, u_\alpha^2, \theta_\alpha)}{\partial(u_\beta^1, u_\beta^2, \theta_\beta)} = \begin{vmatrix} \dfrac{\partial(u_\alpha^1, u_\alpha^2)}{\partial(u_\beta^1, u_\beta^2)} & 0 \\ * & 1 \end{vmatrix} = \frac{\partial(u_\alpha^1, u_\alpha^2)}{\partial(u_\beta^1, u_\beta^2)}$$

which means that $\{(U_\alpha \times I_\alpha, \phi_\alpha \times \psi_\alpha)\}$ form a covering of coherently oriented neighborhoods. Hence, M is orientable.

3322

Let $F : M \to N$ be a local isometry between connected Riemannian manifolds M and N. Show that if M is complete, so is N and F is a covering map.

(*Indiana*)

Solution.

Because F is a local isometry, $F(M)$ is open in N. If y is a limit point of $F(M)$ in N, then there is a point $x \in M$ such that there exists a geodesic in N connecting $F(x)$ and y. The local isometry of F and the completeness of M imply that the above geodesic can be uniquely lifted to a geodesic in M starting from x, and the image of its end point under F must be y.

Therefore, $F(M)$ is also closed in N. Thus, the connectedness of N implies that $F(M) = N$, i.e., F is surjective. Besides, since F maps every geodesic of M into a geodesic of N, the Hopf–Rinow Theorem means that N is complete, too.

Next, we show that F is a covering map. For every $x' \in N$, take $\delta > 0$ so small that $\exp_{x'} : B'(\delta) \to B'_\delta$ is a diffeomorphism, where $B'(\delta) = \{v \in T_{x'}(N) : |v| < \delta\}$ and $B'_\delta = \{y' \in N : d(y', x') < \delta\}$. Since F is a local isometry, $F^{-1}(x')$ is discrete. Denote $F^{-1}(x') = \{x_\alpha\} \subset M$ and set $B^\alpha(\delta) = \{v \in T_{x_\alpha}(M) : |v| < \delta\}$ and $B^\alpha_\delta = \{y \in M : d(y, x_\alpha) < \delta\}$. Then, we claim that B'_δ is an admissible neighborhood of x' and F is a covering map.

Firstly, we claim that $F^{-1}(B'_\delta) = \bigcup_\alpha B^\alpha_\delta$. In fact, if $z \in F^{-1}(B'_\delta)$, then there is a unique geodesic $\gamma : [0, 1] \to B'_\delta$ such that $\gamma(0) = F(z)$ and $\gamma(1) = x'$. Since F is a local isometry, there exists a geodesic $\tilde{\gamma} : [0, 1] \to M$ such that $F(\tilde{\gamma}(t)) = \gamma(t)$, $\forall t$. Hence $F(\tilde{\gamma}(1)) = x'$ and $\tilde{\gamma}(1) = x_\alpha$ for some α. Besides, $L(\tilde{\gamma}) = L(\gamma) < \delta$ means that $z = \tilde{\gamma}(0) \in B^\alpha_\delta$, i.e., $F^{-1}(B'_\delta) \subset \bigcup_\alpha B^\alpha_\delta$. On the other hand, $\bigcup_\alpha B^\alpha_\delta \subset F^{-1}(B'_\delta)$ is obvious. Thus the claim is proved.

Secondly, we say that for any α, $F : B^\alpha_\delta \to B'_\delta$ is a diffeomorphism. Since M is complete, we have the following commutative diagram

$$
\begin{array}{ccc}
B^a(\delta) & \xrightarrow{\ dF\ } & B'(\delta) \\
\Big\downarrow{\scriptstyle \exp_{x_a}} & & \Big\downarrow{\scriptstyle \exp_{x'}} \\
B^a_\delta & \xrightarrow{\hspace{2cm}} & B'_\delta
\end{array}
$$

and $\exp_{x_\alpha}$ is surjective. Besides, we know that dF, $\exp_{x'}$ are diffeomorphisms. Therefore, $F \circ \exp_{x_\alpha} = \exp_{x'} \circ dF$ is a diffeomorphism and hence $\exp_{x_\alpha}$ is an immersion. So, $\exp_{x_\alpha}$ is also a diffeomorphism. Hence $F = \exp_{x'} \circ dF \circ (\exp_{x_\alpha})^{-1}$ is a diffeomorphism.

Thirdly, we claim that if $\alpha \neq \beta$, then $B_\delta^\alpha \cap B_\delta^\beta = \emptyset$. Otherwise, if there is $z \in B_\delta^\alpha \cap B_\delta^\beta$, then there exist unique geodesics γ_α, γ_β connecting z with x_α, x_β respectively. Let γ be the unique geodesic in B_δ' connecting $F(z)$ and x'. Because both $F : B_\delta^\alpha \to B_\delta'$ and $F : B_\delta^\beta \to B_\delta'$ are isometric, we have $F(\gamma_\alpha) = \gamma = F(\gamma_\beta)$. In other words, γ_α, γ_β are the lifts of γ through z. The uniqueness implies $\gamma_\alpha = \gamma_\beta$. In particular, $x_\alpha = \gamma_\alpha(1) = \gamma_\beta(1) = x_\beta$, which contradicts $\alpha \neq \beta$.

3323

Let $F : M \to M$ be an isometry of a Riemannian manifold M.

(i) Show that each component of $X = \{x \in M : F(x) = x\}$ is an embedded totally geodesic submanifold of M.

(Hint. Use exponential coordinates.)

(ii) Give an example in which the components of X have different dimensions.

(Indiana)

Solution.

(i) First, we show that X has submanifold structures. For $x \in X \subset M$, set $B(\delta) = \{\nu \in T_x(M) : |\nu| < \delta\}$ and $B_\delta = \{y \in M : d(x,y) < \delta\}$, where δ is so small that $\exp_x : B(\delta) \to B_\delta$ is a diffeomorphism. Define $V = \{\nu \in T_x(M) : dF(\nu) = \nu\}$. Thus, V is a subspace of $T_x(M)$. Then we claim that $X \cap B_\delta = \exp_x(V \cap B(\delta))$. If this is proved, because $\exp_x(V \cap B(\delta))$ is obviously a submanifold of M, we can assert that X has submanifold structures. In order to prove the claim, we first assume that $y = X \cap B_\delta$ and $\nu \in B(\delta)$ such that $\exp_x \nu = y$. Let $\gamma : [0,1] \to M$ be the unique shortest geodesic $\gamma(t) = \exp_x(t\nu)$ connecting x and y. Since $x, y \in X$, and F is an isometry, $F(\gamma)$ is also a shortest geodesic connecting $F(x) = x$ and $F(y) = y$. Thus the uniqueness implies $F(\gamma) = \gamma$. In particular, $dF(\dot\gamma(0)) = \dot\gamma(0)$, namely, $dF(\nu) = \nu$. Therefore, $\nu \in V$ which means that $y \in \exp_x(V \cap B(\delta))$, i.e., $X \cap B_\delta \subset \exp_x(V \cap B(\delta))$. On the other hand, suppose that $\nu \in V \cap B(\delta)$ and $y = \exp_x \nu$. Let the geodesic $\gamma : [0, 1] \to M$ be defined by $\gamma(t) = \exp_x(t\nu)$. From $dF(\nu) = \nu$ follows $dF(\dot{\gamma}(0)) = \dot{\gamma}(0)$. Then, that F is an isometry implies $F(\gamma) = \gamma$. In particular, $F(y) = F(\gamma(1)) = \gamma(1) = y$, which means that $y \in X \cap B_\delta$, i.e., $\exp_x(V \cap B(\delta)) \subset X \cap B_\delta$.

Next, we show that every geodesic $\gamma : (a, b) \to X$ parameterized by arclength is also a geodesic of M. For any $s_0 \in (a, b)$, let $\zeta(s)$ be a geodesic of M such that $\zeta(s_0) = \gamma(s_0)$, $\dot{\zeta}(s_0) = \dot{\gamma}(s_0)$. Since $F(\zeta(s_0)) = \zeta(s_0)$, $dF(\dot{\zeta}(s_0)) = \dot{\zeta}(s_0)$ and F is an isometry, then $F(\zeta)$ and ζ are two geodesics of M which satisfy the same initial conditions. Therefore, $F(\zeta) = \zeta$, i.e., ζ lies in X. Besides, ζ is naturally a geodesic of X. Thus, in a neighborhood of s_0, $\zeta = \gamma$. Because s_0 is arbitrarily chosen. γ is a geodesic of M. Hence, X is totally geodesic.

(ii) Let

$$M = \{(x, y, z) \in \mathbb{R}^3 : y > 0, z = 0\}$$
$$\cup \{(x, y, z) \in \mathbb{R}^3 : x = 0, y < 0\},$$

and F be a reflection with respect to the plane $z = 0$, i.e., $F(x, y, z) = (x, y, -z)$. Then M is a 2-dimensional manifold, F is an isometry, and

$$X = \{(x, y, z) \in \mathbb{R}^3 : y > 0, z = 0\}$$
$$\cup \{(x, y, z) \in \mathbb{R}^3 : x = 0, y < 0, z = 0\}.$$

3324

Compute the de Rham cohomology groups of the circle S^1. Do so directly; i.e., without citing the de Rham Theorem.

(Indiana)

Solution.

Since $B^0(S^1, d) = 0$ and S^1 is connected, then

$$H^0(S^1, d) = Z^0(S^1, d) = \{f \in C^\infty(S^1, \mathbb{R}^1) | df = 0\} \cong \mathbb{R}^1.$$

For $H^k(S^1,d), k>1$, since there are no non-vanishing k-forms on S^1, we have $Z^k(S^1,d)=0$ and hence $H^k(S^1,d)=0$.

Besides, observe that

$$Z^1(S^1,d)=C^\infty(S^1,\Lambda^1(S^1)),\ B^1(S^1,d)=\{df|f\in C^\infty(S^1,\mathbb{R}^1)\}.$$

Let θ be the polar coordinate characterizing S^1. Then $\frac{\partial}{\partial\theta}$ is a non-vanishing vector field on S^1. Let $d\theta$ be its dual non-vanishing 1-form on S^1 (Caution. Here $d\theta$ is only a formal symbol, because θ in usual sense is not a globally well-defined function on S^1), and it is not exact. For every $\omega = g(\theta)d\theta \in C^\infty(S^1,\Lambda^1(S^1))$, define a function

$$\Omega(\theta)=\int_0^\theta g(\sigma)-\left(\frac{1}{2\pi}\int_0^{2\pi}g(\sigma)d\sigma\right)\theta.$$

Because $\Omega(0)=\Omega(2\pi)$, Ω is globally well-defined on S^1. Hence, denoting

$$C=\frac{1}{2\pi}\int_0^{2\pi}g(\sigma)d\sigma,$$

we see that $\omega - Cd\theta = d\Omega$, i.e., $\omega - Cd\theta$ is exact. Therefore,

$$H^1(S^1,d)=Z^1(S^1,d)/B^1(S^1,d)=\{Cd\theta|C\in\mathbb{R}^1\}\cong\mathbb{R}^1.$$

3325

Let X denote a submanifold of Euclidean space $\mathbb{E}^n$, and set

$$U_\varepsilon X:=\{x+\nu:x\in X,\nu\in N_xX,|\nu|<\varepsilon\},$$
$$B(X,\varepsilon):=\{y\in\mathbb{E}^n:|y-x|<\varepsilon\text{ for some }x\in X\}.$$

Show that $U_\varepsilon X\subset B(X,\varepsilon)$ for all ε. Show that the two are not generally equal. (Consider examples of 1-dimensional submanifold in $\mathbb{E}^2$.) Can you give conditions which imply equality?

(*Indiana*)

Solution.

Setting $y = x + \nu$ immediately implies that $U_\varepsilon X \subset B(X, \varepsilon)$.

Let X be an open line segment in $\mathbb{E}^2$. Then considering the boundary of X, i.e., the end points, can show that $U_\varepsilon X$ is a proper subset of $B(X, \varepsilon)$.

If X has no boundary, e.g., either compact or complete, then $U_\varepsilon X = B(X, \varepsilon)$.

3326

Consider a Riemannian manifold (M, g). Call a vector field Z on M a Killing vector field if Z generates a 1-parameter group of isometries of M.

(i) Show that when Z is Killing, we have $L_z g = 0$, i.e.,

$$Z(g(X, Y)) = g(L_Z X, Y) + g(X, L_Z Y) \qquad (**)$$

for all vector fields X and Y on M. Here $L_z g$ denotes the Lie derivative along Z.

(ii) Show that the expression $(**)$ above is equivalent to

$$g(\nabla_X Z, Y) = -g(\nabla_Y Z, X),$$

where ∇ denotes the Levi–Civita connection for (M, g).

(Indiana)

Solution.

In local coordinates $(x^1, \ldots, x^m)$ of (M, g), let the 1-parameter group of isometries generated by a vector field $Z = \sum_{i=1}^m z^i \frac{\partial}{\partial x_i}$ be expressed by $\bar{x}^i = \bar{x}^i(x^1, \ldots, x^m; t) := \bar{x}^i(x, t)$ such that

$$\sum_{i,j=1}^m g_{ij}(x) dx^i dx^j = \sum_{k,l=1}^m g_{kl}(\bar{x}) d\bar{x}^k d\bar{x}^l,$$

where $g_{ij}(x) = g(\frac{\partial}{\partial x^i}, \frac{\partial}{\partial x^j})$, and $\bar{x}^i = \bar{x}^i(x, t)$ satisfies

$$\bar{x}^i(x, 0) = x^i, \qquad \frac{d\bar{x}^i}{dt}\Big|_{t=0} = z^i.$$

Thus we have

$$g_{ij}(x) = \sum_{k,l=1}^{m} g_{kl}(\bar{x}) \frac{\partial \bar{x}^k}{\partial x^i} \frac{\partial \bar{x}^l}{\partial x^j}.$$

Differentiating the obtained equality with respect to t and then setting $t = 0$, we obtain

$$\sum_{k=1}^{m} \left(\frac{\partial g_{ij}}{\partial x^k} z^k + g_{kj} \frac{\partial z^k}{\partial x^i} + g_{ik} \frac{\partial z^k}{\partial x^j} \right) = 0,$$

which can be written as

$$g_{kj} z_{,i}^k + g_{ik} z_{,j}^k = 0,$$

or equivalently

$$g \left(\nabla_{\frac{\partial}{\partial x^i}} Z, \frac{\partial}{\partial x^j} \right) + g \left(\frac{\partial}{\partial x^i}, \nabla_{\frac{\partial}{\partial x^j}} Z \right) = 0.$$

From this follows what we desire in (ii).

Noting that the Levi–Civita connection ∇ satisfies

$$Z(g(X,Y)) = g(\nabla_Z X, Y) + g(X, \nabla_Z Y)$$

and the Lie derivative satisfies $L_Z X = [Z, X]$, we easily obtain $(**)$.

3327

Let M^2 be a connected Riemannian manifold and X, Y complete vector fields on M. Assume that the flows X_t and Y_t are isometries of M for all t.

(i) Show that the integral curves of X are curves of constant geodesic curvature.

(ii) Assume that X and Y are linearly independent at all $x \in M^2$ and that their flows commute $X_t \circ Y_s = Y_s \circ X_t$. Conclude that $\langle X, X \rangle$, $\langle X, Y \rangle$ and $\langle Y, Y \rangle$ are constant on M.

(*Indiana*)

Solution.

(i) For every $p \in M$, take a local coordinate neighborhood about p such that the coordinate curves are the integral curves of X and their orthogonal trajectories. Namely, we may assume that $X = X^1 \frac{\partial}{\partial x^1}$ and $g_{12} = \langle \frac{\partial}{\partial x^1}, \frac{\partial}{\partial x^2} \rangle = 0$. Because the flow X_t of X for every t is an isometry, the vector field X should satisfy the Killing equation $X_{i,j} + X_{j,i} = 0$; $i, j = 1, 2$, or equivalently,

$$\sum_{k=1}^{2} \left(X^k \frac{\partial g_{ij}}{\partial x^k} + g_{ik} \frac{\partial X^k}{\partial x^j} + g_{jk} \frac{\partial X^k}{\partial x^i} \right) = 0, \quad i, j = 1, 2.$$

Taking $i = 1$, $j = 2$, we obtain $g_{11} \frac{\partial X^1}{\partial x^2} = 0$, i.e., $X^1 = X^1(x^1)$. Now, make the following coordinate transformation

$$\begin{cases} \tilde{x}^1 = \displaystyle\int \frac{1}{X^1(x^1)}, \\ \tilde{x}^2 = x^2, \end{cases} \quad \frac{\partial(\tilde{x}^1, \tilde{x}^2)}{\partial(x^1, x^2)} \neq 0.$$

If we still adopt the original notations, then the vector field $X = \frac{\partial}{\partial x^1}$. Hence from the corresponding killing equation, taking $i = j = 1$ and $i = j = 2$, we obtain

$$\frac{\partial g_{11}}{\partial x^1} = 0, \quad \frac{\partial g_{22}}{\partial x^1} = 0,$$

which means that the metric of M^2 about p can be written as

$$ds^2 = g_{11}(x^2)(dx^1)^2 + g_{22}(x^2)(dx^2)^2.$$

Then, using the Liouville formula, we can compute the geodesic curvature of the x^1-coordinate curve as follows:

$$k_g = \frac{d\theta}{ds} - \frac{1}{2\sqrt{g_{22}}} \frac{\partial \ln g_{11}}{\partial x^2} \cos\theta + \frac{1}{2\sqrt{g_{11}}} \frac{\partial \ln g_{22}}{\partial x^1} \sin\theta$$

$$= -\frac{1}{2\sqrt{g_{22}}} \frac{\partial \ln g_{11}}{\partial x^2} = k_g(x^2),$$

where $\theta = 0$. Therefore, along every integral curve of X, $\frac{\partial k_g}{\partial x^1} = 0$, i.e., $k_g = \text{const}$.

(ii) Analogously, about p, take the integral curves of X and Y as the x^1 and x^2 coordinate curves, respectively. Then, we have locally $X = X^1 \frac{\partial}{\partial x^1}$, $Y = Y^2 \frac{\partial}{\partial x^2}$. Because their flows are commutative $X_t \circ Y_s = Y_s \circ X_t$, that is equivalent to

$$[X, Y] = X^1 \frac{\partial Y^2}{\partial x^1} \frac{\partial}{\partial x^2} - Y^2 \frac{\partial X^1}{\partial x^2} \frac{\partial}{\partial x^1} = 0,$$

we immediately have

$$\frac{\partial Y^2}{\partial x^1} = 0, \quad \frac{\partial X^1}{\partial x^2} = 0.$$

Therefore, we can make a suitable coordinate transformation and then, if adopting the original notations, $X = \frac{\partial}{\partial x^1}$, $Y = \frac{\partial}{\partial x^2}$. Again, by the corresponding Killing equations, we can obtain $\frac{\partial g_{ij}}{\partial x^k} = 0$; $i, j, k = 1, 2$, i.e., all g_{ij}'s are constants. Noting the expressions of X and Y, we obtain

$$\langle X, X \rangle = g_{11} = \text{const}, \quad \langle X, Y \rangle = g_{12} = \text{const},$$

$$\langle Y, Y \rangle = g_{22} = \text{const}.$$

3328

Let M be a compact Riemannian manifold without boundary. For any $f \in C^\infty(M)$, define $\nabla f \in \mathcal{X}(M)$ and $\Delta f \in C^\infty(M)$ as follows:

At any $p \in M$, choose an orthonormal frame field $\{e_1, \ldots, e_n\}$ around p and then define

$$(\nabla f)(p) = \sum_i (e_i f)(e_i)$$

and

$$(\Delta f)(p) = -\sum_i [e_i(e_i f) - (\nabla_{e_i} e_i) f].$$

Verify first that ∇f and Δf are well defined (i.e., they do not depend on the choice of orthonormal frame) and then show that

$$\int f\Delta f = \int \|\nabla f\|^2.$$

(Indiana)

Solution.

Let $\{e_1^*, \ldots, e_n^*\}$ be another orthonormal frame field around p. Then we may suppose that

$$e_i = \sum_j a_i^j e_j^* \quad i = 1, \ldots, n$$

for suitable functions a_i^j, $i, j = 1, \ldots, n$. Noting

$$\langle e_i, e_j \rangle = \langle e_i^*, e_j^* \rangle = \delta_{ij},$$

we have

$$\sum_j a_i^j a_k^j = \delta_{ik}, \quad \sum_j a_j^i a_j^k = \delta^{ik}.$$

Using these equalities and the properties of Riemannian connection, we can easily obtain

$$\sum_i (e_i f) e_i = \sum_i (e_i^* f) e_i^*,$$

$$\sum_i [e_i(e_i f) - (\nabla_{e_i} e_i)f] = \sum_i [e_i^*(e_i^* f) - (\nabla_{e_i^*} e_i^*)f].$$

Hence ∇f and Δf are all well defined.

Using the definition of the Laplacian here, through direct calculation, we have $\operatorname{div}(f\nabla f) = \|\nabla f\|^2 - f\Delta f$. Then Green's Theorem implies

$$\int f\Delta f = \int \|\nabla f\|^2.$$

3329

Let $p(x_1, x_2, x_3, x_4) = (x_1^2 + x_2^2)(x_3^2 + x_4^2)$ and $q(x_1, x_2, x_3, x_4) = x_1^2 + x_2^2 + x_3^2 + x_4^2$. Define

$$S_{a,b} = \{x \in \mathbb{R}^4 | p(x) = a \quad \text{and} \quad q(x) = b\}.$$

For what $a, b \geq 0$, is $S_{a,b}$ a manifold? Explain.

(Indiana)

Solution.

Denote $\alpha = x_1^2 + x_2^2$ and $\beta = x_3^2 + x_4^2$. Then $\alpha\beta = a$ and $\alpha + \beta = b$ means that α and β are two roots of the equation $\lambda^2 - b\lambda + a = 0$. Therefore, $b^2 - 4a \geq 0$ is the prerequisite condition.

(1) If $a = b = 0$, then $x_1 = x_2 = x_3 = x_4 = 0$. Thus $S_{0,0}$ is a 0-dimensional manifold.

(2) If $a = 0$, $b \neq 0$, then

$$S_{0,b} = \{x \in \mathbb{R}^4 \mid x_1 = x_2 = 0, x_3^2 + x_4^2 = b \text{ or}$$
$$x_1^2 + x_2^2 = b, x_3 = x_4 = 0\}.$$

Using the theorem of closed regular submanifolds proved by rank theorem, we can easily show that $S_{0,b}$ is a 1-dimensional submanifold of $\mathbb{R}^4$.

(3) If $a \neq 0$, $b \neq 0$, then when $b^2 - 4a > 0$

$$S_{a,b} = \left\{x \in \mathbb{R}^4 \Big| x_1^2 + x_2^2 = \frac{b + \sqrt{b^2 - 4a}}{2}, \right.$$

$$x_3^2 + x_4^2 = \frac{b - \sqrt{b^2 - 4a}}{2} \text{ or } x_1^2 + x_2^2 = \frac{b - \sqrt{b^2 - 4a}}{2},$$

$$\left. x_3^2 + x_4^2 = \frac{b + \sqrt{b^2 - 4a}}{2} \right\}$$

and when $b^2 - 4a = 0$

$$S_{a,b} = \left\{x \in \mathbb{R}^4 \mid x_1^2 + x_2^2 = x_3^2 + x_4^2 = \frac{b}{2}\right\}.$$

Analogously, we can show that they are all 2-dimensional sub-manifolds of $\mathbb{R}^4$.

3330

Let $F : M \to N$ be a C^∞ map between two C^∞ manifolds. Assume that F is onto. Let X be a smooth vector field on M.

(i) Show by an example that $dF(X)$ may not be a vector field on N.

(ii) Suppose $Y = dF(X)$ is a smooth vector field on N. Show that F takes integral curves of X into integral curves of Y.

(iii) Suppose X_1, Y_1, and X_2, Y_2 are related as X and Y in (ii) above. Show that $dF([X_1, X_2]) = [Y_1, Y_2]$.

(*Indiana*)

Solution.

(i) Let $M = \{(x, y) : x, y \in \mathbb{R}\}$, $N = \{x : x \in \mathbb{R}\}$, $X = (x^2 + y^2)\frac{\partial}{\partial x}$, and $F : M \to N$ be defined by $F(x, y) = x$. Then, if $y_1 \neq y_2$, we have $dF(X(x, y_1)) \neq dF(X(x, y_2))$. Therefore, as a vector field, $dF(X)$ is not well defined in every point of N.

(ii) Let $\alpha(s)$ be an arbitrary integral curve of X. Then, along α, we have $d\alpha(\frac{d}{ds}) = X$. Hence,

$$d(F \circ \alpha)\left(\frac{d}{ds}\right) = (dF \circ d\alpha)\left(\frac{d}{ds}\right) = dF(X) = Y,$$

which means that F takes integral curves of X into integral curves of Y.

(iii) Using local coordinates, by direct computation, one will obtain the desired equality.

3331

Let M^n be a Riemannian manifold. Show that whenever $f : M \to \mathbb{R}$ is a smooth function, there is a unique vector field ∇f

(called the gradient of f on M) such that

$$\langle \nabla f(p), \dot{\gamma}(t) \rangle = \frac{d}{dt} f(\gamma(t))$$

whenever γ is a smooth curve in M with $\gamma(t) = p$.

(Indiana)

Solution.

Let $(x^1, \ldots, x^n)$ be the local coordinates about p and set $p : (x_0^1, \ldots, x_0^n)$. If $\gamma(t)$ is the i-th coordinate curve that passes p, i.e., $\gamma(t)$ has the following expression

$$x^i = t, \quad x^j = x_0^j \quad (j \neq i),$$

then, by

$$\langle \nabla f(p), \dot{\gamma}(t) \rangle = \frac{d}{dt} f(\gamma(t)) = \frac{\partial f}{\partial x^i}$$

we have the expression

$$\nabla f(p) = \sum_{i,j=1}^{n} \left(\frac{\partial f}{\partial x^i} g^{ij} \right)_p \frac{\partial}{\partial x^j},$$

where g^{ij}'s are the components of the matrix $[g_{ij}]^{-1} = [\langle \frac{\partial}{\partial x^i}, \frac{\partial}{\partial x^j} \rangle]^{-1}$. It is easy to verify that the above expression of $\nabla f(p)$ is independent of the choice of local coordinates.

3332

The space $\mathbb{R}^{n \times n}$ of $n \times n$ real matrices forms an n^2-dimensional Euclidean space, in which the dot product between $A = [a_{ij}]$ and $B = [b_{ij}]$ is given by

$$\langle A, B \rangle := \sum_{i=1}^{n} \sum_{j=1}^{n} a_{ij} b_{ij}.$$

Let S^{n-1} be the unit sphere in $\mathbb{R}^n$, and define $\sigma : S^{n-1} \to \mathbb{R}^{n \times n}$ as the map sending $\mathbf{x} = (x_1, \ldots, x_n)$ in S^{n-1} to the symmetric matrix $\sigma(\mathbf{x}) = \frac{1}{\sqrt{2}} [x_i x_j]$.

(i) Show that σ maps S^{n-1} into the sphere of radius $\frac{1}{\sqrt{2}}$ centered at the origin in $\mathbb{R}^{n \times n}$.

(ii) Prove that σ is a local isometry (i.e., the pull-back via σ of the dotproduct metric defined above on $\mathbb{R}^{n \times n}$ is the standard one on S^{n-1}).

$$(Indiana)$$

Solution.

(i) Let $\mathbf{x} = (x_1, \ldots, x_n) \in S^{n-1}$. Then $\sum_{i=1}^{n} x_i^2 = 1$. Hence

$$\langle \sigma(\mathbf{x}), \sigma(\mathbf{x}) \rangle = \frac{1}{2} \sum_{i,j=1}^{n} x_i^2 x_j^2 = \frac{1}{2},$$

which means that $\sigma(\mathbf{x})$ is on the sphere of radius $\frac{1}{\sqrt{2}}$ centered at the origin in $\mathbb{R}^{n \times n}$.

(ii) Let $i : S^{n-1} \to \mathbb{R}^n$ be the standard inclusion map, and $\tau : \mathbb{R}^n \to \mathbb{R}^{n \times n}$ be defined by $(x_1, \ldots, x_n) \to [y_{ij}] = \frac{1}{\sqrt{2}}[x_i x_j]$. Suppose that

$$\mathbf{v} = \sum_{i=1}^{n} v^i \frac{\partial}{\partial x_i}, \quad \mathbf{w} = \sum_{i=1}^{n} w^i \frac{\partial}{\partial x^i}$$

belong to $T_{\mathbf{x}}(S^{n-1})$. Then we have

$$\sum_{i=1}^{n} v^i x_i = \sum_{i=1}^{n} w^i x_i = 0.$$

Therefore

$$d\sigma(\mathbf{v}) = d\tau \circ di \left(\sum_{k=1}^{n} v^k \frac{\partial}{\partial x_k} \right)$$

$$= \sum_{i,j,k=1}^{n} v^k \frac{\partial y_{ij}}{\partial x_k} \frac{\partial}{\partial y_{ij}}$$

$$= \sum_{i,j,k=1}^{n} v^k \frac{1}{\sqrt{2}} (\delta_{ik} x_j + x_i \delta_{jk}) \frac{\partial}{\partial y_{ij}}$$

$$= \sum_{i,j=1}^{n} \frac{1}{\sqrt{2}} (v^i x_j + v^j x_i) \frac{\partial}{\partial y_{ij}}.$$

Hence we have

$$\langle d\sigma(\mathbf{v}), d\sigma(\mathbf{w}) \rangle = \frac{1}{2} \sum_{i,j=1}^{n} (v^i x_j + v^j x_i)(w^i x_j + w^j x_i)$$

$$= \sum_{i=1}^{n} v^i w^i = \langle \mathbf{v}, \mathbf{w} \rangle.$$

3333

Let M be the Riemannian manifold obtained by equipping $\mathbb{R}^n$ with a metric conformal to its usual one; i.e., a metric of the form $[g_{ij}] = e^{2f}[\delta_{ij}]$, where $f : \mathbb{R}^n \to \mathbb{R}$ is a smooth function, and δ_{ij} is the Kronecker delta. Let $e_i = \frac{\partial}{\partial x_i}$ denote the standard coordinate basis vector fields.

(i) Show that for arbitrary indices $i, j, k \in \{1, 2, \ldots, n\}$

$$\langle \nabla_{e_i} e_j, e_k \rangle = e^{2f} \left(\frac{\partial f}{\partial x_i} \delta_{jk} - \frac{\partial f}{\partial x_k} \delta_{ij} + \frac{\partial f}{\partial x_j} \delta_{ki} \right).$$

(ii) Show that when $n = 2$, the sectional curvature of M along an e_1, e_2 plane at p is given by

$$-e^{-2f(p)} \left(\frac{\partial^2 f}{\partial x_1^2} + \frac{\partial^2 f}{\partial x_2^2} \right) (p).$$

(*Indiana*)

Solution.

(i) Set

$$\nabla_{e_i} e_j = \sum_{l=1}^{n} \Gamma_{ij}^l e_l.$$

From $g_{ij} = e^{2f}\delta_{ij}$ it follows that

$$\langle \nabla_{e_i} e_j, e_k \rangle = \sum_{l=1}^{n} \Gamma_{ij}^l g_{lk} = \frac{1}{2} \left(\frac{\partial g_{jk}}{\partial x_i} - \frac{\partial g_{ij}}{\partial x_k} + \frac{\partial g_{ik}}{\partial x_j} \right)$$

$$= e^{2f} \left(\frac{\partial f}{\partial x_i} \delta_{jk} - \frac{\partial f}{\partial x_k} \delta_{ij} + \frac{\partial f}{\partial x_j} \delta_{ki} \right).$$

(ii) The sectional curvature of M along an e_1, e_2 plane at p is just the Gauss curvature of M at p. Noting that $E = G = e^{2f}$, $F = 0$, from the Gauss equation we obtain

$$K(p) = -\frac{1}{\sqrt{EG}} \left[\left(\frac{(\sqrt{E})_2}{\sqrt{G}} \right)_2 + \left(\frac{(\sqrt{G})_1}{\sqrt{E}} \right)_1 \right]$$

$$= -e^{-2f(p)} \left(\frac{\partial^2 f}{\partial x_1^2} + \frac{\partial^2 f}{\partial x_2^2} \right)(p).$$

3334

Let X, Y be complete vector fields on a manifold M and let X_t, Y_t be the flows induced by them.

(i) Show that $X_s \circ Y_t = Y_t \circ X_s$ for all $s, t \in \mathbb{R}$ implies $[X, Y] = 0$.

(ii) Prove the converse to (i).

(*Indiana*)

Solution.

Observe that for a diffeomorphism $F : M \to M$, the complete vector field Y is F-invariant, i.e., $F_* Y = Y$, if and only if $F \circ Y_t = Y_t \circ F, \forall t \in \mathbb{R}$.

(i) Suppose that $X_s \circ Y_t = Y_t \circ X_s$ for all $s, t \in \mathbb{R}$. Since X_s is a diffeomorphism for each s, then Y is X_s-invariant. Thus

$$[X, Y] = L_X Y = \lim_{s \to 0} \frac{1}{s} [Y - X_{s*} Y] = 0.$$

(ii) Now suppose that $[X, Y] = 0$. Then

$$0 = X_{s*}[X, Y] = [X_{s*}X, X_{s*}Y]$$
$$= [X, X_{s*}Y] = L_X(X_{s*}Y).$$

Hence for each $p \in M$ and any $f \in C^\infty(p)$, we have

$$0 = (L_X(X_{s*}Y))_p f = \lim_{\Delta s \to 0} \frac{1}{\Delta s}[(X_{s*}Y)_p f - (X_{\Delta s*}(X_{s*}Y))_p f]$$

$$= \lim_{\Delta s \to 0} \frac{1}{\Delta s}[(X_{s*}Y)_p f - (X_{(s+\Delta s)*}Y)_p f] = -\frac{d}{ds}(X_{s*}Y)_p f$$

for all $s \in \mathbb{R}$. Therefore,

$$(X_{s*}Y)_p f = (X_{0*}Y)_p f = Y_p f.$$

Since f is arbitrary, we obtain $X_{s*}Y = Y$, namely, Y is X_s-invariant. Hence $X_s \circ Y_t = Y_t \circ X_s$ for all $s, t \in \mathbb{R}$.

3335

If $\phi : G_1 \to G_2$ is a Lie group homomorphism, show that for all $v \in LG_1$, we have

$$\phi(\exp(v)) = \exp(\phi_* v).$$

(*Indiana*)

Solution.
Note that $\exp tv$, $t \in \mathbb{R}^1$ is a 1-parameter subgroup of G_1 generated by $v \in LG_1$. Since ϕ is a Lie group homomorphism, then $\phi(\exp tv)$ is also a 1-parameter subgroup of G_2 generated by a suitable $w \in LG_2$, namely, $\phi(\exp tv) = \exp tw$.

Let $(x^1, \ldots, x^m)$ and $(y^1, \ldots, y^n)$ be local coordinates of G_1 and G_2 about the identities e_1 and e_2, respectively. Locally, ϕ can be expressed by

$$\phi(x^1, \ldots, x^m) = (\phi^1(x^1, \ldots, x^m), \ldots, \phi^n(x^1, \ldots, x^m));$$

and $\exp tv$ and $\exp tw$ can be denoted by

$$(x^1(t), \ldots, x^m(t))$$

and

$$(\phi^1(x^1(t), \ldots, x^m(t)), \ldots, \phi^n(x^1(t), \ldots, x^m(t))),$$

respectively. Assuming that

$$v_{e_1} = \sum_{i=1}^m v^i \frac{\partial}{\partial x^i}, \quad w_{e_2} = \sum_{\alpha=1}^n w^\alpha \frac{\partial}{\partial y^\alpha},$$

we have

$$v^i = \left(\frac{dx^i(t)}{dt}\right)_{t=0}, \quad w^\alpha = \left(\frac{d\phi^\alpha}{dt}\right)_{t=0}.$$

Therefore,

$$d\phi(v_{e_1}) = \phi_* v_{e_1} = \sum_{i=1}^m v^i \sum_{\alpha=1}^n \left(\frac{\partial \phi^\alpha}{\partial x^i}\right)_{x=x(0)} \frac{\partial}{\partial y^\alpha}$$

$$= \sum_{i=1}^m \sum_{\alpha=1}^n \left(\frac{dx^i}{dt}\right)_{t=0} \left(\frac{\partial \phi^\alpha}{\partial x^i}\right)_{x=x(0)} \frac{\partial}{\partial y^\alpha}$$

$$= \sum_{\alpha=1}^n \left(\frac{d\phi^\alpha}{dt}\right)_{t=0} \frac{\partial}{\partial y^\alpha} = w_{e_2}.$$

Furthermore, by left translation, we have $d\phi(v) = w$, i.e., $\phi_* v = w$. Hence $\phi(\exp tv) = \exp t\phi_* w$. Taking $t = 1$ completes the proof.

3336

Consider the linearly independent vector fields $\mathbf{r}$ and $\mathbf{v}$ on $U := \mathbb{R}^4 \backslash 0$ whose values at $\mathbf{x} = (x_1, x_2, x_3, x_4) \in \mathbb{R}^4$ are given by

$$\mathbf{r}_x := (x_1, x_2, x_3, x_4),$$

$$\mathbf{v}_x := (-x_2, x_1, -x_4, x_3).$$

(a) Is the rank-2 distribution defined by these two vector fields in U completely integrable?

(b) Is the rank-2 distribution orthogonal to these two vector fields completely integrable?

(*Indiana*)

Solution.

(a) Direct calculation gives $[\mathbf{r}, \mathbf{v}] = 0$. Thus the Frobenius Theorem guarantees that the rank-2 distribution defined by $\mathbf{r}$ and $\mathbf{v}$ in U is completely integrable.

(b) Construct the following linear algebraic equations about y_1, y_2, y_3 and y_4

$$\begin{cases} x_1 y_1 + x_2 y_2 + x_3 y_3 + x_4 y_4 = 0, \\ -x_2 y_1 + x_1 y_2 - x_4 y_3 + x_3 y_4 = 0, \end{cases}$$

which are equivalent to

$$\begin{pmatrix} x_1 & x_2 \\ -x_2 & x_1 \end{pmatrix} \begin{pmatrix} y_1 \\ y_2 \end{pmatrix} + \begin{pmatrix} x_3 & x_4 \\ -x_4 & x_3 \end{pmatrix} \begin{pmatrix} y_3 \\ y_4 \end{pmatrix} = 0.$$

Because $(x_1, x_2, x_3, x_4) \in \mathbb{R}^4 \backslash 0 = U$, without loss of generality, we may assume $x_1 \neq 0$. Therefore, we obtain

$$\begin{pmatrix} y_1 \\ y_2 \end{pmatrix} = \frac{-1}{x_1^2 + x_2^2} \begin{pmatrix} x_1 x_3 + x_2 x_4 & x_1 x_4 - x_2 x_3 \\ x_2 x_3 - x_1 x_4 & x_2 x_4 + x_1 x_3 \end{pmatrix} \begin{pmatrix} y_3 \\ y_4 \end{pmatrix}.$$

Setting

$$\begin{pmatrix} y_3 \\ y_4 \end{pmatrix} = \begin{pmatrix} -(x_1^2 + x_2^2) \\ 0 \end{pmatrix}$$

and

$$\begin{pmatrix} 0 \\ -(x_1^2 + x_2^2) \end{pmatrix}$$

respectively, we obtain the following two linearly independent vector fields α and β which define a rank-2 distribution orthogonal to the above one and whose values at x are given by

$$\alpha = (x_1 x_3 + x_2 x_4, \quad x_2 x_3 - x_1 x_4, \quad -(x_1^2 + x_2^2), \quad 0),$$
$$\beta = (x_1 x_2 - x_2 x_3, \quad x_2 x_4 + x_1 x_3, \quad 0, \quad -(x_1^2 + x_2^2)).$$

By a long but straightforward calculation, we have $[\alpha, \beta] = -2(x_1^2 + x_2^2)\mathbf{v}$ which does not belong to the distribution defined by α and β. Thus the Frobenius Theorem tells us that this distribution is not completely integrable.

3337

Let G be a Riemannian manifold with a global frame-field $\{e_i\}_{i=1}^n$.

(a) Show that any connection on G is competely determined by its effect on the frame field, i.e., by the vector fields $\nabla_{e_i} e_j, i, j = 1, \ldots, n$.

(b) Show that when G is a Lie group with a bi-invariant metric $\langle , \rangle$, and the frame-field is left-invariant, we characterize the Levi–Civita connection on $(G, \langle , \rangle)$ by setting,

$$\nabla_{e_i} e_j = \frac{1}{2}[e_i, e_j]$$

for all $i, j = 1, \ldots, n$.

(Indiana)

Solution.

(a) For arbitrary smooth vector fields X and Y on G, we have

$$X = \sum_{i=1}^n x^i e_i, \quad Y = \sum_{i=1}^n y^i e_i.$$

Then, motivated by the properties of connection, we can well define

$$\nabla_X Y = \nabla_{\sum_i x^i e^i} \left(\sum_j y^i e_j \right) = \sum_{i,j=1}^n x^i e_i(y^j) \cdot e_j + \sum_{i,j=1}^n x^i y^j \nabla_{e_i} e_j.$$

Thus defined operator ∇ certainly satisfies all properties of a linear connection on G.

(b) For every left invariant vector field X on G, let $g(t)$ denote the unique 1-parameter subgroup of G such that $g(0) = e$, $\frac{dg(0)}{dt} = X_e$, where e is the unit element of G. Noting that $g(t)$ is a geodesic of G and $X_{g(t)} = \frac{dg(t)}{dt}$, we have obviously

$$\nabla_{X_e} X = \frac{DX_{g(t)}}{dt}\Big|_{t=0} = \frac{D}{dt}\left(\frac{dg(t)}{dt}\right)\Big|_{t=0} = 0.$$

Besides, because the metric of G is bi-invariant, we know that $\nabla_X X = 0$ is valid everywhere. Especially, if $X = e_i + e_j$, then, $\nabla_{e_i+e_j}(e_i + e_j) = 0$ implies that

$$\nabla_{e_i} e_j + \nabla_{e_j} e_i = 0.$$

On the other hand, the Levi–Civita connection ∇ satisfies

$$\nabla_{e_i} e_j - \nabla_{e_j} e_i = [e_i, e_j].$$

Thus, we obtain

$$\nabla_{e_i} e_j = \frac{1}{2}[e_i, e_j].$$

3338

Let $\phi : \mathbb{R}^n \to \mathbb{R}^n$ be any isometry, that is, a map such that the Euclidean distance between any two points $x, y \in \mathbb{R}^n$ is equal to the distance between their images $\phi(x)$, $\phi(y)$. Show that ϕ is *affine linear*, that is, there exists a vector $b \in \mathbb{R}^n$ and an orthogonal matrix $A \in O(n)$ such that for all $x \in \mathbb{R}^n$, $\phi(x) = Ax + b$.

(*Harvard*)

Solution.

First, we show that ϕ maps every straight line onto a straight line. For two fixed points x, y on a straight line l, every point z on the straight line l can be uniquely expressed by $z = \lambda x +$

$(1-\lambda)y$ for a real number $\lambda \in \mathbb{R}$; or equivalently, the position of z on the line l is uniquely determined by the magnitudes of the distances $d(x, z), d(y, z)$ and $d(x, y)$. Because ϕ is an isometry, we have that

$$d(\phi(x), \phi(z)) = d(x, z), \quad d(\phi(y), \phi(z)) = d(y, z),$$
$$d(\phi(x), \phi(y)) = d(x, y).$$

Therefore the position relations of $\phi(z)$ with respect to $\phi(x)$ and $\phi(y)$ are same as that of z with respect to x and y, which means that the image $\phi(z)$ is on the straight line determined by the images $\phi(x)$ and $\phi(y)$, and the map ϕ preserves the position relations among x, y and z. Hence $\phi(z) = \lambda\phi(x) + (1 - \lambda)\phi(y)$, namely, ϕ has the property that $\phi(\lambda x + (1-\lambda)y) = \lambda\phi(x) + (1-\lambda)\phi(y)$, $\lambda \in \mathbb{R}$. Conversely, using a similar argument, it is easy to show that ϕ is surjective.

Next, if we denote the zero point by θ, then we show that the map $L : \mathbb{R}^n \to \mathbb{R}^n$ defined by $L(x) = \phi(x) - \phi(\theta)$ is linear.

(i) For arbitrary $\lambda \in \mathbb{R}$ and $x \in \mathbb{R}^n$, we have that

$$L(\lambda x) = \phi(\lambda x) - \phi(\theta) = \phi(\lambda x + (1 - \lambda)\theta) - \phi(\theta)$$
$$= \lambda\phi(x) + (1 - \lambda)\phi(\theta) - \phi(\theta)$$
$$= \lambda\phi(x) - \lambda\phi(\theta) = \lambda L(x).$$

(ii) In order to prove that $L(x+y) = L(x)+L(y)$, $\forall x, y \in \mathbb{R}^n$, again, we need to use the above mentioned property of the map ϕ, which can be rewritten as

$$\phi(\lambda x + \mu y) = \lambda\phi(x) + \mu\phi(y), \quad \lambda + \mu = 1.$$

This relation can be easily generalized to $\phi(\lambda x + \mu y + vz) = \lambda\phi(x) + \mu\phi(y) + v\phi(z)$, for $x, y, z \in \mathbb{R}^n$, $\lambda, \mu, v \in \mathbb{R}$ such that $\lambda + \mu + v = 1$. Then,

$$L(x + y) = \phi(x + y) - \phi(\theta) = \phi(x + y - \theta) - \phi(\theta)$$
$$= \phi(x) + \phi(y) - \phi(\theta) - \phi(\theta) = L(x) + L(y)$$

which shows that $L : \mathbb{R}^n \to \mathbb{R}^n$ is a linear map.

For the final assertion, we take an orthonormal basis $\{e_1, e_2, \ldots, e_n\}$ at the origin of $\mathbb{R}^n$. Assuming $L(e_i) = \sum_k A_{ik}e_k$, we have

$$\langle L(e_i), L(e_j) \rangle = \sum_{k,l} A_{ik}A_{jl}\langle e_k, e_l \rangle = \sum_k A_{ik}A_{jk}.$$

On the other hand, by the definitions of L and ϕ,

$$\begin{aligned}
\langle L(e_i), L(e_j) \rangle &= \langle \phi(e_i) - \phi(\theta), \phi(e_j) - \phi(\theta) \rangle \\
&= \langle \phi(e_i), \phi(e_j) \rangle - \langle \phi(e_i), \phi(\theta) \rangle \\
&\quad - \langle \phi(\theta), \phi(e_j) \rangle + \langle \phi(\theta), \phi(\theta) \rangle \\
&= \langle e_i, e_j \rangle = \delta_{ij}.
\end{aligned}$$

Then

$$\sum_k A_{ik}A_{jk} = \delta_{ij},$$

which implies that the matrix $A = [A_{ij}]^T$ is an orthogonal matrix. By denoting $L(x) = Ax$ and $\phi(\theta) = b$, we obtain the conclusion $\phi(x) = Ax + b$, $A \in O(n)$.

3339

Show that the sphere S^{2n} is not the underlying topological space of any Lie group.

(Harvard)

Solution.

We give a proof by contradiction to this assertion. First, assume that the sphere S^{2n} can be defined as a $2n$-dimensional Lie group. Then, for any element a of the Lie group S^{2n}, we can define a right translation R_a by $R_a x = xa$, $\forall x \in S^{2n}$, which is a C^∞ homeomorphism on S^{2n}. Then, by the Poincaré–Brouwer fixed point theorem for S^{2n}, there exists a point b in S^{2n} such that either b is a fixed point of R_a, or R_a maps b to its antipodal point $-b$. That is, either $ba = b$, or $ba = -b$, which implies that either $a = e$, or $a = -e$, because S^{2n} is assumed to be a Lie group. This result means that S^{2n} has at most two elements, i.e., the unit element e and its antipodal $-e$, which contradicts the dimension of S^{2n} when $n > 0$.

3340

Let $U \subset \mathbb{R}^2$ be an open set.

(a) Define a *Riemannian metric* on U.

(b) In terms of your definition, define the *distance* between two points $p, q \in U$.

(c) Let $\Delta = \{(x, y) : x^2 + y^2 < 1\}$ be the open unit disc in $\mathbb{R}^2$, and consider the metric on Δ given by

$$ds^2 = \frac{dx^2 + dy^2}{(1 - x^2 - y^2)^2}.$$

Show that Δ is complete with respect to this metric.

(Harvard)

Solution.

(a) Denote the Descartes coordinates on the $\mathbb{R}^2$ by (x^1, x^2). Given any smooth symmetric positive definite quadratic differential form $ds^2 = \sum_{i,j} g_{ij}(x^1, x^2) dx^i dx^j$ on $\mathbb{R}^2$, then $ds^2|_U$ is a *Riemannian metric* on the open set U.

(b) For any pair of points p, q in U, the length of an arbitrary piecewise smooth curve C which connects p, q and completely lies in U can be calculated by

$$L_C = \int_{t_0}^{t_1} \sqrt{\sum_{i,j} g_{ij}(x^1(t), x^2(t)) \frac{dx^i}{dt} \frac{dx^j}{dt}} \, dt$$

where $x^1 = x^1(t)$, $x^2 = x^2(t)$ are the parametric equations of C, and t_0, t_1 correspond to p, q respectively. Then the *distance* between two points $p, q \in U$ is defined by

$$d(p, q) = \inf_{C \subset U} \{L_C\}.$$

(c) By the Hopf–Rinow Theorem, it suffices to prove that for a fixed point, say the origin $O(0,0)$, the exponential map $\exp_O$ is well-defined on the entire tangent space $T_O(\Delta)$. Obviously, this is equivalent to prove that every geodesic $\gamma : [0, \varepsilon) \to \Delta$ which starts form the point O along an arbitrary direction can be infinitely extended to a geodesic $\bar{\gamma} : [0, \infty) \to \Delta$. For convenience, in the following, we use the polar coordinates (ρ, θ) instead of the Descartes coordinates (x, y), and note that O is not a substantially singular point in the polar coordinates. We can still well define various geometric objects at O by continuity.

Firstly, we assert that ρ-curves are geodesics in Δ. Noticing that the metric is now of the form $ds^2 = \frac{d\rho^2 + \rho^2 d\theta^2}{(1-\rho^2)^2}$, or $E = \frac{1}{(1-\rho^2)^2}$, $F = 0$, $G = \frac{\rho^2}{(1-\rho^2)^2}$, we can easily get the Riemannian connection coefficients, namely the Christoffel symbols as follows:

$$\Gamma_{11}^1 = \frac{E_1}{2E} = \frac{2\rho}{1-\rho^2}, \quad \Gamma_{12}^1 = \Gamma_{21}^1 = \frac{E_2}{2E} = 0,$$

$$\Gamma_{22}^1 = -\frac{G_1}{2E} = -\frac{\rho(1+\rho^2)}{1-\rho^2};$$

$$\Gamma_{11}^2 = -\frac{E_2}{2G} = 0, \quad \Gamma_{12}^2 = \Gamma_{21}^2 = \frac{G_1}{2G} = \frac{1+\rho^2}{\rho(1-\rho^2)}, \quad \Gamma_{22}^2 = \frac{G_2}{2G} = 0.$$

Thus, in the polar coordinates (ρ, θ), by using the arclength as the parameter, the equations for geodesics in Δ

$$\frac{d^2\rho}{ds^2} + \Gamma_{11}^1 \left(\frac{d\rho}{ds}\right)^2 + 2\Gamma_{12}^1 \frac{d\rho}{ds}\frac{d\theta}{ds} + \Gamma_{22}^1 \left(\frac{d\theta}{ds}\right)^2 = 0,$$

$$\frac{d^2\theta}{ds^2} + \Gamma_{11}^2 \left(\frac{d\rho}{ds}\right)^2 + 2\Gamma_{12}^2 \frac{d\rho}{ds}\frac{d\theta}{ds} + \Gamma_{22}^2 \left(\frac{d\theta}{ds}\right)^2 = 0$$

are simplified to

$$\frac{d^2\rho}{ds^2} + \frac{2\rho}{1-\rho^2} \left(\frac{d\rho}{ds}\right)^2 - \frac{\rho(1+\rho^2)}{1-\rho^2} \left(\frac{d\theta}{ds}\right)^2 = 0,$$

$$\frac{d^2\theta}{ds^2} + \frac{2(1+\rho^2)}{\rho(1-\rho^2)} \frac{d\rho}{ds}\frac{d\theta}{ds} = 0.$$

In order to prove that every ρ-curve is a geodesic, we need to notice that the arclength of a curve is

$$s = \int_0^s \sqrt{\left[\left(\frac{d\rho}{ds}\right)^2 + \rho^2 \left(\frac{d\theta}{ds}\right)^2\right] \Big/ (1-\rho^2)^2} ds.$$

For a ρ-curve, $\theta =$ const; hence, taking the derivatives with respect to s of both sides yields $\frac{d\rho}{ds} = 1 - \rho^2$ and then $\frac{d^2\rho}{ds^2} = -2\rho(1-\rho^2)$. Substituting these relations into the left-hand sides of the above simplified geodesic equations, they are all vanishing. Thus every ρ-curve is a geodesic in Δ.

Next, we show that every ρ-curve is well-defined for the arclength $s \in [0, +\infty)$. In fact, from the above arclength formula, we have that, for a ρ-curve,

$$s = \int_0^\rho \frac{d\rho}{1-\rho^2} = \ln\sqrt{\frac{1+\rho}{1-\rho}}.$$

Hence, when $\rho \to 1$, $s \to +\infty$ which shows that the geodesic, starting form the point O along the direction $\theta =$ const, can be infinitely extended. Therefore, the exponential map $\exp_O$ is well-defined on the entire tangent space $T_O(\Delta)$, that concludes our proof.

3341

Let $H = \{(u, v) \in \mathbb{R}^2 | v > 0\}$ and $B = \{(x, y) \in \mathbb{R}^2 | x^2 + y^2 < 1\}$. For $e_2 = (0, 1) \in \mathbb{R}^2$, map H to B by the following diffeomorphism

$$\mathbf{v} \mapsto \mathbf{x} = -e_2 + \frac{2(\mathbf{v} + e_2)}{\|\mathbf{v} + e_2\|^2}.$$

(i) Verify that the image of the above map is indeed B. (Hint. Think of the standard inversion in the circle.)

(ii) Consider the following metric on B:

$$g = \frac{dx^2 + dy^2}{(1 - \|\mathbf{x}\|^2)^2}.$$

Put a metric on H such that the above map is an isometry.
(iii) Show that H is complete.

(*Harvard*)

Solution.

(i) We denote the given map $\mathbf{v} \mapsto \mathbf{x}$ by ϕ and rewrite $\mathbf{x} = -e_2 + \frac{2(\mathbf{v}+e_2)}{\|\mathbf{v}+e_2\|^2}$ as $\mathbf{x}+e_2 = \frac{2(\mathbf{v}+e_2)}{\|\mathbf{v}+e_2\|^2}$, which yields $\|\mathbf{x}+e_2\|\|\mathbf{v}+e_2\| = 2$. From the last two equations we know that, in fact, ϕ is an inversion with respect to the circle centered at $J(0,-1)$ wit radius $\sqrt{2}$. Hence we can easily get the inverse of ϕ as follows:

$$\phi^{-1} : \mathbf{x} \mapsto \mathbf{v} = -e_2 + \frac{2(\mathbf{x}+e_2)}{\|\mathbf{x}+e_2\|^2},$$

namely, $\phi^{-1} = \phi$ and $\phi : \mathbb{R}^2 \to \mathbb{R}^2$ is a diffeomorphism.

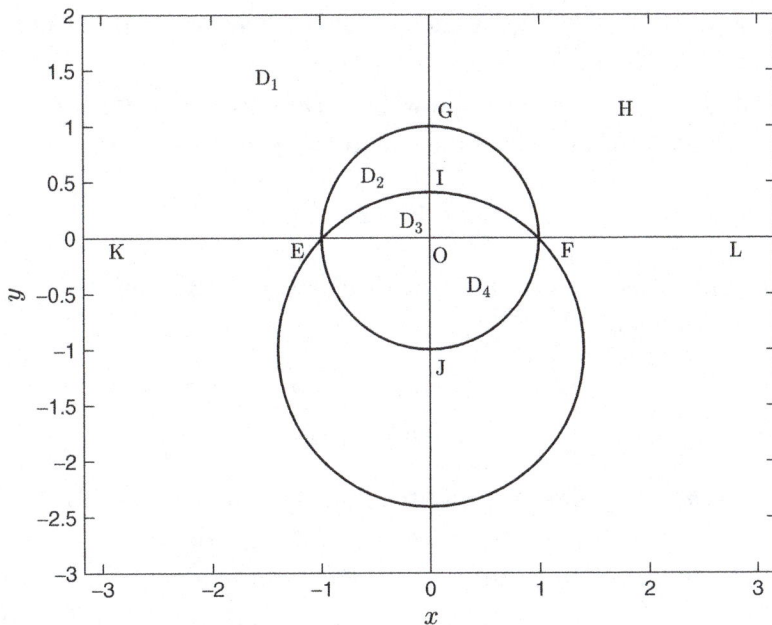

For convenience, in the above illustration, we assume that all the domains D_1, D_2, D_3, D_4 are open sets; all the arc segments arc(EGF), arc(EIF), arc(EJF) and the line segment EOF are open without their end points; and EK and FL are rays. Because the unit circle with the center $O(0,0)$ passes the center $J(0,-1)$ of the inversion circle, its image under the inversion must be a straight line defined by the intersection points E and F. Noticing that $\phi(\text{arc}(EGF)) = EOF$, $\phi(\text{arc}(EIF)) = \text{arc}(EIF)$, $\phi(EK \cup FL) = \text{arc}(EJF)$, it follows that $\phi(D_1) = D_4$, $\phi(D_2) = D_3$ and $\phi(D_3) = D_2$, which implies that the image of $H = \{(u,v) \in \mathbb{R}^2 | v > 0\}$ under the map ϕ is really $B = \{(x,y) \in \mathbb{R}^2 | x^2 + y^2 < 1\}$.

(ii) For avoiding tedious calculations, we make a coordinate transformation on the half plane H by using the diffeomorphism ϕ as follows. For an arbitrary point $p(u,v)$ of H, we take the coordinates (x,y) of its image point $\phi(p)$ as the new coordinates of p. Thus the point p and its image $\phi(p)$ have the same coordinates, then the map ϕ is an identity map, i.e., $\phi = \text{id}$. Hence $\phi^* g = \text{id}^*(g) = \frac{dx^2 + dy^2}{(1 - \|x\|^2)^2}$ is the metric defined on the point $p(x,y)$ in H, and ϕ is obviously an isometry.

(iii) It is equivalent to show that B is complete with respect to the metric g. See the solution to the previous problem.

3342

Define a metric on the unit disc $\{(x.y) \in \mathbb{R}^2 : x^2 + y^2 < 1\}$ by the line element

$$ds^2 = \frac{d\rho^2 + \rho^2 d\theta^2}{(1 - \rho^2)^p}.$$

Here (ρ, θ) are polar coordinates and p is any real number.
   (a) For which p is the circle $\rho = 1/2$ a geodesic?
   (b) Compute the Gaussian curvature of this metric.

(Harvard)

Solution.

(a) From the line element, the coefficients of the first fundamental form are

$$E = \frac{1}{(1-\rho^2)^p}, \quad F = 0, \quad G = \frac{\rho^2}{(1-\rho^2)^p}.$$

In this orthogonal coordinates, the Christoffel symbols (or the Riemannian connection coefficients) can be calculated as follows:

$$\Gamma_{11}^1 = \frac{E_1}{2E} = \frac{p\rho}{1-\rho^2}, \quad \Gamma_{12}^1 = \frac{E_2}{2E} = 0,$$

$$\Gamma_{22}^1 = -\frac{G_1}{2E} = \frac{\rho^3(1-p)-\rho}{1-\rho^2};$$

$$\Gamma_{11}^2 = -\frac{E_2}{2G} = 0, \quad \Gamma_{12}^2 = \frac{G_1}{2G} = \frac{1-\rho^2(1-p)}{\rho(1-\rho^2)}, \quad \Gamma_{22}^2 = \frac{G_2}{2G} = 0.$$

Thus, the equations of geodesics are

$$\frac{d^2\rho}{ds^2} + \frac{p\rho}{1-\rho^2}\left(\frac{d\rho}{ds}\right)^2 + \frac{\rho^3(1-p)-\rho}{1-\rho^2}\left(\frac{d\theta}{ds}\right)^2 = 0,$$

$$\frac{d^2\theta}{ds^2} + 2\cdot\frac{1-\rho^2(1-p)}{\rho(1-\rho^2)}\frac{d\rho}{ds}\frac{d\theta}{ds} = 0.$$

Therefore a θ-curve is a geodesic if and only if

$$\frac{\rho^3(1-p)-\rho}{1-\rho^2}\left(\frac{d\theta}{ds}\right)^2 = 0 \quad \text{and} \quad \frac{d^2\theta}{ds^2} = 0.$$

Now that the θ-curve is defined by $\rho = 1/2$, because $\frac{d\theta}{ds} \neq 0$, we have the restrictions:

$$\left.\frac{\rho^3(1-p)-\rho}{1-\rho^2}\right|_{\rho=1/2} = 0 \quad \text{and} \quad \theta = ks + c.$$

Thus, $\frac{1}{4}(1-p) - 1 = 0$ gives $p = -3$.

(b) In order to compute the Gaussian curvature for the metric when $p = -3$, first, we show that, in an isothermal coordinate system (u, v), if $ds^2 = f^2(du^2 + dv^2)$, $f > 0$, then the Gaussian curvature $K = -\frac{1}{f^2}\Delta \ln f$. In fact, by the Gaussian curvature formula in orthogonal coordinates, we have that

$$K = -\frac{1}{\sqrt{EG}}\left\{\left[\frac{(\sqrt{E})_v}{\sqrt{G}}\right]_v + \left[\frac{(\sqrt{G})_u}{\sqrt{E}}\right]_u\right\}$$

$$= -\frac{1}{f^2}\left[\left(\frac{f_v}{f}\right)_v + \left(\frac{f_u}{f}\right)_u\right]$$

$$= -\frac{1}{f^2}[(\ln f)_{vv} + (\ln f)_{uu}] = -\frac{1}{f^2}\Delta \ln f.$$

In order to use this formula, we convert the polar coordinate system to the Descartes coordinate system and get $ds^2 = (1 - x^2 - y^2)^3(dx^2 + dy^2)$, $f = (1 - x^2 - y^2)^{3/2}$. Hence

$$(\ln f)_x = \frac{3}{2}\cdot\frac{-2x}{1 - x^2 - y^2} = -\frac{3x}{1 - x^2 - y^2},$$

$$(\ln f)_{xx} = \frac{-3(1 - x^2 - y^2) + 3x(-2x)}{(1 - x^2 - y^2)^2} = \frac{-3 - 3x^2 + 3y^2}{(1 - x^2 - y^2)^2};$$

similarly,

$$(\ln f)_{yy} = \frac{-3(1 - x^2 - y^2) + 3y(-2y)}{(1 - x^2 - y^2)^2} = \frac{-3 + 3x^2 - 3y^2}{(1 - x^2 - y^2)^2}.$$

Therefore

$$K = -\frac{1}{(1 - x^2 - y^2)^3}\cdot\frac{-6}{(1 - x^2 - y^2)^2} = \frac{6}{(1 - x^2 - y^2)^5}.$$

Also, we can directly use the above first formula for Gaussian curvature in the given polar coordinates to calculate K. Then we have $K = \frac{6}{(1-\rho^2)^5}$.

3343

Let $(\mathbb{H}^2, g)$ be the two-dimensional hyperbolic space, where

$$\mathbb{H}^2 = \{(x, y) \in \mathbb{R}^2 : y > 0\}$$

is the upper half plane of $\mathbb{R}^2 = \mathbb{C}$ and the metric g is given by

$$g = \frac{dx^2 + dy^2}{y^2}.$$

(i) Suppose a, b, c and d are real numbers such that $ad - bc = 1$. Define

$$\phi(z) = \frac{az + b}{cz + d}$$

for any $z = x + \sqrt{-1}y$. Prove that ϕ is an isometry for $(\mathbb{H}^2, g)$.

(ii) Prove that $(\mathbb{H}^2, g)$ has constant Gaussian curvature.

(Harvard)

Solution.

(i) By differentiation, we get that

$$d\phi(z) = \frac{a(dz)(cz + d) - c(dz)(az + b)}{(cz + d)^2}$$

and

$$\overline{d\phi(z)} = \frac{a(d\bar{z})(c\bar{z} + d) - c(d\bar{z})(a\bar{z} + b)}{(c\bar{z} + d)^2}.$$

Hence, by direct calculation, we have that

$$d\phi(z) \cdot \overline{d\phi(z)}$$
$$= \frac{a^2(dz \cdot d\bar{z})(cz + d)(c\bar{z} + d) - ac(dz \cdot d\bar{z})(az + b)(c\bar{z} + d)}{(cz + d)^2(c\bar{z} + d)^2}$$
$$+ \frac{-ac(dz \cdot d\bar{z})(a\bar{z} + b)(cz + d) + c^2(dz \cdot d\bar{z})(az + b)(a\bar{z} + b)}{(cz + d)^2(c\bar{z} + d)^2}$$
$$= \frac{\mathrm{I} + \mathrm{II} + \mathrm{III} + \mathrm{IV}}{(cz + d)^2(c\bar{z} + d)^2},$$

where

$$\mathrm{I} = (dx^2 + dy^2)(a^2c^2x^2 + 2a^2cdx + a^2d^2 + a^2c^2y^2),$$
$$\mathrm{II} + \mathrm{III} = -ac(dx^2 + dy^2)[2ac(x^2 + y^2) + bc(z + \bar{z})$$
$$+ ad(z + \bar{z}) + 2bd]$$

$$= -(dx^2 + dy^2)(2a^2c^2x^2 + 2a^2c^2y^2 + 2abc^2x$$
$$+ 2a^2cdx + 2abcd),$$
$$\text{IV} = (dx^2 + dy^2)(a^2c^2x^2 + 2abc^2x + b^2c^2 + a^2c^2y^2).$$

Thus,

$$d\phi(z) \cdot \overline{d\phi(z)} = \frac{(ad - bc)^2(dx^2 + dy^2)}{[(cx + d)^2 + c^2y^2]^2} = \frac{dx^2 + dy^2}{[(cx + d)^2 + c^2y^2]^2}.$$

If we notice that

$$\phi(z) = \frac{(az + b)(c\bar{z} + d)}{(cz + d)(c\bar{z} + d)} = \frac{ac(x^2 + y^2) + (ad + bc)x + bd}{(cx + d)^2 + c^2y^2}$$
$$+ \sqrt{-1}\frac{y}{(cx + d)^2 + c^2y^2},$$

then

$$\|d\phi(z)\|^2 = \frac{d\phi(z) \cdot \overline{d\phi(z)}}{[\text{Im}\phi(z)]^2}$$
$$= \frac{dx^2 + dy^2}{[(cx + d)^2 + c^2y^2]^2} \cdot \frac{[(cx + d)^2 + c^2y^2]^2}{y^2}$$
$$= \frac{dx^2 + dy^2}{y^2} = \|dz\|^2,$$

which shows that ϕ is an isometry for $(\mathbb{H}^2, g)$.

(ii) By the formula for the Gaussian curvature in isothermal coordinates, it follows that

$$K = -\frac{1}{y^{-2}}\Delta \ln y^{-1} = y^2\frac{d^2}{dy^2} \ln y = -1,$$

which proves the assertion.

3344

Suppose that ∇ is a connection on a Riemannian manifold M. Define the torsion tensor τ via $\tau(X, Y) = \nabla_X Y - \nabla_Y X - [X, Y]$, where X, Y are vector fields on M. ∇ is called symmetric if the torsion tensor vanishes. Show that ∇ is symmetric if and only if the Christoffel symbols with respect to any coordinate frame are symmetric, i.e., $\Gamma_{ij}^k = \Gamma_{ji}^k$. Remember that if $\{E_i\}$ is a coordinate frame, and ∇ is a connection, the Christoffel symbols are defined via

$$\nabla_{E_i} E_j = \sum_k \Gamma_{ij}^k E_k.$$

(Harvard)

Solution.

It is easy to check that $\forall \lambda \in \mathbb{R}$ and $\forall X, Y, Z \in \Gamma(TM)$,

$$\tau(X + Z, Y) = \tau(X, Y) + \tau(Z, Y), \quad \tau(\lambda X, Y) = \lambda \tau(X, Y).$$

Hence

$$\tau(X, Y + Z) = \tau(X, Y) + \tau(X, Z), \quad \tau(X, \lambda Y) = \lambda \tau(X, Y),$$

because $\tau(X, Y) = -\tau(Y, X)$. Thus $\tau(X, Y)$ is $\mathbb{R}$-bilinear with respect to X and Y. Further, we are going to prove that $\tau(X, Y)$ is $C^\infty(M)$-bilinear with respect to X and Y. It suffices to prove that $\forall f \in C^\infty(M)$ and $\forall X, Y \in \Gamma(TM)$,

$$\tau(fX, Y) = f\tau(X, Y).$$

In fact, by the definition, we can get

$$\begin{aligned}
\tau(fX, Y) &= \nabla_{fX} Y - \nabla_Y (fX) - [fX, Y] \\
&= f\nabla_X Y - (Yf)X - f\nabla_Y X - f[X, Y] + (Yf)X \\
&= f\{\nabla_X Y - \nabla_Y X - [X, Y]\}.
\end{aligned}$$

Now we fix a coordinate system $(x^1, x^2, \ldots, x^n)$, then for the coordinate frame $\left(\frac{\partial}{\partial x^1}, \frac{\partial}{\partial x^2}, \ldots, \frac{\partial}{\partial x^n}\right)$, we have that

$$\begin{aligned}
\tau\left(\frac{\partial}{\partial x^i}, \frac{\partial}{\partial x^j}\right) &= \nabla_{\frac{\partial}{\partial x^i}} \frac{\partial}{\partial x^j} - \nabla_{\frac{\partial}{\partial x^j}} \frac{\partial}{\partial x^i} - \left[\frac{\partial}{\partial x^i}, \frac{\partial}{\partial x^j}\right] \\
&= \sum_k \left(\Gamma_{ij}^k - \Gamma_{ji}^k\right) \frac{\partial}{\partial x^k}.
\end{aligned}$$

Hence, the fact that

$$\tau\left(\frac{\partial}{\partial x^i}, \frac{\partial}{\partial x^j}\right) = 0, \quad \forall i, j = 1, 2, \ldots, n$$

is equivalent to that $\Gamma^k_{ij} = \Gamma^k_{ji}$, $\forall i, j, k = 1, 2, \ldots, n$. In the coordinate domain where the coordinate system $(x^1, x^2, \ldots, x^n)$ is defined, vector fields X and Y can be expressed as

$$X = \sum_i X^i \frac{\partial}{\partial x^i}, \quad Y = \sum_i Y^i \frac{\partial}{\partial x^i},$$

where $X^i = X^i(x^1, x^2, \ldots, x^n)$, $Y^i = Y^i(x^1, x^2, \ldots, x^n)$ are component functions. Then, using the $C^\infty(M)$-bilinearity of the torsion tensor proved in the above, the claim that $\tau(X, Y) = 0$ for any X and Y is equivalent to $\tau\left(\frac{\partial}{\partial x^i}, \frac{\partial}{\partial x^j}\right) = 0$, $\forall i, j = 1, 2, \ldots, n$. This concludes our proof.

3345

(i) Consider $\mathbb{R}^n$ with the standard Euclidean metric and let $p \in \mathbb{R}^n$ be an arbitrary point. For any $x \in \mathbb{R}^n$ let $\rho_p(x)$ be the distance from p to x. Viewing $\rho_p(x)$ as a smooth function of x away from p, verify that $|\text{grad}(\rho_p(x))|^2 = 1$ and that the integral curves of $\text{grad}(\rho_p(x))$ are straight lines. (Here $\text{grad}(\rho_p(x))$ refers to the usual gradient vector field of the function $\rho_p(x)$.)

(ii) More generally, given a smooth function f on a Riemannian manifold (M, g_{ij}), define $\text{grad}(f)$ to be the vector field given locally by

$$\sum_{i,j} \left(g^{ij} \frac{\partial f}{\partial x^j}\right) \frac{\partial}{\partial x^i}$$

show that if $|\text{grad}(f)|^2 = 1$, then the integral curves of the vector field $\text{grad}(f)$ are geodesics.

(Harvard)

Solution.

(i) In $\mathbb{R}^n$, using the Descartes coordinates $(x^1, x^2, \ldots, x^n)$ and noting

$$\rho_p(x) = \sqrt{\sum_i (x^i - p^i)^2},$$

we have that

$$\operatorname{grad} \rho_p(x) = \sum_j \frac{\partial \rho_p(x)}{\partial x^j} \frac{\partial}{\partial x^j} = \sum_j \frac{x^j - p^j}{\rho_p(x)} \frac{\partial}{\partial x^j}.$$

Then

$$|\operatorname{grad}(\rho_p(x))|^2 = \left\langle \sum_i \frac{x^i - p^i}{\rho_p(x)} \frac{\partial}{\partial x^i}, \sum_j \frac{x^j - p^j}{\rho_p(x)} \frac{\partial}{\partial x^j} \right\rangle$$

$$= \rho_p^{-2}(x) \sum_{i,j} (x^i - p^i)(x^j - p^l)\delta_{ij} = 1.$$

(ii) In a Riemannian manifold (M, g_{ij}), if a function f satisfies the condition $|\operatorname{grad}(f)|^2 = 1$, then by the definition of $\operatorname{grad}(f)$, we obtain that

$$|\operatorname{grad}(f)|^2 = \left\langle \sum_{i,k} \left(g^{ik} \frac{\partial f}{\partial x^k} \right) \frac{\partial}{\partial x^i}, \sum_{j,l} \left(g^{jl} \frac{\partial f}{\partial x^l} \right) \frac{\partial}{\partial x^j} \right\rangle$$

$$= \sum_{i,j,k,l} g^{ik} g^{jl} \frac{\partial f}{\partial x^k} \frac{\partial f}{\partial x^l} \left\langle \frac{\partial}{\partial x^i}, \frac{\partial}{\partial x^j} \right\rangle$$

$$= \sum_{k,l} g^{kl} \frac{\partial f}{\partial x^k} \frac{\partial f}{\partial x^l} = 1.$$

Differentiating the above final equation with respect to x^m gives that

$$\sum_{k,l} \frac{\partial g^{kl}}{\partial x^m} \frac{\partial f}{\partial x^k} \frac{\partial f}{\partial x^l} + 2 \sum_{k,l} g^{kl} \frac{\partial^2 f}{\partial x^m \partial x^k} \frac{\partial f}{\partial x^l} = 0 \quad (\forall m = 1, 2, \ldots, n)$$

which, by using the following identities related to the Christoffel symbols

$$\frac{\partial g^{kl}}{\partial x^m} = -\sum_i \left(g^{ki}\Gamma^l_{im} + g^{li}\Gamma^k_{im} \right) \quad (\forall k, l, m = 1, 2, \ldots, n),$$

could be rewritten as

$$\sum_{k,l} g^{kl} \frac{\partial^2 f}{\partial x^m \partial x^k} \frac{\partial f}{\partial x^l} = \frac{1}{2} \sum_{i,k,l} (g^{ki}\Gamma^l_{im} + g^{li}\Gamma^k_{im}) \frac{\partial f}{\partial x^k} \frac{\partial f}{\partial x^l}$$

$$= \sum_{i,k,l} g^{ki}\Gamma^l_{im} \frac{\partial f}{\partial x^k} \frac{\partial f}{\partial x^l}. \tag{3.2}$$

We will use these equations later.

Now let us return to discuss the integral curves $x^i = x^i(t)$ of the vector field $\mathrm{grad}(f)$, where $i = 1, 2, \ldots, n$. First, we have

$$\frac{dx^k}{dt} = \sum_i g^{ki} \frac{\partial f}{\partial x^i}, \text{ or equivalently,}$$

$$\frac{\partial f}{\partial x^k} = \sum_i g_{ki} \frac{dx^i}{dt}, \quad k = 1, 2, \ldots, n. \tag{3.3}$$

Then, differentiating with respect to t again, we get

$$\frac{d^2 x^k}{dt^2} = \sum_{i,j} \left(\frac{\partial g^{ki}}{\partial x^j} \frac{dx^j}{dt} \frac{\partial f}{\partial x^i} + g^{ki} \frac{\partial^2 f}{\partial x^j \partial x^i} \frac{dx^j}{dt} \right)$$

$$= \sum_{i,j} \left[\sum_l \left(-g^{kl}\Gamma^i_{lj} - g^{il}\Gamma^k_{lj} \right) \frac{dx^j}{dt} \frac{\partial f}{\partial x^i} + g^{ki} \frac{\partial^2 f}{\partial x^j \partial x^i} \frac{dx^j}{dt} \right]$$

$$= \sum_{i,j,l} \left(-g^{kl}\Gamma^i_{lj} \frac{dx^j}{dt} \frac{\partial f}{\partial x^i} \right) + \sum_{i,j,l} \left(-g^{il}\Gamma^k_{lj} \frac{dx^j}{dt} \frac{\partial f}{\partial x^i} \right)$$

$$+ \sum_{i,j} g^{ki} \frac{\partial^2 f}{\partial x^j \partial x^i} \frac{dx^j}{dt}$$

$$= \mathrm{I} + \mathrm{II} + \mathrm{III}.$$

Using (3.3), we rewrite the first and second summations as

$$\mathrm{I} = -\sum_{i,j,l} g^{kl}\Gamma^i_{lj}\left(\sum_m g^{jm}\frac{\partial f}{\partial x^m}\right)\frac{\partial f}{\partial x^i} = -\sum_{i,j,l,m} g^{kl}g^{jm}\Gamma^i_{lj}\frac{\partial f}{\partial x^m}\frac{\partial f}{\partial x^i}.$$

$$\mathrm{II} = -\sum_{j,l}\Gamma^k_{jl}\frac{dx^j}{dt}\left(\sum_i g^{il}\frac{\partial f}{\partial x^i}\right) = -\sum_{j,l}\Gamma^k_{jl}\frac{dx^j}{dt}\frac{dx^l}{dt}.$$

Noticing (3.3) and (3.2), the third summation becomes

$$\mathrm{III} = \sum_{i,j} g^{ki}\frac{\partial^2 f}{\partial x^j \partial x^i}\left(\sum_l g^{jl}\frac{\partial f}{\partial x^l}\right) = \sum_i g^{ki}\left(\sum_{j,l} g^{jl}\frac{\partial^2 f}{\partial x^j \partial x^i}\frac{\partial f}{\partial x^l}\right)$$

$$= \sum_i g^{ki}\sum_{j,l,m} g^{jm}\Gamma^l_{mi}\frac{\partial f}{\partial x^j}\frac{\partial f}{\partial x^l} = \sum_{i,j,l,m} g^{ki}g^{jm}\Gamma^l_{mi}\frac{\partial f}{\partial x^j}\frac{\partial f}{\partial x^l}.$$

Therefore, we finally obtain that

$$\frac{d^2 x^k}{dt^2} + \sum_{j,l}\Gamma^k_{jl}\frac{dx^j}{dt}\frac{dx^l}{dt} = 0, \quad k = 1, 2, \ldots, n.$$

Besides, $|\mathrm{grad}(f)|^2 = 1$ is equivalent to $\sum_{i,j} g_{ij}\frac{dx^i}{dt}\frac{dx^j}{dt} = 1$, hence the integral curves of the vector field $\mathrm{grad}(f)$ are geodesics.

PART 4

Real Analysis

Section 1

Measurability and Measure

4101

Let $S \subset [0,1]$ be the set defined by the property that $x \in S$ if and only if in the decimal representation of x the first appearance of the digit 2 precedes the first appearance of the digit 3. Prove that S is Lebesgue measurable and find its measure.

(*Stanford*)

Solution.

For any $x \in [0,1)$ there is a unique sequence $\{p_n(x)\}_{n=1}^{\infty}$ of integers satisfying the following properties

(1) $0 \le p_n(x) \le 9$,   (2) $\forall n, \exists m \ge n : p_m(x) \le 8$ and

(3) $x = \sum_{n=1}^{\infty} \frac{p_n(x)}{10^n}$.

Then

$$S = \{1\} \cup \{x \in [0,1) \mid \forall n, p_n(x) \ne 2 \quad \text{and} \quad p_n(x) \ne 3\}$$
$$\cup \{x \in [0,1) \mid \exists n, p_n(x) = 2, \forall i < n, p_i(x) \ne 2 \quad \text{and}$$
$$p_i(x) \ne 3\}.$$

Let

$$A = \{x \in [0,1) \mid \forall n, p_n(x) \ne 2 \quad \text{and} \quad p_n(x) \ne 3\}$$

and

$$B = \{x \in [0,1) \mid \exists n, p_n(x) = 2, \forall i < n, p_i(x) \neq 2 \quad \text{and}$$
$$p_i(x) \neq 3\}.$$

Then

$$[0,1)\backslash A = \bigcup_{n=1}^{\infty} \bigcup_{k_i \neq 2,3} \left[\frac{k_1}{10} + \cdots + \frac{k_{n-1}}{10^{n-1}} + \frac{2}{10^n}, \right.$$
$$\left. \frac{k_1}{10} + \cdots + \frac{k_{n-1}}{10^{n-1}} + \frac{4}{10^n} \right)$$

and

$$B = \bigcup_{n=1}^{\infty} \bigcup_{k_i \neq 2,3} \left[\frac{k_1}{10} + \cdots + \frac{k_{n-1}}{10^{n-1}} + \frac{2}{10^n}, \right.$$
$$\left. \frac{k_1}{10} + \cdots + \frac{k_{n-1}}{10^{n-1}} + \frac{3}{10^n} \right).$$

It follows that both A and B are measurable and therefore is S.
Since

$$m([0,1)\backslash A) = \sum_{n=1}^{\infty} \frac{2}{10^n} \cdot 8^{n-1} = 1$$

and

$$m(B) = \sum_{n=1}^{\infty} \frac{1}{10^n} \cdot 8^{n-1} = \frac{1}{2},$$

we have

$$m(S) = m(A) + m(B) = \frac{1}{2}.$$

4102

Define $S^1 = \mathbb{R}/\mathbb{Z}$ endowed with the natural Lebesgue measure. Consider on S^1 the equivalence relation: $[x] \sim [y] \Leftrightarrow x - y \in \mathbb{Q}$, for $x, y \in \mathbb{R}$ and $[\,]$ denoting the class in S^1. For each

$\xi \in S^1/\sim$, choose a representative $\bar{\xi} \in S^1$, and let $E \subset S^1$ be the set of these points, i.e.,

$$E = \{\bar{\xi} \mid \bar{\xi} \in \xi, \ \forall \xi \in S^1/\sim\}.$$

Show that E is not measurable.

(Stanford)

Solution.

S^1 is an abelian group under the binary operation $([x], [y]) \mapsto [x+y]$ and the inverse operation $[x] \mapsto [-x]$.

For any $[x] \in S^1$ there is by the construction of E a unique element $[y] \in E$ such that $x - y \in \mathbb{Q}$. Then $[x] = [y] + [r]$, where $r \in [0,1) \cap \mathbb{Q}$ is such that $r \equiv x - y \mod \mathbb{Z}$. So we conclude that $S^1 = \bigcup_{r \in [0,1) \cap \mathbb{Q}} (E + [r])$.

If $r, s \in [0,1) \cap \mathbb{Q}$ are such that $(E+[r]) \cap (E+[s]) \neq \emptyset$ there are $[x], [y] \in E$ such that $[x] + [r] = [y] + [s]$. It follows that $x \equiv y \mod \mathbb{Q}$ and therefore $[x] = [y]$ by the construction of E. Thus $[r] = [s]$, which then implies that $r = s$ since $-1 < r - s < 1$.

It follows that $\{E + [r] \mid r \in [0,1) \cap \mathbb{Q}\}$ is a partition of S^1.

Let m denote the natural Lebesgue measure on S^1 such that $m(S^1) = 1$. If E is measurable then

$$m(S^1) = \sum_{r \in [0,1) \cap \mathbb{Q}} m(E + [r]) = \sum_{r \in [0,1) \cap \mathbb{Q}} m(E).$$

If $m(E) = 0$ then $m(S^1) = 0$, a contradiction; Otherwise $m(S^1) = \infty$, a contradiction, too.

4103

Let X be a set and $\mathcal{D} \subset \mathcal{P}(X)$, $\mathcal{D}$ closed under finite intersection. Denote by $\mathcal{R}$ the ring generated by $\mathcal{D}$. Futhermore, let π be the smallest system, $\mathcal{D} \subset \pi \subset \mathcal{P}(X)$ such that π is closed under the following operations:

(i) finite disjoint unions;
(ii) differences $A \backslash B$, $B \subset A$.

Prove that $\pi = \mathcal{R}$.

(Iowa)

Solution.

Obviously, $\pi \subseteq \mathcal{R}$. We have

$$\{A \in \mathcal{R} \mid A \cap B \in \pi, \quad B \in \mathcal{D}\} = \mathcal{R} \qquad (1)$$

since the former is a ring containing $\mathcal{D}$. Also we have

$$\{B \in \mathcal{R} \mid A \cap B \in \pi, \quad A \in \mathcal{R}\} = \mathcal{R}, \qquad (2)$$

since by the equality (1) the former is a ring containing $\mathcal{D}$. By equality (2), $\mathcal{R} = \pi$ since for any $A \in \mathcal{R}$, $A = A \cap A$.

4104

Let μ^* be the Lebesgue outer measure on $\mathbb{R}$ and A, B subsets of $\mathbb{R}$ such that

$$\inf\{|x - y| \mid x \in A, y \in B\} > 0.$$

Prove or disprove that

$$\mu^*(A \cup B) = \mu^*(A) + \mu^*(B).$$

(Iowa)

Solution.

We will show the equality. Let $r = d(A, B) > 0$. Let

$$U = \left\{x \in \mathbb{R} \,\middle|\, d(x, A) < \frac{r}{2}\right\}$$

and

$$V = \left\{x \in \mathbb{R} \,\middle|\, d(x, B) < \frac{r}{2}\right\}.$$

Then U and V are disjoint open sets containing A and B respectively. We have

$$\mu^*(A \cup B) = \inf\{\mu(W) \mid W \text{ open}, A \cup B \subseteq W\}$$
$$= \inf\{\mu(W) \mid W \text{ open}, A \cup B \subseteq W \subseteq U \cup V\}$$

$$= \inf\{\mu(W_1) + \mu(W_2) \mid W_i \text{ open}, A \subseteq W_1 \subseteq U,$$
$$B \subseteq W_2 \subseteq V\}$$
$$\overset{(1)}{=} \inf\{\mu(W_1) \mid W_1 \text{ open}, \ A \subseteq W_1 \subseteq U\}$$
$$+ \inf\{\mu(W_2) \mid W_2 \text{ open}, \ B \subseteq W_2 \subseteq V\}$$
$$= \mu^*(A) + \mu^*(B),$$

where the equality (1) follows from the following equality

$$\inf(X + Y) = \inf X + \inf Y, \quad X, Y \subseteq \mathbb{R}.$$

4105

(a) Consider a measurable space (X, μ) with a finite, positive, finitely additive measure μ. Finite additivity means that whenever $\{B_i\}$ is a finite collection of mutually disjoint measurable sets, then $\mu(\cup B_i) = \sum \mu(B_i)$. Prove that μ is countably additive if and only if it satisfies the following condition:

If A_n is a decreasing sequence of sets with empty intersection then

$$\lim_{r \to \infty} \mu(A_n) = 0.$$

(b) Now that suppose X is a locally compact Hausdorff space, that $\mathcal{B}$ is the Borel σ-algebra, and that μ is a finite, positive, finitely additive measure on $\mathcal{B}$. Suppose moreover that μ is regular, that is for each $B \in \mathcal{B}$,

$$\mu(B) = \sup\{\mu(K) \mid K \subseteq B \text{ and } K \text{ is compact}\}.$$

Prove that μ is countably additive.

(*Iowa*)

Solution.

(a) The sufficiency. Let $\{B_n\}$ be countably many measurable sets which are mutually disjoint. Let $A_n = \bigcup_{i=n+1}^{\infty} B_i$. Then

$\bigcap_{n=1}^{\infty} A_n = \emptyset$. We have

$$\mu\left(\bigcup_{n=1}^{\infty} B_n\right) = \lim_{n \to \infty}\left(\mu\left(\bigcup_{i=1}^{n} B_i\right) + \mu\left(\bigcup_{i=n+1}^{\infty} B_i\right)\right)$$

$$= \lim_{n \to \infty}\left(\sum_{i=1}^{n} \mu(B_i) + \mu(A_n)\right) = \sum_{i=1}^{\infty} \mu(B_i).$$

Therefore μ is a measure. The necessity is obvious.

(b) By (a) it sufficies to show any decreasing sequence $\{A_n\}$ of measurable sets such that $C := \inf_{n \geq 1} \mu_{A_n} > 0$ has a nonvoid intersection.

For each n there is a compact K_n contained in A_n such that

$$\mu(A_n) < \mu(K_n) + \frac{1}{2^{n+1}}c.$$

It suffices to show $\{K_n\}$ has a nonvoid intersection. However,

$$\mu\left(A_n \backslash \bigcap_{i=1}^{n} K_i\right) \leq \sum_{i=1}^{n} \mu(A_i \backslash K_i) < \frac{1}{2}c,$$

which implies that

$$\mu\left(\bigcap_{i=1}^{n} K_i\right) \neq 0$$

and therefore

$$\bigcap_{i=1}^{n} K_i \neq \emptyset.$$

Thus $\{\bigcap_{i=1}^{n} K_i \mid n \in \mathbb{N}\}$ is a decreasing sequence of non-empty compact subsets in the compact space K_1. So $\bigcap_{n=1}^{\infty} K_n \neq \emptyset$.

4106

For $f : [0, 1] \to \mathbb{R}$, let $E \subset \{x \mid f'(x) \text{ exists}\}$. If $m(E) = 0$, show that $m(f(E)) = 0$.

(Indiana–Purdue)

Solution.

Denote

$$F = \{x \mid x \in E, |f'(x)| < M\},$$

where M is any positive number. It suffices to prove $m(f(F)) = 0$. Set

$$F_n = \left\{ x \mid x \in F, |f(y) - f(x)| \leq M|y - x| \text{ if } |y - x| < \frac{1}{n} \right\}.$$

Then

$$F_1 \subset F_2 \subset \cdots, m(F_n) = 0,$$

and

$$f(F) \subset \cup f(F_n) = \lim_{n \to \infty} f(F_n).$$

For any $\varepsilon > 0$, take a sequence $\{I_{n,k}\}$ of open intervals such that

$$F_n \subset \bigcup_k I_{n,k}, m(I_{n,k}) < \frac{1}{n}, \quad \sum_k m(I_{n,k}) < \varepsilon.$$

For $x, y \in F_n \cap I_{n,k}$, we have

$$|f(x) - f(y)| \leq M m(I_{n,k}).$$

Therefore

$$m^*(f(F_n)) = m^* \left(f \left(F_n \cap \left(\bigcup_k I_{n,k} \right) \right) \right)$$

$$\leq \sum_k m^*(f(F_n \cap I_{n,k}))$$

$$\leq M \sum m(I_{n,k}) < M\varepsilon.$$

Thus $m^*(f(F_n)) = 0$, which implies that $m(f(F)) = 0$.

4107

Suppose $A \subset \mathbb{R}$ is Lebesgue measurable and assume that

$$m(A \cap (a,b)) \leq \frac{b-a}{2}$$

for any $a, b \in \mathbb{R}$, $a < b$. Prove that $m(A) = 0$.

(Iowa)

Solution.

If $m(A) \neq 0$ there is an n such that $m(A \cap (n, n+1)) \neq 0$. There is an open subset U in $(n, n+1)$ such that

$$A \cap (n, n+1) \subseteq U \subseteq (n, n+1)$$

and

$$m(U) < m(A \cap (n, n+1)) + \varepsilon,$$

where $\varepsilon < m(A \cap (n, n+1))$.

There are at most countably many disjoint intervals (a_j, b_j)'s such that

$$U = \bigcup_j (a_j, b_j).$$

Then

$$A \cap (n, n+1) = \bigcup_j A \cap (a_j, b_j).$$

We have

$$m(A \cap (n, n+1)) = \sum_j m(A \cap (a_j, b_j))$$

$$\leq \sum_j \frac{b_j - a_j}{2}$$

$$= \frac{1}{2} m(U) < \frac{1}{2} (m(A \cap (n, n+1)) + \varepsilon)$$

which deduces that

$$m(A \cap (n, n+1)) < \varepsilon,$$

a contradiction.

4108

Choose $0 < \lambda < 1$ and construct the Cantor set K_λ as follows: Remove from $[0, 1]$ its middle part of length λ; we are left with two intervals I_1 and I_2. Remove from each of them their middle parts of lengths $\lambda|I_i|$, etc. and keep doing this ad infinitum. We are left with the set K_λ. Prove that the set K_λ has Lebesgue measure zero.

(Stanford)

Solution.

Claim. *For any* positive *interger* n, *the total length of intervals removed in the n-th step is* $\lambda(1-\lambda)^{n-1}$.

The claim holds for $n = 1$. Assume that it holds for any $n \leq k$. Then the total length of intervals removed in the $(k+1)$-th step is

$$\lambda \left(1 - \sum_{i=1}^{k} \lambda(1-\lambda)^{i-1} \right) = \lambda(1-\lambda)^k.$$

By induction the claim holds for any $n \in \mathbb{N}$.

It follows that the Lebesgue measure of K_λ is

$$1 - \sum_{n=1}^{\infty} \lambda(1-\lambda)^{n-1} = 0.$$

4109

Suppose μ is a positive Borel measure on $\mathbb{R}$ such that
(i) $\mu([0, 1]) = 1$, and (ii) $\mu(E) = \mu(x + E)$ for any Borel set E of $\mathbb{R}$ and every $x \in \mathbb{R}$. Does this imply that μ is the Lebesgue measure? Justify your answer.

(Iowa)

Solution.

Yes, μ is the Lebesgue measure. For any $x \in \mathbb{R}$ define

$$g(x) = \begin{cases} \mu((0, x]), & x \geq 0, \\ -\mu((x, 0]), & x < 0. \end{cases}$$

Then $g : \mathbb{R} \to \mathbb{R}$ is nondecreasing and right-continuous. Moreover, for any $x, y \in \mathbb{R}$ with $x < y$, $\mu(x, y] = g(y) - g(x)$. It follows that the measure μ is induced by g.

For any $x, y \in \mathbb{R}$, from either

$$\mu(x, x + y] = \mu(0, y], \quad y \geq 0$$

or

$$\mu(x + y, x] = \mu(y, 0], \quad y < 0$$

we have

$$g(x + y) = g(x) + g(y).$$

From the right-continuity of g, we conclude that $g(x) = xg(1)$. However,

$$g(1) = \mu((0, 1]) = \mu([0, 1]) - \mu(\{0\})$$

$$= 1 - \lim_{n \to \infty} \mu\left(\left(-\frac{1}{n}, 0\right]\right)$$

$$= 1 - \lim_{n \to \infty} \left(g(0) - g\left(-\frac{1}{n}\right)\right) = 1.$$

It follows that $g(x) \equiv x$ and therefore μ is the Lebesgue measure.

4110

Let $\mathcal{U}$ be a σ-algebra of subsets of a set X and $\mu_n : \mathcal{U} \to \mathbb{R}$ be signed measures such that $\mu(E) = \lim_{n \to \infty} \mu_n(E)$ exists for every $E \in \mathcal{U}$. Prove that μ is a signed measure.

(Iowa)

Solution.

Let, for each $E \in \mathcal{U}$,

$$\tilde{\mu}(E) = \sum_{n=1}^{\infty} \frac{1}{n^2} \frac{|\mu_n|(E)}{1 + |\mu_n|(X)}.$$

Then $\tilde{\mu}$ is a finite measure and μ_n's are absolutely continuous with respect to $\tilde{\mu}$. Given any mutually disjoint measurable sets E_n's, by the Vitali–Hahn–Saks Theorem one has

$$\lim_{n \to \infty} \sup_m \left| \mu_m \left(\bigcup_{k=n}^{\infty} E_k \right) \right| = 0.$$

Then

$$\mu \left(\bigcup_{k=1}^{\infty} E_k \right) = \lim_{m \to \infty} \mu_m \left(\bigcup_{k=1}^{\infty} E_k \right)$$

$$= \lim_{n \to \infty} \lim_{m \to \infty} \left(\mu_m \left(\bigcup_{k=1}^{n} E_k \right) + \mu_m \left(\bigcup_{k=n+1}^{\infty} E_k \right) \right)$$

$$= \lim_{n \to \infty} \left(\sum_{k=1}^{n} \mu(E_k) + \mu_m \left(\bigcup_{k=n+1}^{\infty} E_k \right) \right)$$

$$= \sum_{k=1}^{\infty} \mu(E_k),$$

which shows that μ is a generalized measure and therefore a signed measure.

4111

Let λ be Lebesgue measure on $\mathbb{R}$. Show that for any Lebesgue measurable set $E \subset \mathbb{R}$ with $\lambda(E) = 1$, there is a Lebesgue measurable set $A \subset E$ with $\lambda(A) = \frac{1}{2}$.

(*Iowa*)

Solution.

Define the function $f : \mathbb{R} \to [0, 1]$ by

$$f(x) = \lambda(E \cap (-\infty, x]), \quad x \in \mathbb{R}.$$

It is continuous by the following inequality

$$|f(x) - f(y)| \leq |x - y|, \quad x, y \in \mathbb{R}.$$

Since $\lim_{x \to -\infty} f(x) = 0$ and $\lim_{x \to \infty} f(x) = 1$, there is a point $x_0 \in \mathbb{R}$ such that $f(x_0) = \frac{1}{2}$. Put $A = E \cap (-\infty, x_0]$, as desired.

4112

Let μ be a complex Borel measure on $[0, \infty)$. Show that if

$$\int_0^\infty e^{-nx} d\mu(x) = 0$$

for all $n = 0, 1, 2, \ldots$, then $\mu = 0$.

(Iowa)

Solution.
  Let

$$S = \operatorname{span}\{e^{-nx} \mid n = 0, 1, 2, \ldots\}$$

then S is a self-adjoint subalgebra of $C([0, \infty])$ separating the points of $[0, \infty]$. It follows from the Stone–Weierstrass Theorem, S is dense in $C([0, \infty])$ under the supremum norm topology. Therefore for any $f \in C([0, \infty])$,

$$\int_{[0,\infty]} f d\mu = 0,$$

which then implies that $\mu = 0$.

4113

Let $A \subset [0, 1]$ be a measurable set of positive measure. Show that there exist two points $x' \neq x''$ in A with $x' - x''$ rational.

(Indiana–Purdue)

Solution.
  Denote all the rational numbers in $[-1, 1]$ by $r_1, r_2, \ldots, r_n, \ldots$. Denote

$$A_n = \{x + r_n \mid x \in A\}.$$

Then $m(A_n) = m(A) > 0$. $A_n \subset [-1, 2]$. Thus

$$\bigcup_{n=1}^{\infty} A_n \subset [-1, 2].$$

Suppose that $A_n \cap A_m = \emptyset$ if $n \neq m$. Then

$$\sum_{n=1}^{\infty} m(A_n) \leq m([-1, 2]) = 3,$$

which contradicts $m(A) > 0$. Therefore there must be some m, n such that $A_n \cap A_m \neq \emptyset$. Take $z \in A_n \cap A_m$. Then we can find x', $x'' \in A$ such that

$$z = x' + r_n = x'' + r_m.$$

Thus

$$x' - x'' = r_m - r_n.$$

4114

Let $E = \{\sum_{n=1}^{\infty} \frac{x_n}{3^n} | x_n = 0 \text{ or } 1\}$. Prove that E is Lebesgue measurable with Lebesgue measure 0. Also show that $E + E = [0, 1]$.

(*Iowa*)

Solution.

Obviously, $E \subset [0, 1)$. We will show next that

$$[0, 1) \backslash E = \dot{\bigcup} \left\{ \left[\frac{x_1}{3} + \cdots + \frac{x_{n-1}}{3^{n-1}} + \frac{2}{3^n}, \frac{x_1}{3} + \cdots + \frac{x_{n-1} + 1}{3^{n-1}} \right) \right. $$
$$\left. \left| 0 \leq x_1, \ldots, x_{n-1} \leq 1, n \in \mathbb{N} \right\} \right. . \tag{1}$$

Indeed, for any $x \in [0, 1) \backslash E$, $x = \sum_{n=1}^{\infty} \frac{x_n}{3^n}$ (where $0 \leq x_n \leq 2$ and for any n there is an $m \geq n$ such that $x_m \leq 1$), there is at least one n such that $x_n = 2$. Let $n = \min\{k | x_k = 2\}$ then $0 \leq x_1, \ldots, x_{n-1} \leq 1$ and therefore

$$x \in \left[\frac{x_1}{3} + \cdots + \frac{x_{n-1}}{3^{n-1}} + \frac{2}{3^n}, \frac{x_1}{3} + \cdots + \frac{x_{n-1} + 1}{3^{n-1}} \right).$$

The converse inclusion is obvious.

By equality (1), E is measurable and since

$$m([0,1)\backslash E) = \sum_{n=1}^{\infty} \sum_{0 \le x_1, \dots, x_{n-1} \le 1} \frac{1}{3^n} = \sum_{n=1}^{\infty} \frac{2^{n-1}}{3^n} = 1,$$

$m(E) = 0$. Since $E \subset [0, \frac{1}{2}]$, $E + E \subseteq [0,1]$. For any $x \in [0,1)$ say $x = \sum_{n=1}^{\infty} \frac{x_n}{3^n}$ (where $0 \le x_n \le 2$, and for any n, there is an $m \ge n$ such that $x_m \ne 2$). Define

$$x_n' = \begin{cases} x_n, & 0 \le x_n \le 1, \\ 1, & x_n = 2, \end{cases} \quad n \in \mathbb{N},$$

and

$$x_n'' = \begin{cases} 0, & 0 \le x_n \le 1, \\ 1, & x_n = 2, \end{cases} \quad n \in \mathbb{N}.$$

Then $x' = \sum_{n=1}^{\infty} \frac{x_n'}{3^n}$ and $x'' = \sum_{n=1}^{\infty} \frac{x_n''}{3^n}$ belong to E so that $x' + x'' = x$. Obviously,

$$1 = \left(\sum_{n=1}^{\infty} \frac{1}{3^n} \right) + \left(\sum_{n=1}^{\infty} \frac{1}{3^n} \right) \in E + E.$$

This completes the proof.

4115

Let $f : \mathbb{R} \to \mathbb{R}_+$ be measurable, and let $\varepsilon > 0$. Show that there exists $g : \mathbb{R} \to \mathbb{R}_+$ measurable such that (i) $\|f - g\|_\infty \le \varepsilon$ (ii) for every $r \in \mathbb{R}$, $|\{x \,|\, g(x) = r\}| = 0$.

(*Indiana–Purdue*)

Solution.

Take $\{r_n\}$ such that $0 < r_1 < r_2 < \cdots < r_n < \cdots$, $\lim r_n = +\infty$, $r_{n+1} - r_n < \varepsilon$, for all n. Set $f_1(x) = f(x) + \operatorname{arcctg} x$, Denote

$$E_n = \{x \,|\, r_{n-1} < f_1(x) \le r_n\}.$$

Set $g_1 = \sum r_n \chi_{E_n}$. Then

$$g_1(x) \geq \operatorname{arcctg} x, \quad \|f_1 - g_1\|_\infty \leq \varepsilon.$$

Set $g(x) = g_1(x) - \operatorname{arcctg} x$. We have $g \geq 0$ and $\|f - g\|_\infty = \|f_1 - g_1\|_\infty \leq \varepsilon$. For every $r \in \mathbb{R}$,

$$\{x \mid g(x) = r\} = \{x \mid g_1(x) - \operatorname{arcctg} x = r\}$$
$$= \bigcup_n \{x \mid x \in E_n, \ \operatorname{arcctg} x = r_n - r\}.$$

Since $|\{x \mid \operatorname{arcctg} x = r_n - r\}| = 0$, $|\{x \mid g(x) = r\}| = 0$.

4116

Let f be the function on $[0, 1]$ defined as follows $f(x) = 0$ if x is a point on the Cantor ternary set and $f(x) = \frac{1}{p}$ if x is in one of the complementary intervals of length 3^{-p}.

(a) Prove that f is measurable.
(b) Evaluate $\int_0^1 f(x)dx$.

(*Stanford*)

Solution.

Let K denote the Cantor ternary set. Then

$$[0,1] \backslash K = \bigcup_{p \geq 1} \bigcup_{a_i = 0,2} (0 \cdot a_1 \cdots a_{p-1}1, 0 \cdot a_1 a_2 \cdots a_{p-1}2).$$

By the definition, we have

$$f = \sum_{p \geq 1} \sum_{a_i = 0,2} \frac{1}{p} \chi_{(0 \cdot a_1 \cdots a_{p-1}1, 0 \cdot a_1 a_2 \cdots a_{p-1}2)},$$

which then is measurable, and

$$\int_0^1 f(x)dx = \sum_{p \geq 1} \sum_{a_i = 0,2} \frac{1}{p} \cdot \frac{1}{3^p} = \sum_{p=1}^\infty \frac{1}{p} \frac{2^{p-1}}{3^p} = \ln \sqrt{3}.$$

4117

Let E be a Lebesgue measurable subset of $\mathbb{R}$ with $m(E) < \infty$ and let

$$f(x) = m((E + x) \cap E).$$

Show that

(a) $f(x)$ is a continuous function on $\mathbb{R}$.

(b) $\lim_{x \to +\infty} f(x) = 0$.

(*Illinois*)

Solution.

We will show first that for any Lebesgue measurable set E,

$$\lim_{h \to 0} m((E + h) \cap E) = m(E).$$

If E is of the form $\dot\bigcup_{i=1}^{n}(\alpha_i, \beta_i)$, where (α_i, β_i)'s are finitely many mutually disjoint open intervals of finite length, then

$$m(E) = \sum_{i=1}^{n}(\beta_i - \alpha_i)$$

$$= \lim_{h \to \infty} \sum_{i=1}^{n}(\beta_i \wedge (\beta_i + h) - \alpha_i \vee (\alpha_i + h))$$

$$= \lim_{h \to 0} \sum_{i=1}^{n} m((\alpha_i, \beta_i) \cap (\alpha_i + h, \beta_i + h))$$

$$= \lim_{h \to 0} m\left(\bigcup_{i=1}^{n}(\alpha_i, \beta_i) \cap (\alpha_i + h, \beta_i + h)\right)$$

$$\le \varliminf_{h \to 0} m(E \cap (E + h))$$

$$\le \varlimsup_{h \to 0} m(E \cap (E + h)) \le m(E).$$

So

$$\lim_{h \to 0} m((E + h) \cap E) = m(E).$$

If E is a compact set, then there is, for any $\varepsilon > 0$, an open set E' of the above form such that $E \subset E'$ and $m(E'\backslash E) < \varepsilon$. Let $F = E'\backslash E$ one has

$$m(E) < m(E') = \lim_{h \to 0} m((E' + h) \cap E')$$

$$= \lim_{h \to 0} m(((E + h) \cap E) \dot{\cup} ((E + h) \cap F)$$

$$\dot{\cup} ((F + h) \cap E) \dot{\cup} ((F + h) \cap F))$$

$$\leq \lim_{h \to 0} m((E + h) \cap E) + 3\varepsilon$$

$$\leq \overline{\lim_{h \to 0}} \, m((E + h) \cap E) + 3\varepsilon \leq m(E) + 3\varepsilon.$$

It follows that

$$m(E) = \lim_{h \to 0} m((E + h) \cap E).$$

In general, there is an increasing sequence $\{E_n\}$ of compact sets such that $E_n \subseteq E$ and

$$\lim_{n \to \infty} m(E_n) = m(E).$$

Then

$$m(E) = \lim_{n \to \infty} m(E_n) = \lim_{n \to \infty} \lim_{h \to 0} m((E_n + h) \cap E_n)$$

$$\leq \lim_{n \to \infty} \lim_{h \to 0} m((E + h) \cap E) = \lim_{h \to 0} m((E + h) \cap E).$$

So

$$\lim_{h \to 0} m((E_n + h) \cap E) = m(E).$$

(a) For any $y, x \in \mathbb{R}$

$$|f(y) - f(x)| = |m(((E + y)\backslash(E + x)) \cap E)$$

$$- m((E + x)\backslash(E + y))|$$

$$\leq m((E + y)\backslash(E + x)) + M((E + x)\backslash(E + y))$$

$$= m(E + (y - x)\backslash E) + m(E + (x - y)\backslash E)$$
$$= m(E) - m(E + (y - x)) \cap E) + m(E)$$
$$- m((E + (x - y)) \cap E) \to 0, \quad \text{as} \quad y \to x.$$

(b) If E is compact, then there is an $r > 0$ such that for any $x > r$, $(x + E) \cap E = \emptyset$. The claim follows.

In general there is an increasing sequence $\{E_n\}$ of compact sets such that $E_n \subseteq E$ and $\lim_{n\to\infty} m(E_n) = m(E)$. Then

$$\lim_{x\to+\infty} m((E + x) \cap E)$$

$$= \lim_{x\to+\infty} m((E + x) \cap E) - m((E_n + x) \cap E_n)$$

$$+ m((E_n + x) \cap E_n)$$

$$= \lim_{n\to\infty} \lim_{x\to+\infty} m(((E + x) \cap E)\backslash((E_n + x) \cap E_n))$$

$$+ m((E_n + x) \cap E_n)$$

$$\leq \lim_{n\to\infty} (m((E + x)\backslash(E_n + x)) + m(E\backslash E_n)) = 0.$$

4118

Let $A \subset \mathbb{R}$ be a set of positive Lebesgue measure. Prove that

$$\varphi(x) = \int \chi_A(tx)\chi_A(t)dt$$

is continuous at $x = 1$. Use this result to prove that there exists an $\varepsilon > 0$ such that for any $m \in \mathbb{R}$ with $|m - 1| < \varepsilon$, the line $y = mx$ has a non-void intersection with $A \times A$.

(*Iowa*)

Solution.

If B is a bounded open set, say $\bigcup_{i<m}(a_i, b_i)$, where $2 \leq m \leq \infty$ and (a_i, b_i)'s are mutually disjoint open intervals, then for

any $n < m$ and any $x > 0$ we have

$$\bigcup_{i=1}^{n}(xa_i, xb_i) \cap (a_i, b_i) \subseteq xB \cap B.$$

We have

$$\mu(B) = \sup_{n<m} \sum_{i=1}^{n}(b_i - a_i)$$

$$= \sup_{n<m} \lim_{x \to 1} \sum_{i=1}^{n} \mu((xa_i, xb_i) \cap (a_i, b_i))$$

$$\leq \lim_{x \to 1} \mu(xB \cap B)$$

$$\leq \overline{\lim_{x \to 1}} \mu(xB \cap B) \leq \mu(B), \tag{1}$$

where μ is the Lebesgue measure. If K is a compact set, there is a decreasing sequence $\{B_n\}$ of bounded open sets such that $K = \bigcap_{n=1}^{\infty} B_n$. Since

$$\mu(xB_n \cap B_n) - \mu(xK \cap K) \leq \mu(x(B_n \backslash K)) + \mu(B_n \backslash K)$$

we have by (1)

$$\mu(K) = \lim_{n \to \infty} \mu(B_n) = \lim_{n \to \infty} \lim_{x \to 1} \mu(xB_n \cap B_n)$$

$$\leq \lim_{n \to \infty} \lim_{x \to 1} (\mu(xK \cap K) + \mu(x(B_n \backslash K)) + \mu(B_n \backslash K))$$

$$= \lim_{n \to \infty} \left(\lim_{x \to 1} \mu(xK \cap K) + 2\mu(B_n \cap K) \right)$$

$$= \lim_{x \to 1} \mu(xK \cap K) \leq \overline{\lim_{x \to 1}} \mu(xK \cap K) \leq \mu(K). \tag{2}$$

In general, there is an increasing sequence $\{K_n\}$ of compact sets such that $\bigcup_{n=1}^{\infty} K_n \subseteq A$ and $\mu(A \backslash \bigcup_{n=1}^{\infty} K_n) = 0$. By (2) we have

$$\mu(A) = \lim_{n \to \infty} \mu(K_n) = \lim_{n \to \infty} \lim_{x \to 1} \mu(xK_n \cap K_n)$$

$$\leq \lim_{x \to 1} \mu(xA \cap A) \leq \overline{\lim_{x \to 1}} \mu(xA \cap A) \leq \mu(A). \tag{3}$$

Since

$$|\varphi(x) - \varphi(1)| = \left| \int_{-\infty}^{+\infty} (\chi_{\frac{1}{x}A}(t)\chi_A(t) - \chi_A(t))dt \right|$$

$$= \mu(A) - \mu\left(\frac{1}{x}A \cap A\right)$$

we have by (3) $\lim_{x \to 1} \varphi(x) = \varphi(1)$.

If there is no $\varepsilon > 0$ with the property mentioned in the question, then for any $n \in \mathbb{N}$ there is an $m_n \in \mathbb{R}$ such that $|m_n - 1| < \frac{1}{n}$ and the line $y = m_n x$ has a void intersection with $A \times A$. However, this implies that $m_n(A \cap A) = \emptyset$, and therefore

$$\mu(A) = \lim_{n \to \infty} \mu(m_n(A \cap A)) = 0,$$

a contradiction.

4119

Let μ be a countably additive measure on a set S with $\mu(S) < +\infty$ that is without atoms, i.e., if A is a measurable set with $\mu(A) > 0$, then there is a measurable set $B \subset A$ such that $0 < \mu(B) < \mu(A)$. Prove that the range of μ is the closed interval $[0, \mu(S)]$.

(*Courant Inst.*)

Solution.
If there is a $t_0 \in (0, \mu(S))$ not in the range of μ, let

$$\mathcal{P} = \{A \mid A \text{ measurable and } \mu(A) < t_0\}/\sim,$$

where $\sim$ is an equivalence relation: $A \sim B$ if $\mu(A\backslash B) = \mu(B\backslash A) = 0$. Then $\mathcal{P}$ is a partially ordered set: $[A] \leq [B]$ if $\mu(A\backslash B) = 0$. Given a totally ordered set $\mathcal{Q}$ of $\mathcal{P}$, let $\beta = \sup\{\mu(A) \mid [A] \in \mathcal{Q}\}$. Then $\beta \leq t_0$ and there is an increasing sequence $\{[A_n]\}$ of $\mathcal{Q}$ with $\{\mu(A_n)\}$ increasing to β.

Let $A = \bigcup_{n=1}^{\infty} A_n$, then

$$\beta \leq \mu(A) = \lim_{n \to \infty} \mu\left(\bigcup_{i=1}^{n} A_i\right)$$

$$= \lim_{n \to \infty} \left(\mu\left(\bigcup_{i=1}^{n} A_i \backslash A_n\right) + \mu(A_n)\right)$$

$$= \lim_{n \to \infty} \mu(A_n) = \beta \leq t_0.$$

So $\mu(A) = \beta < t_0$ and therefore $[A] \in \mathcal{P}$. For any $[B] \in \mathcal{Q}$, if $[B] \leq [A_n]$ for some n, then $[B] \leq [A]$; otherwise, $[B] = [A]$ since

$$\mu(A \backslash B) \leq \sum_{n=1}^{\infty} \mu(A_n \backslash B) = 0$$

and

$$\mu(B \backslash A) = \mu(B) - \mu(A \cap B) = \beta - \beta = 0.$$

It follows that $[A]$ is an upper bound of $\mathcal{Q}$ in $\mathcal{P}$. There is by Zorn's Lemma a maximal element $[A_0]$ of $\mathcal{P}$. It follows that $\mu(B) > t_0 - \mu(A_0)$ whenever $B \subseteq S \backslash A_0$ and $\mu(B) > 0$. (For example, $\mu(S \backslash A) > t_0 - \mu(A_0)$). Let

$$\mathcal{R} = \{B \mid B \subseteq S \backslash A_0, \mu(B) > 0\}/\sim,$$

where $\sim$ is defined as above. Equip $\mathcal{R}$ with the partial order $\leq$ as above. Given a totally ordered set $\mathcal{S}$ of $\mathcal{R}$. Let

$$\alpha = \inf\{\mu(B) \mid [B] \in \mathcal{S}\},$$

then $\alpha \geq t_0 - \mu(A)$ and there is a decreasing sequence $\{[B_n]\}$ of $\mathcal{S}$ with $\{\mu(B_n)\}$ decreasing to α. Let

$$B = \bigcap_{n=1}^{\infty} B_n,$$

then

$$\alpha = \lim_{n\to\infty} \mu(B_n) = \lim_{n\to\infty}\left(\mu\left(B_n\backslash\bigcap_{i=1}^{n} B_i\right) + \mu\left(\bigcap_{i=1}^{n} B_i\right)\right)$$

$$\leq \lim_{n\to\infty}\left(\sum_{i=1}^{n}\mu(B_n\backslash B_i) + \mu\left(\bigcap_{i=1}^{n} B_i\right)\right) = \mu(B) \leq \alpha.$$

So $\mu(B) = \alpha > 0$. Therefore $\mu(B) > t_0 - \mu(A_0)$. $[B] \in \mathcal{R}$. For any $[C] \in \mathcal{S}$, if $[B_n] \leq [C]$ for some n, then $[B] \leq [C]$; otherwise $[B] = [C]$ since

$$\mu(C\backslash B) \leq \sum_{n=1}^{\infty}\mu(C\backslash B_n) = 0$$

and

$$\mu(B\backslash C) = \mu(B) - \mu(B\cap C) = \alpha - \alpha = 0.$$

It follows that $[B]$ is a lower bound of $\mathcal{S}$ in $\mathcal{R}$. There is by Zorn's Lemma a minimal element $[B_0]$ of $\mathcal{R}$. However, by the assumption that there is a subset C_0 of B_0 with $0 < \mu(C_0) < \mu(B_0)$, $[C_0] \leq [B_0]$ and $[C_0] \neq [B_0]$, a contradiction.

4120

Let X be a compact Hausdorff space. Consider the σ-algebra $\mathcal{B}$ generated by the compact G_δ sets. Show that any positive measure μ on $\mathcal{B}$ which is finite on compact sets is automatically regular.

(*Iowa*)

Solution.
Recall that sets in $\mathcal{B}$ are called Baire sets and each compact Baire set K is a G_δ set. Also recall that E is outer regular if

$$\mu(E) = \inf\{\mu(V) \mid E \subseteq V, \ V \text{ open and } V \in \mathcal{B}\}$$

and E is inner regular if

$$\mu(E) = \sup\{\mu(K) \mid K \subseteq E, \ K \text{ compact and } K \in \mathcal{B}\}.$$

Let $\mathcal{K}$ and $\mathcal{O}$ denote the classes of compact sets and open sets in $\mathcal{B}$, respectively. Let

$$\mathcal{R} = \left\{ \bigcup_{i=1}^{n} (K_i \setminus L_i) \mid K_i, L_i \in \mathcal{K} \right\},$$

where the symbol $\dot\cup$ means disjoint union. Then $\mathcal{R}$ is a ring such that $\mathcal{B} = \mathbf{S}(\mathcal{R})$.

By definition, each set in $\mathcal{K}$ is outer regular. Let us show each set U in $\mathcal{O}$ is inner regular. For any $\varepsilon > 0$, there is a V in $\mathcal{O}$ such that $X \setminus U \subseteq V$ and $\mu(V) < \mu(X \setminus U) + \varepsilon$. Then $X \setminus V \subseteq U$ and $\mu(U) < \mu(X \setminus V) + \varepsilon$.

Let us preceed in five steps to show the regularity of μ.

Step 1. For any pair K, L in $\mathcal{K}$, $K \setminus L$ is regular. For any $\varepsilon > 0$ there are a $B \in \mathcal{O}$ and an $M \in \mathcal{K}$ such that $K \subseteq B$ and $\mu(B \setminus K) < \varepsilon$ and such that $M \subseteq B \setminus L$ and $\mu(B \setminus L) < \mu(M) + \varepsilon$. Then we have

$$K \setminus L \subseteq B \setminus L, \quad \mu(B \setminus L) < \mu(K \setminus L) + \varepsilon;$$
$$K \cap M \subseteq K \setminus L, \quad \mu(K \setminus L) < \mu(K \cap M) + \varepsilon.$$

Thus $K \setminus L$ is regular.

Step 2. If $\{E_i \mid 1 \leq i \leq n\}$ is a finite class of mutually disjoint, regular sets, then $\dot\bigcup_{i=1}^{n} E_i$ is regular.

Obviously

$$\mu\left(\bigcup_{i=1}^{n} E_i\right) \leq \inf\left\{\mu(V)\Big| \bigcup_{i=1}^{n} E_i \subseteq V \in \mathcal{O}\right\}. \tag{1}$$

For any $\varepsilon > 0$, there are $B_1, \ldots, B_n \in \mathcal{O}$ such that $E_i \subseteq B_i$ with

$$\mu(B_i \setminus E_i) < \frac{\varepsilon}{n}, \quad i = 1, \ldots, n.$$

Consequently,

$$\bigcup_{i=1}^{n} E_i \subseteq \bigcup_{i=1}^{n} B_i$$

and

$$\mu\left(\bigcup_{i=1}^{n} B_i\right) < \mu\left(\bigcup_{i=1}^{n} E_i\right) + \varepsilon,$$

which shows the outer regularity of $\bigcup_{i=1}^{n} E_i$. The inner regularity follows from the following inequalities

$$\mu\left(\bigcup_{i=1}^{n} E_i\right) = \sum_{i=1}^{n} \sup\{\mu(K_i) \mid E_i \supseteq K_i \in \mathcal{K}\}$$

$$= \sup\left\{\sum_{i=1}^{n} \mu(K_i) \,\middle|\, E_i \supseteq K_i \in \mathcal{K}\right\}$$

$$\leq \sup\left\{\mu(K) \,\middle|\, \bigcup_{i=1}^{n} \subseteq K \in \mathcal{K}\right\}.$$

Step 3. If $\{E_n \mid n \in \mathbb{N}\}$ is an increasing sequence of regular sets then $\bigcup_{n=1}^{\infty} E_n$ is regular.

Let $E = \bigcup_{n=1}^{\infty} E_n$. Obviously

$$\mu(E) \leq \inf\{\mu(V) \mid E \subseteq V \in \mathcal{O}\}.$$

For any $\varepsilon > 0$ there are $V_1, \ldots, V_n, \ldots \in \mathcal{O}$ such that $E_n \subseteq V_n$ and $\mu(V_n) < \mu(E_n) + \frac{\varepsilon}{2^n}$. Let $V = \bigcup_{n=1}^{\infty} V_n \in \mathcal{O}$, then

$$E \subseteq V, \quad \mu(V) < \mu(E) + \varepsilon,$$

which shows the outer regularity of E, while the inner regularity follows from the following inequalities

$$\mu(E) = \sup_{n} \mu(E_n)$$

$$= \sup_{n} \sup\{\mu(K) \mid K \subseteq E_n, K \in \mathcal{K}\}$$

$$\leq \sup\{\mu(K) \mid E \supseteq K \in \mathcal{K}\}.$$

Step 4. If $\{E_n\}$ is a decreasing sequence of regular sets, then $\bigcap_{n=1}^{\infty} E_n$ is regular.

Let $E = \bigcap_{n=1}^{\infty} E_n$. For any $\varepsilon > 0$, $n \in \mathbb{N}$, there is a $K_n \in \mathcal{R}$ such that $K_n \subseteq E_n$ and $\mu(E_n) < \mu(K_n) + \frac{\varepsilon}{2^n}$, then

$$E \supseteq \bigcap_{n=1}^{\infty} K_n$$

and

$$\mu(E) < \mu\left(\bigcap_{n=1}^{\infty} K_n\right) + \varepsilon,$$

which shows the inner regularity of E. The outer regularity follows from the following inequalities,

$$\inf\{\mu(V) \mid E \subseteq V \in \mathcal{O}\} \leq \inf_n \inf\{\mu(V) \mid E_n \subseteq V \in \mathcal{O}\}$$

$$= \inf_n \mu(E_n) = \mu(E).$$

Step 5. Let

$$S = \{E \in \mathcal{B} \mid E \text{ is regular}\}.$$

By steps 1 and 2, S contains $\mathcal{R}$. By steps 3 and 4, S is a monotone class. It follows that $S = \mathcal{B}$.

4121

Suppose that μ is a non-negative Borel measure on $\mathbb{R}^n$. Let

$$f(r) = \sup\{\mu(B(x, r)) \mid x \in \mathbb{R}^n\}.$$

Assume $f(r)$ is finite for all $r > 0$ and assume

$$\lim_{r \to 0} \frac{f(r)}{r^n} = 0.$$

Prove that μ is identically 0.

(*Indiana*)

Solution.

Let

$$C(x,r) = \{y \in \mathbb{R}^n \mid -r < y_i - x_i \leq r, \quad i = 1, \ldots, n\}$$

and let

$$g(r) = \sup\{\mu(C(x,r)) \mid x \in \mathbb{R}^n\}.$$

Since $C(x,r) \subset B(x, \sqrt{n+1}r)$,

$$\lim_{r \to 0} \frac{g(r)}{r^n} = 0.$$

Given a compact set K of $\mathbb{R}^n$, let $s > 0$ be such that $K \subset C(0,s)$. For any $\varepsilon > 0$ there exists an $r > 0$ such that $g(r) < (2r)^n \varepsilon$. There exist finitely many $x^1, \ldots, x^k \in \mathbb{R}^n$ such that

$$K \subset \bigcup_{i=1}^{k} C(x^i, r) \subset C(0, 2s)$$

and $C(x^i, r)$'s are mutually disjoint. Then

$$\mu(K) \leq \sum_{i=1}^{k} \mu(C(x^i, r)) \leq \sum_{i=1}^{k} (2r)^n \varepsilon$$

$$= \sum_{i=1}^{k} \lambda(C(x^i, r))\varepsilon \leq \lambda(C(0, 2s))\varepsilon.$$

Letting $\varepsilon \to 0$, we get that $\mu(K) = 0$ and therefore μ is identically 0.

4122

Let μ be a finite Borel measure defined on $\mathbb{R}^n$, and define a function f on $\mathbb{R}^n$ by $f(x) = \mu(B(x,1))$ where $B(x,1)$ denotes the open ball centered at x of radius 1. Prove that f attains its minimum on each compact set of $\mathbb{R}^n$.

(Indiana)

Solution.

Let K be any non-empty compact set of $\mathbb{R}^n$ and let $\alpha = \inf f(K)$. There exists a sequence $\{x_k\}$ of K such that

$$\lim_{k \to \infty} f(x_k) = \alpha.$$

Assume without loss of generality that $x_k \to x$ in K. It is easy to show that

$$B(x, 1) \subseteq \lim_{k \to \infty} B(x_k, 1).$$

Hence

$$\mu(B(x, 1)) \leq \mu \left(\lim_{k \to \infty} B(x_k, 1) \right) \leq \lim_{k \to 0} \mu(B(x_k, 1)) = \alpha,$$

which shows that

$$\mu(B(x, 1)) = \alpha.$$

Hence α is finite and f attains its minimum on K.

4123

Let E be the set of all numbers in $[0, 1]$ which can be written in a decimal expansion with no sevens appearing. Thus,

$$\frac{1}{3} = 0.333 \cdots , \quad \frac{27}{100} = 0.2699 \cdots , \quad \frac{28}{100} = 0.2800 \cdots \in E.$$

(i) Compute the Lebesgue measure of E.

(ii) Determine whether E is a Borel set.

(Indiana)

Solution.

For any $x \in [0, 1] \backslash E$, write $x = 0.a_1 a_2 \cdots a_n \cdots$. Let

$$n = \min\{k \geq 1 \,|\, a_k = 7\}.$$

If $a_k = 0$ for all $k > n$ then

$$x = 0.a_1 a_2 \cdots a_{n-1} 7 = 0.a_1 a_2 \cdots a_{n-1} 699 \cdots \in E,$$

a contradiction. It follows that

$$0.a_1a_2 \cdots a_{n-1}7 < x < 0.a_1a_2 \cdots a_{n-1}8.$$

and therefore

$$[0,1] \backslash E \subseteq \cup\{(0.a_1a_2 \cdots a_{n-1}7, \ 0.a_1a_2 \cdots a_{n-1}8)$$

$$\mid a_1, \ldots, a_{n-1} \neq 7, \ n = 1, 2, \ldots\}.$$

The reverse inclusion is obvious. So E is Borel measurable and

$$m(E) = 1 - \sum_{n=1}^{\infty} 9^{n-1} \times \frac{1}{10^n} = 1 - 1 = 0.$$

4124

Let $f : \mathbb{R} \to \mathbb{R}^n$ be a function such that for all $x, y \in \mathbb{R}$

$$|f(x) - f(y)|^n \leq e^{|x|+|y|}|x - y|. \tag{$*$}$$

Show that if $E \subset \mathbb{R}$ is a measurable set with $m_1(E) = 0$ then $m_n(f(E)) = 0$.

(Indiana)

Solution.

Assume without loss of generality that E is bounded. Let $E \subseteq (-r, r)$. For any $\varepsilon > 0$ there exists an open set U such that $E \subseteq U \subseteq (-r, r)$ and $m_1(U \backslash E) < \varepsilon$. Write $U = \bigcup_i (a_i, b_i)$, where (a_i, b_i)'s are mutually disjoint.

Then condition $(*)$ implies that

$$f((a_i, b_i)) \subseteq B(f(a_i), (e^{2r}|b_i - a_i|)^{\frac{1}{n}}).$$

It follows that

$$m_n(f(U)) \leq \sum_i C_n e^{2r}|b_i - a_i| < C_n e^{2r}\varepsilon,$$

which implies that $m_n^*(f(E)) = 0$ and therefore $f(E)$ is measurable. Hence $m_n(f(E)) = 0$.

4125

Let $f : \mathbb{R}^n \to \mathbb{R}$ be an arbitrary function having the property that for each $\varepsilon > 0$, there is an open set U with $\lambda(U) < \varepsilon$ such that f is continuous on $\mathbb{R}^n \backslash U$ (in the relative topology). Prove that f is measurable.

(Indiana)

Solution.

Let U_k be an open set such that $\lambda(U_k) < \frac{1}{k}$ and f is continuous on $\mathbb{R}^n \backslash U_k$. Let $f_k = f \chi_{\mathbb{R}^n \backslash U_k}$, then f_k is measurable. For any $\varepsilon > 0$,

$$m^*(\{x \mid |f_k - f|(x) \geq \varepsilon\}) = m^*(\{x \in U_k \mid |f(x)| \geq \varepsilon\}) \leq \frac{1}{k}.$$

It follows that $\{f_k\}$ converges to f in measure. Since the Lebesgue measure is complete f is measurable.

4126

Let $A \subset \mathbb{R}$ ba Lebesgue measurable set and let $rA = \{rx \mid x \in A\}$ where r is a real number. Assume that $rA = A$ for every non-zero rational number r. Prove that either A or $\mathbb{R} \backslash A$ has Lebesgue measure zero.

(Indiana)

Solution.

Let $\mathbb{R}^* = \mathbb{R} \backslash \{0\}$ and $B = A \backslash \{0\}$. Then $rB = B$ and $r(\mathbb{R}^* \backslash B) = \mathbb{R}^* \backslash B$ for every non-zero rational number r. If $m(\mathbb{R}^* \backslash B) = m(\mathbb{R} \backslash A) > 0$ there exists a compact subset K of $\mathbb{R}^* \backslash B$ with positive Lebesgue measure. For any compact subset L of B, define function $f : \mathbb{R}^* \to [0, \infty)$ by

$$f(x) = \int_{\mathbb{R}^*} \chi_K(y) \chi_{L^{-1}}(y^{-1}x) \frac{dy}{y}, \quad x \in \mathbb{R}^*.$$

Then f is continuous and for any non-zero rational number r,

$$f(r) = \int_{\mathbb{R}^*} \chi_k(y) \chi_{rL}(y) \frac{dy}{y} \leq \int_{\mathbb{R}^*} \chi_{\mathbb{R}^* \backslash B}(y) \chi_B(y) \frac{dy}{y} = 0.$$

Hence $f(x) = 0$ for any $x \in \mathbb{R}^*$. Since

$$\int_{\mathbb{R}^*} \chi_K(y) \frac{dy}{y} \int_{\mathbb{R}^*} \chi_{L^{-1}}(y) \frac{dy}{y} = \int_{\mathbb{R}^*} f(x) \frac{dx}{x} = 0,$$

we conclude that $m(L^{-1}) = 0$ and therefore $m(L) = 0$. It follows that $m(B) = 0$ and therefore $m(A) = 0$.

4127

Let μ be a σ-finite measure on the measure space (X, m). Prove that there exists a probability measure ν on (X, m) ($\nu(x) = 1$) such that μ is absolutely continuous with respect to ν, and ν is absolutely continuous with respect to μ.

(Indiana)

Solution.

There exists a sequence $\{E_n\}_{n=1}^{\infty}$ of mutually disjoint measurable sets of finite and positive measure such that $X = \bigcup_{n=1}^{\infty} E_n$. For each measurable set E let

$$\nu(E) = \sum_{n=1}^{\infty} \frac{\mu(E \cap E_n)}{2^n \mu(E_n)},$$

then ν is the desired probability measure.

4128

Give an example of first category subset of $[0,1]$ which is of Lebesgue measure 1.

(Auburn)

Solution.

For each interval $[a, b]$ in $\mathbb{R}$ and $0 < r < 1$, we can construct a set E of Cantor type in $[a, b]$ such that $m(E) = r(b - a)$. Indeed, let $s = (1 - r)/(1 + 2(1 - r))$ and remove from $[a, b]$ its middle part of length $s(b - a)$; we are left with two intervals I_{11} and I_{12}. Remove from each of them their middle parts of lengths $s^2 |I_{1j}|$, and keep doing this ad infinitum. We are left with a compact set E whose Lebesgue measure is then

$$(b-a)\left(1 - \sum_{n=1}^{\infty} 2^{n-1} s^n\right) = r(b-a).$$

Now E contains no interval of positive length. Indeed, if $[x, y] \subseteq E$ then $[x, y]$ is contained in one of the intervals left in the n-th step above and thus

$$0 \le y - x \le \frac{b-a}{2^n} : n \ge 1.$$

Passing to the limits yields that $x = y$.

It follows that we get a generalized Cantor set $E_n \subseteq [0, 1]$ with Lebesgue measure $1 - \frac{1}{n}$. Let $E = \cup_{n \ge 1} E_n$ then E is of first category and of Lebesgue measure 1.

4129

Let $(X, \mathcal{F}, \mu)$ be a measure space and suppose $(f_n : X \to \mathbb{C})$ is a sequence of measurable functions with the property that for all $n \ge 1$ and all $\lambda > 0$,

$$\mu\{x \in X : |f_n(x)| \ge \lambda\} \le ce^{-\lambda^2/n}$$

where c is a constant independent of n. Let $n_k = 2^k$. Prove that

$$\varlimsup_{k \to \infty} \frac{|f_{n_k}|}{\sqrt{n_k \log \log n_k}} \le 1, a.e.$$

(*Purdue*)

Solution.

Let $g_k = \frac{|f_{n_k}|}{\sqrt{n_k \log \log n_k}}$ then for any $r > 1$ we have

$$\{x \in X : \varlimsup_{k \to \infty} g_k(x) > r\} \subseteq \bigcap_{k \ge 1} \bigcup_{l \ge k} \{x \in X : g_l(x) \ge r\},$$

$$\mu\{x \in X : g_l(x) \ge r\} \le c \exp\left(-\frac{(r\sqrt{n_l \log \log n_l})^2}{n_l}\right)$$

$$= \frac{c}{(l \log 2)^{r^2}}$$

and thus $\mu\{x \in X \mid \overline{\lim}\, g_k(x) > r\} = 0$. It remains to notice that

$$\{x \in X \mid \varlimsup_{k\to\infty} g_k(x) > 1\}$$
$$= \bigcup_{l\in\mathbb{N}}\{x \in X \mid \varlimsup_{k\to\infty} g_k(x) > 1 + 2^{-l}\}.$$

4130

(1) Let $(X, \mathcal{F}, \mu)$ be a finite measure space. Let f_n be a sequence of measurable functions. Prove that $f_n \to f$ in measure if and only if every subsequence $\{f_{n_k}\}$ contains a further subsequence $\{f_{n_{k_j}}\}$ that converges almost everywhere to f.

(2) Let $(X, \mathcal{F}, \mu)$ be a finite measure space. Let $F : \mathbb{R} \to \mathbb{R}$ be a continuous function and $f_n \to f$ in measure. Prove that $F(f_n) \to F(f)$ in measure. (You may assume, of course, that $f_n, f, F(f_n)$ and $F(f)$ are all measurable.)

(*Purdue*)

Solution.

(1) The necessity follows from the Riesz Theorem: Each sequence convergent in measure contains a subsequence convergent almost everywhere.

Sufficiency: For any $\varepsilon > 0$ let $c = \varlimsup_{n\to\infty}\mu(|f_n - f| \geq \varepsilon)$ and we have to show that $c = 0$. Passing to a subsequence we may assume $c = \lim_{n\to\infty}\mu(|f_n - f| \geq \varepsilon)$ and again passing to a subsequence we may assume (f_n) converges almost everywhere to f. By Lebesgue Theorem, (f_n) converges to f in measure and thus $c = 0$.

(2) For any subseqeunce $\{F(f_{n_k})\}$, the sequence $\{f_{n_k}\}$ admits a subsequence $\{f_{n_{k_j}}\}$ that converges to f almost everywhere by (1) so the sequence $\{F(f_{n_{k_j}})\}$ converges to $F(f)$ almost everywhere. Again by (1), $F(f_n)$ converges to $F(f)$ in measure.

4131

Let $f \in L^1[0,1]$ and let $g(x) = \int_0^x f(t)dt$. If E is a measurable subset of $[0,1]$, show that (i) $g(E) = \{y : \exists x \in E \text{ with } y = g(x)\}$ is measurable. (ii) $m(g(E)) \le \int_E |f(t)|dt$.

(*Purdue*)

Solution.

(1) If K is an F_σ-subset of $[0,1]$, namely $K = \cup_{n \ge 1} K_n$ for some sequence (K_n) of compact sets, then $g(K)$ is of F_σ and thus measurable.

(2) If $[a,b] \subseteq [0,1]$, take $\{x_0, x_1\} \subset [a,b]$ such that

$$g(x_0) = \min\{g(t) \,|\, a \le t \le b\},$$
$$g(x_1) = \max\{g(t) \,|\, a \le t \le b\}$$

then $g([a,b]) = [g(x_0), g(x_1)]$ and

$$|g(x_1) - g(x_0)| = \left| \int_{x_0}^{x_1} f(t)dt \right| \le \int_a^b |f(t)|dt.$$

Consequently, $m(g(A)) \le \int_A |f(t)|dt$ whenever A is a subinterval of $[0,1]$.

Now for any $\epsilon > 0$ take a $\delta > 0$ such that $\int_{E'} |f(t)|dt < \epsilon$ for any measurable set E' with $m(E') < \delta$. Given any sequence $(A_i)_{i \ge 1}$ of disjoint subintervals of $[0,1]$ such that $E \subseteq \coprod_{i \ge 1} A_i$ and $\sum_{i \ge 1} m(A_i) < m(E) + \delta$ we have

$$m^*(g(E)) \le m\left(\bigcup_{i \ge 1} g(A_i) \right) \le \sum_{i \ge 1} m(g(A_i))$$

$$\le \sum_{i \ge 1} \int_{A_i} |f(t)|dt < \int_E |f(t)|dt + \epsilon$$

and thus $m^*(g(E)) \le \int_E |f(t)|dt$. Especially, $m(g(E)) = 0$ if $m(E) = 0$. In general, E is of the form $K \cup E_0$ where K is of F_σ and E_0 is of measure zero so that $g(E)$ is of the form $g(K) \cup g(E_0)$ and thus measurable.

4132

Given a function $f : \mathbb{R} \to \mathbb{Z}$, show that $\{x : f$ is not continuous at $x\}$ is a Borel set.

(*Purdue*)

Solution.
We will show a stronger result: Suppose that X is a topological space and Y is a metric space. For any function $f : X \to Y$ define

$$\omega(x) = \inf_V \sup\{d(f(x'), f(x'')) \,|\, x', x'' \in V\}$$

for any $x \in X$, where V runs through open neighborhoods of x, then f is continuous at x iff $\omega(x) = 0$. Moreover, $(\omega < b)$ is an open set of X. Indeed, if $\omega(x) < b$ then

$$\sup\{d(f(x'), f(x'')) \,|\, x', x'' \in V\} < b$$

for some open neighborhood V of x and thus $V \subseteq (\omega < b)$.

It follows that the set of $x \in X$ at which f is not continuous is exactly $(\omega > 0)$ and thus a Borel set.

4133

Let A and B be (not necessarily Lebesgue measurable) subsets of $\mathbb{R}$ and let $|\cdot|_e$ stand for Lebesgue outer measure. Prove that if $|A|_e = 1$ and $|B|_e = 1$ and $|A \cup B|_e = 2$ then $|A \cap B|_e = 0$.

(*Purdue*)

Solution.
It suffices to show that

$$|A \cup B|_e + |A \cap B|_e \le |A|_e + |B|_e$$

for any subsets A and B of $\mathbb{R}$. To this end, take a Borel set E of $\mathbb{R}$ such that $A \cap B \subseteq E$ and $|A \cap B|_e = |E|_e$. Now $A \cap B \subseteq B \cap E$ and

$$|(A \cup B) \cap E|_e \le |E|_e = |A \cap B \cap E|_e \le |A \cap E|_e.$$

By Carathéodory condition we get that

$$
\begin{aligned}
|A \cup B|_e &+ |A \cap B|_e \\
&= |(A \cup B) \cap E|_e + |(A \cup B) \backslash E|_e + |A \cap B|_e \\
&\le |A \cap E|_e + |A \backslash E|_e + |B \backslash E|_e + |B \cap E|_e \\
&= |A|_e + |B|_e.
\end{aligned}
$$

4134

Let f be a bounded Lebesgue measurable function on $\mathbb{R}$. Put

$$g(x) = \sup\{a \in \mathbb{R} : |\{y \in (x, x+1) \,|\, f(y) > a\}| > 0\},$$

where $|\cdot|$ is the Lebesgue measure (i.e., $g(x)$ equals the essential supremum of f over $(x, x+1)$). Prove $\underline{\lim}_{x \to 0} g(x) \ge g(0)$.

(*Purdue*)

Solution.

It suffices to show $(g > b)$ is open for any $b \in \mathbb{R}$. Let $h_a(x) = |(x, x+1) \cap (f > a)|$ then h_a is uniformly continuous since

$$|h_a(x') - h_a(x)| \le |x' - x|.$$

Now $g(x) > b$ iff $h_a(x) > 0$ for some $a > b$. It follows that

$$(g > b) = \bigcup_{a > b} (h_a > 0)$$

and thus $(g > b)$ is open.

4135

(1) Prove that a continuous function $T : \mathbb{R}^n \to \mathbb{R}^n$ maps sets F_σ into sets F_σ.

(2) Prove that a Lipschitz map $T : \mathbb{R}^n \to \mathbb{R}^n$, i.e., a map satisfying $|T(x) - T(y)| \leq L|x - y|$ for every $x, y \in \mathbb{R}^n$, and for some $L > 0$, maps measurable sets into measurable sets. You can take for granted here that a Lipschitz map preserves sets of measure zero.

(*Purdue*)

Solution.

(1) It suffices to note that any F_σ-set in $\mathbb{R}^n$ is the union of countably many compact sets and continuous functions map compact sets into compact sets.

(2) Since any Lebesgue set is the union of an F_σ-set and a set of measure zero it suffices to show $m(T(E)) = 0$ if $m(E) = 0$. So given $a \in \mathbb{R}^n$ and $r > 0$, write $O(a, r)$ for the open ball $\{x \in \mathbb{R}^n : |x - a| < r\}$ then $T(O(a, r)) \subseteq O(T(a), Lr)$. For any $\delta > 0$ we have a countable set F of E such that $E \subseteq \cup_{a \in F} O(a, r)$ and $\sum_{a \in F} |O(a, r)| < \delta$. Now $T(E) \subseteq \cup_{a \in F} O(T(a), Lr)$ and $\sum_{a \in F} |O(T(a), Lr)| < L^n \delta$. It follows that $|T(E)| = 0$.

4136

Show that if $E \subseteq \mathbb{R}$ and if $|\cdot|_e$ stands for outer measure then

$$|E|_e = \sum_{n=-\infty}^{+\infty} |E \cap [n, n+1]|_e.$$

Also prove that if $O_i, i \geq 1$, are open subsets of $\mathbb{R}$ satisfying $\cup_{i=1}^\infty O_i = \mathbb{R}$ then

$$\lim_{k \to \infty} \left| \left(\bigcup_{i=1}^k O_i \right) \cap E \right|_e = |E|_e.$$

(*Purdue*)

Solution.

Let $E_{2k} = [k, k+1), E_{2k+1} = [-k-1, -k)$ for $k \geq 0$, and $S_n = \cup_{i=0}^n E_i$ then

$$|E|_e = |E \cap E_0| + |E \backslash S_0|_e,$$
$$|E \backslash S_0|_e = |E \cap E_1|_e + |E \backslash S_1|_e,$$
$$\vdots$$
$$|E \backslash S_n|_e = |E \cap E_{n+1}|_e + |E \backslash S_{n+1}|_e,$$
$$\vdots$$

It follows that for any $n \geq 0$ we have

$$|E|_e = \sum_{i=0}^n |E \cap E_i|_e + |E \backslash S_n|_e \geq \sum_{i=0}^n |E \cap E_i|_e.$$

Passing to the limit we get that $|E| \geq \sum_{i=0}^{\infty} |E \cap E_i|_e$, the reverse being obvious.

Let $F_k = (\cup_{i=1}^k O_i) \cap E$ then (F_k) increases to E. Take Borel sets G_k such that $F_k \subseteq G_k$ and $|G_k|_e = |F_k|_e$. Replacing G_k with $\cap_{l \geq k} G_l$ we may assume that (G_k) increases to some Borel set G containing E. It follows that

$$|E|_e \leq |G|_e = \lim_{k \to \infty} |G_k|_e$$
$$= \lim_{k \to \infty} |F_k|_e \leq \lim_{k \to \infty} |E|_e.$$

4137

If ν and ρ are measures on a measurable space $(X, \mathcal{M})$, we say that $\nu \leq \rho$ if $\nu(A) \leq \rho(A)$ for all $A \in \mathcal{M}$. Assume that (ν_n) is a sequence of measures on $(X, \mathcal{M})$ such that $\nu_1 \leq \nu_2 \leq \cdots$. Define $\nu(A) = \lim_{n \to \infty} \nu_n(A), A \in \mathcal{M}$.

(1) Show that ν defines a measure.

(2) Assume there exists a measure μ so that for each $n, \nu_n \ll \mu$. Show that $\nu \ll \mu$ also and $\frac{d\nu}{d\mu} = \lim_{n \to \infty} \frac{d\nu_n}{d\mu}, \mu$-a.e.

Here $\nu \ll \mu$ means that if $\mu(A) = 0$ then $\nu(A) = 0$ for all measurable set A.

(Rutgers)

Solution.

(1) It suffices to show $\nu(A) = \sum_{k \in \mathbb{N}} \nu(A_k)$ whenever A is the union of disjoint measurable sets A_k ($k \in \mathbb{N}$). However, regard the series as the integral with respect to the counting measure then applying the Monotone Convergence Theorem to the equalities $\nu_n(A) = \sum_{k \in \mathbb{N}} \nu_n(A_k)$ yields the required equality.

(2) Write $\nu_n = f_n \, d\mu$ then (we may assume) $f_n \le f_{n+1}$. Let $f = \lim f_n$ then applying the Monotone Convergence Theorem to the equalities $\nu_n(E) = \int_E f_n \, d\mu$ yields that $\nu(E) = \int_E f \, d\mu$.

4138

Define $f : [0,1] \to \mathbb{R}$ as follows: if $x \in [0,1]$ has decimal expansion $x = 0.a_1 a_2 a_3 \cdots$ then $f(x) = \max_i a_i$. Prove that the function f is measurable.

(Rutgers)

Solution.

It suffices to show that f_i is measurable for any i, where f_i is defined by $f_i(x) = a_i$. However,

$$(f_i = a) = \bigcup \{[0.a_1 \cdots a_{k-1}n, 0.a_1 \cdots a_{k-1}(n+1)] |$$
$$a_j \in \{0, 1, \ldots, 9\}\}.$$

4139

Let m and μ be, respectively, the Lebesgue measure and the counting measure on the Borel σ-algebra on $[0, 1]$.

(1) Show that $m \ll \mu$ but $dm \ne f \, d\mu$ for any f.

(2) State the Lebesgue–Radon–Nikodym Theorem for signed measures, and explain why part (1) is not a counter-example!

(Rutgers)

Solution.

(1) If $\mu(E) = 0$ then E is empty so that $m(E) = 0$. If $dm = f\,d\mu$ for some measurable function $f : \mathbb{R} \to [0, +\infty]$ then

$$f(x) = \int_{\{x\}} f(t)\mu(dt) = m\{x\} = 0$$

for any $x \in \mathbb{R}$, a contradiction.

(2) Suppose that μ and ν are signed measures on the measurable space $(X, \mathfrak{M})$ and μ is σ-finite. If $\nu \ll \mu$ then $d\nu = f\,d\mu$ for some measurable function f on X, unique in the sense that if $d\nu = g\,d\mu$ then

$$|\mu|(f \neq g) = 0.$$

The reason why the part (1) is not a counter-example lies in the fact that the counting measure on the Borel σ-alegbra is not σ-finite.

Section 2

Integral

4201

Prove or disprove that the composition of any two Lebesgue integrable functions with compact support $f, g : \mathbb{R} \to \mathbb{R}$ is still integrable.

(Stanford)

Solution.

It is not true. For example, let

$$f(x) = \chi_{\{0\}}(x) \quad \text{and} \quad g(x) = \chi_{\{0,1\}}(x).$$

Then f and g are integrable functions with compact support. However, since $g \circ f(x) \equiv 1$, the function $g \circ f$ is not integrable.

4202

Let $f \in L_1(0,1)$. Assume that for any $x \in (0,1)$ and every $\varepsilon > 0$, there is an open interval $J_x \subseteq (0,1)$ such that

$$x \in J_x, \ m(J_x) < \varepsilon, \ \text{and} \ \int_{J_x} f \, dm = 0.$$

Prove that for every open interval $I \subseteq (0,1)$

$$\int_I f \, dm = 0.$$

(Illinois)

Solution.

There is a measurable set E of measure zero such that any $x \in (0,1)\backslash E$ is a Lebesgue point of f, i.e.,

$$\lim_{\substack{\alpha \leq x \leq \beta \\ \beta - \alpha > 0 \\ \beta - \alpha \to 0}} \frac{1}{\beta - \alpha} \int_{\alpha}^{\beta} f(t)dt = f(x). \tag{1}$$

For any $x \in (0,1)\backslash E$ and for any n there is an open interval $J_n \subseteq (0,1)$ such that $x \in J_n, m(J_n) < \frac{1}{n}$ and

$$\int_{J_n} f(t)dt = 0.$$

It follows by equality (1) that $f(x) = 0$, i.e., $f(x) = 0$ a.e.. Therefore

$$\int_I f\,dm = 0.$$

4203

Let $(X, \mathcal{M}, \mu)$ be a positive measure space with $\mu(X) < \infty$. Show that a measurable function $f : X \to [0, \infty)$ is integrable (i.e., one has $\int_X f\,d\mu < \infty$) if and only if the series

$$\sum_{n=0}^{\infty} \mu(\{x | f(x) \geq n\})$$

converges.

(*Iowa*)

Solution.

Suppose that f is integrable. Then

$$\sum_{n=0}^{\infty} \mu(\{x | f(x) \geq n\}) = \sum_{n=0}^{\infty} \sum_{m=n}^{\infty} \mu(\{x | m \leq f(x) < m + 1\})$$

$$= \sum_{m=0}^{\infty} \sum_{n=0}^{m} \mu(\{x | m \leq f(x) < m + 1\})$$

$$= \sum_{m=0}^{\infty} (m + 1)\mu(\{x | m \leq f(x) < m + 1\})$$

$$= \sum_{m=0}^{\infty} m\mu(\{x|m \le f(x) < m+1\})$$

$$+ \sum_{m=0}^{\infty} \mu(\{x|m \le f(x) < m+1\})$$

$$\le \sum_{m=0}^{\infty} \int_{\{x|m \le f(x) < m+1\}} f(x)d\mu(x) + \mu(X)$$

$$= \int_X (f+1)d\mu < \infty.$$

Conversely,

$$\int_X f\,d\mu = \sum_{m=0}^{\infty} \int_{\{x|m \le f(x) < m+1\}} f\,d\mu$$

$$\le \sum_{m=0}^{\infty} (m+1)\mu(\{x|m \le f(x) < m+1\})$$

$$= \sum_{n=0}^{\infty} \mu(\{x|f(x) \ge n\}) < \infty,$$

which shows that f is integrable.

4204

(a) Is there a Borel measure μ (positive or complex) on $\mathbb{R}$ with the property that

$$\int_{\mathbb{R}} f\,d\mu = f(0)$$

for all continuous $f : \mathbb{R} \to \mathbb{C}$ of compact support? Justify.

(b) Is there a Borel measure μ (positive or complex) on $\mathbb{R}$ with the property that

$$\int_{\mathbb{R}} f\,d\mu = f'(0)$$

for all continuously differentiable $f : \mathbb{R} \to \mathbb{C}$ of compact support? Justify.

(Iowa)

Solution.

(a) Yes. Let $\mu(E) = \chi_E(0)$ for any Borel set E.

(b) No. If there were such a Borel measure, let $\varphi \geq 0$ be a continuously differentiable function of compact support, taking value one on $[-1, 1]$. Then a contradiction occurs from the following limits

$$\lim_{n \to \infty} \int_{\mathbb{R}} \varphi(t) e^{\frac{t}{n}} \, dt = \int_{\mathbb{R}} \varphi(t) dt > 0$$

and

$$\lim_{n \to \infty} (\phi(t) e^{\frac{t}{n}})'|_{t=0} = \lim_{n \to \infty} \frac{e^{\frac{t}{n}}}{n} \bigg|_{t=0} = 0.$$

4205

Let E be a Banach space, (X, π, μ) a probability space, and $f : X \to E$ such that $g \circ f$ is μ-integrable for every $g \in E'$.

Define

$$L : E' \to \mathbb{R}, \quad L(g) = \int g \circ f \, d\mu.$$

Does $L \in E''$? Justify your answer.

(Iowa)

Solution.

It is ture that $L \in E''$. Define the linear operator

$$T : E' \to L^1(X), \quad \varphi \mapsto \varphi \circ f.$$

Assume that $\varphi_n \to \varphi$ in E' and $T(\varphi_n) \to h$ in $L^1(X)$. It follows that $\varphi_n \circ f$ converges to $\varphi \circ f$ everywhere and to h in measure and therefore $h = \varphi \circ f$ in $L^1(X)$. By the closed graph theorem,

T is bounded. We have

$$|L(g)| \leq \int_X |g \circ f| d\mu \leq \int_X \|T\| \|g\| d\mu \leq \|T\| \mu(X) \|g\|.$$

So L is bounded.

4206

Let $(X, \mathcal{M}, \mu)$ be a positive measure space with $\mu(X) < \infty$, and let f and g be real-valued measurable functions with

$$\int_X f \, d\mu = \int_X g \, d\mu.$$

Show that either (a) $f = g$ a.e., or (b) there exists an $E \in \mathcal{M}$ such that

$$\int_E f \, d\mu > \int_E g \, d\mu.$$

(*Iowa*)

Solution.

If (b) does not hold, then for any $E \in \mathcal{M}$,

$$\int_E f \, d\mu \leq \int_E g \, d\mu.$$

Since

$$\int_X f \, d\mu = \int_X g \, d\mu,$$

for any $E \in \mathcal{M}$,

$$\int_E f \, d\mu = \int_E g \, d\mu.$$

For any $\varepsilon > 0$ the sets

$$E_+ = \{x | f(x) \geq g(x) + \varepsilon\}$$

and

$$E_- = \{x | f(x) \leq g(x) - \varepsilon\}$$

are measurable. From the equalities

$$\int_{E+} f\,d\mu = \int_{E+} g\,d\mu$$

and

$$\int_{E-} f\,d\mu = \int_{E-} g\,d\mu,$$

we conclude that $\mu(E_+) = \mu(E_-) = 0$. It follows that

$$\mu(\{x|f(x) \neq g(x)\}) = \mu\left(\bigcup_{n=1}^{\infty}\left\{x\,\Big|\,|g(x)-f(x)| \geq \frac{1}{n}\right\}\right) = 0.$$

Therefore (a) holds.

4207

Let $(X, \mathcal{M}, \mu)$ be a positive measure space, and S a closed set in $\mathbb{C}$. Suppose that

$$\frac{1}{\mu(E)}\int_E f\,d\mu \in S$$

whenever $E \in \mathcal{M}$ and $\mu(E) > 0$. Show that $\{x \in X | f(x) \notin S\}$ has measure 0.

(Iowa)

Solution.

Since $\mathbb{C}$ is second countable, one finds that $\mathbb{C}\backslash S$ is the union of countably many closed balls $\{z \in \mathbb{C} | |z - \lambda_n| \leq \varepsilon_n\}$'s. If

$$\mu(\{x \in X | f(x) \notin S\}) \neq 0,$$

there is at least an n such that

$$\mu(\{x \in X | |f(x) - \lambda_n| \leq \varepsilon_n\}) \neq 0.$$

But then $\frac{1}{\mu(E)}\int_E f\,d\mu$ belongs to $\{z \in \mathbb{C} | |z-\lambda_n| \leq \varepsilon_n\}$, not to S, where $E = \{x \in X | |f(x) - \lambda_n| \leq \varepsilon_n\}$), a contradiction.

4208

Let $f : [0, 1] \to (0, \infty)$ and let $0 < \alpha \le 1$. Show that

$$\inf \left\{ \int_E f \right\} > 0,$$

where inf is extended over all measurable $E \subset [0, 1]$ with $m(E) \ge \alpha$.

(*Indiana–Purdue*)

Solution.

Obviously,

$$[0, 1] = (f \ge 1) \cup \left(\bigcup_{n=1}^{\infty} \left(\frac{1}{n} > f \ge \frac{1}{n+1} \right) \right).$$

Thus

$$m((f \ge 1)) + \sum_{n=1}^{\infty} m\left(\left(\frac{1}{n} > f \ge \frac{1}{n+1} \right) \right) = 1.$$

Take an N such that

$$\sum_{n=N}^{\infty} m\left(\left(\frac{1}{n} > f \ge \frac{1}{n+1} \right) \right) < \frac{\alpha}{2}.$$

Suppose that $E \subset [0, 1]$, $m(E) > \alpha$. Denote $E_1 = E \ (f \ge \frac{1}{N})$, $E_2 = E \backslash E_1$. Then $m(E_2) < \frac{\alpha}{2}$, so $m(E_1) > \frac{\alpha}{2}$. Therefore

$$\int_E f > \int_{E_1} f \ge \frac{\alpha}{2} \cdot \frac{1}{N}.$$

Thus

$$\inf_{m(E) > \alpha} \left\{ \int_E f \right\} \ge \frac{\alpha}{2N} > 0.$$

4209

Let $\{f_n\}$ be a sequence of real-valued functions in $L^1(\mathbb{R})$ and suppose that for some $f \in L^1(\mathbb{R})$,

$$\int_{-\infty}^{+\infty} |f_n(t) - f(t)|dt \leq \frac{1}{n^2}, \quad n \geq 1.$$

Prove that $f_n \to f$ almost everywhere with respect to Lebesgue measure.

(*Illinois*)

Solution.

Since

$$\sup_n \int \sum_{k=1}^n |f_{k+1} - f_k|(t)dt \leq \sum_{k=1}^{\infty} \left(\frac{1}{(k+1)^2} + \frac{1}{k^2} \right) < \infty,$$

there is, by Levi's Lemma, a measurable set E of measure zero such that for any $t \in \mathbb{R} \backslash E$,

$$\sup_n \sum_{k=1}^n |f_{k+1} - f_k|(t) < \infty.$$

Therefore for any $t \in \mathbb{R} \backslash E$,

$$f_n(t) = f_1(t) + \sum_{k=2}^n (f_k - f_{k-1})(t)$$

converges. It follows that $f_n \to f$ almost everywhere.

4210

Let μ be a finite measure on $\mathbb{R}$, and define

$$f(x) = \int_{\mathbb{R}} \frac{\ln|x - t|}{|x - t|^{1/2}} d\mu(t), \quad x \in \mathbb{R}.$$

Show that $f(x)$ is finite a.e. with respect to the Lebesgue measure on $\mathbb{R}$.

(*Indiana*)

Solution.

Let

$$g(x) = \begin{cases} \dfrac{\ln|x|}{|x|^{1/2}}, & x \neq 0, \\ 0, & x = 0, \end{cases}$$

then $g \in L^1(\mathbb{R}, d\nu)$, where

$$d\nu(x) = \frac{dx}{1+x^2},$$

and

$$f(x) = \int_{\mathbb{R}} g(x-t) d\mu(t).$$

Since

$$\int_{\mathbb{R}} \left(\int_{\mathbb{R}} |g(x-t)| d\nu(t) \right) d\mu(t)$$

$$= \int_{\mathbb{R}} \left(\int_{|x-t|\leq 1} \frac{|\ln|x-t||}{|x-t|^{1/2}} \frac{dx}{1+x^2} \right.$$

$$\left. + \int_{|x-t|\geq 1} \frac{|\ln|x-t||}{|x-t|^{1/2}} \frac{dx}{1+x^2} \right) d\mu(t)$$

$$\leq \int_{\mathbb{R}} \left(\int_{|x|\leq 1} \frac{|\ln|x||}{|x|^{1/2}} dx + \int_{\mathbb{R}} \frac{dx}{1+x^2} \right) d\mu(t)$$

$$< +\infty,$$

by Fubini's Theorem, the function

$$x \mapsto \int_{\mathbb{R}} g(x-t) d\mu(t)$$

is finite a.e. with respect to the measure ν. The conclusion follows from that the measure ν and the Lebesgue measure are equivalent.

4211

Let $(X, \mathcal{M}, \mu)$ be a positive measure space, $f_n : X \to [0, \infty]$ a sequence of integrable functions, and $f : X \to [0, \infty]$ an integrable function. Suppose that $f_n \to f$ a.e. $[\mu]$, and that

$$\int_X f_n \, d\mu \to \int_X f \, d\mu.$$

Prove that

$$\int_x |f - f_n| d\mu \to 0.$$

(Iowa)

Solution.

Let $g_n = f - f_n$ then $g_n = g_n^+ - g_n^-$ and $|g_n| = g_n^+ + g_n^-$. Now $g_n^+ \le f$ and $\lim_{n\to\infty} g_n^+ = g^+$ a.e. $[\mu]$ so that applying the Dominated Convergence Theorem yields that $\lim_{n\to\infty} \int_X g_n^+ \, d\mu = \int_X g_n^- \, d\mu$, which together with the condition $\lim_{n\to\infty} \int_X g_n \, d\mu = 0$ implies $\lim_{n\to\infty} \int_X g_n^- \, d\mu = \int_X g^- \, d\mu$. It follows that $\lim_{n\to\infty} \int |g_n| d\mu = 0$, as desired.

4212

Let μ be a σ-finite, positive measure on a σ-algebra $\mathcal{M}$ in a set X.

(a) Show that there exists $W \in L^1(\mu)$ which takes its values in the open interval $(0,1)$. Show also that

$$\tilde{\mu}(E) = \int_E W \, d\mu$$

is a positive, finite measure on $\mathcal{M}$, and that

$$\tilde{\mu}(E) = 0 \Leftrightarrow \mu(E) = 0$$

for $E \in \mathcal{M}$.

(b) Show that if f is a complex function on X which is measurable with respect to $\mathcal{M}$, then

$$\int_X f \, d\tilde{\mu} = \int_X f W \, d\mu.$$

(Iowa)

Solution.

(a) Let $\{X_n\}$ be a disjoint sequence of $\mathcal{M}$ such that $X = \cup_{n=1}^{\infty} X_n$ and $\mu(X_n) > 0$. Let

$$W = \sum_{n=1}^{\infty} \frac{1}{2^n} \frac{\chi_{X_n}}{1 + \mu(X_n)}$$

as desired.

The set function $\tilde{\mu}$ is of course a positive, finite measure on $\mathcal{M}$. If $\mu(E) = 0$ then $\tilde{\mu}(E) = 0$. Conversely,

$$\int_E W \, d\mu = 0$$

implies that

$$\mu(\{x | x \in E, W(x) \neq 0\}) = 0,$$

i.e., $\mu(E) = 0$.

(b) If f is a simple function, i.e.,

$$f = \sum_{i=1}^{n} a_i \chi_{E_i},$$

where $\mu(E_i) < +\infty, i = 1, \ldots, n$, then both $\int_X f \, d\tilde{\mu}$ and $\int_X fW \, d\mu$ equal to $\sum_{i=1}^{n} a_i \tilde{\mu}(E_i)$.

In general, if f is integrable with respect to the measure $\tilde{\mu}$, there is a sequence $\{f_n\}$ of simple functions integrable with respect to $\tilde{\mu}$ such that

$$\lim_{n \to \infty} \int_X |f_n - f| d\tilde{\mu} = 0.$$

Then $f_n W$'s are integrable with respect to μ, and since

$$\lim_{\substack{n \to \infty \\ m \to \infty}} \int_X |f_n W - f_m W| d\mu = \lim_{\substack{n \to \infty \\ m \to \infty}} \int_X |f_n - f_m| d\tilde{\mu} = 0,$$

$$\lim_{n \to \infty} f_n W = fW,$$

we see that fW is integrable with respect to μ. We have

$$\lim_{n\to\infty} \int_X f_n W \, d\mu = \int_X fW \, d\mu$$

and therefore

$$\int_X f \, d\tilde{\mu} = \int_X fW \, d\mu.$$

Conversely, if fW is integrable with respect to μ, assuming without loss of generality that f is positive-valued, there is a sequence of simple functions $\{f_n\}$ such that $|f_n| \leq f$ and $\lim_{n\to\infty} f_n = f$. Then $f_n W$ is integrable with respect to μ and hence f_n is integrable with respect to $\tilde{\mu}$. Since

$$\lim_{m,n\to\infty} \int_X |f_m - f_n| d\tilde{\mu} = 0$$

and $\lim_{n\to\infty} f_n = f$, f is integrable with respect to $\tilde{\mu}$. Accordingly,

$$\int_X f \, d\tilde{\mu} = \int_X fW \, d\mu.$$

4213

Let m denote the Lebesgue measure on $[0,1]$ and let (f_n) be a sequence in $L^1(m)$ and h a non-negative element of $L^1(m)$. Suppose that

(i) $\int f_n g \, dm \to 0$ for each $g \in C([0,1])$ and
(ii) $|f_n| \leq h$ for all n.

Show that

$$\int_A f_n \, dm \to 0$$

for each Borel subset $A \subseteq [0,1]$.

(Iowa)

Solution.

For any $\varepsilon > 0$, there is a $\delta > 0$ such that

$$\int_E h\,dm < \varepsilon$$

whenever $m(E) < \delta$. For such a δ there are a compact set K and an open set U such that (1) $K \subseteq A \subseteq U$ and (2) $m(U \backslash K) < \delta$. There is a continuous function $g : [0, 1] \to \mathbb{R}$ such that (3) $0 \leq g \leq 1$, (4) $g = 1$ on K and (5) $g = 0$ outside U. Then we have

$$\varlimsup_{n\to\infty} \left| \int_A f_n\,dm \right| = \varlimsup_{n\to\infty} \left| \int_0^1 f_n \chi_A\,dm \right|$$

$$\leq \varlimsup_{n\to\infty} \left| \int_A f_n g\,dm \right| + \left| \int_0^1 f_n(\chi_A - g)\,dm \right|$$

$$\leq \varlimsup_{n\to\infty} \left(\left| \int_A f_n g\,dm \right| + \int_0^1 h\chi_{U\backslash K}\,dm \right)$$

$$\leq \varepsilon.$$

It follows that

$$\lim_{n\to\infty} \int_A f_n\,dm = 0.$$

4214

Let $\{f_n\}$ be a sequence of non-negative measurable functions in $L^P(R)$ for some $1 < p < \infty$. Show that $f_n \to f(L^p)$ if and only if $f_n^p \to f^p(L^1)$.

(Indiana–Purdue)

Solution.

Suppose that $\|f_n - f\|_p \to 0$. Then $\|f_n\|_p \to \|f\|_p$, i.e.,

$$\int f_n^p \to \int f^p.$$

Denote

$$\tilde{f}_n = \min(f_n, f), \quad \bar{f}_n = \max(f_n, f).$$

Then $\tilde{f}_n \leq f$ and

$$|\tilde{f}_n - f| \leq |f_n - f|$$

which implies that $\|\tilde{f}_n - f\|_p \to 0$. Just as above, we have

$$\int \tilde{f}_n^p \to \int f^p.$$

Since

$$\tilde{f}_n^p + \bar{f}_n^p = f_n^p + f^p,$$

so

$$\int \bar{f}_n^p \to \int f^p.$$

Therefore

$$\int |f_n^p - f^p| = \int (\bar{f}_n^p - \tilde{f}_n^p) \to \int f^p - \int f^p = 0.$$

Conversely, suppose that

$$\int |f_n^p - f^p| \to 0.$$

Then

$$f_n \xrightarrow{m} f, \quad f \geq 0.$$

Since

$$\left| \int (f_n^p - f^p) \right| \leq \int |f_n^p - f^p| \to 0,$$

it follows that

$$\int f_n^p \to \int f^p.$$

For any $\varepsilon > 0$, take N_1 such that

$$\left| \int_\varepsilon (f_n^p - f^p) \right| < \varepsilon p/2$$

for $n > N_1$ and any measurable set E. Take A, a measurable set, such that $m(A) < +\infty$ and

$$\int_{A^c} f^p < \varepsilon p/2.$$

Then

$$\int_{A^c} f_n^p < \varepsilon^p$$

for $n > N_1$. Take $\delta > 0$ such that

$$\int_e f^p < \frac{\varepsilon^p}{2}$$

if $m(e) < \delta$. Thus

$$\int_e f_n^p < \varepsilon^p$$

if $m(e) < \delta$ and $n > N_1$. Denote $\eta = \varepsilon/m(A)^{\frac{1}{p}}$. Take N such that $N \geq N_1$ and $m(|f_n - f| \geq \eta) < \delta$ if $n > N$. We have

$$\left(\int_R |f_n - f|^p \right)^{1/p} \leq \left(\int_{|f_n - f| \geq \eta} |f_n - f|^p \right)^{1/p}$$

$$+ \left(\int_{A(|f_n - f| < \eta)} |f_n - f|^p \right)^{1/p} + \left(\int_{A^c} |f_n - f|^p \right)^{1/p}$$

$$\leq \left(\int_{|f_n - f| \geq \eta} f_n^p \right)^{1/p} + \left(\int_{|f_n - f| \geq \eta} f^p \right)^{1/p} + \left(\int_{A^c} f_n^p \right)^{1/p}$$

$$+ \left(\int_{A^c} f^p \right)^{1/p} + \left(\int_{A(|f_n - f| < \eta)} |f_n - f|^p \right)^{1/p}$$

$$< 4\varepsilon + (\eta^p m(A))^{1/p} = 5\varepsilon,$$

for any $n > N$.

4215

Let $1 \leq p < \infty$. All parts refer to Lebesgue measure on R.

(a) Give an example where $\{f_n\}$ converges to f pointwise, $\|f_n\|_p \leq M$, $\forall n$ and $\|f_n - f\|_p \nrightarrow 0$.

(b) If $\{f_n\}$ converges to f pointwise and $\|f_n\|_p \to M < +\infty$, what can you conclude about $\|f\|_p$? Justify this conclusion.

(c) Show that if $\{f_n\}$ converges to f pointwise and $\|f_n\|_p \to \|f\|_p$, then $\|f_n - f\|_p \to 0$.

(Indiana–Purdue)

Solution.

(a) Consider $L[0, 1]$. Set $f_n = n\chi_{(0,\frac{1}{n})}$, then

$$\lim_{n\to\infty} f_n(x) = f(x) = 0, \quad x \in [0, 1]$$

and

$$\|f_n - f\| = \|f_n\| = 1.$$

(b) We conclude that $\|f\|_p \leq M$. By the Fatou's Lemma,

$$\int |f|^p \, dx \leq \varliminf \int |f_n|^p \, dx = M^p.$$

(c) Set

$$g_n = 2^p(|f_n|^p + |f|^p) - |f_n - f|^p.$$

Then $g_n \geq 0$, and

$$\lim_{n\to\infty} g_n(x) = 2^{p+1}|f|^p$$

pointwise. Using the Fatou Lemma, we have

$$2^{p+1}\int |f|^p \, dx \leq \varliminf \int g_n(x) \, dx$$

$$= 2^{p+1}\int |f|^p \, dx - \varlimsup \int |f_n - f|^p \, dx.$$

Therefore

$$\lim_{n\to\infty} \int |f_n - f|^p dx = 0.$$

4216

Suppose f_n is a sequence of measurable functions on $[0,\ 1]$ with

$$\int_0^1 |f_n(x)|^2 dx \le 10$$

and $f_n \to 0$ a.e. on $[0,\ 1]$. Prove

$$\int_0^1 |f_n|\ dx \to 0.$$

(Hint. Use Egorov and Cauchy–Schwartz.)

(*Stanford*)

Solution.

For any $\varepsilon > 0$ there is by Egorov's Theorem a measurable set $E \subset [0,1]$ such that (1) $m([0,1]|E) < \varepsilon$ and (2) f_n converges to 0 uniformly on $[0,\ 1]\backslash E$. We have by the Cauchy–Schwartz Inequality

$$\int_0^1 |f_n| dx = \int_{[0,1]\backslash E}^1 |f_n| dx + \int_E |f_n| dx$$

$$\le \left(\int_{[0,1]\backslash E} |f_n|^2 dx \right)^{1/2} \mu([0,1]\backslash E)^{1/2} + \int_E |f_n| dx$$

and therefore

$$\overline{\lim_{n\to\infty}} \int_0^1 |f_n| dx \le \sqrt{10}\mu([0,1]\backslash E)^{1/2} < \sqrt{10\varepsilon}.$$

Let $\varepsilon \to 0$, and we have

$$\lim_{n\to\infty} \int_0^1 |f_n| dx = 0.$$

4217

Let $(X, \mathcal{M}, \mu)$ be a probability space (a positive measure space with $\mu(X) = 1$). Show that if f is an integrable function with values in $[1, \infty)$, then

$$\lim_{p \downarrow 0} \left(\int_X f^p \, d\mu \right)^{1/p} = \exp\left(\int_X \log f \, d\mu \right).$$

(Iowa)

Solution.

It suffices to show that

$$\lim_{p \downarrow 0} \frac{\log(\int_X f \, d\mu)}{p} = \int_X \log f \, d\mu. \tag{1}$$

If $f = 1$ a.e., equality (1) holds; Otherwise, for any $0 < p < 1$ and any $x \in X$

$$0 \leq \frac{f(x)^p - 1}{p} = f(x)^{\xi p} \log f(x) \leq f(x), \quad (0 < \xi < 1).$$

By the Dominated Convergence Theorem we have

$$\lim_{n \to \infty} \frac{\log(\int_X f^{p_n} \, d\mu)}{p_n}$$

$$= \lim_{n \to \infty} \frac{\log(\int_X 1 + (f^{p_n} - 1) d\mu)}{p_n}$$

$$= \lim_{n \to \infty} \left(\frac{\log(1 + \int_X (f^{p_n} - 1) d\mu)}{\int_X (f^{p_n} - 1) d\mu} \cdot \frac{\int_X (f^{p_n} - 1) d\mu}{p_n} \right)$$

$$= \int_X \log f \, d\mu,$$

where $\{p_n\}$ is any sequence of $(0, 1)$ decreasing to 0.

It follows that

$$\lim_{p \downarrow 0} \frac{\log(\int_X f^p \, d\mu)}{p} = \int_X \log f \, d\mu.$$

4218

Let $\{a_n\}_{n\geq 2}$ be a sequence of real numbers with $|a_n| < \log n$. Consider

$$\sum_{n=2}^{\infty} a_n n^{-x}, \quad \forall\, 2 \leq x < \infty.$$

(a) Prove that this series converges in $L^1[2, \infty)$.

(b) Prove that

$$\sum_{n=2}^{\infty} \int_2^{\infty} a_n n^{-x}\, dx = \int_2^{\infty} \sum_{n=2}^{\infty} a_n n^{-x}\, dx,$$

where the sum on the right is the pointwise limit (you need not prove that this pointwise limit exists).

(Stanford)

Solution.

(a) Since

$$\sum_{n=2}^{\infty} \int_2^{\infty} |a_n| n^{-x}\, dx = \sum_{n=2}^{\infty} \frac{|a_n|}{\log n} n^{-2} < \sum_{n=2}^{\infty} \frac{1}{n^2} < \infty.$$

The series $\sum_{n=2}^{\infty} a_n n^{-x}$ converges in $L^1[2, \infty)$.

(b) Let

$$f_n(x) = \sum_{k=2}^{n} a_k k^{-x}, \quad x \geq 2.$$

Then

(i) $\lim_{n\to\infty} f_n(x) = \sum_{k=2}^{\infty} a_k k^{-x}$ and

(ii) $|f_n(x)| \leq \sum_{k=2}^{\infty} |a_n| n^{-x}$.

By (a), $x \mapsto \sum_{k=2}^{\infty} |a_k| k^{-x}$ is integrable, and by the Dominated Convergence Theorem,

$$\sum_{n=2}^{\infty} \int_2^{\infty} a_n n^{-x}\, dx$$

$$= \lim_{n\to\infty} \int_2^{\infty} f_n(x)\, dx = \int_2^{\infty} \lim_{n\to\infty} f_n(x)\, dx = \int_2^{\infty} \sum_{n=2}^{\infty} a_n n^{-x} dx.$$

4219

Let S be a bounded Lebesgue measurable set in $\mathbb{R}$, and let $\{c_n\}$ be a sequence in $\mathbb{R}$. Show that

$$\lim_{n\to\infty} \int_S \cos^2(nt + c_n)d\lambda(t) = \frac{1}{2}\lambda(S),$$

where λ is Lebesgue measure.

(Iowa)

Solution.

By the Riemann–Lebesgue Lemma, we have

$$\lim_{n\to\infty} \int_S \cos^2(nt + c_n)d\lambda(t)$$

$$= \lim_{n\to 0} \int_{\mathbb{R}} \frac{1 + \cos(2nt + 2c_n)}{2}\chi_S(t)d\lambda(t)$$

$$= \lim_{n\to\infty} \left(\frac{1}{2}\lambda(S) + \frac{\cos 2c_n}{2}\int_{\mathbb{R}} \cos 2nt\chi_S(t)dt\right.$$

$$\left. -\frac{\sin 2c_n}{2}\int_{\mathbb{R}} \sin 2nt\chi_S(t)dt\right)$$

$$= \frac{1}{2}\lambda(S).$$

4220

(a) Is the function

$$f(x, y) = \begin{cases} \dfrac{x^2 - y^2}{(x^2 + y^2)^2}, & (x, y) \neq (0, 0), \\ 0, & (x, y) = (0, 0). \end{cases}$$

Lebesgue integrable on the unit square $0 \leq x \leq 1$, $0 \leq y \leq 1$?

(b) Compute the repeated integrals in the two orders.

(c) Does the integral

$$\int_0^1 \left(\int_0^1 |f(x, y)|dx\right) dy$$

exist? (If you use a theorem, be explicit!)

(*Iowa*)

Solution.
(a) The function f is not integrable on the unit square, for

$$\int_0^1 \int_0^1 |f(x,y)| dx\, dy \geq \int_{\{(x,y)|x,y\geq 0,\ x^2+y^2\leq 1\}} |f(x,y)| dx\, dy$$

$$= \int_0^1 \int_0^{\frac{\pi}{2}} \frac{\cos 2\theta}{r^2} r\, dr\, d\theta$$

$$\geq \int_0^1 \int_0^{\frac{\pi}{4}} \frac{\cos 2\theta}{r} dr\, d\theta$$

$$\geq \int_0^1 \frac{dr}{r} \frac{\sqrt{2}}{2} \frac{\pi}{4} = \infty.$$

(b) We have

$$\int_{(0,1]} \left(\int_{(0,1]} \frac{x^2 - y^2}{(x^2+y^2)^2} dy \right) dx$$

$$= \int_{(0,1]} \int_{(0,\operatorname{arctg}\frac{1}{x}]} \frac{x^2(1 - \operatorname{tg}^2 t)}{\sec^4 t} \sec^2 t\, dt\, dx$$

$$= \int_{(0,x]} \int_{(0,\operatorname{arctg}\frac{1}{x}]} x^2 \cos 2\theta\, dt\, dx$$

$$= \int_{(0,1]} \frac{x^3}{1+x^2} dx = \frac{1 - \ln 2}{2}$$

and

$$\int_{(0,1]} \left(\int_{(0,1]} \frac{x^2 - y^2}{(x^2+y^2)^2} dx \right) dy = \frac{\ln 2 - 1}{2}.$$

(c) If $\int_0^1 (\int_0^1 |f(x,y)| dx) dy$ did exist, then $\int_{(0,1]\times(0,1]} \|f(x,y)| d(x,y)$ would exist by Fubini Theorem, which contradicts (a). So $\int_0^1 (\int_0^1 |f(x,y)| dx) dy$ does not exist.

4221

Let $(X, \mathcal{M}, \mu)$ be a measure space and $f \in L(\mu)$. Evaluate

$$\lim_{n \to \infty} \int_X n \ln\left(1 + \left(\frac{|f|}{n}\right)^2\right) d\mu.$$

(Iowa)

Solution.

It is easy to show that for any $x \geq 0$

$$\ln(1 + x^2) \leq x.$$

It follows that for any $n \in \mathbb{N}$ the function $n \ln(1 + \frac{|f|^2}{n^2})$ is dominated by $|f|$ and therefore integrable. By the Dominated Convergence Theorem

$$\lim_{n \to \infty} \int_X n \ln\left(1 + \left(\frac{|f|}{n}\right)^2\right) d\mu$$

$$= \int_X \lim_{n \to \infty} n \ln\left(1 + \left(\frac{|f|}{n}\right)^2\right) d\mu$$

$$= \int_X 0 \, d\mu = 0.$$

4222

Suppose that f is a measurable real valued function on $\mathbb{R}$ such that $t \mapsto e^{tx} f(t)$ is in $L^1(\mathbb{R})$ for all $x \in (-1, 1)$. Define the function

$$\varphi(x) = \int_{-\infty}^{+\infty} e^{tx} f(t) dt$$

for all $x \in (-1, 1)$. Prove that φ is differentiable on $(-1, 1)$.

(Iowa)

Solution.

Since for any $x \in (-1, 1)$,

$$|e^{tx}tf(t)| \leq e^{tx}\frac{2}{1-|x|}e^{\frac{|t|(1-|x|)}{2}}|f(t)|$$

$$= \frac{2}{1-|x|}e^{t\left(x+\text{sign}\,t\frac{1-|x|}{2}\right)}|f(t)|,$$

and

$$\left|x + \text{sign}\,t\frac{1-|x|}{2}\right| = \left|x \pm \frac{1-|x|}{2}\right| < 1,$$

we conclude that $t \mapsto e^{tx}tf(t)$ is also in $L^1(\mathbb{R})$, $x \in (-1, 1)$.

Next we show that

$$\frac{d}{dx}\int_{-\infty}^{+\infty}e^{tx}f(t)dt = \int_{-\infty}^{+\infty}e^{tx}tf(t)dt.$$

For any fixed $x \in (-1, 1)$, for any $y > x$,

$$\left|\frac{e^{ty}f(t) - e^{tx}f(t)}{y - x} - e^{tx}tf(t)\right|$$

$$= |(e^{tz} - e^{tx})tf(t)|, \quad (x < z < y)$$

$$= |(e^{tz''}t^2f(t)(z - x)| \quad (x < z' < z)$$

$$\leq \begin{cases} |y - x|e^{t\frac{1+x}{2}}t^2f(t), & t \geq 0 \text{ and } x < y < \dfrac{1+x}{2}, \\ |y - x|e^{tx}t^2f(t), & t < 0 \text{ and } x < y < \dfrac{1+x}{2}. \end{cases}$$

By the same method as above, we see that $t \mapsto e^{tx}t^2f(t)$ is integrable. It follows that

$$\lim_{y\downarrow x}\int_{-\infty}^{+\infty}\left(\frac{e^{ty}f(t) - e^{tx}f(t)}{y - x} - e^{tx}tf(t)\right)dt = 0$$

and

$$\lim_{y\uparrow x}\int_{-\infty}^{+\infty}\left(\frac{e^{ty}f(t) - e^{tx}f(t)}{y - x} - e^{tx}tf(t)\right)dt = 0.$$

4223

Evaluate

$$\lim_{n\to\infty} \int_0^n \left(1+\frac{x}{n}\right)^n e^{-2x}\, dx,$$

justifying any interchange of limits you use.

(Hint. First show that $(1+\frac{x}{n})^n \le e^x$ for $x \ge 0$.)

(Stanford)

Solution.

Since for any $n \in \mathbb{N}$ and $x \ge 0$,

$$\left(1+\frac{x}{n}\right)^n = \sum_{i=0}^n \binom{n}{i}\left(\frac{x}{n}\right)^i$$

$$= \sum_{i=0}^n \frac{n!}{(n-i)!n^i}\frac{x^i}{i!}$$

$$= 1 + \sum_{i=1}^n \left(1-\frac{0}{n}\right)\left(1-\frac{1}{n}\right)\cdots\left(1-\frac{i-1}{n}\right)\frac{x^i}{i!}$$

$$\le 1 + \sum_{i=1}^n \left(1-\frac{0}{n+1}\right)\left(1-\frac{1}{n+1}\right)\cdots$$

$$\times \left(1-\frac{i-1}{n+1}\right)\frac{x^i}{i!} + \left(\frac{x}{n+1}\right)^{n+1}$$

$$= 1 + \sum_{i=1}^{n+1} \binom{n+1}{i}\left(\frac{x}{n+1}\right)^i = \left(1+\frac{x}{n+1}\right)^{n+1}$$

and

$$\lim_{n\to\infty}\left(1+\frac{x}{n}\right)^n = e^x,$$

we have

$$\left(1+\frac{x}{n}\right)^n \le e^x.$$

By the Dominated Convergence Theorem

$$\lim_{n \to \infty} \int_0^n \left(1 + \frac{x}{n}\right)^n e^{-2x}\, dx$$

$$= \lim_{n \to \infty} \int_0^\infty \chi_{[0,n]}(x)\left(1 + \frac{x}{n}\right)^n e^{-2x}\, dx$$

$$= \int_0^\infty \lim_{n \to \infty} x_{[0,n]}(x)\left(1 + \frac{x}{n}\right)^n e^{-2x}\, dx$$

$$= \int_0^\infty e^{-x}\, dx = 1.$$

4224

Let $(X, \mathcal{A}, \mu)$ be a probability space and $f : X \to [1, \infty)$ a measurable function. Prove or disprove:

$$\int f \ln f\, d\mu \geq \int f\, d\mu \int \ln f\, d\mu.$$

(Iowa)

Solution.

Under the condition that $f \ln f$ is integrable we conclude that both f and $\ln f$ are integrable, and

$$\int_X f\, d\mu \int_X \ln f\, d\mu \leq \int_X f \ln f\, d\mu.$$

Indeed, since

$$0 \leq f \leq \begin{cases} e, & f(x) \leq e, \\ f\ln f, & \text{otherwise} \end{cases}$$

and $0 \leq \ln f \leq f \ln f$ we see that both f and $\ln f$ are integrable. Since $\ln t \leq t \ln t$ for any $t > 0$,

$$\int_X f\, d\mu \int_X \ln f\, d\mu - \int_X f \ln f\, d\mu$$

$$= \frac{1}{2} \int_{X \times X} (f(x) \ln f(y) + f(y) \ln f(x)) d\mu(x) d\mu(y)$$

$$-\frac{1}{2}\int_{X\times X}(f(x)\ln f(x)+f(y)\ln(y))d\mu(x)d\mu(y)$$

$$=\frac{1}{2}\int_{X\times X}f(x)\left(\ln\frac{f(y)}{f(x)}-\frac{f(y)}{f(x)}\ln\frac{f(y)}{f(x)}\right)d\mu(x)d\mu(y)$$

$$\leq 0.$$

4225

For $i = 1, 2$, let $X_i = \mathbb{N}$ (the natural numbers), Let $M_i = 2^{\mathbb{N}}$ (the σ-algebras of all subsets of $\mathbb{N}$) and let μ_i be the counting measure. For the function $f : X_1 \times X_2 \to \mathbb{R}$ defined by

$$f(i, j) = \begin{cases} -2^{-i}, & j = i, \\ 2^{-i}, & j = i+1, \\ 0, & \text{otherwise.} \end{cases}$$

Compute the iterated integrals

$$\int_{X_1}\left(\int_{X_2}f\,d\mu_2\right)d\mu_1$$

and

$$\int_{X_2}\left(\int_{X_1}f\,d\mu_1\right)d\mu_2.$$

How do you reconcile your answers with Fubini's Theorem?

(*Iowa*)

Solution.

For any $i \in X_1, j \mapsto f(i, j)$ is integrable and

$$\int_{X_2}f(i, j)d\mu_2(j) = -2^{-i} + 2^{-i} = 0.$$

Therefore

$$\int_{X_1}\left(\int_{X_2}f\,d\mu_2\right)d\mu_1 = 0.$$

For any $j \in X_2$, $i \mapsto f(i, j)$ is integrable and

$$\int_{X_1} f(i, 1)d\mu_1(i) = -\frac{1}{2},$$

$$\int_{X_1} f(i, j)d\mu_1(i) = 2^{-j} \ (j \geq 2).$$

We have

$$\int_{X_2} \left(\int_{X_1} f(i, j)d\mu_1(i) \right) d\mu_2(j) = -\frac{1}{2} + \sum_{j=2}^{\infty} \frac{1}{2^j} = 0.$$

Since f is integrable ($\sum_{i,j \in \mathbb{N}} |f|(i, j) = 2$), the two iterated integrals exist and coincide by the Fubini's Theorem.

4226

Let (Ω, μ) be any measure space. For $f \in L^1(\Omega, \mu)$, and for $\lambda > 0$ define

$$\varphi(\lambda) = \mu(\{x \in \Omega | f(x) > \lambda\})$$

and

$$\psi(\lambda) = \mu(\{x \in \Omega | f(x) > -\lambda\}).$$

Show that the functions φ and ψ are Borel measurable and that

$$\|f\|_1 = \int_0^\infty (\varphi(\lambda) + \psi(\lambda))d\lambda.$$

(Hint. As usual, it may be helpful to consider f positive first.)

(*Iowa*)

Solution.

The measurability of φ and ψ follows from that they are monotone.

The function $(x, \lambda) \mapsto \chi_{(0,\infty)}(|f|(x) - \lambda)$ is measurable. On one hand,

$$\int_0^\infty \left(\int_\Omega \chi_{(0,\infty)}(|f|(x) - \lambda)d\mu(x) \right) d\lambda$$

$$= \int_0^\infty \left(\int_\Omega \chi_{\{x \| f(x)| > \lambda\}}(x)d\mu(x) \right) d\lambda$$

$$= \int_0^\infty \mu(\{x||f(x)| > \lambda\})\, d\lambda$$

$$= \int_0^\infty (\varphi(\lambda) + \psi(\lambda))d\lambda;$$

but on the other,

$$\int_\Omega \left(\int_0^\infty \chi_{(0,\infty)}(|f|(x) - \lambda)d\lambda\right)d\mu(x) = \int_\Omega \int_0^{|f|(x)} d\lambda\, d\mu(x)$$

$$= \int_\Omega |f|(x)d\mu(x) = \|f\|_1.$$

By Fubini's Theorem we have

$$\|f\|_1 = \int_0^\infty (\varphi(\lambda) + \psi(\lambda))d\lambda,$$

provided that either side exists.

4227

Let $f \in L^2(0,1)$. Set

$$F(x) = \int_0^x f(t)dt.$$

Show that

$$\left(\int_0^{1-h} \left|\frac{F(x+h) - F(x)}{h}\right|^2 dx\right)^{1/2} \le c\|f\|_2$$

for each $h, 0 < h < 1$ where c is a positive number independent of f.

(*UC, Irvine*)

Solution.
We have

$$\left(\int_0^{1-h} \left|\frac{F(x+h) - F(x)}{h}\right|^2 dx\right)^{1/2}$$

$$= \frac{1}{h}\left(\int_0^{1-h} \left|\int_x^{x+h} f(t)dt\right|^2 dx\right)^{1/2}$$

$$\leq \frac{1}{h} \left(\int_0^{1-h} \int_x^{x+h} |f(t)|^2 dt \cdot h\, dx \right)^{1/2}$$

$$= \frac{1}{\sqrt{h}} \left(\int_0^{1-h} dx \int_x^{x+h} |f(t)|^2 dt \right)^{1/2}$$

$$= \frac{1}{\sqrt{h}} \left(\int_0^h dt \int_0^t |f(t)|^2 dx + \int_h^{1-h} dt \int_{t-h}^t |f(t)|^2 dx \right.$$

$$\left. + \int_{1-h}^1 dt \int_{t-h}^{1-h} |f(t)|^2 dx \right)^{1/2}$$

$$= \frac{1}{\sqrt{h}} \left(\int_0^h |f(t)|^2 t\, dt + \int_h^{1-h} |f(t)|^2 h\, dt + \int_{1-h}^1 |f(t)|^2 (1-t) dt \right)^{1/2}$$

$$\leq \frac{1}{\sqrt{h}} \left(\int_0^h |f(t)|^2 h\, dt + \int_h^{1-h} |f(t)|^2 h\, dt + \int_{1-h}^1 |f(t)|^2 h\, dt \right)^{1/2}$$

$$= \|f\|_2.$$

4228

Let $f, g \in L^1(0,1)$ and assume that $f(x)g(y) = f(y)g(x)$ for all $x, y \in [0,1]$. Show that

$$\int_0^1 \int_0^1 f(x)g(y) dx\, dy = 2 \int_\Delta f(x)g(y) dA,$$

where

$$\Delta = \{(x,y) | 0 \leq x < y \leq 1\}$$

and dA denotes the planar Lebesgue measure.

(*Iowa*)

Solution.
We have

$$\int_0^1 \int_0^1 f(x)g(y) dx\, dy = \int_{[0,1] \times [0,1]} f(x)g(y) dA$$

$$= \int_\Delta f(x)g(y) dx\, dy + \int_{\{(x,y)|0 \leq x = y \leq 1\}} f(x)g(y) dA$$

$$+ \int_{\{(x,y)|0\leq y<x\leq 1\}} f(x)g(y)dA$$

$$= \int_{\Delta} f(x)g(y)dx\,dy + \int_{\{(x,y)|0\leq x<y\leq 1\}} f(x)g(y)dA$$

$$= 2\int_{\Delta} f(x)g(y)dA.$$

4229

Give an example of a Baire-1 function which is not Riemann integrable. Explain why such a function would have to be Lebesgue integrable.

(*Auburn*)

Solution.

For each rational number r take a positive number a_r such that $\sum_{r\in\mathbb{Q}} a_r = 1$. Let $U = \cup_{r\in\mathbb{Q}}(r - a_r, r + a_r)$ and $E = \mathbb{R}\backslash U$ then U is open, dense in $\mathbb{R}$, while E is closed in $\mathbb{R}$ and $|E|_1 = +\infty$.

Define a function $f : \mathbb{R} \to \mathbb{R}$ such that $f(x) = 0$ for $x \in E$ and

$$f(x) = (x-u)^2(x-v)^2 \sin g(x) : u < x < v$$

for each component (u, v) of U, where

$$g(x) = \frac{1}{(v-u)(x-u)(x-v)} : u < x < v$$

observing that (u, v) is finite since U is of finite measure. Clearly,

$$f'(x) = 2(x-u)(x-v)(2x-u-v)\sin g(x)$$
$$+ \frac{(u+v-2x)\cos g(x)}{v-u} : u < x < v \qquad (*)$$

As for $x \in E$, $f'(x) = 0$. Indeed, if $z \in U$ tends to x, with $u_z < z < v_z$, either $x \leq u_z$ or $v_z \leq x$. It follows that

$$\frac{f(z) - f(x)}{z - x} = \frac{(z-u_z)^2(v_z-z)^2}{z-x} \sin g(z) \to 0.$$

Now $|f'| \le 3$ and E is the set of discontinuity of f'. Indeed, for any $x_0 \in E$. by $(*)$, we see that $\overline{\lim}_{x \to x_0}|f'(x) - f'(x_0)| > 0$.

Since E is of infinite measure, $|E \cap [a,b]|_1 > 0$ for some $a < b$. It follows that f' is not Riemannian integrable on $[a,b]$.

However, f' is a bounded function of Baire-1 so that f' is Lebesgue integrable on each Lebesgue measurable set of finite measure.

4230

Let $(\Omega, \mathcal{A}, \mu)$ be a σ-finite measure space, $f : \Omega \to \mathbb{R}$ measurable. Suppose there is a $c \in \mathbb{R}$ such that for all $X \subset \Omega$ of finite measure $|\int_X f \, d\mu| \le c$ holds. Prove that $f \in L^1(\Omega, \mathcal{A}, \mu)$.

(*Purdue*)

Solution.

Let $\Omega_\pm = (\pm f > 0)$ then $\int_X |f| \le c$ for any $X \subseteq \Omega_\pm$ of finite measure. Take a sequence $(X_n^\pm)$ of measurable subsets of finite measure increasing to $\Omega_\pm$ so that

$$\int_{\Omega_\pm} |f| = \lim_{n \to \infty} \int_{X_n^\pm} |f| \le c$$

and thus f is integrable.

4231

Let $(X, \mathcal{M}, \mu)$ be a measure space. Two functions $f, g : X \to \mathbb{R}$ are said to be comonotone if

$$(f(x) - f(y))(g(x) - g(y)) \ge 0$$

for every $(x, y) \in X \times X$. If $\mu(X) = 1$ and $f, g \in L^1(\mu)$ are comonotone, show that

$$\int_X f \, d\mu \int_X g \, d\mu \ge \int_X fg \, d\mu.$$

(*Purdue*)

Solution.

The condition for being comonotone is equivalent to that

$$f(x)g(x) + f(y)g(y) \geq f(x)g(y) + f(y)g(x)$$

for every $(x, y) \in X \times X$. Integrating over $X \times X$ yields that

$$\int_X \mu(dy) \int_X f(x)g(x)\mu(dx) + \int_X f(y)g(y)\mu(dy) \int_X \mu(dx)$$

$$\geq \int_X g(y)\mu(dy) \int_X f(x)\mu(dx) + \int_X f(y)\mu(dy) \int_X g(x)\mu(dx).$$

It remains to note that $\mu(X) = 1$.

4232

Let $f : \mathbb{R}_+ \to \mathbb{R}_+, \mathbb{R}_+ = \{x \in \mathbb{R} : x \geq 0\}$. Assume that $f \in L^1(\mathbb{R}_+)$ satisfies

$$f(x) \leq c \int_0^x f(t)dt : x > 0$$

with c independent of x. Show that $f(x) = 0$ for every $x \in \mathbb{R}_+$.
(Hint. Write $f(x)$ as iterated integrals.)

(*Purdue*)

Solution.

Let $b = c \int_0^\infty f(t)dt$ then $f(x) \leq cb$. Inductivety,

$$f(x) \leq c^n \int_0^x dt_1 \int_0^{t_1} dt_2 \cdots \int_0^{t_{n-1}} f(t_n)dt_n \leq \frac{b(cx)^n}{n!}$$

and thus $f(x) = 0$ for $x \in \mathbb{R}_+$.

4233

Show that the following limit exists

$$\lim_{n\to\infty} n \int_{1/n}^1 \frac{\cos(x + 1/n) - \cos x}{x^{3/2}} dx.$$

(*Purdue*)

Solution.

With I_n denoted the n-th integral above, integration by part yields that

$$I_n = \int_{1/n}^1 4n \sin\left(x + \frac{1}{2n}\right) \sin\frac{1}{2n} d\frac{1}{\sqrt{x}}$$

$$= \frac{4n}{\sqrt{x}} \sin\left(x + \frac{1}{2n}\right) \sin\frac{1}{2n}\Big|_{1/n}^1$$

$$- \int_{1/n}^1 \frac{4n}{\sqrt{x}} \cos\left(x + \frac{1}{2n}\right) \sin\frac{1}{2n} dx$$

$$= 4n \sin\left(1 + \frac{1}{2n}\right) \sin\frac{1}{2n} - 4n\sqrt{n}\sin\frac{3}{2n}\sin\frac{1}{2n}$$

$$- \int_{1/n}^1 \frac{4n}{\sqrt{x}} \cos\left(x + \frac{1}{2n}\right) \sin\frac{1}{2n} dx.$$

Applying Dominated Convergence Theorem yields that

$$\lim_{n\to\infty} I_n = 2\sin 1 - \int_0^1 \frac{2\cos x}{\sqrt{x}} dx.$$

4234

A Lebesgue integrable function $f : \mathbb{R} \to \mathbb{R}$ has the property that

$$\int_E f(x)dx = 0$$

for all Lebesgue measurable sets $E \subset \mathbb{R}$ with $m(E) = \pi$. Prove or disprove that $f = 0$ a.e.

(*Purdue*)

Solution.

We will show $f = 0$, using the fact that if $f(x) \geq 0$ for $x \in F$ and $\int_F f(x)dx = 0$ then $f(x) = 0$ for almost every $x \in F$.

Let $E_\pm = (\pm f > 0)$ then $m(E_\pm) < \pi$. Otherwise, assume $m(E_+) \geq \pi$ there would be a measurable subset F of E_+ such that $m(F) = \pi$ by a similar argument to the solution of the Problem 4111 in this book. But then $f = 0$ a.e, on F, a contradiction.

Let $E_0 = (f = 0)$ then $m(E_0) = \infty$ and there are measurable subsets $F_\pm$ of E_0 such that $m(F_\pm) = \pi - m(E_\pm)$. It follows that $f = 0$ a.e, on $E_\pm \cup F_\pm$ and thus $m(E_\pm) = 0$.

4235

Prove that the following limit exists

$$\lim_{n \to \infty} \int_0^\infty \frac{\exp(-x) \cos x}{nx^2 + \frac{1}{n}} dx,$$

and find it, justifying all your steps.

(Purdue)

Solution.

Applying the integration by part yields that

$$\int_0^\infty \frac{\exp(-x) \cos x}{(nx)^2 + 1} d(nx) = \int_0^\infty \exp(-x) \cos x \, d \arctan nx$$

$$= \exp(-x) \cos x \arctan nx |_0^\infty - \int_0^\infty \arctan nx \, d(\exp(-x) \cos x).$$

The signed measure $d(\exp(-x) \cos x)$ is finite and $|\arctan nx| \le \pi$ so that applying the Bounded Convergence Theorem yields that

$$\lim_{n \to \infty} \int_n^\infty \frac{\exp(-x) \cos x}{nx^2 + \frac{1}{n}} d(x) = -\int_0^\infty \frac{\pi}{2} d(\exp(-x) \cos x) = \frac{\pi}{2}.$$

4236

Let $(X, \mathcal{F}, \mu)$ be a measure space and let $f_n : X \to \mathbb{R}$ be a sequence of measurable functions on it satisfying

$$\int_X |f_k|^2 \, d\mu \le c, \text{ for all } k,$$

where c is a finite constant independent of k, and

$$\int_X f_j f_k \, d\mu = 0, \text{ for all } j \neq k.$$

For each $n = 1, 2, \ldots$, set $g_n = \sum_{k=1}^{n^2} f_k$. Prove that

$$\lim_{n \to \infty} \frac{g_n}{n^\alpha} = 0, \text{ a.e.}$$

for all $\alpha > 3/2$.

(Purdue)

Solution.

Since $2(\alpha - 1) > 1$, integration in series yields that

$$\int_X \sum_{n=1}^{\infty} \frac{g_n^2}{n^{2\alpha}} d\mu = \sum_{n=1}^{\infty} \sum_{i=1}^{n^2} \int_X \frac{f_i^2}{n^{2\alpha}} d\mu \leq \sum_{n \geq 1} \frac{c}{n^{2(\alpha-1)}} < +\infty$$

so that $\sum_{n \geq 1}^{\infty} \frac{g_n^2(x)}{n^{2\alpha}} < +\infty$ for almost every $x \in X$. It follows that $\lim_{n \to \infty} \frac{g_n}{n^\alpha} = 0$, a.e.

4237

(1) Let $f \in L^1(0, \infty)$. Prove that there exists a sequence x_k increasing to ∞ such that $\lim_{k \to \infty} x_k f(x_k) = 0$.

(2) Let $f \in L^1(\mathbb{R}^n)$ with $n \geq 2$. Prove that there exists a sequence r_k increasing to ∞ such that $\lim_{k \to \infty} r_k \int_{\mathbb{S}(r_k)} |f| d\sigma = 0$, where $\mathbb{S}(r) = \{x \in \mathbb{R}^n : |x| = r\}$, and $d\sigma$ represents the $(n-1)$-dimensional Lebesgue measure on the sphere $\mathbb{S}(r)$.

(Purdue)

Solution.

(1) It suffices to show that $\underline{\lim}_{x \to +\infty} x|f(x)| = 0$. Otherwise, there would be $b > 0$ and $c > 0$ such that $x|f(x)| \geq c$ for any $x \geq b$ and thus

$$\int_0^\infty |f(x)| dx \geq \int_b^\infty \frac{c \, dx}{x} = +\infty,$$

a contradiction.

(2) Let $g(r) = \int_{S(r)} |f| d\sigma$ then by the spherical coordinate transformation.

$$\int_{\mathbb{R}^n} |f(x)| dx = \int_0^\infty \left(\int_S |f(rx)| \sigma(dx) \right) r^{n-1} dr = \int_0^\infty g(r) dr.$$

Applying (1) yields a sequence $(r_k)_{k\geq 1}$ increasing to $+\infty$ such that $\lim_{k\to\infty} r_k g(r_k) = 0$.

4238

Let (S, Q, m) be a measure space satisfying $m(S) < \infty$. Let $f : S \to \mathbb{R}$ be a measurable function satisfying $|f| < 1$. Prove that either $\lim_{n\to\infty} \int_S (1 + f + \cdots + f^n) dm$ exists or that this limit is $+\infty$.

(Purdure)

Solution.

Write $f = f_+ - f_-$ then $f^n = f_+^n + (-f_-)^n$ and

$$\int_S (1 + f + f^2 + \cdots + f^n) dm$$

$$= \int_S \frac{1 - f_+^{n+1}}{1 - f_+} dm - \int_S \frac{f_- - (-f_-)^{n+1}}{1 + f_-} dm$$

$$= \int_S \frac{dm}{1 - f_+} - \int_S \frac{f_- dm}{1 + f_-}.$$

The sequence $\left(\frac{1 - f_+^{n+1}}{1 - f_+} \right)_{n \geq 1}$ of positive measurable functions increases to $\frac{1}{1 - f_+}$ and the sequence $\left(\frac{f_- - (-f_-)^{n+1}}{1 + f_-} \right)_{n \geq 1}$ of measurable functions is dominated by the integrable function f_- and converges to $\frac{f_-}{1 + f_-}$ pointwise. Applying Monotone Convergence Theorem and Bounded Convergence Theorem yields that

$$\lim_{n\to\infty} \int_S \sum_{i=0}^n f^i \, dm = \int_S \frac{dm}{1 - f_+} - \int_S \frac{f - dm}{1 + f_-}$$

which is finite or infinite if $(1 - f_+)^{-1}$ is integrable or not.

4239

If $\{r_k\}$ dentoes the rational numbers in $[0,1]$ and $\{a_k\}$ satisfies $\sum_{k=1}^{\infty} |a_k| < \infty$, prove that $\sum_{k=1}^{\infty} \frac{a_k}{|x-r_k|^{1/2}}$ converges absolutely a.e. $[0,1]$.

(*Temple*)

Solution.

Applying Monotone Convergence Theorem yields that

$$\int_0^1 \sum_{k=1}^{\infty} \frac{|a_k|}{|x-r_k|^{1/2}} dx = \sum_{k=1}^{\infty} \int_0^1 \frac{|a_k|dx}{|x-r_k|^{1/2}}$$

$$= \sum_{k=1}^{\infty} 2(\sqrt{r_k} + \sqrt{1-r_k})|a_k| < +\infty.$$

It follows that $\sum_{k=1}^{\infty} \frac{|a_k|}{|x-r_k|^{1/2}} < +\infty$ for almost every x in $[0,1]$.

4240

Prove that the series

$$\sum_{n=0}^{\infty} \int_0^{\pi/3} (1 - \sqrt{\sin x})^n \cos x \, dx$$

is summable and find its sum.

(*SUNY, Albany*)

Solution.

Since the integrands are positive we see that

$$\sum_{n\geq 0} \int_0^{\pi/3} (1 - \sqrt{\sin x})^n \cos x \, dx = \int_0^{\pi/3} \sum_{n\geq 0} (1 - \sqrt{\sin x})^n \cos x \, dx$$

$$= \int_0^{\pi/3} \frac{\cos x \, dx}{\sqrt{\sin x}} = \sqrt{2\sqrt{3}}.$$

4241

Let $\{a_{n,k}|n, k = 1, 2, \ldots\} \subset \mathbb{R}$ with $|a_{n,k}| \leq 1$ for $n, k \geq 1$. Assume that for each n, $\lim_{k \to \infty} a_{n,k} = 0$. Let $p > 1$. Show that $\lim_{k \to \infty} \sum_{n=1}^{\infty} a_{n,k}/n^p = 0$.

(SUNY, Albany)

Solution.

Regard the series as the integral relative to the counting measure. Since $\left|\frac{a_{n,k}}{n^p}\right| \leq \frac{1}{n^p}$ and $\sum_{n \geq 1} \frac{1}{n^p}$ is summable, applying the Dominated Convergence Theorem yields that

$$\lim_{k \to \infty} \sum_{n \geq 1} \frac{a_{n,k}}{n^p} = \sum_{n \geq 1} \lim_{k \to \infty} \frac{a_{n,k}}{n^p} = 0.$$

4242

Prove that if f and g are positive continuous functions on $(-\infty, +\infty)$ which are periodic of period 1 that

$$\lim_{n \to \infty} \int_0^1 f(x)g(nx)dx = \int_0^1 f(x)dx \int_0^1 g(x)dx.$$

(SUNY, Albany)

Solution.

We can show a stronger result: If $f : \mathbb{R} \to \mathbb{C}$ is integrable and $g : \mathbb{R} \to \mathbb{C}$ is a bounded measurable function with period T such that $\int_0^T g(x)dx = I$ then

$$\lim_{x \to \infty} \int_{\mathbb{R}} f(y)g(xy)dy = \frac{I}{T} \int_{\mathbb{R}} f(y)dy.$$

Define a bounded semi-norm $\psi : L^1\mathbb{R} \to \mathbb{C}$ by

$$\psi(f) = \overline{\lim_{x \to \infty}} \left| \int_{\mathbb{R}} f(y)g(xy)dx - \frac{I}{T} \int_{\mathbb{R}} f(y)dy \right|.$$

It suffices to show that $\psi(f) = 0$ for any function f of the form $\chi_{(a,b]}$ since such functions span a dense subspace of $L^1\mathbb{R}$ and ker ψ is a closed linear subspace of $L^1\mathbb{R}$. However, let k be the integer part of $\frac{bx-ax}{T}$, depending on x, then $\lim_{x\to\infty}\frac{k}{x}=\frac{b-a}{T}$ since $\frac{bx-ax}{T}\le k\le \frac{bx-ax}{T}+1$. Now

$$\int_{\mathbb{R}} f(y)g(xy)dy = \int_a^b g(xy)dy = \int_{ax}^{bx} \frac{g(y)}{x}dy$$

$$= \int_{ax}^{ax+kT} \frac{g(y)}{x}dy + \int_{ax+kT}^{bx} \frac{g(y)}{x}dy = \frac{kI}{x}+\frac{J}{x}.$$

Now $\lim \frac{J}{x}=0$ since $|J|\le \|g\|_\infty T$. The desired result follows.

4243

Given functions $\{f_n : [0,1] \to [-1,1]\}$ such that $\lim_{n\to\infty}$ $\int_a^b f_n(x)dx = 0$ for all $0\le a\le b\le 1$, show that

(1) For every Lebesgue measurable subset $A\subseteq [0,1]$, $\lim_{n\to\infty}$ $\int_A f_n(x)dx = 0$.

(2) For every Lebesgue measurable f on $[0,1]$ such that $\int_0^1 |f(x)|dx < \infty$, $\lim_{n\to\infty}\int_0^1 f(x)f_n(x)dx = 0$.

(Rutgers)

Solution.

We will show (1) and (2) together. For any $f\in L^1[0,1]$ define

$$\phi(f) = \overline{\lim_{n\to\infty}}\left|\int_0^1 f(x)f_n(x)dx\right|$$

then $\phi : L^1[0,1] \to \mathbb{R}$ is a semi-norm such that $\phi(f)\le \|f\|_1$ and $\phi(\chi_{(a,b]}) = 0$ so that $\phi = 0$ since ker ϕ is a closed linear subspace of $L^1[0,1]$ and $\{\chi_{(a,b]}|0\le a\le b\le 1\}$ spans a dense subspace in $L^1[0,1]$.

4244

Let f be a Lebesgue integrable function on $\mathbb{R}^2$, that is $\int_{\mathbb{R}^2} |f(x)|dx < \infty$. Assume that $\int_B f(x)dx = 0$ for all square subsets $B\subset \mathbb{R}^2$. Prove that $f(x)=0$ almost everywhere.

(Rutgers)

Problems and Solutions in Mathematics

Solution.

It suffices to show the claim that $\int_A f(x)dx = 0$ for any measurable set A of finite measure.

In the case A is a dyadic rectangle $(\frac{i}{2^m}, \frac{j}{2^m}] \times (\frac{k}{2^n}, \frac{l}{2^n}]$, increasing m and n if necessary, we may assume $m = n$ so that

$$A = \coprod \left\{ \left(\frac{p-1}{2^n}, \frac{p}{2^n}\right] \times \left(\frac{q-1}{2^n}, \frac{q}{2^n}\right] \middle| \begin{matrix} i < p \le j \\ k < q \le l \end{matrix} \right\}$$

and the claim holds.

In the case A is a rectangle $(a, b] \times (c, d]$, take a sequence (A_u) of dyadic rectrangles increasing to $(a, b) \times (c, d)$, then $\lim_{n \to \infty} |A \backslash A_n| = 0$ and $\int_A f(x)dx = \lim_{n \to \infty} \int_{A_n} f(x)dx$ so that the claim holds.

In general case, take a sequence (A_n) of finite disjoint union of rectangles such that $\lim_{n \to \infty} |A \Delta A_n| = 0$ then it remains to note that

$$\lim_{n \to \infty} \left| \int_A f(x)dx - \int_{A_n} f(x)dx \right| \le \lim_{n \to \infty} \int_{A \Delta A_n} |f(x)|dx = 0.$$

4245

Let $f \in L^1[0, 1]$ and $\epsilon > 0$. Prove the existence of a compact set K and a polynomial p such that the Lebesgue measure of K exceeds $1 - \epsilon$ and

$$|f(x) - p(x)| < \epsilon : \forall x \in K.$$

$$(Rutgers)$$

Solution.

Applying Lusin Theorem yields a closed subset K of $[0, 1]$ and a continuous function $g : [0, 1] \to \mathbb{C}$ such that $m(K) > 1 - \epsilon$ and $f(x) = g(x)$ for $x \in K$. Applying Weierstrass' Approximation Theorem yields a polynomial p such that $|g(x) - p(x)| < \epsilon$ for $x \in [0, 1]$ and thus $|f(x) - p(x)| < \epsilon$ for $x \in K$.

4246

Assume that $f_n \leq g_n \leq h_n$ are real-valued Lebesgue integrable functions converging almost everywhere to Lebesgue integrable functions f, g and h respectively. Assume that $\lim_{n \to \infty} \int f_n(x)dx = \int f(x)dx$ and $\lim_{n \to \infty} \int h_n(x)dx = \int h(x)dx$. Prove that $\lim_{n \to \infty} \int g_n(x)dx = \int g(x)dx$. (Hint. The monotone convergence theorem will not work. Look at $h_n - g_n$ and $g_n - f_n$.)

(*Rutgers*)

Solution.

Applying Fatou's Lemma yields that

$$\int (h(x) - g(x))dx \leq \varliminf_{n \to \infty} \int (h_n(x) - g_n(x))dx$$

$$= \int h(x)dx - \varlimsup_{n \to \infty} \int g_n(x)dx,$$

$$\int (g(x) - f(x))dx \leq \varliminf_{n \to \infty} \int (g_n(x) - f_n(x))dx$$

$$= \varliminf_{n \to \infty} \int g_n(x)dx - \int f(x)dx.$$

It follows that

$$\int g(x)dx \leq \varliminf_{n \to \infty} \int g_n(x)dx \leq \varlimsup_{n \to \infty} \int g_n(x)dx \leq \int g(x)dx,$$

as desired.

4247

Let $f_n : [0,1] \to [0,\infty)$ be Lebesgue integrable for $n = 1, 2, \ldots$, Assume that

$$\int_0^1 f_n(x)dx = 1, \int_{1/n}^1 f_n(x)dx < \frac{1}{n} : n \geq 1.$$

Let $g(x) = \sup\{f_n(x) : n \geq 1\}$. Prove that $\int_0^1 g(x)dx = \infty$.

(*Rutgers*)

Solution.

Otherwise, g would be integrable. However,

$$\int_0^{1/n} g(x)dx \geq \int_0^{1/n} f_n(x)dx \geq 1 - \frac{1}{n}$$

and passing to the limit yields that

$$\int_0^0 g(x)dx = 1,$$

a contradiction.

4248

Let λ be Lebesgue measure and μ be counting measure both regarded as Borel measures on $I = [0,1]$. Let $\Delta = \{(x,x)|x \in I\}$, the diagonal of $I \times I$.

(a) Show that Δ is a Borel subsets of $I \times I$.

(b) Let f be the characteristic function of the diagonal Δ. Compute the integrals $\int_I(\int_I f \, d\lambda)d\mu$ and $\int_I(\int_I f \, d\mu)d\lambda$.

(c) Explain why the result of part (b) does not contradict Fubini's Theorem. As part of the explanation, you should compute the double integral $\int_{I \times I} f(d\mu \times d\lambda)$.

(SUNY, Albany)

Solution.

(a) Δ is a closed subset of $I \times I$.

(b) Since $f(x,y) = \chi_{\{x\}}(y) = \chi_{\{y\}}(x)$ we have that

$$\int_I \mu(dx) \int_I f(x,y)\lambda(dy) = \int_I 0\mu(dx) = 0,$$

$$\int_I \lambda(dy) \int_I f(x,y)\mu(dx) = \int_I 1\lambda(dy) = 1.$$

(c) For any partition $\{E_i\}$ of $[0,1]$ we have $\inf f(E_i \times E_j) = \delta_{ij}$ and $\lambda(E_i) > 0$ implies $\mu(E_i) = +\infty$. Now $\int_{I \times I} f(d\mu \times d\lambda) = +\infty$ since

$$\sum_{i,j \geq 1} \inf f(E_i \times E_j)(\mu \times \lambda)(E_i \times E_j) = \sum_{i \geq 1} \mu(E_i)\lambda(E_i) = +\infty.$$

The reason for the result of part (b) not to contradict Fubini's Theorem is that $(f \neq 0)$ is not σ-finite relative to the product measure $\mu \times \lambda$.

Section 3

Space of Integrable Functions

4301

Let X be a positive measure space of total measure 1. Show that for any $[0, \infty)$-valued measurable function f on X,

$$I(p) := \left(\int_X f^p \right)^{\frac{1}{p}}$$

is a non-decreasing function of $p \in (1, \infty)$. Under what circumstances is $I(p)$ strictly increasing as a function of p?

(Iowa)

Solution.

For any $p, q \in (1, \infty)$ with $p < q$, let $\alpha = \frac{q}{p}$ and $\beta = \frac{q}{q-p}$ then $\frac{1}{\alpha} + \frac{1}{\beta} = 1$. By Hölder's Inequality, we have

$$\left(\int_X f^p \right)^{\frac{1}{p}} \leq \left(\left(\int_X f^{p\alpha} \right)^{\frac{1}{\alpha}} \left(\int_X 1^\beta \right)^{\frac{1}{\beta}} \right)^{\frac{1}{p}} = \left(\int_X f^q \right)^{\frac{1}{q}},$$

which shows that $I(p)$ is a non-decreasing function of $p \in (1, \infty)$. The function I is strictly increasing if and only if f is not almost everywhere equal to any constant function.

4302

Let $(X, \mathcal{M}, \mu)$ be a fixed measure space, let $\{p_i\}_{i=1}^n$ be positive numbers such that $p_i > 1$, $i = 1, \ldots, n$, $p > 1$ and

$$\sum_{i=1}^n \frac{1}{p_i} = \frac{1}{p}.$$

If, for $i \in \{1, \ldots, n\}$, $f_i \in \mathcal{L}^{p_i}(\mu)$, must it be the case that

$$f_1 f_2 \cdots f_n \in \mathcal{L}^p(\mu)?$$

Justify.

(*Iowa*)

Solution.

Yes, $f_1 \cdots f_n \in \mathcal{L}^p(\mu)$. Moreover

$$\|f_1 \cdots f_n\|_p \leq \|f_1\|_{p_1} \cdots \|f_n\|_{p_n}. \tag{1}$$

In case $n = 2$, $|f_1|^p \in \mathcal{L}^{\frac{p_1}{p}}(\mu)$ and $|f_2|^p \in \mathcal{L}^{\frac{p_2}{p}}(\mu)$, and by Hölder's inequality,

$$\left(\int_X |f_1|^p |f_2|^p \right)^{\frac{1}{p}} \leq \left(\left(\int_X (|f_1|^p)^{\frac{p_1}{p}} \right)^{\frac{p}{p_1}} \left(\int_X (|f_2|^p)^{\frac{p_2}{p}} \right)^{\frac{p}{p_2}} \right)^{\frac{1}{p}}$$

$$= \left(\int_X |f_1|^{p_1} \right)^{\frac{1}{p_1}} \left(\int_X |f_2|^{p_2} \right)^{\frac{1}{p_2}}.$$

Assume that the conclusion and the inequality (1) hold for $n = k$. Then for $n = k + 1$,

$$0 < \frac{1}{p_1} + \cdots + \frac{1}{p_k} = \frac{1}{p} - \frac{1}{p_{k+1}} = \frac{1}{\frac{pp_{k+1}}{p_{k+1}-p}} < 1,$$

which shows that $\frac{pp_{k+1}}{p_{k+1}-p} > 1$. By inductive assumption $f_1 \cdots f_k$ belongs to $\mathcal{L}^{\frac{pp_{k+1}}{p_{k+1}-p}}(\mu)$ and

$$\left(\int_X (f_1 \cdots f_k)^{\frac{pp_{k+1}}{p_{k+1}-p}} \right)^{\frac{pp_{k+1}}{p_{k+1}-p}} \leq \|f_1\|_{p_1} \cdots \|f_k\|_{p_k}.$$

Then by the case $n = 2$ we conclude that $(f_1 \cdots f_k)f_{k+1} \in \mathcal{L}^p$ and

$$\|(f_1 \cdots f_n)f_{k+1}\|_p \leq \|f_1 \cdots f_k\|_{\frac{pp_{k+1}}{p_{k+1}-p}} \|f_{k+1}\|_{p_{k+1}}$$

$$\leq \|f_1\|_{p_1} \cdots \|f_k\|_{p_k} \|f_{k+1}\|_{p_{k+1}},$$

which completes the proof.

4303

Let $X = \{a, b, c\}$, let $\mathcal{M}$ be the σ-algebra of all subsets of X, and let μ be the measure determined by

$$\mu(\{a\}) = 0, \quad \mu(\{b\}) = 1, \quad \text{and} \quad \mu(\{c\}) = \infty.$$

What are the dimensions of the vector spaces $L^1(\mu)$, $L^2(\mu)$, and $L^\infty(\mu)$? Justify your answers.

(*Iowa*)

Solution.
We have $\dim L^1(\mu) = \dim L^2(\mu) = 1$ and $\dim L^\infty(\mu) = 2$. Since each function f of $L^1(\mu)$ or $L^2(\mu)$ vanishes at c and two functions taking the same value at b coincide in $L^1(\mu)$ or $L^2(\mu)$, we conclude that $\dim L^1(\mu) = \dim L^2(\mu) = 1$. Since two functions are equal in $L^\infty(\mu)$ if and only if they are identical on $\{b, c\}$ we have $\dim L^\infty(\mu) = 2$.

4304

Let $f, g \in L^2([0, 1])$,

$$\int_0^1 f \, dm = 0.$$

Show that

$$\left(\int_0^1 fg \, dm\right)^2 \leq \left(\int_0^1 f^2 \, dm\right)\left(\int_0^1 g^2 \, dm - \left(\int_0^1 g \, dm\right)^2\right).$$

(*Indiana–Purdue*)

Solution.

Denoting

$$a = \int_0^1 g\, dm,$$

we have

$$\left(\int_0^1 fg\, dm\right)^2 = \left(\int_0^1 f(g-a)dm\right)^2$$

$$\leq \left(\int_0^1 f^2\, dm\right)\left(\int_0^1 (g-a)^2\, dm\right)$$

$$= \left(\int_0^1 f^2\, dm\right)\left(\int_0^1 g^2\, dm - 2a\int_0^1 g\, dm + a^2\right)$$

$$= \left(\int_0^1 f^2\, dm\right)\left(\int_0^1 g^2\, dm - \left(\int_0^1 g\, dm\right)^2\right).$$

4305

Let f be a non-negative function in $L^p(\mathbb{R})$ for some $1 \leq p < \infty$ and let $r + s = p$, $r > 0$, $s > 0$. Also let $f_h(x) = f(x+h)$.
(i) Show that $f_r^h f^s \in L^1(\mathbb{R})$.
(ii) Investigate what happens to $\|f_h^r f^s\|_1$ as $|h| \to \infty$.

(*Indiana–Purdue*)

Solution.

(i) Obviously $f_h \in L^p(\mathbb{R})$. From $f_h^r \in L^{\frac{p}{r}}$, $f^s \in L^{\frac{p}{s}}$, and $\frac{r}{p} + \frac{s}{p} = 1$, it follows that

$$f_h^r f^s \in L^1(\mathbb{R}).$$

(ii) We claim that

$$\lim_{|h| \to \infty} \|f_h^r f^s\|_1 = 0.$$

For any $\varepsilon > 0$, take N such that $\int_{E^c} f^p < \varepsilon$ where $E = [-N, N]$. Now if $|h| > 2N$, then $x + h \notin E$ whenever $x \in E$. Therefore

$$\|f_h^r f^s\|_1 = \int_E |f_h^r f^s| + \int_{E^c} |f_h^r f^s|$$

$$\leq \left(\int_E |f_h|^p \right)^{\frac{r}{p}} \left(\int_E |f|^p \right)^{\frac{r}{p}} + \left(\int_{E^c} |f_h|^p \right)^{\frac{r}{p}} \left(\int_{E_c} |f|^p \right)^{\frac{s}{p}}$$

$$\leq \varepsilon^{\frac{r}{p}} \|f\|_p^s + \varepsilon^{\frac{s}{p}} \|f\|_p^r.$$

4306

Let $f : \mathbb{R} \to \mathbb{R}_+$ be measurable, and let $0 < r < \infty$. Show that

$$\frac{1}{\frac{1}{|I|} \int_I f} \leq \left(\frac{1}{|I|} \int_I \frac{1}{f^r} \right)^{\frac{1}{r}}.$$

(Indiana–Purdue)

Solution.

Set $p = 1 + \frac{1}{r}$. Then $p > 1$ and $\frac{1}{p} + \frac{1}{rp} = 1$. Since

$$|I| = \int_I f^{\frac{1}{p}} f^{-\frac{1}{p}} \leq \left(\int_I f^{\frac{1}{p} p} \right)^{\frac{1}{p}} \left(\int_I f^{-\frac{1}{p} \cdot rp} \right)^{\frac{1}{rp}}$$

$$= \left(\int_I f \right)^{\frac{1}{p}} \left(\int_I f^{-r} \right)^{\frac{1}{rp}},$$

we have

$$|I|^{1+\frac{1}{r}} \leq \left(\int_I f \right) \left(\int_I f^{-r} \right)^{\frac{1}{r}},$$

which implies

$$\frac{1}{\frac{1}{|I|} \int_I f} \leq \left(\frac{1}{|I|} \int_I f^{-r} \right)^{\frac{1}{r}}.$$

4307

Let f be a bounded measurable function on $(0, 1)$. Prove that

$$\lim_{p \to \infty} \left(\int_0^1 |f(x)|^p dx \right)^{\frac{1}{p}} = \operatorname{ess\,sup}|f|,$$

where

$$\operatorname{ess\,sup}|f| = \|f\|_\infty = \inf\{t \mid m(\{|f| > t\}) = 0\}.$$

(*Illinois*)

Solution.

For any $\varepsilon > 0$,

$$m(\{x \mid |f(x)| > \|f\|_\infty - \varepsilon\}) > 0.$$

Then

$$\|f\|_\infty = \lim_{\varepsilon \to 0} \lim_{p \to \infty} (\|f\|_\infty - \varepsilon) m(\{x \mid |f(x)| > \|f\|_\infty - \varepsilon\})^{\frac{1}{p}}$$

$$\leq \lim_{\varepsilon \to 0} \lim_{p \to \infty} \left(\int_{\{x \mid |f(x)| > \|f\|_\infty - \varepsilon\}} |f(x)|^p dx \right)^{\frac{1}{p}}$$

$$\leq \lim_{\varepsilon \to \infty} \lim_{p \to \infty} \left(\int_0^1 |f(x)|^p dx \right)^{\frac{1}{p}}$$

$$= \lim_{p \to \infty} \left(\int_0^1 |f(x)|^p dx \right)^{\frac{1}{p}}$$

$$\leq \overline{\lim_{p \to \infty}} \left(\int_0^1 |f(x)|^p dx \right)^{\frac{1}{p}}$$

$$\leq \|f\|_\infty.$$

Therefore

$$\lim_{p \to \infty} \left(\int_0^1 |f(x)|^p dx \right)^{\frac{1}{p}} = \|f\|_\infty.$$

4308

Let f be a non-negative measurable function on $[0,1]$ satisfying

$$m(\{x \mid f(x) \geq t\}) < \frac{1}{1+t^2}, \quad t > 0.$$

Determine those values of $p, 1 \leq p < \infty$ for which $f \in L^p$ and find the minimum value of p for which f may fail to be in L^p.

(*Illinois*)

Solution.

If $1 \leq p < 2$, then $f \in L^p$. The minimum value of p for which f may fail to be in L^p is 2.

Indeed, for any $p \in [1, 2)$,

$$\sum_{n=1}^{\infty} m(\{x \mid f^p(x) \geq n\}) = \sum_{n=1}^{\infty} m(\{x \mid f(x) \geq n^{\frac{1}{p}}\})$$

$$\leq \sum_{n=1}^{\infty} \frac{1}{1+n^{\frac{2}{p}}} < \sum_{n=1}^{\infty} \frac{1}{n^{\frac{2}{p}}} < \infty,$$

it follows that f^p is integrable. Let $f(x) = \frac{1}{x^{\frac{1}{2}}} - 1$. Then f^2 is not integrable. However for any $t > 0$

$$m(\{x \mid f(x) \geq t\}) = m\left(\left\{x \;\middle|\; \frac{1}{x^{\frac{1}{2}}} - 1 \geq t\right\}\right)$$

$$= m\left(\left\{x \;\middle|\; x \leq \frac{1}{(1+t)^2}\right\}\right)$$

$$= \frac{1}{(1+t)^2} < \frac{1}{1+t^2}.$$

4309

With Lebesgue measure on $[0,1]$, prove that

$$S = \{f \in C[0,1] \mid \|f\|_\infty \leq 1\}$$

is not compact in $L^1[0,1]$.

(*Iowa*)

Solution.

It suffices to show that S is not closed in $L^1([0,1])$. For each $n \in \mathbb{N}$, let

$$
f_n(x) = \begin{cases} 1, & 0 \leq x \leq \frac{1}{2}, \\ 0, & \frac{1}{2} + \frac{1}{4n} \leq x \leq 1, \\ -4nx + (2n+1), & \frac{1}{2} < x < \frac{1}{2} + \frac{1}{4n}. \end{cases}
$$

Then $f_n \in S$ and $\lim_{n \to \infty} f_n = \chi_{[0,\frac{1}{2}]}$ in $L^1[0,1]$. Since $\chi_{[0,\frac{1}{2}]} \notin S$, S is not closed.

4310

Let $\mathcal{H}$ be Hilbert space $L^2(0, 2\pi)$, with inner product defined by

$$
(u, v) = \int_0^{2\pi} u(x)\overline{v(x)}dx, \quad u, v \in \mathcal{H}.
$$

Consider the elements $u_n \in \mathcal{H}$, $n = 1, 2, \ldots$, defined by $u_n(x) = \sin(nx)$ for $x \in (0, 2\pi)$. Show that

(a) the set $\{u_n\}_{n=1}^\infty$ is closed and bounded, but not compact, in the strong (i.e., norm) topology of $\mathcal{H}$.

(b) $u_n \to 0$ as $n \to \infty$ in the weak topology of $\mathcal{H}$, i.e., for every $v \in \mathcal{H}$.

$$
\lim_{n \to \infty} (u_n, v) = 0.
$$

(Stanford)

Solution.

(a) Obviously, the set $\{u_n\}_{n=1}^\infty$ is bounded. For any $m, n \geq 1$ with $m \neq n$

$$
\|u_m - u_n\|_2 = \left(\int_0^{2\pi} (\sin mx - \sin nx)^2 dx \right)^{\frac{1}{2}} = \sqrt{2\pi},
$$

which shows that $\{u_m \mid m \in \mathbb{N}\}$ is closed. Since it admits no convergent subsequence, it is not compact.

(b) For any $v \in \mathcal{H}$, $v \in L^1(0, 2\pi)$ by Hölder's inequality. By Riemann–Lebesgue Lemma,

$$\lim_{n \to \infty} (u_n, v) = \lim_{n \to \infty} \int_0^{2\pi} v(x) \sin nx \, dx = 0.$$

4311

Let H_1 be the Sobolev's space on the unit interval $[0, 1]$, i.e., the Hilbert space consisting of functions $f \in L^2[0, 1]$ such that

$$\|f\|_1^2 = \sum_{n=-\infty}^{+\infty} (1 + n^2)|\hat{f}(n)|^2 < \infty,$$

where

$$\hat{f}(n) = \frac{1}{2\pi} \int_0^1 f(x) \exp(-2\pi inx) dx$$

are Fourier coeffents of f. Show that there exists constant $C > 0$ such that $\|f\|_{L^\infty} \le C\|f\|_1$.

(*Stanford*)

Solution.

Since for any $x \in [0, 1]$

$$\sum_{n=-\infty}^{+\infty} |\hat{f}(n)e^{2\pi inx}| \le \left(\sum_{\substack{n=-\infty \\ n \ne 0}}^{+\infty} n^2|\hat{f}(n)|^2 \right)^{\frac{1}{2}} \left(\sum_{\substack{n=-\infty \\ n \ne 0}}^{+\infty} \frac{1}{n^2} \right) + |\hat{f}(0)|,$$

(1)

the series $\sum_{n=-\infty}^{+\infty} \hat{f}(n)e^{2\pi inx}$ converges to $f(x)$ both in H_1 and in L^∞. It follows from (1) that

$$\|f\|_{L^\infty} \le C\|f\|_1.$$

4312

Let f be a periodic function on $\mathbb{R}$ with period 2π such that $f|_{[0,2\pi]}$ belongs to $L^2(0, 2\pi)$. Suppose

$$f(x) = \sum_{n=-\infty}^{+\infty} a_n e^{inx}.$$

For each $h \in \mathbb{R}$ define the function f_h by $f_h(x) = f(x - h)$.
  (i) Give the Fourier expansion of f_h.
  (ii) Find the L_2-norm $\|f_h - f\|_2$ in terms of the a_n and h.
  (iii) Prove that

$$\lim_{h \to 0} \frac{\|f_h - f\|_2}{|h|} > 0,$$

unless f is constant almost everywhere.

(Stanford)

Solution.
  (i) We have

$$f_h(x) = \sum_{n=-\infty}^{+\infty} a_n e^{in(x-h)}$$

$$= \sum_{n=-\infty}^{+\infty} a_n e^{-inh} e^{inx}.$$

  (ii) We have

$$\|f_h - f\|_2 = \left(\sum_{n=-\infty}^{+\infty} 2\pi |a_n e^{-inh} - a_n|^2 \right)^{\frac{1}{2}}$$

$$= \left(\sum_{n=-\infty}^{+\infty} 8\pi \sin^2 \frac{nh}{2} |a_n|^2 \right)^{\frac{1}{2}}.$$

  (iii) If f is constant almost everywhere, then $a_n = 0$, $n \neq 0$. It follows that $f_h = f$ and $\|f_h - f\|_2 = 0$. If f is not constant almost everywhere, there is an $n \neq 0$ such that $a_n \neq 0$. Then

$$\lim_{h \to 0} \frac{\|f_h - f\|_2}{|h|} \geq \lim_{h \to 0} \frac{(8\pi \sin^2 \frac{nh}{2} |a_n|^2)^{\frac{1}{2}}}{|h|} = \sqrt{2\pi} |na_n| > 0.$$

4313

Let $f_n(x)$ be an orthonormal family of functions in the Hilbert space $L^2(0,1)$. Prove that

$$\sum_{n=1}^{\infty} \left| \int_0^x f_n(t) dt \right|^2 \leq x,$$

for all $x \in [0,1]$, and that this inequality is sharp (equality) if and only if $\{f_n \mid n = 1, \ldots\}$ span a dense subspace of $L^2(0,1)$.

(*Iowa*)

Solution.

By Bessel's inequality

$$\sum_{n=1}^{\infty} \left| \int_0^x f_n(t) dt \right|^2 = \sum_{n=1}^{\infty} \left| \int_0^1 \chi_{(0,x]}(t) f_n(t) \right|^2$$

$$= \sum_{n=1}^{\infty} |(\chi_{(0,x]}, f_n)|^2 \leq \|\chi_{(0,x]}\|_2^2 = x. \quad (1)$$

If $\{f_n \mid n \in \mathbb{N}\}$ span a dense subspace of $L^2(0,1)$, then $\{f_n\}$ is an orthonomal basis of $L^2(0,1)$ and therefore (1) is sharp. Conversely, since

$$\text{span}\{\chi_{(0,x]} \mid 0 \leq x \leq 1\} = \text{span}\{\chi_{(a,b]} \mid 0 \leq a \leq b \leq 1\}$$

is a dense subspace of $L^2(0,1)$ while

$$\left\{ f \in L^2(0,1) \,\Big|\, \|f\|_2^2 = \sum_{n=1}^{\infty} |(f, f_n)|^2 \right\}$$

is a closed subspace of $L^2(0,1)$, containing span $\{\chi_{(0,x]} \mid 0 \leq x \leq 1\}$, it follows that

$$\left\{ f \in L^2(0,1) \Big| \|f\|_2^2 = \sum_{n=1}^{\infty} |(f, f_n)|^2 \right\} = L^2(0,1)$$

and therefore, $\{f_n\}$ span a dense subspace of $L^2(0,1)$.

4314

If f is a function on $\mathbb{R}$, let f_t be the translate $f_t(x) = f(t+x)$. Prove that if f is square integrable with respect to Lebesgue measure, then

$$\lim_{t \to 0} \|f_t - f\|_{L^2(\mathbb{R})} = 0.$$

(Stanford)

Solution.

Suppose that $f : \mathbb{R} \to \mathbb{C}$ be a continuous function with compact support. Then f_t converges to f uniformly. We have

$$\lim_{t \to 0} \|f_t - f\|_2 = \lim_{t \to 0} \left(\int_{-\infty}^{+\infty} |f(t+x) - f(x)|^2 dx \right)^{\frac{1}{2}}$$

$$= \lim_{t \to 0} \left(\int_{K-[-1,1]} |f(t+x) - f(x)|^2 dx \right)^{\frac{1}{2}} = 0,$$

where $K = \mathrm{supp}(f)$ and

$$K - [-1, 1] = \{x - y \mid x \in K, y \in [-1, 1]\}$$

is compact. In general, there is a sequence $\{f_n\}$ of continuous functions with compact support converging to f in $L^2(\mathbb{R})$. Then

$$\overline{\lim_{t \to 0}} \|f_t - f\|_2 \leq \overline{\lim_{t \to 0}}(\|f_t - f_{nt}\|_2 + \|f_{nt} - f_n\|_2 + \|f_n - f\|_2)$$

$$\leq \overline{\lim_{t \to 0}}(2\|f_n - f\|_2 + \|f_{nt} - f_n\|_2) = 2\|f_n - f\|_2.$$

Letting $n \to \infty$, we have

$$\lim_{t \to 0} \|f_t - f\|_2 = 0.$$

4315

A trigonometric polynomial is a function p on $\mathbb{R}$ of the form

$$p(\theta) = \sum_{n=-N}^{N} C_n e^{in\theta},$$

for some $C_n \in \mathbb{C}$, $N \geq 0$. Suppose f is continuous and 2π-periodic on $\mathbb{R}$, and let

$$a_n = \frac{1}{2\pi} \int_0^{2\pi} f(\theta) e^{-in\theta} \, d\theta.$$

Show that for every $\varepsilon > 0$, there exists a trigonometric polynomial

$$p(\theta) = \sum_{n=-N}^{N} C_n e^{in\theta}$$

such that

$$\|f - p\|_\infty < \varepsilon$$

and

$$|C_n| \leq |a_n| \quad \text{for all } n.$$

(Hint. You may wish to think about harmonic functions on the unit disk.)

(Stanford)

Solution.

Regard f as a continuous function on the unit circle S^1 of $\mathbb{C}$. Then

$$a_n = \frac{1}{2\pi} \int_\pi f(z) \bar{z}^n \sigma(dz)$$

where $\sigma(dz)$ is the Harr measure on S^1 such that $\sigma(S^1) = 2\pi$. Let u be the harmonic extension of f to the unit disc $D = \{z \mid$

$|z| \leq 1\}$. Then

$$\lim_{r \to 1-} \sup_{z \in S^1} |u(rz) - f(z)| = 0.$$

Moreover

$$u(rz) = \sum_{n=-\infty}^{+\infty} a_n r^{|n|} z^n, \quad z \in S^1.$$

There is an N such that

$$\sum_{|n|>N} |a_n| r^{|n|} < \varepsilon/2$$

where r is fixed so that

$$\sup_{z \in S^1} |u(rz) - f(z)| < \varepsilon/2.$$

Put

$$p(z) = \sum_{n=-N}^{N} a_n r^{|n|} z^n, \quad z \in S^1,$$

which is required.

4316

Prove or disprove the following statements:

(i) the set of continuous functions on the interval $[0, 1]$ is dense in $L^\infty([0,1])$.

(ii) $L^\infty([0,1])$ is a separable metric space.

(*Stanford*)

Solution.

(i) False. Since $C([0,1])$ is separable while $L^\infty([0,1])$ is not as shown below, $C([0,1])$ is not dense in $L^\infty([0,1])$.

(ii) False. Let

$$S = \left\{ f \in L^\infty([0,1]) \mid f(x) = 0 \text{ or } 1, \right.$$

$$\left. \frac{1}{n+1} < x \leq \frac{1}{n}, n \in \mathbb{N}, f(0) = 0 \right\}.$$

Then as a subset of $L^\infty([0,1])$, S is of cardinal $\aleph$. However, since the distance between any two elements in S is 1, S is not separable.

4317

Let $\{f_n\}_{n=1}^\infty$ be a sequence in $L^p([0,1])$ (with Lebesgue measure) and $\{g_n\}_{n=1}^\infty$ a sequence in $L^q([0,1])$, where $\frac{1}{p}+\frac{1}{q}=1$. If $\lim_{n\to\infty} f_n = f$ in L_p and $\lim_{n\to\infty} g_n = g$ in L^q, is it true that $f_n g_n \to fg$ in measure? Justify.

(Iowa)

Solution.

Yes, $f_n g_n \to fg$ in measure. Indeed, since

$$\|f_n g_n - fg\|_1 \le \|f_n - f\|_p \|g_n\|_q + \|f\|_p \|g_n - g\|_q \to 0 \quad \text{as } n \to \infty,$$

we have for any $\varepsilon > 0$

$$\lambda(\{x \mid |f_n g_n(x) - fg(x)| > \varepsilon\})$$
$$\le \frac{1}{\varepsilon} \int_{\{x \mid |f_n g_n - fg|(x) > \varepsilon\}} |f_n g_n - fg|(x)dx$$
$$\le \frac{\|f_n g_n - fg\|_1}{\varepsilon} \to 0 \quad \text{as } n \to \infty,$$

where λ is the Lebesgue measure.

4318

Let X be a measure space with measure μ and suppose that $\mu(X) < \infty$. Let

$S = \{(\text{equivalence class}) \text{ of measurable complex functions on } X\}$.

(Here, as usual, two measurable complex functions are equivalent if they agree a.e.) For $f \in S$, define

$$\rho(f) = \int_X \frac{|f|}{1+|f|} d\mu.$$

Show that $d(f, g) \doteq p(f - g)$ is a metric on S, and that $f_n \to f$ in this metric if and only if $f_n \to f$ in measure.

(Stanford)

Solution.

Obviously, $d(f, g) = d(g, f) \geq 0$ and $d(f, g) = 0$ iff $\frac{|f-g|}{1+|f-g|} = 0$ a.e., iff $f = g$ in S. For f, g, $h \in S$, we have

$$d(f, h) = \int_X \frac{|f - h|}{1 + |f - h|} d\mu$$

$$\leq \int_X \frac{|f - g| + |g - h|}{1 + |f - g| + |g - h|} d\mu$$

$$\leq \int_X \left(\frac{|f - g|}{1 + |f - g|} + \frac{|g - h|}{1 + |g - h|} \right) d\mu$$

$$= f(f, g) + d(g, h).$$

Therefore d is a metric on S.

If $f_n \to f$ in measure, then $\frac{|f_n - f|}{1 + |f_n - f|} \to 0$ in measure by the equality

$$\left\{ x \in X \,\middle|\, \frac{|f_n - f|}{1 + |f_n - f|} \geq \varepsilon \right\}$$

$$= \left\{ x \in X \,\middle|\, |f_n - f| \geq \frac{\varepsilon}{1 - \varepsilon} \right\} \quad (0 < \varepsilon < 1).$$

By the Dominated Convergence Theorem $d(f_n, f) \to 0$.

If $d(f_n, f) \to 0$, then

$$\mu(\{ x \in x \mid |f_n - f|(x) \geq \varepsilon \})$$

$$= \mu \left(\left\{ x \in X \,\middle|\, \frac{|f_n - f|(x)}{1 + |f_n - f|(x)} \geq \frac{\varepsilon}{1 + \varepsilon} \right\} \right)$$

$$\leq \frac{1 + \varepsilon}{\varepsilon} \int_X \frac{|f_n - f|(x)}{1 + |f_n - f|(x)} dx$$

$$= \frac{1 + \varepsilon}{\varepsilon} d(f_n, f) \to 0 \quad \text{as } n \to \infty.$$

4319

Let g be a measurable function such that $\int_{\mathbb{R}} |fg| < \infty$ for every $f \in L^p(\mathbb{R})$ (fixed $p \geq 1$). Prove that there is a constant M such that

$$\int_{\mathbb{R}} |fg| \leq M|f|_p$$

all $f \in L^p(\mathbb{R})$.

(Stanford)

Solution.
Define for each $n \in \mathbb{N}$ a measurable function g_n

$$g_n(x) = \begin{cases} g(x), & \text{if } |g(x)| \leq n \text{ and } |x| \leq n, \\ 0, & \text{else.} \end{cases}$$

Then $g_n \in L^q(\mathbb{R})$ and for any $f \in L^p(\mathbb{R})$.

$$|g_1 f| \leq |g_2 f| \leq \cdots \leq |g_n f| \leq \cdots \leq |gf|.$$

Moreover, $\lim_{n\to\infty} g_n f = gf$. It follows that the sequence of bounded linear functionals $f \mapsto \int_{\mathbb{R}} f(x)g_n(x)dx$ on $L^p(\mathbb{R})$ converges to $f \mapsto \int_{\mathbb{R}} f(x)g(x)dx$, pointwise. By the Banach–Steinhaus Theorem, $f \mapsto \int_{\mathbb{R}} f(x)g(x)dx$ is continuous on $L^p(\mathbb{R})$ and therefore there is an $h \in L^q(\mathbb{R})$ such that

$$\int_{\mathbb{R}} f(x)g(x) = \int_{\mathbb{R}} f(x)h(x)dx, \quad f \in L^p$$

which then implies that $g \overset{\text{a.e.}}{=} h \in L^q(\mathbb{R})$.

4320

Let S^1 denote the unit circle (the set of complex numbers with modulus one, or the real numbers modulo 2π). The convolution of two functions on S^1 is

$$f * g(\alpha) = \frac{1}{2\pi} \int_0^{2\pi} f(\theta)g(\alpha - \theta)d\theta.$$

Suppose that f is an element of $L^2(S^1)$ with the property that its Fourier coefficient

$$\hat{f}(n) = \frac{1}{2\pi} \int_0^{2\pi} f(\theta)e^{-in\theta}\, d\theta$$

is non-zero for all $n \in \mathbb{Z}$. Show that the linear space $\{f * k \mid k \in L^2(S^1)\}$ is dense in $L^2(S^1)$.

(*Iowa*)

Solution.

Let for any $n \in \mathbb{Z}$, $\chi_n(\theta) = e^{in\theta}$ then $\{\chi_n\}_{n\in\mathbb{Z}}$ is an orthonormal basis of $L^2(S^1)$. We have

$$
\begin{aligned}
(f * \chi_n)(\theta) &= \frac{1}{2\pi} \int_0^{2\pi} f(\alpha)\chi_n(\theta - \alpha)\, d\alpha \\
&= \frac{e^{in\theta}}{2\pi} \int_0^{2\pi} f(\alpha)e^{-in\alpha}\, d\alpha \\
&= \hat{f}(n)e^{in\theta}.
\end{aligned}
$$

Therefore $\{f * k \mid k \in L^2(S^1)\}$ contains $\{\chi_n \mid n \in \mathbb{Z}\}$ and therefore is dense in $L^2(S^1)$.

4321

For functions $f, g \in L^2(\mathbb{R})$, define the convolution $f * g$ by

$$f * g = \int_{-\infty}^{+\infty} f(y)g(x - y)\, dy,$$

where the integral is with respect to Lebesgue measure.

(a) Show that $f * g \in L^\infty(\mathbb{R})$. Do not neglect to check the measurablity of $f * g$.

(b) Suppose that f has the property that for all $g \in L^2(\mathbb{R})$ the convolution $f * g$ is also an element of $L^2(\mathbb{R})$. Define $T_f : L^2(\mathbb{R}) \to L^2(\mathbb{R})$ by $T_f(g) = f * g$. It is evident that T_f is a linear operator; you need not check this. Show that T_f is a bounded linear operator.

(Hint. Closed graph theorem.)

(Iowa)

Solution.

Since by Hölder's Inequality

$$\int_{-\infty}^{+\infty} |f(y)g(x-y)|dy$$

$$\leq \left(\int_{-\infty}^{+\infty} |f(y)|^2 dy\right)^{\frac{1}{2}} \left(\int_{-\infty}^{+\infty} |g(x-y)|^2 dy\right)^{\frac{1}{2}}$$

$$\leq \|f\|_2 \|g\|_2, \tag{1}$$

$f * g(x)$ is well defined.

(a) We will show that $f * g$ is uniformly continuous. Indeed,

$$\lim_{|x_2-x_1|\to 0} |(f*g)(x_1) - (f*g)(x_2)|$$

$$= \lim_{|x_2-x_1|\to 0} \left|\int_{-\infty}^{+\infty} f(y)(g(x_1-y) - g(x_2-y))dy\right|$$

$$\leq \lim_{|x_2-x_1|\to 0} \|f\|_2 \left(\int_{-\infty}^{+\infty} |g(x_1-y) - g(x_2-y)|^2 dy\right)^{\frac{1}{2}}$$

$$= \lim_{|x_2-x_1|\to 0} \|f\|_2 \left(\int_{-\infty}^{+\infty} |g(x_1-x_2+y) - g(y)|^2\right)^{\frac{1}{2}}$$

$$= 0.$$

By (1), $f * g$ is bounded and therefore belongs to $L^\infty(\mathbb{R})$.

(b) Let $\{g_n\}$ be a sequence of $L^2(\mathbb{R})$ such that

$$g_n \to g \text{ and } T_f g_n \to h \quad \text{in } L^2(\mathbb{R}).$$

Since

$$|(f * g_n - f * g)(x)| \leq \|f\|_2 \|g_n - g\|_2 \to 0, \quad x \in \mathbb{R}$$

we see that $h = f * g = T_f g$. By the Closed Graph Theorem, T_f is bounded.

4322

If $f \in L^1(\mathbb{R})$, define

$$\hat{f}(\xi) = \int_{-\infty}^{+\infty} f(x) e^{ix\xi} \, dx.$$

Prove that, for any $f \in L^1(\mathbb{R})$, $\hat{f}(\xi) \to 0$ as $|\xi| \to \infty$.

(Stanford)

Solution.

If f is simple, say

$$f = \sum_{i=1}^{n} a_i \chi(\alpha_i, \beta_i] \tag{1}$$

then

$$\lim_{|\xi| \to \infty} |\hat{f}(\xi)| = \lim_{|\xi| \to \infty} \left| \sum_{k=1}^{n} a_k \frac{e^{i\xi\beta_k} - e^{i\xi\alpha_k}}{i\xi} \right| = 0.$$

In general, there is a sequence $\{f_n\}$ of functions of the form (1) such that

$$\lim_{n \to \infty} \|f_n - f\|_1 = 0.$$

Then

$$\overline{\lim_{|\xi| \to \infty}} |\hat{f}(\xi)| \leq \overline{\lim_{|\xi| \to \infty}} \left(|\hat{f}(\xi) - \hat{f}_n(\xi)| + |\hat{f}_n(\xi)| \right)$$

$$\leq \overline{\lim_{|\xi| \to \infty}} \left(\|f_n - f\|_1 + |\hat{f}_n(\xi)| \right) = \|f_n - f\|_1.$$

Letting $n \to +\infty$, we have

$$\lim_{|\xi| \to \infty} \left(|\hat{f}(\xi)| \right) = 0.$$

4323

Let $f : [0,1] \to [0,\infty)$ be an essentially bounded function, $\|f\|_\infty > 0$. Show that

$$\lim_{n \to \infty} \left(\int_0^1 f(x)^{n+1} dx \right) \bigg/ \left(\int_0^1 f(x)^n dx \right) = \|f\|_\infty.$$

(Indiana)

Solution.

For any α with $0 < \alpha < \|f\|_\infty$, let

$$E_\alpha = \{x \in [0,1] \,|\, f(x) \geq \alpha\}$$

and $F_\alpha = [0,1]\backslash E_\alpha$ then $\lambda(E_\alpha) > 0$, where λ is the Lebesgue measure. For any $k \in \mathbb{N}$, by the Dominated Convergence Theorem,

$$\lim_{n \to \infty} \left(\int_{F_\alpha} f(x)^{n+k} dx \right) \bigg/ \left(\int_{E_\alpha} f(x)^n dx \right)$$

$$\leq \lim_{n \to \infty} \frac{1}{\lambda(E_\alpha)} \int_{F_\alpha} \left(\frac{f(x)}{\alpha} \right)^n \|f\|_\infty^k dx$$

$$= 0.$$

Hence

$$\lim_{n \to \infty} \left(\int_0^1 f(x)^{n+1} dx \right) \bigg/ \left(\int_0^1 f(x)^n dx \right)$$

$$\geq \lim_{n \to \infty} \left(\alpha \int_{E_\alpha} f(x)^n dx + \int_{F_\alpha} f(x)^{n+1} dx \right)$$

$$\bigg/ \left(\int_{E_\alpha} f(x)^n dx + \int_{F_\alpha} f(x)^n dx \right)$$

$$= \alpha.$$

Letting $\alpha \uparrow \|f\|_\infty$, we get that

$$\lim_{n \to \infty} \left(\int_0^1 f(x)^{n+1} dx \right) \bigg/ \left(\int_0^1 f(x)^n dx \right) = \|f\|_\infty.$$

4324

Let $(X, \mathcal{M}, \mu)$ be a measure space for which $\mu(X) < \infty$. Let $1 < p < \infty$. Suppose that $\{f_k\}$ is a sequence in $L^p(X)$ such that $\sup_k \|f_k\|_p < \infty$ and $\lim_{n \to \infty} f_k(x) = f(x)$ exists for μ-a.e. x. Prove that

$$\lim_{k \to \infty} \|f_k - f\|_1 = 0.$$

(Indiana)

Solution.

If on the contrary $\overline{\lim}_{k \to \infty} \|f_k - f\|_1 > 0$ there exists a subsequence of $\{f_k\}$, also denoted $\{f_k\}$, such that $\lim_{k \to \infty} \|f_k - f\|_1 > 0$. Since $\{|f_k|\}$ is bounded in $L^p(X)$ and $L^P(X)$ is reflexive there exists a subsequence $\{|f_{k_l}|\}$ of $\{|f_k| \mid k \in \mathbb{N}\}$, converging to $|f|$ weakly in $L^p(X)$.

By the Vitali–Hahn–Saks Theorem, for any $\varepsilon > 0$ there exists a $\delta > 0$ such that for $k = 1, 2 \ldots,$

$$\int_E |f_{k_l}| d\mu + \int_E |f| d\mu < \varepsilon$$

whenever $\mu(E) < \delta$. For such $\delta > 0$ there exists by Egroff's Theorem a set E such that $\mu(E) < \delta$ and $\{f_k\}$ converges to f uniformly on $X \backslash E$. Then

$$\overline{\lim_{l \to \infty}} \int_X |f_{k_l} - f| d\mu \le \overline{\lim_{l \to \infty}} \int_{X \backslash E} |f_{k_l} - f| d\mu$$

$$+ \overline{\lim_{l \to \infty}} \int_E (|f_{k_l}| + |f|) d\mu \le \varepsilon.$$

Letting $\varepsilon \to 0$, we get that

$$\lim_{l \to \infty} \int_X |f_{k_l} - f| d\mu = 0,$$

a contradiction.

4325

Give an example of a derivative on $[0, 1]$ which is not Lebesgue integrable.

(Auburn)

Solution.

Define the function $f : \mathbb{R} \to \mathbb{R}$ by

$$f(x) = \begin{cases} x^2 \cos \dfrac{1}{x^2}, & x \neq 0, \\ 0, & x = 0. \end{cases}$$

then

$$f'(x) = \begin{cases} 2x \cos \dfrac{1}{x^2} + \dfrac{2}{x} \sin \dfrac{1}{x^2}, & x \neq 0, \\ \displaystyle\lim_{\delta \to 0} \delta \cos \dfrac{1}{\delta^2} = 0, & x = 0. \end{cases}$$

Now $x \mapsto \frac{2}{x} \sin \frac{1}{x^2}$ is not Lebesgue integrable on $[0, 1]$ since

$$\int_0^1 \left| \frac{1}{x} \sin \frac{1}{x^2} \right| dx = \int_1^\infty \frac{|\sin t|}{2t} dt = \infty$$

while the function $x \mapsto 2x \cos \frac{1}{x^2}$ is Lebesgue integrable on $[0, 1]$. It follows that f' is not Lebesgue integrable on $[0, 1]$.

4326

Let $(X, \mathcal{F}, \mu)$ be a finite measure space. Let f_n be a sequence of measurable functions with $f_1 \in L^1(\mu)$ and with the property that

$$\mu\{x \in X : |f_n(x)| > \lambda\} \leq \mu\{x \in X : |f_1(x)| > \lambda\}$$

for all n and all $\lambda > 0$. Prove that

$$\lim_{n \to \infty} \frac{1}{n} \int_X (\max_{1 \leq j \leq n} |f_j|) d\mu = 0.$$

(Hint. You may assume the fart that $\|f\|_1 = \int_0^\infty \mu\{|f(x)| > \lambda\}d\lambda$.)

(*Purdue*)

Solution.

Let $g_n(\lambda) = \frac{1}{n}\mu(\max_{1\leq j\leq n}|f_j| > \lambda)$ for $\lambda > 0$ then

$$g_n(\lambda) = \frac{1}{n}\mu\left(\bigcup_{j=1}^n(|f_j| > \lambda)\right) \leq \min\left\{g_1(\lambda), \frac{\mu(X)}{n}\right\}$$

and g_1 is integrable. Applying the Dominated Convergence Theorem yields that

$$\lim_{n\to\infty}\frac{1}{n}\int_X \max_{1\leq j\leq n}|f_j|d\mu = \lim_{n\to\infty}\int_X g_n(\lambda)d\lambda = 0.$$

4327

Let $g : \mathbb{R}^n \to \mathbb{R}$ be non-negative with $\int_{\mathbb{R}^n} g(y)dy = 1$, let $g_\varepsilon(x) = \varepsilon^{-n}g(x/\varepsilon)$ for $\varepsilon > 0$, and define

$$(g_\varepsilon * f)(x) = \int_{\mathbb{R}^n} g_\varepsilon(y)f(x - y)dy.$$

If $f : \mathbb{R}^n \to \mathbb{R}$ is continuous and bounded, then show that $f_\varepsilon \to f$ as $\varepsilon \to 0$ uniformly on compact subsets of $\mathbb{R}^n$.

(Hint. Change of variables.)

(*Temple*)

Solution.

Given any compact set K and $r > 0$, by change of variables

$$|(g_\varepsilon * f)(x) - f(x)| = \left|\int_{\mathbb{R}^n} g(y)(f(x - \varepsilon y) - f(x))dy\right|$$

$$\leq \int_{|y|\leq r} g(y)|(f(x - \varepsilon y) - f(x))|dy$$

$$+ \int_{|y|>r} g(y)|(f(x - \varepsilon y) - f(x))|dy$$

$$\leq \int_{\mathbb{R}^n} g(y)dy \sup_{|z|\leq r} |f(x - \epsilon z) - f(x)|dy$$

$$+ \int_{|y|>r} 2g(y)\|f\|_\infty dy.$$

Since $\lim_{\epsilon \to 0} \sup_{x \in K} \sup_{z|\leq r} |f(x - \epsilon z) - f(x)| = 0$ we have

$$\lim_{\epsilon \to 0} \sup_{x \in K} |(g_\epsilon * f)(x) - f(x)| \leq 2\|f\|_\infty \int_{|y|>r} g(y)dy.$$

Finally, let $r \to \infty$ as desired.

4328

Fix $p \geq 1$. Assume without proof that the set of continuous functions with compact support is dense in $L^p(\mathbb{R}^n)$. Let $f \in L^p(\mathbb{R}^n)$ and, for $t \in \mathbb{R}^n$, let $f_t(x) = f(x + t)$ be the translate of f by t. Show

(1) The map $t \mapsto f_t$ is continuous from $\mathbb{R}^n$ to $L^p(\mathbb{R}^n)$, and
(2) $g_\varepsilon * f \to f$ in L^p as $\varepsilon \to 0$.

(Temple)

Solution.
(1) Let $\phi(f) = \sup_{t \in \mathbb{R}^n} \overline{\lim}_{s \to t}\|f_s - f_t\|$ then $\phi : L^P(\mathbb{R}^n) \to \mathbb{R}$ is a semi-norm and $\phi(f) \leq 2\|f\|_p$ so that ker ϕ is a closed linear subspace of $L^p(\mathbb{R}^n)$. However, $\phi(f) = 0$ for any continuous function f with compact support and such functions form a dense set in $L^p(\mathbb{R}^n)$. It follows that $\phi = 0$ and thus $t \mapsto f_t$ is continuous from $\mathbb{R}^n$ to $L^p(\mathbb{R}^n)$ for any $f \in L^p(\mathbb{R}^n)$.

(2) By change of variables, we have

$$|(g_\varepsilon * f)(x) - f(x)| = \left|\int_{\mathbb{R}^n} g(y)(f(x - \epsilon y) - f(x))dy\right|$$

$$\leq \int_{|y|\leq r} g(y)|(f(x - \epsilon y) - f(x))|dy$$

$$+ \int_{|y|>r} g(y)|f(x - \epsilon y) - f(x)|dy.$$

By Minkowski's Inequality we have

$$\|g_\epsilon * f - f\|_p \leq \int_{|y| \leq r} g(y) dy \sup_{|z| \leq r} \|f_{-\epsilon z} - f\|_p$$

$$+ \int_{|y| > r} g(y) 2 \|f\|_p dy.$$

It follows that $\lim_{\epsilon \to 0} \|g_\epsilon * f - f\|_p \leq 2 \int_{|y| > r} g(y) dy$. Let $r \to 0$ as desired.

4329

This problem concerns Lebesgue integration of functions from $\mathbb{R}$ to $\mathbb{R}$. For such functions, a trivial form of Hölder's inequality says that if $f \in L^\infty(\mathbb{R})$ and $g \in L^1(\mathbb{R})$ then

$$\left| \int_{\mathbb{R}} fg \right| \leq \|f\|_\infty \|g\|_1. \tag{*}$$

(1) Prove that for each $g \in L^1(\mathbb{R})$ there is an $f \in L^\infty(\mathbb{R})$ such that $\|f\|_\infty = 1$ such that equality holds in (*).

(2) Prove that for each $f \in L^\infty(\mathbb{R})$ and each $\epsilon > 0$ there is a $g \in L^1(\mathbb{R})$ with $\|g\|_1 = 1$ such that

$$\left| \int_{\mathbb{R}} fg \right| \geq \|f\|_\infty - \epsilon.$$

(Rutgers)

Solution.

(1) Assume without loss of generality that $\|g\|_1 = 1$. Let $f(x) = \frac{|g(x)|}{g(x)}$ with the convention that $\frac{0}{0} = 0$ then $\|f\|_\infty = 1$ and $\int_{\mathbb{R}} fg = 1$.

(2) Assume without loss of generality that $\|f\|_\infty = 1$. Take a measurable subset E of $(|f| > 1 - \epsilon)$ such that $0 < m(E) < +\infty$ and define $g(x) = \frac{|f(x)| \chi_E(x)}{f(x) m(E)}$ with the convention that $\frac{0}{0} = 0$ then $\|g_n\|_1 = 1$ and

$$\int_{\mathbb{R}} f g_n = \int \frac{|f|}{m(E)} \chi_E \geq \int_E \frac{1 - \epsilon}{m(E)} = 1 - \epsilon.$$

Section 4

Differential

4401

Let $f : [0, 1] \to \mathbb{R}$ and $g : [0, 1] \times [0, 1] \to \mathbb{R}$ satisfy the inequality

$$f(y) - f(x) \le g(y, x)(y - x) \quad \text{(for all } x, y \in (0, 1)). \qquad (*)$$

Assume also that g is non-decreasing in each variable, i.e., $u \le x, v \le y \Rightarrow g(u, v) \le g(x, y)$. Show that $\lim_{y \to x} g(x, y) = \phi(x)$ exists except in a countable set and that for $0 \le x \le y \le 1$ we have

$$f(y) - f(x) = \int_x^y \phi(t) dt.$$

(Hint. Observe that $(*)$ is equivalent to

$$g(x, x) \le g(y, x) \le \frac{f(y) - f(x)}{y - x}$$
$$\le g(x, y) \le g(y, y), \quad 0 \le x < y \le 1.) \qquad (**)$$

(*Iowa*)

Solution.

Let $\phi(x) = g(x,x)$ then ϕ is non-decreasing and

$$\begin{cases} \phi(x) \le g(y,x) \le \dfrac{f(y)-f(x)}{y-x} \le g(x,y) \le \phi(y), & x < y, \\[2mm] \phi(y) \le g(x,y) \le \dfrac{f(y)-f(x)}{y-x} \le g(y,x) \le \phi(x), & y < x. \end{cases} \quad (1)$$

By (1) we have an inclusion

$$\left\{x \Big| \lim_{y\to x} g(x,y) \ne \phi(x)\right\} \subseteq \left\{x \Big| \lim_{y\to x} \phi(y) \ne \phi(x)\right\}$$

while the latter set is at most countable. So $\lim_{y\to x} g(x,y) \ne \phi(x)$ exists except in a countable set. Again by (1) f is Lipschitz and therefore absolutely continuous. However, since by (1), $f'(x) = \phi(x)$ whenever ϕ is continuous at x, we have

$$f(y) - f(x) = \int_x^y f'(t)dt = \int_x^y \phi(t)dt.$$

4402

Prove that a function $f : [0,1] \to \mathbb{R}$ is Lipschitz, with

$$|f(x) - f(y)| \le M|x-y|$$

for all $x, y \in [0,1]$, if and only if there is a sequence of continuously differentiable functions $f_n : [0,1] \to \mathbb{R}$ such that
(I) $|f_n'(x)| \le M$ for all $x \in [0,1]$;
(II) $f_n(x) \to f(x)$ for all $x \in [0,1]$.
(Hint. There are several different ways to do this problem; one is to use the Fundamental Theorem of Calculus.)

(Stanford)

Solution.

If f is Lipschitz, then f is differentiable almost everywhere and moreover for any $x \in [0,1]$

$$f(x) = f(0) + \int_0^x f'(x)dx.$$

Since $|f'(x)| \le M, x \in [0,1]$ there is a sequence $\{g_n\}$ of continuous functions such that

$$|g_n(x)| \le M, \quad x \in [0,1]$$

and

$$\lim_{n\to\infty} \int_0^1 |g_n(x) - f'(x)| dx = 0.$$

For any $n \in \mathbb{N}$, define

$$f_n(x) = f(0) + \int_0^x g_n(t)dt, \quad x \in [0,1],$$

which is required.

Conversely, by the Mean Value Theorem

$$\begin{aligned}
|f(x) - f(y)| &= \lim_{n\to\infty} |f_n(x) - f_n(y)| \\
&= \lim_{n\to\infty} |f_n'(\xi_n)||x - y| \\
&\le M|x - y|.
\end{aligned}$$

This completes the proof.

4403

Let $\{f_n\}$ be a sequence of absolutely continuous real-valued functions on $[0, 1]$ such that
(a) $f(x) = \sum_{n=1}^{\infty} f_n(x)$ converges for every $x \in [0,1]$.
(b) $\int_0^1 (\sum_{n=1}^{\infty} |f_n'(x)|) dx < +\infty$.
Show that f is absolutely continuous on $[0, 1]$.

(*Illinois*)

Solution.
Let

$$g(x) = \sum_{n=1}^{\infty} f_n'(x)$$

in $L^1[0, 1]$. Then

$$f(x) = \sum_{n=1}^{\infty}(f_n(x) - f_n(0)) + \sum_{n=1}^{\infty} f_n(0)$$

$$= \sum_{n=1}^{\infty} f_n(0) + \sum_{n=1}^{\infty} \int_0^x f_n'(x)dx$$

$$= f(0) + \int_0^x g(t)dt.$$

It follows that f is absolutely continuous.

4404

Assume that $f \in AC(I)$ for every $I \subset R$. If both f and f' are in $L^1(R)$ show that (i) $\int_R f' = 0$, (ii) $f(x) \to 0$ as $|x| \to \infty$.
(*Indiana–Purdue*)

Solution.
Since $f \in AC(I)$, for every $I \subset R$,

$$\int_0^x f'(t)dt = f(x) - f(0).$$

Since $f' \in L^1$, $\lim_{x \to +\infty} \int_0^x f'(t)dt$ exists, which means $\lim_{x \to +\infty} f(x)$ exists. Since $f \in L^1$, we must have $\lim_{x \to +\infty} f(x) = 0$. Therefore

$$\int_0^{+\infty} f'(x)dx = \lim_{x \to +\infty} f(x) - f(0) = -f(0).$$

In the same way, we have $\lim_{x \to -\infty} f(x) = 0$ and

$$\int_{-\infty}^0 f'(x)dx = f(0).$$

Thus

$$\int_R f'(x)dx = \int_{-\infty}^0 f'(x)dx + \int_0^{+\infty} f'(x)dx = 0.$$

4405

Let $\{f_n\} \subset AC([0,1])$, $f_n(0) = 0$ for every n. If $\{f_n'\}$ is Cauchy (L^1), show that there is $f \in AC([0,1])$ such that $f_n \to f$ uniformly on $[0,1]$.

(Indiana–Purdue)

Solution.

Since $f_n \in AC([0,1])$,

$$f_n(x) = \int_0^x f_n'(t)dt.$$

Thus

$$|f_n(x) - f_m(x)| \le \int_0^1 |f_n'(t) - f_m'(t)|dt \to 0.$$

So there exists an $f \in C([0,1])$ such that $f_n \to f$ uniformly on $[0,1]$.

Moreover, there exists $g \in L^1$ such that $f_n' \xrightarrow{L^1} g$. Then

$$\left| \int_0^x [f_n'(t) - g(t)]dt \right| \le \int_0^1 |f_n'(t) - g(t)dt| \to 0.$$

Therefore

$$f(x) = \lim_{n \to \infty} \int_0^x f_n'(t)dt = \int_0^x g(t)dt$$

which implies $f \in AC([0,1])$.

4406

(a) Assume that $f \in AC(I)$ and $f' \in L^\infty(I)$. Show that f is Lipschitz.

(b) Show that the following two statements are equivalent.

(1) $f : I \to R$ is Lipschitz

(2) $\varepsilon > 0 \Rightarrow \exists \delta = \delta(\varepsilon) > 0$ such that $\{I_j = [a_j, b_j]\} \subset I$ with

$$\sum |I_j| \le \delta \Rightarrow \sum |f(b_j) - f(a_j)| \le \varepsilon.$$

(Indiana–Purdue)

Solution.

(a) For any $x_1, x_2 \in I$,

$$|f(x_2) - f(x_1)| = \left| \int_{x_1}^{x^2} f'(x)dx \right| \leq \|f'\|_\infty |x_1 - x_2|.$$

(b) $(1) \Rightarrow (2)$ obviously.

$(2) \Rightarrow (1)$. Take $\delta > 0$ such that $\sum |f(b_j) - f(a_j)| \leq \varepsilon$ for any $I_j = [a_j, b_j] \subset I$, $\sum |I_j| \leq \delta$. For any $x_1, x_2 \in I, x_1 < x_2$, we give the following fact.

Suppose that $x_2 - x_1 \geq \delta$. Take N such that $\frac{\delta}{2} \leq \frac{x_2 - x_1}{N} < \delta$. Take $\{c_j\}$ such that $x_1 = c_0 < c_1 < \cdots < c_N = x_2$. $c_i - c_{i-1} < \delta$. Then

$$|f(x_2) - f(x_1)| \leq \sum |f(c_j) - f(c_{j-1})| < N \leq \frac{2}{\delta}|x_2 - x_1|.$$

Suppose that $x_2 - x_1 < \delta$. Take $N \geq 0$, such that $\frac{\delta}{2} < N(x_2 - x_1) < \delta$. Then $N|f(x_1) - f(x_1)| < 1$. So we have

$$|f(x_2) - f(x_1)| \leq \frac{1}{N} < \frac{2}{\delta}|x_2 - x_1|.$$

Therefore we find a Lipschitz constant $\frac{2}{\delta}$.

4407

Let $G_n, n = 1, 2, \ldots$ be open subsets of $[0, 1]$ such that

$$G_1 \supset G_2 \supset \cdots, \quad |G_n| \leq \frac{1}{2^n}.$$

Let

$$f(x) = \sum_{n \geq 1} |G_n \cap [0, x]|.$$

Show that

(i) $f \in AC([0, 1])$.

(ii) $|f(x') - f(x'')| \leq M|x' - x''|, x', x'' \in [0, 1]$ iff there exists no such that $G_n = \emptyset, n \geq n_0$.

(*Indiana–Purdue*)

Solution.

(i) For any $\varepsilon > 0$, take N such that $\sum_{N+1}^{\infty} \frac{1}{2^n} < \frac{\varepsilon}{2}$. Set $\delta = \frac{\varepsilon}{2N}$. If $\{(a_i, b_i)\}$ is a sequence of disjoint open intervals in $[0, 1]$ and $\sum(b_i - a_i) < \delta$, then

$$\sum_i |f(b_i) - f(a_i)| = \sum_n \left(\sum_i |G_n \cap [a_i, b_i]| \right)$$

$$\leq \sum_{N+1}^{\infty} |G_n| + \sum_1^N \left| \sum_i |G_n \cap [a_i, b_i]| \right|$$

$$< \frac{\varepsilon}{2} + N \sum (b_i - a_i) < \frac{\varepsilon}{2} + \frac{\varepsilon}{2} = \varepsilon,$$

which means $f \in AC([0, 1])$.

(ii) Suppose that there exists n_0 such that $G_n = \emptyset$ for $n \geq n_0$. Then for $x'' > x'$,

$$|f(x'') - f(x')| = \sum_1^{n_0} |G_n \cap [x', x'']| \leq n_0 |x'' - x'|.$$

Conversely, suppose that for any natural number K, there is $k' > K$, $G_{k'} \neq \emptyset$. Then $G_K \neq \emptyset$. Take $x \in G_K, \delta > 0$ such that $(x - \delta, x + \delta) \subset G_K$. Take $a, b, a < b$,

$$a, b \in (x - \delta, x + \delta) \subset G_K = \bigcap_{i=1}^K G_i.$$

Then

$$|f(b) - f(a)| = \sum_{n=1}^{\infty} |G_n \cap [a, b]|$$

$$\geq \sum_{n=1}^K |G_n \cap [a, b]|$$

$$= \sum_1^K |[a, b]|$$

$$= K|b - a|.$$

4408

For what values "a" and "b" is the function

$$f(x) = \begin{cases} 0, & x = 0, \\ |x|^a \sin |x|^b, & x \neq 0 \end{cases}$$

(i) of bounded variation in $(-1, 1)$; (ii) differentiable at "0"

(Iowa)

Solution.

The function f is of bounded variation if and only if

$$\text{(I)} \begin{cases} b = 0 \\ a \geq 0, \end{cases} \text{or} \quad \text{(II)} \begin{cases} b > 0 \\ a + b \geq 0, \end{cases} \text{or} \quad \text{(III)} \begin{cases} b < 0 \\ a + b > 0. \end{cases} \quad (1)$$

To show (1), let us first establish the following equality

$$\bigvee_0^1(f) = \sup_{\varepsilon > 0} \left(|f(\varepsilon)| + \int_\varepsilon^1 |f'(x)| dx \right)$$

$$= \sup_{1 > \varepsilon > 0} \left(|\varepsilon^a \sin \varepsilon^b| + \int_\varepsilon^1 \frac{|a \frac{\sin x^b}{x^b} + b \cos x^b|}{x^{1-a-b}} dx \right) \quad (2)$$

provided that either side is finite.

Indeed, since f is continuously differentiable on $(0, 1]$, for any $\varepsilon > 0$, f is of bounded variation on $[\varepsilon, 1]$ and

$$\bigvee_\varepsilon^1(f) = \int_\varepsilon^1 |f'(x)| dx.$$

Then

$$\sup_{1 > \varepsilon > 0} \left(|f(\varepsilon)| + \int_\varepsilon^1 |f'(x)| dx \right)$$

$$= \sup_{1 > \varepsilon > 0} \left(|f(\varepsilon) - f(0)| + \bigvee_\varepsilon^1(f) \right)$$

$$\leq \sup_{1 > \varepsilon > 0} \left(\bigvee_0^\varepsilon(f) + \bigvee_\varepsilon^1(f) \right) = \bigvee_0^1(f)$$

$$= \sup_{0<\varepsilon<0<x_1<\cdots<x_n=1} (|f(\varepsilon) - f(0)| + |f(x_1) - f(\varepsilon)|$$
$$+ \cdots + |f(1) - f(x_{n-1})|)$$
$$\leq \sup_{0<\varepsilon<1} \left(|f(\varepsilon)| + \int_\varepsilon^1 |f'(x)| dx \right).$$

Case I. $b = 0$. Then $\sup_{0<\varepsilon<1} |f(\varepsilon)| < \infty$ iff $a \geq 0$, while $a \geq 0$ implies that

$$\sup_{0<\varepsilon<1} \int_\varepsilon^1 |f'(x)| dx = \sup_{0<\varepsilon<1} \int_\varepsilon^1 \frac{|a \sin 1|}{x^{1-a}} dx < \infty.$$

Case II. $b > 0$. Then $\sup_{0<\varepsilon<1} |f(\varepsilon)| < \infty$ iff $a + b \geq 0$, for

$$|f(\varepsilon)| = \left| \frac{\sin \varepsilon^b}{\varepsilon^b} \right| |\varepsilon^{a+b}|.$$

Since $a + b = 0$ and $a + b > 0$ imply respectively that

$$\lim_{x \to 0} \frac{|f'(x)|}{x^{2b-1}} = \frac{b}{3}$$

and

$$\lim_{x \to 0} \frac{|f'(x)|}{x^{a+b-1}} = a + b,$$

the function f' is integrable on $(0, 1]$.

Case III. $b < 0$. If $a + b \leq 0$, there is a $\delta > 0$ such that $0 < x \leq \delta$ implies that

$$\frac{\sqrt{3}}{2} |b| - \frac{|a|}{x^b} \geq \frac{|b|}{2}.$$

Let $N = \left[\frac{\delta^b}{2\pi} \right] + 1$. Then

$$\int_0^1 |f'(x)| dx \geq \int_0^\delta \frac{|b||\cos x^b| - |a|x^{-b}}{x^{1-a-b}} dx$$

$$\geq \sum_{k=N}^\infty \int_{(2k\pi + \frac{\pi}{6})^{\frac{1}{b}}}^{(2k\pi)^{\frac{1}{b}}} \frac{|b|}{2} x^{a+b-1} dx$$

$$= \begin{cases} \displaystyle\sum_{k=N}^{\infty} \frac{|b|\left((2k\pi)^{1+\frac{a}{b}} - (2k\pi + \frac{\pi}{6})^{1+\frac{a}{b}}\right)}{2(a+b)}, & a+b < 0, \\[4mm] \displaystyle\sum_{k=N}^{\infty} \frac{|b|\ln(1 + \frac{1}{12k})}{-2b}, & a+b = 0. \end{cases}$$

If $a + b < 0$, since

$$\lim_{k\to\infty} \left(\left(2k\pi + \frac{\pi}{6}\right)^{1+\frac{a}{b}} - (2k\pi)^{1+\frac{a}{b}}\right) \bigg/ k^{\frac{a}{b}} = \frac{(2\pi)^{\frac{a}{b}}\pi(a+b)}{6b}$$

and

$$\sum_{k=1}^{\infty} k^{\frac{a}{b}} = \infty,$$

$$\int_0^1 |f'(x)|dx = +\infty;$$

if $a + b = 0$, since

$$\lim_{k\to\infty} \ln\left(1 + \frac{1}{12k}\right) \bigg/ \frac{1}{12k} = 1$$

and

$$\sum_{k=1}^{\infty} \frac{1}{12k} = \infty,$$

we have

$$\int_0^1 |f'(x)|dx = +\infty.$$

If $a + b > 0$, then

$$\sup_{0<\varepsilon<1} |f(\varepsilon)| < +\infty$$

and

$$|f'(x)| = \frac{|a\frac{\sin x^b}{x^b} + b\cos x^b|}{x^{1-a-b}} \leq (|a| + |b|)x^{a+b-1}$$

and therefore f' is integrable.

(ii) f is differentiable at 0 if and only if

$$\begin{cases} b \geq 0 \\ a+b \geq 1 \end{cases} \quad \text{or} \quad \begin{cases} b < 0 \\ a > 1 \end{cases}$$

4409

Prove or disprove

(a) Let f be a real function on $[0, 2\pi]$ satisfying $|f(x) - f(y)| \leq |x - y|$, all $x, y \in [0, 2\pi]$ and $f'(x) = 0$ a.e. on $[0, 2\pi]$. Then f must be constant. (b) Let f be a real valued function of bounded variation on $[0, 2\pi]$ with $f'(x) = 0$ a.e.. Then f must be constant.

(*Stanford*)

Solution.

(a) The conclusion is ture. Since by the condition that

$$|f(x) - f(y)| \leq |x - y|, \quad x, y \in [0, 2\pi]$$

f is absolutely continuous, f is constant.

(b) is false. Let

$$f(x) = 0 \text{ on } [0, \pi) \text{ and } f(x) = 1 \text{ on } [\pi, 2\pi].$$

Then f is of bounded variation with $f'(x) = 0$ a.e., but f is not constant.

4410

Let $f, g \in L^1(\mathbb{R})$. Show

(a)

$$\lim_{h \to 0} \frac{1}{h} \int_t^{t+h} f(x) dx = f(t), \quad \text{a.e.}$$

(b) if

$$\lim_{h \to 0} \frac{1}{h} \int_a^{a+h} f(x) dx = c,$$

then

$$\lim_{h\to 0}\int_a^{a+t}\frac{f(x+h)-f(x)}{h}dx = f(a+t) - c \quad \text{a.e.}$$

(c) if

$$\lim_{n\to 0}\int_{\mathbb{R}}\left|\frac{f(x+h)-f(x)}{h} - g(x)\right|dx = 0$$

then there are constants a, c such that

$$f(a+t) = \int_a^{a+t} g(x)dx + c \quad \text{a.e.}$$

Can you deduce that $f'(x) = g(x)$ a.e.?
    State explicitly the theorems you use.

(Stanford)

Solution.
    Assume, without loss of generality, that f and g are real-valued. Since the function

$$x \mapsto \int_{-\infty}^x f(t)dt = \int_{-\infty}^x f_+(t)dt - \int_{-\infty}^x f_-(t)dt$$

is of bounded variation, it is differentiable almost everywhere by Lebesgue's Theorem.
    (a) For any $\varepsilon > 0$ there is a continuous function $\tilde h : \mathbb{R} \to \mathbb{R}$ such that

$$\int_{-\infty}^{+\infty}|f(x) - \tilde h(x)|dx < \varepsilon.$$

We have

$$\int_{-\infty}^{+\infty}\left|\frac{d}{dx}\int_{-\infty}^x f(t)dt - f(x)\right|dx$$

$$= \int_{-\infty}^{+\infty}\left|\frac{d}{dx}\int_{-\infty}^x (f(t) - \tilde h(t))dt + \tilde h(x) - f(x)\right|dx$$

$$\leq \int_{-\infty}^{+\infty} \left| \frac{d}{dx} \int_{-\infty}^{x} (f(t) - \tilde{h}(t)) dt \right| + \int_{-\infty}^{+\infty} |\tilde{h}(x) - f(x)| dx$$

$$< \int_{-\infty}^{+\infty} \left| \frac{d}{dx} \left(\int_{-\infty}^{x} (f - \tilde{h})_+(t) dt - \int_{-\infty}^{x} (f - \tilde{h})_-(t) dt \right) \right| dx + \varepsilon$$

$$\leq \int_{-\infty}^{+\infty} \left(\frac{d}{dx} \int_{-\infty}^{x} (f - \tilde{h})_+(t) + \frac{d}{dx} \int_{-\infty}^{x} (f - \tilde{h})_-(t) dt \right) dx + \varepsilon$$

$$\leq \int_{-\infty}^{+\infty} |f - \tilde{h}|(x) dx + \varepsilon < 2\varepsilon.$$

Letting $\varepsilon \to 0$, we see that

$$\int_{-\infty}^{+\infty} \left| \frac{d}{dx} \int_{-\infty}^{x} f(t) dt - f(x) \right| dx = 0,$$

and therefore

$$\frac{d}{dx} \int_{-\infty}^{x} f(t) dt \overset{\text{a.e.}}{=} f(x).$$

(b) We have by (a)

$$\lim_{h \to 0} \int_{a}^{a+h} \frac{f(x+h) - f(x)}{h} dx$$

$$= \lim_{h \to 0} \left(\int_{a}^{a+t} \frac{f(x+h)}{h} dx - \int_{a}^{a+t} \frac{f(x)}{h} dx \right)$$

$$= \lim_{h \to 0} \left(\int_{a+h}^{a+h+t} \frac{f(x)}{h} dx - \int_{a}^{a+t} \frac{f(x)}{h} dx \right)$$

$$= \lim_{h \to 0} \left(\int_{a+t}^{a+t+h} \frac{f(x)}{h} dx - \int_{a}^{a+h} \frac{f(x)}{h} dx \right)$$

$$= f(a+t) - c \quad \text{a.e. } t \in \mathbb{R}.$$

(c) By (a), there is an $a \in \mathbb{R}$ and a c such that

$$\lim_{h \to 0} \frac{1}{h} \int_{a}^{a+h} f(x) dx = c.$$

By (b) and by the assumption, we have

$$f(a + t) - \int_a^{a+t} g(x)dx - c$$

$$= f(a + t) - c - \int_a^{a+t} \frac{f(x+h) - f(x)}{h} dx$$

$$+ \int_a^{a+t} \left(\frac{f(x+h) - f(x)}{h} - g(x) \right) dx$$

$$= \lim_{h \to 0} \left(f(a + t) - c - \int_a^{a+t} \frac{f(x+h) - f(x)}{h} dx \right.$$

$$\left. + \int_a^{a+t} \left(\frac{f(x+h) - f(x)}{h} - g(x) \right) dx \right)$$

$$= 0, \quad \text{a.e. } t \in \mathbb{R}.$$

Therefore

$$f(a + t) = \int_a^{a+t} g(x)dx + c \quad \text{a.e. } t \in \mathbb{R},$$

i.e.,

$$f(x) = \int_a^x g(t)dt + c \quad \text{a.e. } x \in \mathbb{R}.$$

It follows that $f'(x) = g(x)$ a.e..

4411

Let $F : (a, b) \to \mathbb{R}$ be measurable.
(1) Prove that the following two statements are equivalent:
(a) There is an $f \in L^2(a, b)$ such that

$$F(x) = \int_a^x f \, dm \quad (m \text{ the Lebesgue measure});$$

(b) There is an $M > 0$ such that

$$\sum_{k=1}^n \frac{|F(x_k) - F(x_{k-1})|^2}{x_k - x_{k-1}} \leq M$$

for any finite partition $x_0 < x_1 < \cdots < x_n$ of (a, b).

(2) Show that the smallest constant M in (b) is equal to $\|f\|_2^2$.

(*Iowa*)

Solution.

(1) (a) $\Rightarrow$ (b). By the Cauchy–Schwarz inequality

$$
\sum_{i=1}^{n} \frac{|F(x_i) - F(x_{i-1})|^2}{x_i - x_{i-1}}
$$

$$
= \sum_{i=1}^{n} \frac{1}{x_i - x_{i-1}} \left| \int_{x_{i-1}}^{x_i} f(x) dx \right|^2
$$

$$
\leq \sum_{i=1}^{n} \frac{1}{x_i - x_{i-1}} \left(\int_{x_{i-1}}^{x_i} |f(x)|^2 dx \right) \left(\int_{x_{i-1}}^{x_i} 1^2 dx \right) = \|f\|_2^2.
$$

$$(1)$$

(b) $\Rightarrow$ (a) Obviously, for any mutually disjoint intervals (α_n, β_n) of (a, b), $n \in \mathbb{N}$, we have

$$
\sum_{n=1}^{\infty} \frac{|F(\beta_n) - F(\alpha_n)|^2}{\beta_n - \alpha_n} \leq M. \qquad (2)
$$

Since

$$
\sum_{i=1}^{n} |F(\beta_i) - F(\alpha_i)| = \sum_{i=1}^{n} \frac{|F(\beta_i) - F(\alpha_i)|}{\sqrt{\beta_i - \alpha_i}} \sqrt{\beta_i - \alpha_i}
$$

$$
\leq \left(\sum_{i=1}^{n} \frac{|F(\beta_i) - F(\alpha_i)|^2}{\beta_i - \alpha_i} \right)^{\frac{1}{2}} \left(\sum_{i=1}^{n} (\beta_i - \alpha_i) \right)^{\frac{1}{2}}
$$

$$
\leq \sqrt{M \sum_{i=1}^{n} (\beta_i - \alpha_i)},
$$

F is absolutely continuous, and therefore there is an integrable function $f : (a, b) \to \mathbb{R}$ such that

$$
F(x) = \int_a^x f(t) dt + c, \quad x \in (a, b).
$$

We will show that if $E_1, \ldots, E_n$ are any mutually disjoint measurable sets, then for any $\varepsilon > 0$

$$\sum_{i=1}^{n} \frac{1}{m(E_i) + \varepsilon} \left| \int_{E_i} f(x) dx \right|^2 \leq M. \tag{3}$$

If E_i's are open sets of (a, b), say $E_i = \dot{\bigcup}_{j \in \Lambda_i} (\alpha_{ij}, \beta_{ij})$ (Λ_i at most countable), then by inequality (2)

$$\sum_{i=1}^{n} \frac{\left| \int_{E_i} f \right|^2}{m(E_i) + \varepsilon} = \sum_{i=1}^{n} \frac{\left| \sum_{j \in \Lambda_i} \int_{\alpha_{ij}}^{\beta_{ij}} f \right|^2}{m(E_i) + \varepsilon}$$

$$= \sum_{i=1}^{n} \frac{\left| \sum_{j \in \Lambda_i} \left(\frac{1}{\sqrt{\beta_{ij} - \alpha_{ij}}} \int_{\alpha_{ij}}^{\beta_{ij}} f \right) \sqrt{\beta_{ij} - \alpha_{ij}} \right|^2}{m(E_i) + \varepsilon}$$

$$\leq \sum_{i=1}^{n} \frac{\sum_{j \in \Lambda_i} \frac{1}{\beta_{ij} - \alpha_{ij}} \left| \int_{\alpha_{ij}}^{\beta_{ij}} f \right|^2 \sum_{j \in \Lambda_i} (\beta_{ij} - \alpha_{ij})}{m(E_i) + \varepsilon}$$

$$\leq \sum_{i=1}^{n} \sum_{j \in \Lambda_i} \frac{\left| \int_{\alpha_{ij}}^{\beta_{ij}} f \right|^2}{\beta_{ij} - \alpha_{ij}}$$

$$\leq M.$$

If E_i's are compact sets there is for each i a decreasing sequence $\{E_{ij}\}_{j=1}^{\infty}$ of open sets of (a, b) such that $E_i = \bigcap_{j=1}^{\infty} E_{ij}$ and $E_{i1} \cap E_{k1} = \emptyset$ ($i \neq k$). Applying (3) to the mutually disjoint open sets $E_{1j}, \ldots, E_{nj}$ and passing to the limits we see that (3) holds for $E_1, \ldots, E_n$.

In general, there is for each i an increasing sequence $\{E_{ij}\}_{j=1}^{\infty}$ of compact sets such that $\bigcup_{j=1}^{\infty} E_{ij} \subseteq E_i$ and $\lim_{j \to \infty} \mu(E_{ij}) = m(E_i)$. Applying (3) to the mutually disjoint compact sets $E_{1j}, \ldots, E_{nj}$ and passing to the limits we see that (3) holds in general.

To show that f is square integrable, it suffices to show that

$$\sum_{n=1}^{\infty} nm(E_n) < \infty,$$

where $E_n = \{x \mid n \le |f(x)|^2 < n+1\}$. Let

$$E_n^+ = \{x \mid \sqrt{n} \le f(x) < \sqrt{n+1}\}$$

and

$$E_n^- = \{x \mid -\sqrt{n+1} < f(x) \le -\sqrt{n}\}.$$

Then

$$\sum_{n=1}^{\infty} nm(E_n) = \sum_{n=1}^{\infty} (nm(E_n^+) + nm(E_n^-))$$

$$\le \lim_{n\to\infty} \sum_{k=1}^{n} \lim_{\varepsilon\to 0} \left(\frac{1}{m(E_k^+) + \varepsilon} \left| \int_{E_k^+} f \right|^2 + \frac{1}{m(E_k^-) + \varepsilon} \left| \int_{E_k^-} f \right|^2 \right)$$

$$= \lim_{n\to\infty} \lim_{\varepsilon\to\infty} \sum_{k=1}^{n} \left(\frac{1}{m(E_k^+) + \varepsilon} \left| \int_{E_k^+} f \right|^2 + \frac{1}{m(E_k^-) + \varepsilon} \left| \int_{E_k^-} f \right|^2 \right)$$

$$\le M.$$

(2) Define, for each $f \in L^2(a,b)$,

$$\|f\| = \sup_{a=x_0<\cdots<x_n=b} \left(\sum_{i=1}^{n} \frac{1}{x_i - x_{i-1}} \left| \int_{x_{i-1}}^{x_i} f \right|^2 \right)^{\frac{1}{2}},$$

then $\| \cdot \|$ is a norm on $L^2(a,b)$ by (3). Moreover $\|f\| \le \|f\|_2$ by (1). For any $f \in C[a,b]$, there is $\xi_n^i \in (x_n^{i-1}, x_n^i)(x_n = a + \frac{i}{n}(b-a))$ such that

$$\int_{x_n^{i-1}}^{x_n^i} f = f(\xi_n^i)(x_n^i - x_n^{i-1}).$$

Then

$$\|f\|_2^2 = \lim_{n \to \infty} \sum_{i=1}^{n} |f(\xi_n^i)|^2 (x_n^i - x_n^{i-1})$$

$$= \lim_{n \to \infty} \sum_{i=1}^{n} \frac{\left| \int_{x_n^{i-1}}^{x_n^i} f(x)dx \right|^2}{x_n^i - x_n^{i-1}}$$

$$\leq \|f\|^2.$$

It follows that $\|f\| = \|f\|_2$ for any $f \in C[a,b]$. For any $f \in L^2(a,b)$ there is a sequence $\{f_n\}_{n=1}^{\infty}$ of $C[a,b]$ such that $\|f_n - f\|_2 \to 0$, so

$$\lim_{n \to \infty} \|f_n - f\| \leq \lim_{n \to \infty} \|f_n - f\|_2 = 0.$$

We have

$$\|f\| = \lim_{n \to \infty} \|f_n\| = \lim_{n \to \infty} \|f_n\|_2 = \|f_n\|_2.$$

This completes the proof.

4412

Let $X = [0,1]$, M the σ-algebra of Borel subsets of X. Let $\alpha(t) = t^2$, $\beta(t) = t^3$ and define measures μ and ϕ on M by

$$\mu(E) = \int_E 1 \, d\alpha \quad \text{and} \quad \phi(E) = \int_E 1 \, d\beta.$$

Does $\frac{d\mu}{d\phi}$ exist? Does $\frac{d\phi}{d\mu}$ exist? Compute the value of the Radon–Nikodym derivatives that exist. Justify.

(Iowa)

Solution.
For any Borel subset E of X,

$$\mu(E) = \int_E 2t \, dt$$

and

$$\phi(E) = \int_E 3t^2 \, dt.$$

We have

$$\mu(E) = 0 \Leftrightarrow 2t = 0 \text{ a.e. } t \in E \Leftrightarrow \lambda(E) = 0$$
$$\Leftrightarrow 3t^2 = 0 \text{ a.e. } t \in E \Leftrightarrow \lambda(E) = 0$$
$$\Leftrightarrow \phi(E) = 0.$$

It follows that $\mu \sim \phi \sim \lambda$, where λ the Lebesgue measure. We have

$$\frac{d\mu}{d\phi} = \frac{d\mu}{d\lambda} \bigg/ \frac{d\phi}{d\lambda} = 2t/3t^2 = \frac{2}{3t}$$

and

$$\frac{d\phi}{d\mu} = \frac{3t}{2}.$$

4413

Let λ be the Lebesgue measure on $\mathbb{R}$ and μ and ν the Borel measures on $\mathbb{R}$ defined by:

$$\mu(A) = \sum_{n=1}^{\infty} \frac{1}{2^n} \lambda((n, 2n) \cap A),$$

$$\nu(A) = \sum_{n=1}^{\infty} \frac{1}{3^n} \lambda\left(\left(n, \frac{3}{2}n\right) \cap A\right).$$

Is $\mu \ll \nu$ or $\nu \ll \mu$? Find the corresponding derivaties if they exist.

(*Iowa*)

Solution.

It is true that $\nu \ll \mu$ and it is false that $\mu \ll \nu$. Because $\mu(A) = 0$ implies for all $n \in \mathbb{N}$, $\lambda((n, 2n) \cap A) = 0$ and therefore

$\mu((n, \frac{3}{2}n) \cap A) = 0$, i.e., $\nu(A) = 0$; however, let $A = (\frac{3}{2}, 2]$ and then $\nu/(A) = 0$ but $\mu(A) = \frac{1}{4}$.

We show next that

$$\frac{d\nu}{d\mu} = \begin{cases} 0, & x \leq 0, \\ \dfrac{\sum_{n=1}^{\infty} \frac{1}{3^n} \chi_{(n, \frac{3}{2}n]}(x)}{\sum_{n=1}^{\infty} \frac{1}{2^n} \chi_{(n, 2n]}(x)}, & x > 0. \end{cases} \tag{1}$$

Since

$$\mu(A) = \int_A \sum_{n=1}^{\infty} \frac{1}{2^n} \chi_{(n, 2n]}(x) dx$$

and

$$\nu(A) = \int_A \sum_{n=1}^{\infty} \frac{1}{3^n} \chi_{(n, \frac{3}{2}n]}(x) dx,$$

we have

$$\frac{d\mu}{d\nu} = \sum_{n=1}^{\infty} \frac{1}{2^n} \chi_{(n, 2n]}$$

and

$$\frac{d\nu}{d\lambda} = \sum_{n=1}^{\infty} \frac{1}{3^n} \chi_{(n, \frac{3}{2}n]}.$$

Therefore (1) holds.

4414

Let μ be the Lebesgue measure on $[0, \infty]$. Define

$$\mu_1(E) = \sum_{n=1}^{\infty} \frac{1}{n^3} \int_{E \cap [n, n+1]} x \, d\mu$$

$$\mu_2(E) = \int_{E \cap [1, \infty]} \frac{1}{x^2} d\mu$$

for any Lebesgue measurable subset E of $[0, \infty]$. Is $\mu \ll \mu_2$ or/and $\mu_2 \ll \mu_1$? If so, find the corresponding derivatives.

(Iowa)

Solution.

Since

$$\mu_1(E) = \int_E \sum_{n=1}^{\infty} \frac{1}{n^3} \chi_{[n,n+1)}(x) dx,$$

$$\mu_2(E) = \int_E \chi_{[1,\infty)}(x) \frac{1}{x^2} dx,$$

we have

$$\mu_1(E) = 0 \Leftrightarrow \sum_{n=1}^{\infty} \frac{1}{n^3} \chi_{[n,n+1)} \chi_E(x) = 0, \quad \text{a.e. } x \in [0, \infty)$$

$$\Leftrightarrow \forall n \in \mathbb{N}, \quad \chi_{E \cap [n,n+1)}(x) = 0, \quad \text{a.e. } x \in [0, \infty)$$

$$\Leftrightarrow \forall n \in \mathbb{N}, \quad \mu(E \cap [n, n+1)) = 0$$

$$\Leftrightarrow \mu(E \cap [1, \infty)) = 0$$

$$\Leftrightarrow \chi_{E \cap [1,\infty)}(x) \frac{1}{x^2} = 0$$

$$\Leftrightarrow \mu_2(E) = 0.$$

It follows that both $\mu_1 \ll \mu_2$ and $\mu_2 \ll \mu_1$. Moreover

$$\frac{d\mu_1}{d\mu_2}(x) = \begin{cases} \sum_{n=1}^{\infty} \frac{x^2 \chi_{[n,n+1)}(x)}{n^3}, & x \geq 1, \\ 0, & 0 \leq x < 1, \end{cases}$$

$$\frac{d\mu_2}{d\mu_1}(x) = \begin{cases} \dfrac{1}{x^2 \left(\sum_{n=1}^{\infty} \frac{1}{n^3} \chi_{[n,n+1)}(x) \right)}, & x \geq 1, \\ 0, & 0 \leq x < 1. \end{cases}$$

From $\mu_2([0,1)) = 0$ and $\mu([0,1)) = 1$, $\mu \ll \mu_2$ does not hold.

4415

Let $\phi : \mathbb{R} \to \mathbb{R}$ be a bounded, continuously differentiable function with a bounded derivative, and assume $g \in L(\mathbb{R})$. Define

$$f(t) = \int_{\mathbb{R}} \phi(tg(x))dx, \quad t \in \mathbb{R}.$$

(a) What additional assumption on ϕ will insure that f is well defined? (That is, that $\phi(tg(\cdot)) \in L(\mathbb{R})$ for all $t \in \mathbb{R}$).

(b) Under the additional assumption in (a) above, show that f is differentiable, i.e., $f'(t)$ exists for all $t \in \mathbb{R}$.

(Indiana)

Solution.

(a) Assume $\phi(0) = 0$ then f is well defined. Indeed, for any t, x there exists an s with $0 < s < 1$ such that

$$\phi(tg(x)) = tg(x)\phi'(y)|_{y=stg(x)}.$$

Since

$$\sup_{y\in\mathbb{R}} |\phi'(y)| < \infty \quad \phi(tg(\cdot)) \in L(\mathbb{R}).$$

(b) We will show that

$$f'(t) = \int_{\mathbb{R}} \phi'(tg(x)g(x))dx.$$

Indeed, for any $s, t \in \mathbb{R}\ (s \neq t)$,

$$\left| \frac{\phi(sg(x)) - \phi(tg(x))}{s - t} - \phi'(tg(x))g(x) \right|$$
$$= |\phi'(rg(x)) - \phi'(tg(x))g(x)|$$
$$\leq 2\sup_{y\in\mathbb{R}} |\phi'(y)||g(x)|$$

and

$$\lim_{s\to t} \left| \frac{\phi(sg(x)) - \phi(tg(x))}{s - t} - \phi'(tg(x))g(x) \right| = 0.$$

The conclusion follows from the Dominated Convergence Theorem.

4416

Let $\{f_k\}$ denote a sequence of non-decreasing functions defined on $(0,1)$ with the property that $\lim_{k\to\infty} f_k(x) = 1$ for almost every $x \in (0,1)$. Prove that $\varliminf_{k\to\infty} f_k'(x) = 0$ for almost every $x \in (0,1)$.

(Indiana)

Solution.

There are sequences $\{a_n\}$ and $\{b_n\}$ such that $0 < a_n < b_n < 1$,

$$\lim_{n\to\infty} (b_n - a_n) = 1$$

and

$$\lim_{k\to\infty} f_k(a_n) = \lim_{k\to\infty} f_k(b_n) = 1$$

for any $n \in \mathbb{N}$. By Fatou's Lemma, for any $n \in \mathbb{N}$

$$\int_{a_n}^{b_n} \varliminf_{k\to\infty} f_k'(x)dx \leq \varliminf_{k\to\infty} \int_{a_n}^{b_n} f_k'(x)dx$$
$$\leq \varliminf_{k\to\infty} (f_k(b_n) - f_k(a_n)) = 0.$$

It follows that $\varliminf_{k\to\infty} f_k'$ is integrable and

$$\int_0^1 \varliminf_{k\to\infty} f_k'(x)dx = 0.$$

Hence $\varliminf_{k\to\infty} f_k'(x) = 0$ for almost every $x \in (0,1)$.

4417

Describe a function which is of bounded variation on $[0,1]$ but not monotone on any subinterval of $[0,1]$.

(Auburn)

Solution.

We first construct a Borel set A such that $m((a,b) \cap A) > 0$ and $m((a, b) \backslash A)) > 0$ whenever a $< b$.

For $0 < c < 1$ set $s = c/(1+c)$. Suppose that I is an open interval of finite length. Remove from I its middle part of length $sm(I)$ and we are left with two intervals I_{11} and I_{12}. Remove from I_{1j} its middle part of length $s^2 m(I)$ and keep doing this ad infinitum. All open intervals removed from I form an open set $I(c)$. Let $I[c] = I \backslash I(c)$. Since

$$s + 2s^2 + \cdots + 2^{n-1}s^n + \cdots = c$$

we see that $m(I(c)) = cm(I)$ and $m(I[c]) = (1-c)m(I)$. Clearly, the largest length of components of $I(c)$ is $sm(I)$.

Now given any open set V whose compoents I are all of finite length, let $V(c) = \cup_I I(c)$.

Let $c_n = 1 - 2^{-n}$ and form the open set $V = \bigcup_{k \in \mathbb{Z}}(k-1, k)$. Clearly $\Pi_{n \geq 1} c_{n+k} \neq 0$. Form the open sets $V_1 = V(c_1)$ and $V_n = V_{n-1}(c_n)$ inductively, and define the Borel set $A = \bigcap_{n \geq 1} V_n$.

The largest length of components of V_1 is $c_1/(1+c_1)$ and $\mathbb{R} \backslash V_1$ is nowhere dense. Inductively, the largest length of components of V_n is $\Pi_{i=1}^n \frac{c_i}{1+c_i}$ and $\mathbb{R} \backslash V_n$ is nowhere dense, which shows that $(a,b) \cap V_n \neq \emptyset$. It follows that (a,b) contains some component I of some V_n. Inductively, $(a,b) \cap V_{n+k} \supseteq I(c_{n+1}) \cdots (c_{n+k})$ and so that

$$m((a,b) \cap A) = \lim_{k \to \infty} m((a,b) \cap V_{n+k}) = m(I) \prod_{k=1}^{\infty} c_{n+k} > 0.$$

However, $I[c_{n+1}] \subseteq (a,b) \backslash V_{n+1} \subseteq (a,b) \backslash A$ so that

$$m((a,b) \backslash A) > 0.$$

Now define an absolutely continuous function $f : \mathbb{R} \to \mathbb{R}$ by

$$f(x) = \int_0^x (2\chi_A(t) - 1)dt.$$

Clearly, $f'(x) \doteq 2\chi_A(x) - 1$. For any subinterval E we have

$$m(E(f' > 0)) = m(E(2\chi_A > 1)) = m(E \cap A) > 0,$$
$$m(E(f' < 0)) = m(E(2\chi_A < 1)) = m(E \backslash A) > 0.$$

It follows that f is neither increasing nor decreasing on E.

4418

Let $f : [0, 1] \to \mathbb{R}$ be strictly increasing and

$$g(x) = \lim_{h \to 0} \frac{h}{f(x + h) - f(x - h)}, \quad 0 < x < 1.$$

Prove that if $A = \{x \in (0, 1) : g(x) < 1\}$, then $m^*(A) \le 2f(1) - 2f(0)$, where m^* denotes the outer measure of the Lebesgue measure m.

(Texas A & M)

Solution.

We will show a stronger result: If $E \subseteq \{x \in (0, 1) : g(x) < 1\}$, then $m^*(E) \le 2m^*f(E)$. Indeed, $x \in E$ implies

$$\varlimsup_{h \to 0+} \frac{f(x + h) - f(x - h)}{h} > 1.$$

Since $m^*(f(E)) = \inf_W m(W)$ where W runs through open sets containing $f(E)$, whose components intersect with E, it suffices to show that $m^*(E) \le 2m(W)$ for any such W. Given any component I of W, the preimage $f^{-1}(I)$ is an interval. Let $E_1 = E \cap f^{-1}(I)$. For any $\delta > 0$ and $x \in E_I$ there is an $h \in (0, \delta)$ such that the interval $[x - h, x + h]$ lies in $f^{-1}(I)$ and $f(x + h) - f(x - h) > h$. Applying Vitali's Covering Theorem yields finitely many such intervals $[x_i - h_i, x_i + h_i] : i = 1, \ldots, n$ such that

$$m^* \left(E_I \backslash \bigcup_{i=1}^n [x_i - h_i, x_i + h_i] \right) < 2\delta.$$

It follows that

$$m^*(E_I) \leq \sum_{i=1}^{n} m[x_i - h_i, x_i + h_i] < 2\delta$$

$$\leq \sum_{i=1}^{n} 2(f(x_i + h_i) - f(x_i - h_i) + \delta) \leq 2(m^*(I) + \delta)$$

and thus $m^*(E_I) \leq 2m(I)$ so that $m^*(E) \leq 2m(W)$.

4419

Find all the functions $f : [0,1] \to \mathbb{R}$ of bounded variation satisfying

$$f(x) + \left(\bigvee_0^x f \right)^{\frac{1}{2}} = 1, \quad \forall x \in [0,1],$$

and

$$\int_0^1 f(x)dx = \frac{1}{3}.$$

(Hint. Prove first that f is monotonic.)

(Purdue)

Solution.
Since $f(x) = 1 - \left(\bigvee_0^x f \right)^{1/2}$ the function f is decreasing and $f(0) = 1$ so that $\bigvee_0^x f = f(0) - f(x)$ and thus $f(x) + \sqrt{1 - f(x)} = 1$. Clearly, $f(x) = 0$ or 1. Let $I = (f = 1)$, an interval containing 0, then $f = \chi_I$ and $m(I) = 1/3$. It follows that $I = \left[0, \frac{1}{3}\right)$ or $I = \left[0, \frac{1}{3}\right]$.

4420

Let $c(x)$ be the Cantor function on $[0,1]$ (so $c(x)$ is a continuous non-decreasing function which equals $\frac{1}{2}$ on $\left(\frac{1}{3}, \frac{2}{3}\right)$, equals $\frac{1}{4}$ on $\left(\frac{1}{9}, \frac{2}{9}\right)$, equals $\frac{3}{4}$ on $\left(\frac{7}{9}, \frac{8}{9}\right)$, etc.) Clearly $c'(x) = 0$ for $x \in (0,1), x \notin C$, where C, the Cantor set, is the complement of the union of $\left(\frac{1}{3}, \frac{2}{3}\right)$, $\left(\frac{1}{9}, \frac{2}{9}\right)$, $\left(\frac{7}{9}, \frac{8}{9}\right)$ and all the rest of the middle thirds. Are there any $x \in (0,1) \cap C$ such that $c'(x)$ exists and is finite? Prove your answer.

(Purdue)

Solution.

There are no $x \in (0,1) \cap C$ such that $c'(x)$ exists and is finite. Indeed, we will show that $\overline{\lim}_{z \to x} \frac{c(z) - c(x)}{z - x} = +\infty$ for such x. To this end, write $x = \sum_{n \geq 1} \frac{a_n}{3^n}$ such that $\{a_n | n \geq 1\} \subseteq \{0, 2\}$ then $c(x) = \sum_{n \geq 1} \frac{a_n}{2^{n+1}}$ by definition. Let $I = \{i | a_n = 0\}$. If I is infinite then for any $n \in I$, with $y = \sum_{i < n} \frac{a_i}{3^i} + \frac{2}{3^n} + \sum_{i > n} \frac{a_i}{3^i}$ we have

$$\frac{c(y) - c(x)}{y - x} = \frac{\frac{2}{2^{n+1}} - \frac{0}{2^{n+1}}}{\frac{2}{3^n} - \frac{0}{3^n}} = \frac{1}{2}\left(\frac{3}{2}\right)^n.$$

Let $n \in I$ tend to $+\infty$ as desired. If I is finite then $a_n = 2$ for any $n > \max I$ and with $y = \sum_{i < n} \frac{a_i}{3^i} + \frac{0}{3^n} + \sum_{i > n} \frac{a_i}{3^i}$ we have

$$\frac{c(y) - c(x)}{y - x} = \frac{\frac{0}{2^{n+1}} - \frac{2}{2^{n+1}}}{\frac{0}{3^n} - \frac{2}{3^n}} = \frac{1}{2}\left(\frac{3}{2}\right)^n.$$

Let n tend to infinity as desired.

4421

(1) Prove that if f is a continuous function on $[0,1]$ such that $f'(x) = 0$ for all but perhaps a finite number of x then f is a constant function, i.e. $f(x) = f(y), x, y \in [0,1]$.

(2) Prove that if g is a continuous function on $[0,1]$ satisfying $g(s) \leq g(t)$ if $s \leq t$ such that $g'(x) = 0$ for all but perhaps a countable number of x then g is a constant function.

(Purdue)

Solution.

(1) Write $0 = t_0 < t_1 < \cdots < t_n = 1$ such that $f'(x) = 0$ for $x \in \bigcup_{i=1}^{n}(t_{i-1}, t_i)$. It follows that f is constant on each (t_{i-1}, t_i) and thus on $[0,1]$.

(2) It suffices to show that g is constant on $(0, 1)$. Otherwise, there would be $\{x_1, x_2\} \subseteq (0, 1)$ such that $x_1 < x_2$ and $g(x_1) < g(x_2)$. Let $d = \frac{g(x_2) - g(x_1)}{2}$ and define a continuous function $h_c : [0, 1] \to \mathbb{R}$ by

$$h_c(x) = g(x) - g(x_1) - c(x - x_1)$$

whenever $0 < c < \frac{d}{x_2 - x_1}$. Clearly, $h_c(x_1) = 0$ and $h_c(x_2) > d$ so that $h_c(z) = d$ for some $z \in (x_1, x_2)$. Let $\xi(c)$ be the supremum of such z then $h(\xi(c)) = d$ and $h(\eta) > d$ if $\xi(c) < \eta < x_2$. It follows that

$$0 \le D_+ h(\xi(c)) = D_+ g(\xi(c)) - c$$

and thus $D_+ g(\xi_c) \ge c$ so that ξ_c lies in $(D_+ g > 0)$, where

$$D_+ g(x) = \lim_{z \to x+} \frac{g(z) - g(x)}{z - x}.$$

If $\xi(c) = \xi(c')$ then $c = c'$ follows from the following equalities

$$g(\xi(c)) - g(x_1) - c(\xi(c) - x_1) = d,$$
$$g(\xi(c')) - g(x_1) - c'(\xi(c') - x_1) = d.$$

Now $(D_+ g > 0) \supseteq \{\xi(c) | 0 < c < d/(x_2 - x_1)\}$ and the latter is uncountable, a contradiction.

Section 5

Miscellaneous Problems

4501

Let $S \subset R$ be a set of real numbers with the property that $|S_1 + \cdots + S_n| \leq 1$ for every finite subset $\{S_1, \ldots, S_n\} \subset S$. Show that S is countable.

(Indiana–Purdue)

Solution.

Since $S \cap (\frac{1}{n}, n)$ and $S \cap (-n, -\frac{1}{n})$ must be finite and

$$S = S \cap \left(\bigcup_{n=1}^{\infty} \left(\left(\frac{1}{n}, n \right) \cup \left(-n, -\frac{1}{n} \right) \right) \cup \{0\} \right),$$

S is countable.

4502

Let $\{I_\alpha\}$, $\alpha \in \Gamma$, be a collection of closed intervals in R. Show that

$$\bigcup_{\alpha \in \Gamma} I_\alpha \setminus \bigcup_{\alpha \in \Gamma} I_\alpha^0$$

is countable.

(Indiana–Purdue)

Solution.

Obviously $\bigcup_{\alpha \in \Gamma} I_\alpha^0$ is an open set. We have

$$\bigcup_{\alpha \in \Gamma} I_\alpha^0 = \bigcup_{n=1}^{\infty} (\alpha_n, \beta_n),$$

where $\alpha_n, \beta_n \bar{\in} \bigcup_{\alpha \in \Gamma} I_\alpha^0$. For any $x \in \bigcup_\alpha I_\alpha \setminus \bigcup_\alpha I_\alpha^0$, there must be α such that $x \in I_\alpha$, but x cannot be in I_α^0. Take n such that

$$I_\alpha^0 \subset (\alpha_n, \beta_n).$$

If $I_\alpha \subset (\alpha_n, \beta_n)$, then $x \notin I_\alpha$, which is a contradiction. Thus we must have

$$x = \alpha_n \quad \text{or} \quad \beta_n.$$

Therefore

$$\bigcup_\alpha I_\alpha \setminus \bigcup_\alpha I_\alpha^0 \subset \bigcup_n \{\alpha_n, \beta_n\}.$$

which implies $\bigcup_\alpha I_\alpha \setminus \bigcup_\alpha I_\alpha^0$ is countable.

4503

Show that any infinite set of non-empty, mutually disjoint, open sets in a separable metric space X is countable.

(*Stanford*)

Solution.

Let $\{U_i \mid i \in I\}$ be such an infinite set. For each $i \in I$ there is an $a_i \in D$ such that $a_i \in U_i$, where D is a countable dense subset of X. Then $a_i \neq a_j$ whenever $i \neq j$. It follows that the map, $i \mapsto a_i$ of I to D is injective and therefore I is countable.

4504

Prove that for almost every $x \in [0, 2\pi]$

$$\overline{\lim_{n \to \infty}} \sin(nx) = 1.$$

(*Stanford*)

Solution.

Let

$$A = \left\{ x \in (0, 2\pi) \,\middle|\, \frac{x}{\pi} \text{ is irrational} \right\}.$$

Then A is a measurable set of measure 2π. Moreover, for any $x \in A$,

$$\overline{\lim_{n \to \infty}} \sin(nx) = 1.$$

Indeed for any $x \in A$, since $\{k\frac{x}{\pi} - 2l \mid k, l \in \mathbb{Z}\}$ is a dense subgroup of $\mathbb{R}$, there are sequences $\{k_n\}$ and $\{l_n\}$ of $\mathbb{Z}$ such that

$$\lim_{n \to \infty} \left(k_n \frac{x}{\pi} - 2l_n \right) = \frac{1}{2}.$$

Since $\frac{1}{2} \notin \{k\frac{x}{\pi} - 2l \mid k, l \in \mathbb{Z}\}$, $\{k_n\}$ admits a subsequence $\{k'_n\}$ either increasing to $+\infty$ or decreasing to $-\infty$. If $\lim_{n \to \infty} k'_n = +\infty$ then

$$\lim_{n \to \infty} \sin(k'_n x) = \lim_{n \to \infty} \sin(k'_n x - 2l'_n \pi) = 1;$$

Otherwise $\lim_{n \to \infty} (-3k'_n) = +\infty$ and

$$\lim_{n \to \infty} \left((-3k'_n) \frac{x}{\pi} + 2(3l'_n + 1) \right) = \frac{1}{2}$$

and therefore

$$\lim_{n \to \infty} \sin(-3k'_n x) = \lim_{n \to \infty} \sin(-3k'_n x + 2(3l'_n + 1)\pi) = 1.$$

4505

Let $f \in C(I)$. Show that there exists a sequence of polynomials $\{p_n\}$ such that $p_n \to f$ uniformly on I and $p_1(x) \leq p_2(x) \leq \cdots$ for every $x \in I$.

(Indiana–Purdue)

Solution.

For any n, take a polynomial p_n such that

$$\left| p_n(x) - \left[f(x) - \frac{3}{2^{n+2}} \right] \right| < \frac{1}{2^{n+2}}, \quad x \in I.$$

Then

$$f(x) - \frac{1}{2^{n+1}} > p_n(x) > f(x) - \frac{1}{2^n}, \quad x \in I.$$

Obviously, $p_1(x) \le p_2(x) \le \cdots$ and $p_n \to f$ uniformly on I.

4506

Let $f \in C([0,1])$. Show that there is a sequence of odd polynomials $p_n(x)$ with $p_n \to f$ uniformly on $[0,1]$ iff $f(0) = 0$.

(Indiana–Purdue)

Solution.

Suppose that there is a sequence of odd polynomials p_n with $p_n \to f$. Then

$$f(0) = \lim_{n \to \infty} p_n(0) = 0.$$

Conversely, suppose that $f(0) = 0$. Set

$$f(x) = -f(-x), \quad x \in [-1, 0].$$

Then f is a continuous function on $[-1,1]$. Take a sequence of polynomials $p_n(x)$ with $p_n \to f$ uniformly on $[-1,1]$. Set

$$P_n(x) = \frac{1}{2}[p_n(x) - p_n(-x)].$$

Then P_n is an odd polynomial for any n. We have

$$\lim_{n \to \infty} P_n(x) = \lim \frac{1}{2}[p_n(x) - p_n(-x)]$$

$$= \frac{1}{2}[f(x) - f(-x)] = f(x)$$

uniformly on $[0,1]$.

4507

(a) Show that the mapping $I : C[0, 1] \to C[0, 1]$

$$(If)(x) = e^x + \frac{1}{2} \int_0^{x^2} f(t)dt$$

is a contracting mapping on $C[0, 1]$, with the supremum norm.

(b) Show that there exists one and only one smooth function f on [0,1] satisfying the conditions:

$$\begin{cases} \dfrac{d}{dx}f(x) = e^x + xf(x^2), \\ f(0) = 1. \end{cases}$$

(*Stanford*)

Solution.

(a) For any f and g in $C([0, 1])$,

$$\|If - Ig\| = \max_{x \in [0,1]} \frac{1}{2} \int_0^{x^2} |f(t) - g(t)|dt \leq \frac{1}{2}\|f - g\|.$$

(b)

By the Banach fixed point theorem, there is only one function $f \in C[0, 1]$ such that $I(f) = f$, i.e.,

$$f(x) = e^x + \frac{1}{2} \int_0^{x^2} f(t)dt. \tag{1}$$

By (1) f is smooth and satisfies the conditions:

$$\begin{cases} \dfrac{d}{dx}f(x) = e^x + xf(x^2), \\ f(0) = 1. \end{cases} \tag{2}$$

If g is another smooth function satisfying (2), then

$$g(x) = e^x + \frac{1}{2} \int_0^{x^2} g(x^2)dx,$$

i.e., $I(g) = g$, by the uniqueness of $f, g = f$.

4508

Given $f : R \to R$ bounded and uniformly continuous and $\{K_n\}$, $n = 1, 2, 3, \ldots$, $K_n \in L^1$ such that

(i) $\|K_n\|_1 \leq M < \infty$, $n = 1, 2, 3, \ldots$.

(ii) $\int K_n \to 1$ as $n \to \infty$.

(iii) $\int_{\{x:|x|>\delta\}} |K_n| \to 0$ as $n \to \infty$ for all $\delta > 0$.

Show $K_n * f \to f$ uniformly.

(UC, Irvine)

Solution.

Take $M_1 > 0$ such that $|f(x)| < M_1$ for all $x \in R$. For any $\varepsilon > 0$, by (ii) there is an N_1 such that

$$\left| \int K_n(y) dy - 1 \right| < \frac{\varepsilon}{2M_1}.$$

If $n > N_1$. Thus

$$\left| \int K_n(y) f(x) dy - f(x) \right| < M_1 \cdot \frac{\varepsilon}{2M_1} = \frac{\varepsilon}{2}. \qquad (1)$$

Take $\delta > 0$ such that

$$|f(x - y) - f(x)| < \frac{\varepsilon}{4M}$$

holds for all $x \in R$ any $|y| < \delta$. For the above δ, take N_2 such that

$$\int_{\{y:|y|>\delta\}} |K_n(y)| dy < \frac{\varepsilon}{8M_1},$$

if $n > N_2$. Therefore

$$\left| (K_n * f)(x) - \int K_n(y) f(x) dy \right|$$

$$\leq \int |K_n(y)| |f(x - y) - f(x)| dy$$

$$= \int_{\{y:|y|>\delta\}} |K_n(y)| |f(x - y) - f(x)| dy$$

$$+ \int_{\{y:|y|\le\delta\}} |K_n(y)||f(x-y) - f(x)|dy$$

$$\le 2M_1 \int_{|y|>\delta} |K_n(y)|dy + \int_{|y|\le\delta} |K_n(y)|dy \cdot \frac{\varepsilon}{4M}$$

$$\le 2M_1 \cdot \frac{\varepsilon}{8M_1} + \|K_n\|_1 \cdot \frac{\varepsilon}{4M}$$

$$< \frac{\varepsilon}{4} + M \cdot \frac{\varepsilon}{4M} = \frac{\varepsilon}{2}. \tag{2}$$

Set $N = \max(N_1, N_2)$. It follows from (1), (2) that

$$|(K_n * f)(x) - f(x)| \le \left|(K_n * f)(x) - \int K_n(y)f(x)dy\right|$$

$$+ \left|\int K_n(y)f(x)dy - f(x)\right|$$

$$< \frac{\varepsilon}{2} + \frac{\varepsilon}{2} = \varepsilon$$

holds for all $x \in R$ if $n > N$.

4509

Let $f : R \to (-\infty, \infty)$ be upper semicontinuous and define $m : R \to [-\infty, \infty)$ by $m(x) = \liminf_{y\to x} f(y)$. Let $S = \{x \mid f(x) - m(x) \ge 1\}$.

(a) Show S is closed.

(b) Show: If I is an open interval contained in S, then $m(x) = -\infty$ on I.

(c) Show that S is nowhere dense.

(*UC, Irvine*)

Solution.

(a) For any $x_0 \in S^c$, there exists $\varepsilon > 0$ such that

$$f(x_0) - m(x_0) < 1 - \varepsilon < 1.$$

Take $\delta > 0$ such that

$$f(x) < f(x_0) + \frac{\varepsilon}{2}, \quad f(x) > m(x_0) - \frac{\varepsilon}{2}$$

for $x \in O(x_0, \delta)$. Therefore

$$m(x) = \liminf_{y \to x} f(y) \geq m(x_0) - \frac{\varepsilon}{2}$$

holds for $x \in O(x_0, \delta)$. Thus

$$f(x) - m(x) < f(x_0) + \frac{\varepsilon}{2} - m(x_0) + \frac{\varepsilon}{2} < 1 - \varepsilon + \varepsilon = 1$$

if $x \in O(x_0, \delta)$, i.e., $x \in S^c$. So S^c is open.

(b) Suppose that there is $x_0 \in I$ such that $m(x_0) > -\infty$. As in (a), we can find $\delta > 0$ such that $m(x) > m(x_0) - \frac{1}{2}$ for $x \in O(x_0, \delta) \subset I$. By the definition of $m(x)$, there must be $\bar{x} \in O(x_0, \delta)$ such that $f(\bar{x}) < m(x_0) + \frac{1}{2}$. Therefore $f(\bar{x}) - m(\bar{x}) < 1$, i.e., $\bar{x} \notin S$ which contradicts $\bar{x} \in I \subset S$.

(c) Suppose that S is not nowhere dense. Since S is closed, there is an open interval $I \subset S$. From (b), $m(x) = -\infty$ for $x \in I$. Denote

$$A_1 = \{x \in I, f(x) < -1\}.$$

A_1 is a non-empty open set. Take an open interval I_1 such that $\bar{I}_1 \subset A_1$. In the same way, we can take an open interval I_2 satisfying $\bar{I}_2 \subset I_1$ and $f(x) < -2$ for $x \in \bar{I}_2$. By induction, we obtain a sequence of open intervals $\{I_n\}$ such that $\bar{I}_n \subset I_{n-1}, f(x) < -n$ for $x \in \bar{I}_n$. Obviously $f(x) = -\infty$ for $x \in \cap \bar{I}_n$, which is a contradiction.

4510

Prove that any topological metric space is homeomorphic to a bounded metric space.

(Stanford)

Solution.

Suppose that (X, d) is a metric space. Define another metric $\tilde{d}$ on X by

$$\tilde{d}(x, y) = \frac{d(x, y)}{1 + d(x, y)}, \quad x, y \in X.$$

It is easy to show that the identity mapping of X is a homeomorphism of (X, d) to $(X, \tilde{d})$.

4511

Let M be a metric space with distance function d, suppose $A : M \to M$ is a distance non-increasing periodic map of order 3, i.e.,

$$A \circ A \circ A = Id \quad \text{and} \quad d(Ax, Ay) \leq d(x, y).$$

(i) Show that A is a continuous bijective isometry.

(ii) Give an example of a complete metric space M and an isometry A on M, periodic of order 3, which has exactly two fixed points.

(*Stanford*)

Solution.

(i) Since $A^{-1} = A^2$, A^{-1} is distance non-increasing, too. Therefore A is an isometry and therefore continuous.

(ii) Let M be the 2-dimensional sphere

$$M = \{(z, x) \in \mathbb{C} \times \mathbb{R} \mid |z|^2 + x^2 = 1\}.$$

The isometry $A : M \to M$ defined by

$$A(z, x) = (e^{\frac{2}{3}\pi i} z, x), \quad (z, x) \in M,$$

is a periodic map of order 3, having exactly two fixed points $(0, 1)$ and $(0, -1)$.

4512

Let H be a Hilbert space and let $f : H \to \mathbb{R}$ be a continuous convex function such that $f(x_n) \to \infty$ whenever $\|x_n\| \to \infty$. Prove that f attains a minimum.

(SUNY, Stony Brook)

Solution.

Let $\alpha = \inf f(H)$. There is a sequence $\{x_n\}$ of H such that

$$\lim_{n \to \infty} f(x_n) = \alpha.$$

If

$$\sup_n \|x_n\| = +\infty,$$

there is a subsequence $\{x_{n_k}\}_{k=1}^{\infty}$ such that $\|x_{n_k}\| \to +\infty$, then

$$\alpha = \lim_{k \to \infty} f(x_{n_k}) = +\infty,$$

a contradiction. So $\sup_n \|x_n\| < +\infty$. By the weak compactness of the closed ball of H, there is a subsequence $\{x_{n_k}\}$ converging weakly to a point $x \in H$.

For any $\beta > \alpha$ there is an $N \in \mathbb{N}$ such that for any $k > N$, $f(x_{n_k}) < \beta$. Since

$$x \in \overline{\{x_{n_k} \mid k > N\}}^W \subseteq \overline{\mathrm{cov}\{x_{n_k} \mid k > N\}}^W$$

$$= \overline{\mathrm{cov}\{x_{n_k} \mid k > N\}}^{\|\cdot\|}$$

and for any $y \in \mathrm{cov}\{x_{n_k} \mid k > N\}$, $f(y) < \beta$, one has $f(x) \leq \beta$. Therefore

$$\alpha \leq f(x) \leq \lim_{\beta \downarrow \alpha} \beta = \alpha.$$

It follows that α is finite and f attains its minimum at x.

4513

A is the subset of $L^1(\mathbb{R})$ consisting of all functions f satistying $|f(x)| \leq 1$ a.e. on $\mathbb{R}$. Prove that A is closed in the norm topology of $L^1(\mathbb{R})$.

(Stanford)

Solution.

If $\{f_n\}$ is a sequence of A such that $f = \lim_{n \to \infty} f_n$ exists in $L^1(\mathbb{R})$. Then $\{f_n\}$ converges to f in measure and therefore by F. Riesz Theorem there is a subsequence $\{f_{n_k}\}$ converging to f almost everywhere. It follows that $|f(x)| \leq 1$ a.e. on $\mathbb{R}$. So A is closed.

4514

Let A be a bounded linear operator on Hilbert space H. Recall that the adjoint A^* is the unique bounded linear operator on H such that $(Ax, y) = (x, A^*y)$ for all $x, y \in H$.

(a) Show that $\|A^*\| = \|A\|$, where $\|A\|$ is the norm of A.

(b) Show that $AA^* - A^*A$ cannot be the identity on H. (You may wish to use (a) prove this.)

(Stanford)

Solution.

(a) Indeed, since

$$\|A^*\| = \sup_{\|y\| \leq 1} \|A^*y\| = \sup_{\|y\| \leq 1} \sup_{\|x\| \leq 1} |(x, A^*y)|$$

$$= \sup_{\|y\| \leq 1} \sup_{\|x\| \leq 1} |(Ax, y)| \leq \|A\|$$

and a fortiori $\|A\| = \|A^*\|$.

(b) First, we will show that

$$\|A^*A\| = \|A\|^2 = \sup_{\|x\| \leq 1} (A^*Ax, x). \tag{1}$$

Indeed equalities (1) follow from the following inequalities

$$\|A\|^2 = \sup_{\|x\|\le 1}(A^*Ax,x) \le \sup_{\|x\|\le 1}\|A^*Ax\|\,\|x\|$$

$$= \|A^*A\| \le \|A^*\|\,\|A\| = \|A\|^2.$$

If for some bounded linear operator A on $H, AA^* - A^*A = I$ then

$$\|AA^*\| = \sup_{\|x\|=1}(AA^*x,x)$$

$$= \sup_{\|x\|=1}(A^*Ax + x,x)$$

$$= \sup_{\|x\|=1}(A^*Ax,x) + 1$$

$$= \|A^*A\| + 1.$$

It follows that $\|A\|^2 = \|A\|^2 + 1$, a contradiction.

4515

Suppose that $A, B \subset \mathbb{R}$ are Lebesgue measurable, with $m(A) > 0$, $m(B) > 0$. Show that

$$A + B = \{x \in \mathbb{R} \mid x = a + b, a \in A, b \in B\}$$

contains an interval of positive Lebesgue measure.

(*Indiana*)

Solution.

Assume without less of generality that A and B are compact sets. Since

$$\int_{\mathbb{R}} m((x-B)\cap A)dx = \int_{\mathbb{R}}\left(\int_{\mathbb{R}} \chi_{(x-B)\cap A}(y)dy\right)dx$$

$$= \int_{\mathbb{R}}\left(\int_{\mathbb{R}} \chi_B(x-y)\chi_A(y)dy\right)dx$$

$$= m(A)m(B) > 0,$$

there exists an $x_0 \in \mathbb{R}$ such that $m((x_0 - B) \cap A) > 0$. Since $x \mapsto m((x - B) \cap A)$ is continuous, the set $\{x \mid m((x - B) \cap A) > 0\}$ is non-empty and open. Then the conclusion follows from the following inclusion

$$\{x \mid m((x - B) \cap A) > 0\} \subset A + B.$$

4516

Let X be a compact Hausdorff topological space and let μ be a finite regular Borel measure on X. Is it true that if $f : X \to \mathbb{R}$ is μ-measurable then there exists a sequence $\{f_n\}_{n=1}^{\infty}$ of continuous real-valued functions such that $\lim_{n \to 0} f_n = f$ a.e. $[\mu]$? Justify.

(*Iowa*)

Solution.

Yes. For any $n \in \mathbb{N}$ there is a compact subset F_n of X such that $\mu(X \backslash F_n) < \frac{1}{n}$ and f is continuous on F_n. Let $X_n = \bigcup_{i=1}^{n} F_i$. Then (1) $\mu(X \backslash X_n) < \frac{1}{n}$, (2) f is continuous on X_n, and (3) $\mu(X \backslash \bigcup_{n>1}^{\infty} X_n) = 0$. There is for each $n \in \mathbb{N}$ a real-valued continuous function $f_n : X \to \mathbb{R}$ such that $f_n = f$ on X_n. Then $\lim_{n \to \infty} f_n = f$ a.e. $[\mu]$.

4517

Consider a complete metric space (X, d) and a sequence $F_1 \supseteq F_2 \supseteq \cdots$ of non-empty set and closed subsets of X. Assume that for each n, the set F_n can be covered by a finite number of balls of radius $1/n$. For each n, select a point $x_n \in F_n$. Prove that the sequence $(x_n)_{n=1}^{\infty}$ has a convergent subsequence.

(*Indiana*)

Solution.

Note that $(x_n)_{n \geq k}$ is contained in F_k for any $k \geq 1$. Since $F_1 \subseteq \bigcup \{O(y, 1) \mid y \in E_1\}$ for a finite subset E_1 of X, the sequence $(x_n)_{n \geq 1}$ has a subsequence $(x_n)_{n \in J_1}$ contained in $O(y_1, 1)$ for some $y_1 \in E_1$, where J_1 is an infinite subset of $\mathbb{N}$ such that

$\min J_1 > 1$. Since $F_2 \subseteq \bigcup \{O(y, \frac{1}{2}) \mid y \in E_2\}$ for a finite subset E_2 of X, the sequence $(x_n)_{n \in J_1}$ has a subsequence $(x_n)_{n \in J_2}$ contained in $O(y_2, \frac{1}{2})$ for some $y_2 \in E_2$, where J_2 is an infinite subset of J_1 such that $\min J_2 > \min J_1$. By induction, we get a sequence $(y_k)_{k \geq 1}$ of X and a decreasing sequence $(J_n)_{n \geq 1}$ of subsets of $\mathbb{N}$ such that $(x_n)_{n \in J_k}$ is contained in $O(y_k, \frac{1}{k})$ and $p_k > p_{k-1}$, where $p_k = \min J_k$. Since

$$d(x_{p_l}, x_{p_k}) \leq d(x_{p_l}, y_k) + d(y_k, x_{p_k}) < \frac{2}{k}$$

for $l \geq k$, the subsequence $(x_{p_k})_{k \geq 1}$ of $(x_n)_{n \geq 1}$ converges in X.

4518

Let $u : \mathbb{R} \to \mathbb{R}$ be arbitrary, and $v(x) = \inf_{|x-y|<1} u(y)$. Show that v is upper semi-continuous.

(*Purdue*)

Solution.

For any $b \in \mathbb{R}$, if $v(x) < b$ then $u(y) < b$ for some $y \in \mathbb{R}$ with $|x - y| < 1$ and thus $v(x') < b$ for any $x' \in \mathbb{R}$ with $|x' - y| < 1$. It follows that $(v < b)$ is an open set and thus v is upper semi-continuous.

4519

Let X be a Banach space and Y a closed subspace of X. Show that every extreme point g of the unit ball of the dual Y^* extends to an extreme point of the unit ball of X^*.

(*Texas A & M*)

Solution.

Let $[X]_1$ be the unit ball of the norm space X. Let

$$E = \{f \in [X^*]_1 : f|_Y = g\},$$

then E is non-empty by Hahn–Banach Theorem, convex and weakly*-closed. Applying Krein–Milman Theorem produces an extreme point f_0 of E. We have to show that f_0 is an extreme point of $[X^*]_1$. Write $f_0 = (f_1 + f_2)/2$ with $\{f_1, f_2\} \subseteq [X^*]_1$ then

$$g = f_0|_Y = (f_1|_Y + f_2|_Y)/2$$

and thus $g = f_i|_Y$ for $i = 1, 2$ since g is one extreme point of $[Y^*]_1$. It follows that f_1 and f_2 lie in E and thus $f_0 = f_i$ for $i = 1, 2$.

4520

Prove that span$\{x^{2n} : n \geq 0\}$ is norm dense in $C[0, 1]$. Deduce that the norm closure of span$\{x^{2n+1} | n \geq 0\}$ is $\{f \in C[0, 1] : f(0) = 0\}$.

(*Texas A & M*)

Solution.
(1) For any $f \in C[0, 1]$, applying Weierstrass Theorem produces a sequence $(f_n(x))_{n \geq 1}$ of polynomials convergent to $f(\sqrt{x})$ uniformly, and thus $f_n(x^2)$ converges to $f(x)$ uniformly.
(2) If $f(x) = 0$ extends f to an odd function on $[-1, 1]$ and we find a sequence (g_n) of polynomials converges to f uniformly on $[-1, 1]$. Let

$$f_n(x) = \frac{g_n(x) - g_n(-x)}{2} : n \geq 1$$

then (f_n) converges to f uniformly. Clearly each $f_n(x)$ lies in span$\{x^{2n-1} | n \geq 0\}$.

4521

Let $f \in L^1[0, 1]$, $E_n \subseteq [0, 1]$ be measurable subsets, and

$$g_n(x) = \int_0^x \chi_{E_n}(t) f(t) dt.$$

Prove that there exists $\{n_k\}$ and a continuous function $g(x)$ on $[0, 1]$ such that $g_{n_k}(x)$ converges uniformly to $g(x)$ on $[0, 1]$.

(*Rutgers*)

Solution.

Clearly, $|g_n(x)| \leq \|f\|_1$ for $x \in [0,1]$. By Arzela–Ascoli Theorem it suffices to show that g_n's are equicontinuous. For any $\epsilon > 0$ there is a $\delta > 0$ such that $|E| < \delta$ implies $\int_E |f(t)|dt < \epsilon$. If $0 \leq x < x' \leq 1$ such that $x' - x < \delta$ then

$$|g_n(x') - g_n(x)| = \left| \int_x^{x'} \chi_{E_n}(t)f(t)dt \right| \leq \int_x^{x'} |f(t)|dt < \epsilon.$$

4522

Prove or disprove: if K and M are metric spaces, $f : K \times M \to \mathbb{R}$ is a continuous function, and K is compact, then the function $g : M \to \mathbb{R}$ defined by

$$g(x) = \max\{f(k,x)|\, k \in K\}$$

is continuous.

(Rutgers)

Solution.

Clearly, g is lower semi-continuous. To show it is upper semi-continuous, it suffices to show $(g \geq c)$ is closed for any $c \in \mathbb{R}$. To this end, let $\pi : K \times M \to M$ be the natural projection then it is well known that π is a closed map. It remains to note that

$$(g \geq c) = \{x \in M|\, \exists k \in K : f(k,x) \geq c\} = \pi(f \geq c).$$

PART 5

Complex Analysis

Section 1

Analytic and Harmonic Functions

5101

True-False. If the assertion is true, quote a relevant theorem or reason; if false, give a counterexample or other justification.

(a) If $f(z) = u + iv$ is continuous at $z = 0$, and the partials u_x, u_y, v_x, v_y exist at $z = 0$ with $u_x = v_y$ and $u_y = -v_x$ at $z = 0$, then $f'(0)$ exists.

(b) If $f(z)$ is analytic in Ω and has infinitely many zeros in Ω, then $f \equiv 0$.

(c) If f and g are analytic in Ω and $f(z) \cdot g(z) \equiv 0$ in Ω, then either $f \equiv 0$ or $g \equiv 0$.

(d) If $f(z)$ is analytic in $\Omega = \{z; \operatorname{Re} z > 0\}$, continuous on $\bar{\Omega}$ with $|f(iy)| \leq 1$ ($-\infty < y < +\infty$), then $|f(z)| \leq 1$ ($z \in \Omega$).

(e) If $\sum a_n z^n$ has radius of convergence exactly R, then $\sum n^3 a_n z^n$ has radius of convergence exactly R.

(f) $\sin \sqrt{z}$ is an entire function.

(*Indiana–Purdue*)

Solution.

(a) False. A counterexample is $f(x, y) = \sqrt{|xy|}$. f satisfies Cauchy–Riemann equations at $z = 0$, but $f'(0)$ does not exist.

(b) False. A counterexample is $f(z) = \sin\frac{1}{1-z}$. f is analytic in $\Omega = \{z : |z| < 1\}$, and has zeros $z = 1 - \frac{1}{n\pi}$, $n = 1, 2, \ldots$. But f is not identically zero in Ω.

(c) True. If neither of f and g is identically zero in Ω, then both f and g have at most countably many zeros in Ω, and the zeros have no limit point in Ω. Then $f(z)\cdot g(z)$ is not identically zero in Ω.

(d) False. A counterexample is $f(z) = e^z$, which is analytic in Ω, and continuous on $\bar{\Omega}$ with $|f(iy)| \equiv 1$. But $f(z)$ is not bounded in Ω.

(e) True. Because $\lim_{n\to\infty} \sqrt[n]{n^3} = 1$, it follows from

$$\varlimsup_{n\to\infty} \sqrt[n]{|a_n|} = \frac{1}{R}$$

that

$$\varlimsup_{n\to\infty} \sqrt[n]{n^3|a_n|} = \frac{1}{R}.$$

(f) False. $\sin\sqrt{z}$ is not analytic at $z = 0$. Actually, $z = 0$ is a branch point of $\sin\sqrt{z}$.

5102

(a) Let $f(z)$ be a complex-valued function of a complex variable. If both $f(z)$ and $zf(z)$ are harmonic in a domain Ω, prove that f is analytic there.

(b) Suppose that f is analytic with $|f(z)| < 1$ in $|z| < 1$ and that $f(\pm a) = 0$ where a is a complex number with $0 < |a| < 1$. Show that $|f(0)| \le a^2$. What can you conclude if this holds with equality.

(c) Determine all entire function f that $|f'(z)| < |f(z)|$.

(*Stanford*)

Solution.

(a) It is well known that the Laplacian can be written as

$$\Delta = \frac{\partial^2}{\partial x^2} + \frac{\partial^2}{\partial y^2} = 4\frac{\partial^2}{\partial z\partial\bar{z}}.$$

Because

$$\frac{\partial^2}{\partial z \partial \bar{z}}(zf(z)) = \frac{\partial}{\partial \bar{z}}f(z) + z\frac{\partial^2}{\partial z \partial \bar{z}}f(z),$$

it follows from

$$\frac{\partial^2}{\partial z \partial \bar{z}}f(z) = 0$$

and

$$\frac{\partial^2}{\partial z \partial \bar{z}}(zf(z)) = 0$$

that

$$\frac{\partial}{\partial \bar{z}}f(z) = 0,$$

which implies that $f(z)$ is analytic in Ω.

(b) Define

$$F(z) = f(z) \cdot \frac{1 - \bar{a}z}{z - a} \cdot \frac{1 + \bar{a}z}{z + a},$$

then $F(z)$ is analytic in $\{|z| < 1\}$. When $|z| = 1$,

$$\left| \frac{1 - \bar{a}z}{z - a} \cdot \frac{1 + \bar{a}z}{z + a} \right| = 1,$$

hence

$$\overline{\lim_{|z| \to 1}} |F(z)| \le 1,$$

which implies that $|F(z)| \le 1$ for $|z| < 1$. Take $z = 0$, we obtain

$$|f(0)| \le |a|^2.$$

When it holds with equality, we have $F(z) \equiv e^{i\theta}$, which is equivalent to

$$f(z) = e^{i\theta} \frac{z - a}{1 - \bar{a}z} \cdot \frac{z + a}{1 + \bar{a}z}.$$

(c) From $|f'(z)| < |f(z)|$, we know that f has no zero in $\mathbb{C}$, which implies that $\frac{f'(z)}{f(z)}$ is also an entire function. It follows from $\left|\frac{f'(z)}{f(z)}\right| < 1$ that $\frac{f'(z)}{f(z)} = c$, $|c| < 1$. Integrating on both sides, we obtain $\log f(z) = cz + d$. Hence $f(z) = c'e^{cz}$, where c and c' are constants and $|c| < 1$.

5103

Let G be a region in $\mathbb{C}$ and suppose $u : G \to \mathbb{R}$ is a harmonic function.

(a) Show that $\frac{\partial u}{\partial x} - i\frac{\partial u}{\partial y}$ is an analytic function on G.

(b) Show that u has a harmonic conjugate on G if and only if $\frac{\partial u}{\partial x} - i\frac{\partial u}{\partial y}$ has a primitive (anti-derivative) on G.

(*Indiana*)

Solution.

(a) Let

$$P(x, y) = \frac{\partial u}{\partial x}, \quad Q(x, y) = -\frac{\partial u}{\partial y}.$$

Because u is a harmonic function, we have

$$\frac{\partial P}{\partial x} - \frac{\partial Q}{\partial y} = \frac{\partial^2 u}{\partial x^2} + \frac{\partial^2 u}{\partial y^2} = 0.$$

We also have

$$\frac{\partial P}{\partial y} + \frac{\partial Q}{\partial x} = \frac{\partial^2 u}{\partial x \partial y} - \frac{\partial^2 u}{\partial x \partial y} = 0.$$

So $P(x, y)$ and $Q(x, y)$ satisfy the Cauchy–Riemann equations, hence

$$P + iQ = \frac{\partial u}{\partial x} - i\frac{\partial u}{\partial y}$$

is analytic on G.

(b) If $\frac{\partial u}{\partial x} - i\frac{\partial u}{\partial y}$ has a primtive, then for any closed curve $c \subset G$, the integral

$$\int_c \left(\frac{\partial u}{\partial x} - i\frac{\partial u}{\partial y} \right) dz = \int_c \left(\frac{\partial u}{\partial x} dx + \frac{\partial u}{\partial y} dy \right)$$
$$+ i \left(-\frac{\partial u}{\partial y} dx + \frac{\partial u}{\partial x} dy \right) = 0.$$

It follows that

$$\int_c -\frac{\partial u}{\partial y} dx + \frac{\partial u}{\partial x} dy = 0$$

holds for any closed curve $c \subset G$. Hence we can define a single-valued function $v(x, y)$ on G:

$$v(x, y) = \int_{z_0}^{z} -\frac{\partial u}{\partial y} dx + \frac{\partial u}{\partial x} dy,$$

where $z_0, z \in G$, and the integral is taken along any curve connecting z_0 and z in G. Because

$$\frac{\partial v}{\partial x} = -\frac{\partial u}{\partial y}, \quad \frac{\partial v}{\partial y} = \frac{\partial u}{\partial x},$$

we know that $v(x, y)$ is a harmonic conjugate of $u(x, y)$ on G.

On the contrary, if u has a harmonic conjugate v on G, then

$$dv = \frac{\partial v}{\partial x} dx + \frac{\partial v}{\partial y} dy = -\frac{\partial u}{\partial y} dx + \frac{\partial u}{\partial x} dy.$$

For any closed curve $c \subset G$, we have

$$\int_c \left(\frac{\partial u}{\partial x} - i\frac{\partial u}{\partial y} \right) dz = \int_c \left(\frac{\partial u}{\partial x} dx + \frac{\partial u}{\partial y} dy \right) + i \left(-\frac{\partial u}{\partial y} dx + \frac{\partial u}{\partial x} dy \right)$$
$$= \int_c \left(\frac{\partial u}{\partial x} dx + \frac{\partial u}{\partial y} dy \right) + i \left(\frac{\partial v}{\partial x} + \frac{\partial v}{\partial y} dy \right)$$
$$= \int_c d(u + iv) = 0.$$

Hence $\frac{\partial u}{\partial x} - i\frac{\partial u}{\partial y}$ has a primitive $\int_{z_0}^{z} \left(\frac{\partial u}{\partial x} - i\frac{\partial u}{\partial y} \right) dz$ on G.

5104

Suppose that u and v are real valued harmonic functions on a domain Ω such that u and v satisfy the Cauchy–Riemann equations on a subset S of Ω which has a limit point in Ω. Prove that $u + iv$ must be analytic on Ω.

(*Indiana–Purdue*)

Solution.

Because u and v are harmonic functions, $f_1 = \frac{\partial u}{\partial x} - i\frac{\partial u}{\partial y}$ and $f_2 = \frac{\partial v}{\partial x} - i\frac{\partial v}{\partial y}$ are analytic functions on Ω. The reason lies on the fact that the real and imaginary parts of f_1 and f_2 satisfy the Cauchy–Riemann equations respectively (see 5103 (a)).

By the assumption of the problem,

$$\frac{\partial u}{\partial x} = \frac{\partial v}{\partial y}, \quad \frac{\partial u}{\partial y} = -\frac{\partial v}{\partial x}$$

when $z \in S \subset \Omega$. Hence

$$f_1 = \frac{\partial u}{\partial x} - i\frac{\partial u}{\partial y} = if_2 = i\left(\frac{\partial v}{\partial x} - i\frac{\partial v}{\partial y}\right)$$

when $z \in S$. Because the subset S has a limit point in Ω, by the uniqueness theorem of analytic functions, we know that $f_1 = if_2$ holds for all $z \in \Omega$. It follows from $f_1 = if_2$ for $z \in \Omega$ that

$$\frac{\partial u}{\partial x} = \frac{\partial v}{\partial y}, \quad \frac{\partial u}{\partial y} = -\frac{\partial v}{\partial x}$$

for $z \in \Omega$, which implies that $u + iv$ is analytic on Ω.

5105

Let $Q = [0,1] \times [0,1] \subset \mathbb{C}$ be the unit square, and let f be holomorphic in a neighborhood of Q. Suppose that

$$f(z+1) - f(z) \text{ is real and } \geq 0 \quad \text{for } z \in [0,i],$$
$$f(z+i) - f(z) \text{ is real and } \geq 0 \quad \text{for } z \in [0,1].$$

Show that f is constant.

(Indiana)

Solution.

Because f is holomorphic on the closed unit square Q, by Cauchy integral theorem, we have

$$\int_{\partial Q} f(z)dz = \int_0^1 f(x)dx + \int_0^1 f(1+yi)i\,dy$$

$$- \int_0^1 f(x+i)dx - \int_0^1 f(yi)i\,dy$$

$$= \int_0^1 (f(x) - f(x+i))dx$$

$$+ i \int_0^1 (f(1+yi) - f(yi))dy$$

$$= 0.$$

As

$$f(x) - f(x+i) \leq 0$$

for $0 \leq x \leq 1$ and

$$f(1+yi) - f(yi) \geq 0$$

for $0 \leq y \leq 1$, by comparing the real and imaginary parts in the above identity, we obtain that $f(x+i) = f(x)$ for $0 \leq x \leq 1$ and $f(1+yi) = f(yi)$ for $0 \leq y \leq 1$. Hence $f(z)$ can be analytically extended to a double-periodic function by

$$f(z) = f(z+1) = f(z+i),$$

which is holomorphic in $\mathbb{C}$ and satisfies

$$|f(z)| \leq \max_{z \in Q}\{|f(z)|\} < +\infty.$$

This shows that $f(z)$ must be a constant.

5106

Let f be continuous on the closure $\bar{S}$ of the unit square

$$S = \{z = x + iy \in \mathbb{C} : 0 < x < 1, 0 < y < 1\},$$

and let f be analytic on S. If $\mathbf{R}f = 0$ on $\bar{S} \cap (\{y = 0\} \cup \{y = 1\})$, and if $\mathbf{I}f = 0$ on $\bar{S} \cap (\{x = 0\} \cup \{x = 1\})$, prove that $f = 0$ everywhere on S.

(Indiana)

Solution.

Define $F(z) = \int_0^z f(z)dz$, where the integral is taken along any curve in $\bar{S}$ which has endpoints 0 and z. Then $F(z)$ is analytic in S and continuous on $\bar{S}$. For $z \in \partial S$, we choose the integral path on ∂S and consider the differential form $f(z)dz$ in the integral. Let $f = u + iv$, then

$$f(z)dz = (u\,dx - v\,dy) + i(v\,dx + u\,dy).$$

On $\bar{S} \cap (\{y = 0\} \cup \{y = 1\})$ we have $u = 0$ and $dy = 0$, and on $\bar{S} \cap (\{x = 0\} \cup \{x = 1\})$ we have $v = 0$ and $dx = 0$. Hence we obtain $\mathrm{Re}(f(z)dz) = 0$ on ∂S which implies $\mathrm{Re}\, F(z) = 0$ when $z \in \partial S$.

Let $G(z) = e^{F(z)}$. Then $G(z)$ is analytic in S and $|G(z)| = 1$ when $z \in \partial S$. Because $G(z)$ has no zeros in S, so $1/G(z)$ is also analytic in S and $|1/G(z)| = 1$ when $z \in \partial S$. Apply the maximum modulus principle to both $G(z)$ and $1/G(z)$, we obtain $|G(z)| \equiv 1$ for $z \in S$, which implies that $G(z)$ is a constant of modulus 1. It follows from $G(z) = e^{F(z)}$ that $F(z)$ is also a constant. Hence $f(z) = F'(z) \equiv 0$.

5107

(a) Find the constant c such that the function

$$f(z) = \frac{1}{z^4 + z^3 + z^2 + z - 4} - \frac{c}{z - 1}$$

is holomorphic in a neighborhood of $z = 1$.

(b) Show that the function f is holomorphic on an open set containing the closed disk $\{z : |z| \le 1\}$.

(Iowa)

Solution.

(a) As

$$\lim_{z \to 1} (z - 1) \cdot \frac{1}{z^4 + z^3 + z^2 + z - 4}$$

$$= \lim_{z \to 1} \frac{1}{(z^4 + z^3 + z^2 + z - 4)'}$$

$$= \lim_{z \to 1} \frac{1}{4z^3 + 3z^2 + 2z + 1}$$

$$= \frac{1}{10},$$

we know that $z = 1$ is a simple pole of $\frac{1}{z^4+z^3+z^2+z-4}$ with residue equal to $\frac{1}{10}$. Hence when $c = \frac{1}{10}$, $f(z)$ is holomorphic in a neighborhood of $z = 1$.

(b) When $|z| \le 1$, we have

$$|z^4 + z^3 + z^2 + z - 4| \ge 4 - |z^4 + z^3 + z^2 + z|$$

$$\ge 4 - |z|^4 - |z|^3 - |z|^2 - |z| \ge 0,$$

and the equalities hold if and only if $z = 1$, which shows that $z = 1$ is the only zero of $z^4+z^3+z^2+z-4$ in $\{z : |z| \le 1\}$. By (a), we obtain that $f(z)$ has no singular point in $\{z : |z| \le 1\}$, hence $f(z)$ is holomorphic on an open set containing $\{z : |z| \le 1\}$.

5108

Let $P(z)$ be a polynomial of degree d with simple roots $z_1, z_2, \ldots, z_d$. A "partial fractions" expression of $\frac{1}{p}$ has the form:

$$\frac{1}{P(z)} = \sum_{n=1}^{d} \frac{c_n}{z - z_n}. \tag{$*$}$$

(a) Give a direct formula for c_n in terms of P.

(b) Show that $\frac{1}{P(z)}$ really has a representation of the form $(*)$.

(c) Give a formula similar to $(*)$ that works when $z_1 = z_2$ but all other roots are simple.

(Courant Inst.)

Solution.

(a)

$$c_n = \lim_{z \to z_n} \frac{z - z_n}{P(z)} = \frac{1}{P'(z_n)},$$

which is the residue of $\frac{1}{P(z)}$ at $z = z_n$.

(b) Let

$$f(z) = \frac{1}{P(z)} - \sum_{n=1}^{d} \frac{c_n}{z - z_n}.$$

Then $f(z)$ is analytic on $\mathbb{C}$ and $\lim_{z \to \infty} f(z) = 0$. By Liouville's Theorem, $f(z)$ is identically equal to zero, hence

$$\frac{1}{P(z)} = \sum_{n=1}^{d} \frac{c_n}{z - z_n}.$$

(c) Denote the Taylor expansion of $P(z)$ at $z = z_1$ $(= z_2)$ by

$$P(z) = \sum_{n=2}^{\infty} a_n (z - z_1)^n.$$

Then the Laurent expansion of $\frac{1}{P(z)}$ at $z = z_1$ is

$$\frac{1}{P(z)} = \frac{c_1'}{(z - z_1)^2} + \frac{c_2'}{z - z_1} + \sum_{n=0}^{\infty} b_n (z - z_1)^n,$$

where

$$c_1' = \frac{1}{a_2} = \frac{2}{P''(z_1)}, \quad c_2' = -\frac{a_3}{a_2^2} = -\frac{2P'''(z_1)}{3P''(z_1)^2}.$$

Hence $\frac{1}{P(z)}$ has the form:

$$\frac{1}{P(z)} = \frac{c_1'}{(z-z_1)^2} + \frac{c_2'}{z-z_1} + \sum_{n=3}^{d} \frac{c_n}{z-z_n}.$$

5109

Suppose f is meromorphic in a neighborhood of $\bar{D}$ ($D = \{|z| < 1\}$) whose only pole is a simple one at $z = a \in D$. If $f(\partial D) \subseteq \mathbb{R}$, show that there is a complex constant A and a real constant B such that

$$f(z) = \frac{Az^2 + Bz + \bar{A}}{(z-a)(1-\bar{a}z)}.$$

(Indiana)

Solution.
Assume that the residue of f at $z = a$ is A_1. Define

$$g(z) = f(z) - \frac{A_1}{z-a} - \frac{\bar{A}_1 z}{1-\bar{a}z}.$$

It is obvious that $g(z)$ is analytic on $\bar{D}$ and $g(\partial D) \subseteq \mathbb{R}$. By the reflection principle, $g(z)$ can be extended to an analytic function on the Riemann sphere $\bar{\mathbb{C}}$, hence $g(z)$ must be a constant. Suppose $g(z) \equiv B_1$, then B_1 is real and

$$f(z) = \frac{A_1}{z-a} + \frac{\bar{A}_1 z}{1-\bar{a}z} + B_1 = \frac{Az^2 + Bz + \bar{A}}{(z-a)(1-\bar{a}z)},$$

where $A = \bar{A}_1 - \bar{a}B_1$, $B = -(\bar{a}A_1 + a\bar{A}_1) + B_1(1 + |a|^2) \in \mathbb{R}$.

5110

Let $K_1, K_2, \ldots, K_n$ be pairwise disjoint disks in $\mathbb{C}$, and let f be an analytic function in $\mathbb{C} \backslash \bigcup_{j=1}^{n} K_j$. Show that there exist functions $f_1, f_2, \ldots, f_n$ such that
  (a) f_j is analytic in $\mathbb{C} \backslash K_j$, and

(b) $f(z) = \sum_{j=1}^{n} f_j(z)$ for $z \in \mathbb{C} \backslash \bigcup_{j=1}^{n} K_j$.

(Indiana)

Solution.

Assume $K_1 = \{z; |z - z_1| \leq r_1\}$. Choose $\varepsilon_1 > 0$ sufficiently small, such that

$$\Sigma_1 = \{z; r_1 < |z - z_1| < r_1 + \varepsilon_1\} \subset \mathbb{C} \backslash \bigcup_{j=1}^{n} K_j.$$

In Σ_1, $f(z)$ has the Laurent expansion

$$f(z) = \sum_{k=-\infty}^{+\infty} a_k^{(1)} (z - z_1)^k.$$

Set

$$f_1(z) = \sum_{k=-\infty}^{0} a_k^{(1)} (z - z_1)^k.$$

$f_1(z)$ is analytic in $\mathbb{C} \backslash K_1$. Because $f(z) - f_1(z)$ has an analytic continuation to K_1, $f(z) - f_1(z)$ is analytic in $\mathbb{C} \backslash \bigcup_{j=2}^{n} K_j$.

Assume $K_2 = \{z; |z - z_2| \leq r_2\}$. Choose $\varepsilon_2 > 0$ sufficiently small, such that $\Sigma_2 = \{z; r_2 < |z - z_2| < r_2 + \varepsilon_2\} \subset \mathbb{C} \backslash \bigcup_{j=2}^{n} K_j$. In Σ_2, $f(z) - f_1(z)$ has the Laurent expansion

$$f(z) - f_1(z) = \sum_{k=-\infty}^{+\infty} a_k^{(2)} (z - z_2)^k.$$

Set $f_2(z) = \sum_{k=-\infty}^{0} a_k^{(2)} (z - z_2)^k$. $f_2(z)$ is analytic in $\mathbb{C} \backslash K_2$. Because $f(z) - f_1(z) - f_2(z)$ has an analytic continuation to K_2, $f(z) - f_1(z) - f_2(z)$ is analytic in $\mathbb{C} \backslash \bigcup_{j=3}^{n} K_j$.

Repeat the above procedure $n - 1$ times, we get a function $f(z) - f_1(z) - f_2(z) - \cdots - f_{n-1}(z)$, which is analytic $\mathbb{C} \backslash K_n$. Set

$$f_n(z) = f(z) - f_1(z) - f_2(z) - \cdots - f_{n-1}(z).$$

Then we have

$$f(z) = \sum_{j=1}^{n} f_j(z),$$

where $f_j(z)$ is analytic in $\mathbb{C} \backslash K_j$, and the above identity holds for $z \in \mathbb{C} \backslash \bigcup_{j=1}^{n} K_j$.

5111

Recall that a divisor D_f of a rational function $f(z)$ on $\mathbb{C}$ is a set $\{p \in \mathbb{C} \cup \{\infty\}\}$, consisting of zeros and poles p of $f(z)$ (including the point ∞), counted with their multiplicities $n_p \in Z$. Let f and g be two rational functions with disjoint divisors. Prove that

$$\prod_{p \in D_f} g(p)^{n_p} = \prod_{q \in D_g} f(q)^{n_q}.$$

(*SUNY, Stony Brook*)

Solution.
Let p_i ($i = 1, 2, \ldots, n$) be all the zeros and poles of $f(z)$ with multiplicities n_{p_i} respectively. It should be noted that p_i is a zero of f when $n_{p_i} > 0$ and a pole of f when $n_{p_i} < 0$. By the property of rational functions, we have $\sum_{i=1}^{n} n_{p_i} = 0$. Similarly, let q_j ($j = 1, 2, \ldots, m$) be all the zeros and poles of $g(z)$ with multiplicities m_{q_j} respectively, then we have $\sum_{j=1}^{m} m_{q_j} = 0$.

First we assume that the point ∞ is not a zero or a pole of f or g, then f and g can be represented by

$$f(z) = A \prod_{i=1}^{n} (z - p_i)^{n_{p_i}}$$

and

$$g(z) = B \prod_{j=1}^{m} (z - q_j)^{m_{q_j}}.$$

Then

$$\prod_{p \in D_f} g(p)^{n_p} = \prod_{i=1}^{n} g(p_i)^{n_{p_i}} = \prod_{i=1}^{n} B^{n_{p_i}} \cdot \prod_{i=1}^{n} \prod_{j=1}^{m} (p_i - q_j)^{n_{p_i} m_{q_j}}$$

$$= \prod_{i=1}^{n} \prod_{j=1}^{m} (p_i - q_j)^{n_{p_i} m_{q_j}},$$

and

$$\prod_{q \in D_g} f(q)^{n_q} = \prod_{j=1}^{m} f(q_j)^{m_{q_j}} = \prod_{j=1}^{m} A^{m_{q_j}} \prod_{j=1}^{m} \prod_{i=1}^{n} (q_j - p_i)^{n_{p_i} m_{q_j}}$$

$$= \prod_{i=1}^{n} \prod_{j=1}^{m} (p_i - q_j)^{n_{p_i} m_{q_j}}.$$

In case the point ∞ is a zero or a pole of f or g, we may assume $p_n = \infty$ without loss of generality. Then

$$f(z) = A \prod_{i=1}^{n-1} (z - p_i)^{n_{p_i}}$$

and

$$g(z) = B \prod_{j=1}^{m} (z - q_j)^{m_{q_j}}.$$

Since $\sum_{j=1}^{m} m_{q_j} = 0$, we may assume that $g(p_n) = g(\infty) = B$. Hence

$$\prod_{p \in D_f} g(p)^{n_p} = \prod_{i=1}^{n-1} g(p_i)^{n_{p_i}} \cdot B^{n_{p_n}}$$

$$= \left(\prod_{i=1}^{n-1} B^{n_{p_i}} \right) \cdot B^{n_{p_n}} \cdot \prod_{i=1}^{n-1} \prod_{j=1}^{m} (p_i - q_j)^{n_{p_i} m_{q_j}}$$

$$= \prod_{i=1}^{n-1} \prod_{j=1}^{m} (p_i - q_j)^{n_{p_i} m_{q_j}},$$

and

$$\prod_{q \in D_g} f(q)^{n_q} = \prod_{j=1}^{m} f(q_j)^{m_{q_j}} = \prod_{j=1}^{m} A^{m_{q_j}} \cdot \prod_{j=1}^{m} \prod_{i=1}^{n-1} (q_j - p_i)^{n_{p_i} m_{q_j}}$$

$$= \prod_{i=1}^{n-1} \prod_{j=1}^{m} (p_i - q_j)^{n_{p_i} m_{q_j}},$$

which completes the proof of the problem.

5112

Let $f(z)$ be the "branch" of $\log z$ defined off the negative real axis so that $f(1) = 0$.

(a) Find the Taylor polynomial of f of degree 2 at $-4 + 3i$, simplifying the coefficients.

(b) Find the radius of convergence R of the Taylor series $T(z)$ of $f(z)$ at $-4 + 3i$.

(c) Identify on a picture any points z where $T(z)$ converges but $T(z) \neq f(z)$, and describe the relationship between f and T at such points. If there are no such points, is this something special to this example, or a general impossibility? Explain and/or give examples.

(*Minnesota*)

Solution.

(a) When z is in the neighborhood of $z_0 = -4 + 3i$, we have

$$f(z) = \log z = \log[(-4 + 3i) + (z + 4 - 3i)]$$

$$= \log(-4 + 3i) + \log\left[1 + \frac{z + 4 - 3i}{-4 + 3i}\right]$$

$$= \log 5 + i\left(\pi - \arcsin\frac{3}{5}\right) + \frac{z + 4 - 3i}{-4 + 3i}$$

$$- \frac{1}{2}\left(\frac{z + 4 - 3i}{-4 + 3i}\right)^2 + \cdots.$$

Hence the Taylor polynomial of f of degree 2 at $-4 + 3i$ is

$$c_0 + c_1(z + 4 - 3i) + c_2(z + 4 - 3i)^2,$$

where

$$c_0 = \log 5 + i\left(\pi - \arcsin\frac{3}{5}\right),$$

$$c_1 = -\frac{4 + 3i}{25},$$

and

$$c_2 = -\frac{25 + 24i}{1250}.$$

(b) Denote the Taylor series of $f(z)$ at $-4 + 3i$ by $T(z)$. Because $\log z$ has only $z = 0$ and $z = \infty$ as its branch points, and has no other singular point, the radius of convergence R of $T(z)$ is equal to the distance between $z = -4 + 3i$ and $z = 0$. Hence $R = 5$.

(c) Denote the shaded domain shown in Fig. 5.1 by Ω. When $z \in \Omega = \{z : |z + 4 - 3i| < 5, \operatorname{Im} z < 0\}$, $T(z) \neq f(z)$. It is because $T(z)$ in Ω is the continuation of $\log z$ at $-4 + 3i$ in the disk $\{z : |z + 4 - 3i| < 5\}$, while $f(z)$ in Ω is the continuation of $\log z$ at $-4 + 3i$ in the slit plane $\mathbb{C}\backslash(-\infty, 0]$. Hence the difference is $2\pi i$, i.e., $T(z) = f(z) + 2\pi i$.

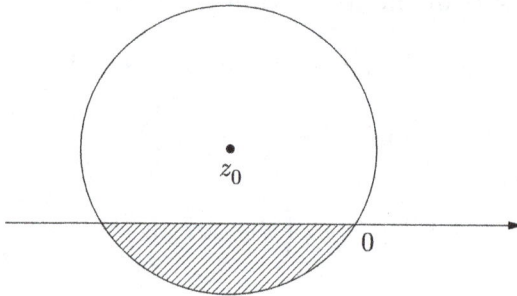

Fig. 5.1.

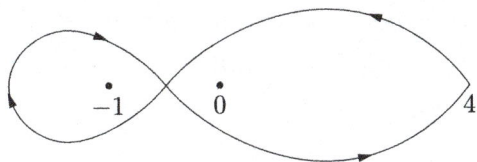

Fig. 5.2.

5113

Let f be the analytic function defined in the disk $\Delta = \{z : |z-4| < 4\}$ so that $f(z) = z^{\frac{1}{3}}(z+1)^{\frac{1}{2}}$ in Δ and $f(x)$ is positive for $0 < x < 8$. An analytic function g in Δ is obtained from f by analytic continuation along the path starting and ending at $z = 4$ (see Fig. 5.2). Express g in terms of f.

(Indiana)

Solution.

Denote the closed path in Fig. 5.2 by Γ, and denote the change of $\phi(z)$ when z goes along Γ from the start point to the end point by $\Delta_\Gamma \phi(z)$. Then

$$g(z) = |g(z)|e^{i \arg g(z)} = |f(z)|e^{i(\arg f(z)+\Delta_\Gamma \arg f(z))}.$$

We have

$$\Delta_\Gamma \arg f(z) = \frac{1}{3}\Delta_\Gamma \arg z + \frac{1}{2}\Delta_\Gamma \arg(z+1) = \frac{1}{3}(2\pi) + \frac{1}{3}(-2\pi)$$

$$= -\frac{\pi}{3}.$$

Hence

$$g(z) = e^{-\frac{\pi}{3}i}f(z).$$

5114

Define

$$f(z) = \frac{e^{\sqrt{z}} - e^{-\sqrt{z}}}{\sin \sqrt{z}}.$$

(a) Where is f single-valued and analytic?

(b) Classify the singularities of f.

(c) Evaluate $\int_{|z|=25} f(z)dz$.

(Indiana)

Solution.

(a) It is known that $z = 0$ and $z = \infty$ are the branch points of function $\sqrt{z}$. Let $\Gamma = \{z : |z| = r\}$, and when z goes along Γ once in the counterclockwise sense, $\sqrt{z}$ is changed to $-\sqrt{z}$, while $f(z)$ is changed to

$$\frac{e^{-\sqrt{z}} - e^{\sqrt{z}}}{\sin(-\sqrt{z})} = \frac{e^{\sqrt{z}} - e^{-\sqrt{z}}}{\sin\sqrt{z}}$$

which is still $f(z)$. Hence $z = 0$ and $z = \infty$ are no longer the branch points of $f(z)$.

When z is in the small neighborhood of $z = 0$, $f(z)$ can be represented by

$$f(z) = \frac{e^{\sqrt{z}} - e^{-\sqrt{z}}}{\sin\sqrt{z}} = \frac{\sum_{n=0}^{\infty}\frac{1}{n!}z^{\frac{n}{2}} - \sum_{n=0}^{\infty}\frac{(-1)^n}{n!}z^{\frac{n}{2}}}{\sum_{n=0}^{\infty}\frac{(-1)^n}{(2n+1)!}z^{n+\frac{1}{2}}}$$

$$= \frac{2\sum_{n=0}^{\infty}\frac{1}{(2n+1)!}z^n}{\sum_{n=0}^{\infty}\frac{(-1)^n}{(2n+1)!}z^n},$$

which implies that $z = 0$ is a removable singular point of $f(z)$. It is obvious that $z = n^2\pi^2$ $(n = 1, 2, \ldots)$ are poles of $f(z)$. Hence $f(z)$ is single-valued and analytic in $\mathbb{C}\backslash\{z = n^2\pi^2 : n = 1, 2, \ldots\}$.

(b) We have

$$\lim_{z \to n^2\pi^2} \frac{e^{\sqrt{z}} - e^{-\sqrt{z}}}{(\sin\sqrt{z})'} = \frac{e^{n\pi} - e^{-n\pi}}{\frac{\cos(n\pi)}{2n\pi}} = 2n\pi(-1)^n(e^{n\pi} - e^{-n\pi}),$$

which shows that $z = n^2\pi^2$ are simple poles of $f(z)$ with residues

$$2n\pi(-1)^n(e^{n\pi} - e^{-n\pi}).$$

As to $z = \infty$, it is the limit point of the poles of $f(z)$, and hence is a non-isolated singular point of $f(z)$.

(c) $f(z)$ has only one pole $z = \pi^2$ in the disk $\{z : |z| < 25\}$. Hence

$$\int_{|z|<25} f(z)dz = 2\pi i \operatorname{Res}(f, \pi^2) = -4\pi^2 i(e^\pi - e^{-\pi}).$$

5115

Let Ω be the plane with the segment $\{-1 \le x \le 1, y = 0\}$ deleted. For which of the multi-valued functions

(a) $f(z) = \frac{z}{\sqrt{1-z^2}}$,

(b) $g(z) = \frac{1}{\sqrt{1-z^2}}$,

can we choose single-valued branches which are holomorphic in Ω. Which of these branches are (is) the derivative of a single-valued holomorphic function in Ω. Why?

(*Indiana–Purdue*)

Solution.

Let Γ be an arbitrary simple closed curve in Ω, and denote by $\Delta_\Gamma \phi(z)$ the change of $\phi(z)$ when z goes continuously along Γ counterclockwise once. It is known that f and g can be represented by

$$f(z) = \frac{z}{\sqrt{1-z^2}} = e^{\{\log z - \frac{1}{2}\log(1+z) - \frac{1}{2}\log(1-z)\}}$$

$$= \left|\frac{z}{\sqrt{1-z^2}}\right| e^{i[\arg z - \frac{1}{2}\arg(1+z) - \frac{1}{2}\arg(1-z)]}$$

and

$$g(z) = \frac{1}{\sqrt{1-z^2}} = e^{\{-\frac{1}{2}\log(1+z) - \frac{1}{2}\log(1-z)\}}$$

$$= \left|\frac{1}{\sqrt{1-z^2}}\right| e^{i[-\frac{1}{2}\arg(1+z) - \frac{1}{2}\arg(1-z)]}.$$

Because

$$\Delta_\Gamma \left[\arg z - \frac{1}{2} \arg(1 + z) - \frac{1}{2} \arg(1 - z) \right] = 0$$

and

$$\Delta_\Gamma \left[-\frac{1}{2} \arg(1 + z) - \frac{1}{2} \arg(1 - z) \right]$$
$$= \begin{cases} 0, & \{-1 \le x \le 1, y = 0\} \text{ not inside } \Gamma, \\ -2\pi, & \{-1 \le x \le 1, y = 0\} \text{ inside } \Gamma, \end{cases}$$

we have $\Delta_\Gamma f(z) = 0$ and $\Delta_\Gamma g(z) = 0$. Hence both $f(z)$ and $g(z)$ have single-valued branches which are holomorphic in Ω, and each of f and g has two single-valued branches.

In order to know which of f and g has a single-valued primitive in Ω, we consider the integrals $\int_\Gamma f(z)dz$ and $\int_\Gamma g(z)dz$. If the segment $\{-1 \le x \le 1, y = 0\}$ is not inside Γ, it is obvious that $\int_\Gamma f(z)dz = 0$ and $\int_\Gamma g(z)dz = 0$. If the segment $\{-1 \le x \le 1, y = 0\}$ is inside Γ, we consider the Laurent expansion of f and g about $z = \infty$:

$$f(z) = \pm i \left(1 - \frac{1}{z^2} \right)^{-\frac{1}{2}} = \pm i \left(1 + \frac{a_2}{z^2} + \frac{a_4}{z^4} + \cdots \right),$$

$$g(z) = \pm \frac{i}{z} \left(1 - \frac{1}{z^2} \right)^{-\frac{1}{2}} = \pm i \left(\frac{1}{z} + \frac{b_3}{z^3} + \frac{b_5}{z^5} + \cdots \right).$$

It follows that $\int_\Gamma f(z)dz = 0$ and $\int_\Gamma g(z)dz = \pm 2\pi$. Hence we obtain that both of the single-valued branches of f are the derivatives of single-valued holomorphic functions in Ω, and the primitives are $\int_{z_0}^z f(z)dz + c$, where the integral is taken along any curve connecting z_0 and z in Ω. But neither of the branches of g is the derivative of a single-valued holomorphic function in Ω.

5116

(a) Let $D \subset \mathbb{C}$ be the complement of the simply connected closed set $\{e^{\theta + i\theta} \mid \theta \in \mathbb{R}\} \cup \{0\}$. Let log be a branch of the logarithm on D such that $\log e = 1$. Find $\log e^{15}$. Justify your answer.

(b) Let γ denote the unit circle, oriented counterclockwise. By lifting the integration to an appropriate covering space, give a precise meaning to the integral $\int_\gamma (\log z)^2 dz$ and find all possible values which can be assigned to it.

(Harvard)

Solution.

(a) The set $\{e^{\theta + i\theta} \mid \theta \in \mathbb{R}\} \cup \{0\}$ is a spiral which intersects the positive real axis at $\{e^{2n\pi} : n = 0, \pm 1, \pm 2, \ldots\}$. The single-valued branch of $\log z$ is defined by $\log e = 1$. Hence $\log e^{15} = \log e + \Delta_\Gamma \log z$, where Γ is a continuous curve connecting $z = e$ and $z = e^{15}$ in D and $\Delta_\Gamma \log z$ is the change of $\log z$ when z goes continuously along Γ from $z = e$ to $z = e^{15}$. It follows that $\Delta_\Gamma \log z = \Delta_\Gamma \log |z| + i\Delta_\Gamma \arg z$, and $\Delta_\Gamma \log |z| = 15 - 1 = 14$. Because $e \in (e^0, e^{2\pi}), e^{15} \in (e^{4\pi}, e^{6\pi})$, we know that when Γ connects e and e^{15} in D, $\Delta_\Gamma \arg z$ must be 4π. Hence $\log e^{15} = 1 + (14 + 4\pi i) = 15 + 4\pi i$.

(b) Define the lift mapping by $w = \log z$ which lifts the unit circle γ one-to-one onto a segment with length 2π on the imaginary axis of w-plane. Because both the starting point of γ and the single-valued branch of $\log z$ on γ can be arbitrarily chosen, the segment on w-plane can be denoted by $[it, i(t + 2\pi))$, where t can be any real number. Hence we have

$$\int_\gamma (\log z)^2 dz = \int_{it}^{i(t+2\pi)} w^2 e^w dw$$

$$= (w^2 e^w)|_{it}^{i(t+2\pi)} - 2 \int_{it}^{i(t+2\pi)} w e^w dw$$

$$= e^{it}(-4\pi t - 4\pi^2) - (2we^w)|_{it}^{i(t+2\pi)}$$

$$+ 2\int_{it}^{i(t+2\pi)} e^w dw$$

$$= -4\pi(t + \pi + i)e^{it} = 4\pi(t + \pi + i)e^{i(t+\pi)},$$

which implies that the set of values being assigned to the integral $\int_\gamma (\log z)^2 dz$ is a spiral $\{4\pi(s + i)e^{is} : s \in \mathbb{R}\}$.

5117

Find the most general harmonic function of the form $f(|z|)$, $z \in \mathbb{C}\backslash 0$. Which of these $f(|z|)$ have a single valued harmonic conjugate?

(*Indiana*)

Solution.

Because $f(|z|)$ is harmonic, we have reason to assume that the function f (with real variable t) has continuous derivatives $f'(t)$ and $f''(t)$. Note that the Laplacian

$$\Delta = \frac{\partial^2}{\partial x^2} + \frac{\partial^2}{\partial y^2} = 4\frac{\partial^2}{\partial z \partial \bar{z}},$$

and

$$\frac{\partial}{\partial z}f(|z|) = \frac{\partial}{\partial z}f(\sqrt{z\bar{z}}) = \frac{1}{2}f'(|z|) \cdot \sqrt{\frac{\bar{z}}{z}},$$

$$\frac{\partial^2}{\partial z \partial \bar{z}}f(|z|) = \frac{1}{4}f''(|z|) + \frac{1}{4|z|}f'(|z|),$$

we obtian

$$f''(t) + \frac{f'(t)}{t} = 0,$$

where $t = |z|$. This differential equation is easy to solve, and the solution is $f(t) = \alpha \log t + \beta$, where α, β are two real constants.

Hence the most general harmonic function of the form $f(|z|)$ in $\mathbb{C}\backslash 0$ is $\alpha \log|z| + \beta$.

Since $\log|z|$ has no single-valued harmonic conjugate in $\mathbb{C}\backslash 0$, we know that when $f(|z|)$ has a single-valued harmonic conjugate in $\mathbb{C}\backslash 0$, it must be a constant.

5118

Consider the regular pentagram centered at the origin in the complex plane. Let u be the harmonic function in the interior of the pentagram which has boundary values 1 on the two segments shown and 0 on the rest of the boundary. What is the value of u at the origin? Justify your claim.

(*Stanford*)

Solution.
Denote the interior domain of the pentagram shown in Fig. 5.3 by D, and the ten segments of the boundary by $l_1, l_2, \ldots, l_{10}$, put in order of counterclockwise.

Then denote the harmonic function on D with boundary values 1 on l_k and 0 on the rest of the boundary by $u_k(z)$,

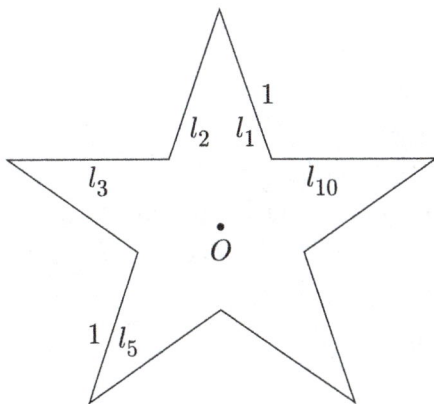

Fig. 5.3.

$k = 1, 2, \ldots, 10$. By the symmetry of domain D, we have

$$u_2(z) = u_1(-\bar{z}),$$

$$u_3(z) = u_1(e^{-\frac{2\pi}{5}i}z),$$

$$u_4(z) = u_2(e^{-\frac{2\pi}{5}i}z),$$

$$u_5(z) = u_1(e^{-\frac{4\pi}{5}i}z),$$

$$\cdots$$

$$u_{10}(z) = u_2(e^{-\frac{8\pi}{5}i}z).$$

It follows from

$$\sum_{k=1}^{10} u_k(z) \equiv 1$$

and $u_1(0) = u_2(0) = \cdots = u_{10}(0)$ that $u_k(0) = \frac{1}{10}$ for $k = 1, 2, \ldots, 10$. Hence

$$u(0) = u_1(0) + u_5(0) = \frac{1}{5}.$$

5119

Suppose G is a region in $\mathbb{C}$, $[0,1] \subset G$, and $h : G \to \mathbb{R}$ is continuous. $h|_{G\setminus[0,1]}$ is harmonic, does this implies that h is harmonic on G?

(*Iowa*)

Solution.

The answer is No.

A counterexample is $h(z) = \operatorname{Re}\sqrt{z(z-1)}$, where the single-valued branch of $\sqrt{z(z-1)}$ is chosen by $\sqrt{z(z-1)}|_{z=2} = \sqrt{2}$. Since $\sqrt{z(z-1)}$ is analytic in $\mathbb{C}\setminus[0,1]$, $h(z)$ is harmonic there. When $0 \le x \le 1$,

$$\lim_{\substack{z=x+yi\to x \\ y>0}} \sqrt{z(z-1)} = \sqrt{x(1-x)}i,$$

$$\lim_{\substack{z=x+yi\to x \\ y<0}} \sqrt{z(z-1)} = -\sqrt{x(1-x)}i.$$

Hence $h(z) = 0$ when $z = x$, $0 \le x \le 1$, and $h(z)$ is continuous on $\mathbb{C}$. But $h(z)$ is not harmonic on $\mathbb{C}$, because $z = 0$ and $z = 1$ are branch points of $\sqrt{z(z-1)}$.

Remark. If the problem is changed to $h : G \to \mathbb{C}$ is continuous and $h|_{G \setminus [0,1]}$ is holomorphic, then h must be holomorphic on G.

5120

Let γ be an arc of the unit circle. Suppose that u and v are harmonic in $D = \{z : |z| < 1\}$ and continuously differentiable on $D \cup \gamma$. If the boundary values satisfy $u = v$ on γ and the radial derivatives satisfy $\frac{\partial u}{\partial r} = \frac{\partial v}{\partial r}$ on γ, prove that $u = v$ in D.

(*Indiana*)

Solution.
Let u^* be a conjugate harmonic function of u in D and v^* be a conjugate harmonic function of v in D. We know that a variation of Cauchy–Riemann equations for $f = u + iu^*$ and $g = v + iv^*$ are

$$r \frac{\partial u}{\partial r} = \frac{\partial u^*}{\partial \theta}, \quad \frac{\partial u}{\partial \theta} = -r \frac{\partial u^*}{\partial r}$$

and

$$r \frac{\partial v}{\partial r} = \frac{\partial v^*}{\partial \theta}, \quad \frac{\partial v}{\partial \theta} = -r \frac{\partial v^*}{\partial r}.$$

It follows from the continuous differentiability of u and v on $D \cup \gamma$ that u^* and v^* can be continuously extended to $D \cup \gamma$ and then are also continuously differentiable on $D \cup \gamma$. Let z_0 be a fixed point on γ, and for $z \in \gamma$ denote the subarc of γ from z_0 to z by γ_z. Without loss of generality, we may assume that $u^*(z_0) = v^*(z_0) = 0$. Then for $z \in \gamma$,

$$u^*(z) = \int_{\gamma_z} \frac{\partial u^*}{\partial r} dr + \frac{\partial u^*}{\partial \theta} d\theta = \int_{\gamma_z} \frac{\partial u^*}{\partial \theta} d\theta = \int_{\gamma_z} \frac{\partial u}{\partial r} d\theta$$

$$= \int_{\gamma_z} \frac{\partial v}{\partial r} d\theta = v^*(z).$$

Hence we obtain two functions $f = u + iu^*$ and $g = v + iv^*$ which are analytic in D and continuous on $D \cup \gamma$, such that $f = g$ on γ. Let $F = f - g$. Then by the reflection principle, F can be analytically extended to an analytic function on $D \cup \gamma \cup D^*$, where $D^* = \{z : |z| > 1\}$. Since $F = 0$ on γ, we obtain $F \equiv 0$ on $D \cup \gamma \cup D^*$, which implies $u = v$ in D.

<h3 style="text-align:center">5121</h3>

Use conformal mapping to find a harmonic function $U(z)$ defined on the unit disc $\{|z| < 1\}$ such that

$$\lim_{r \to 1-} U(re^{i\theta}) = \begin{cases} +1, & \text{for} \quad 0 < \theta < \pi, \\ -1, & \text{for} \quad \pi < \theta < 2\pi. \end{cases}$$

Give the correct determination of any multiple-valued functions appearing in your answer.

(Courant Inst.)

Solution.
It is easy to know that $w = -i\frac{z+1}{z-1}$ is a conformal mapping of the unit disc $D = \{z : |z| < 1\}$ onto the upper half plane $H = \{w : \operatorname{Im} w > 0\}$. The boundary correspondence is that the negative real axis $\{w : -\infty < w < 0\}$ corresponds to the arc $\Gamma_1 = \{z = e^{i\theta} : 0 < \theta < \pi\}$ and the positive real axis $\{w : 0 < w < +\infty\}$ corresponds to the arc $\Gamma_2 = \{z = e^{i\theta} : \pi < \theta < 2\pi\}$.
It is well known that $u(w) = \frac{2}{\pi} \arg w - 1$ is a harmonic function in H and assume $+1$ on the negative real axis and -1 on the positive real axis. Hence

$$U(z) = u\left(-i\frac{z+1}{z-1}\right) = \frac{2}{\pi} \arg\left(\frac{z+1}{z-1}\right) - 2,$$

where the single-valued branch of $\arg(\frac{z+1}{z-1})$ is defined by $\arg(\frac{z+1}{z-1})|_{z=0} = \pi$, is a harmonic function in $D = \{z : |z| < 1\}$ with the boundary values $+1$ on Γ_1 and -1 on Γ_2.

Remark. This problem can be solved directly from the Poisson formula as follows:

$$
\begin{aligned}
U(z) &= \frac{1}{2\pi} \int_{|\zeta|=1} U(\zeta) \mathrm{Re}\left(\frac{\zeta+z}{\zeta-z}\right) \frac{d\zeta}{i\zeta} \\
&= \frac{1}{2\pi} \int_{\Gamma_1} \mathrm{Re}\left(\frac{\zeta+z}{\zeta-z}\right) \frac{d\zeta}{i\zeta} - \frac{1}{2\pi} \int_{\Gamma_2} \mathrm{Re}\left(\frac{\zeta+z}{\zeta-z}\right) \frac{d\zeta}{i\zeta} \\
&= \frac{1}{\pi} \int_{\Gamma_1} \mathrm{Re}\left(\frac{\zeta+z}{\zeta-z}\right) \frac{d\zeta}{i\zeta} - 1 \\
&= \frac{1}{\pi}\mathrm{Re}\left\{\int_{\Gamma_1} \frac{\zeta+z}{i\zeta(\zeta-z)} d\zeta\right\} - 1 \\
&= \frac{1}{\pi}\mathrm{Im}\left\{\int_{\Gamma_1} \left(\frac{2}{\zeta-z} - \frac{1}{\zeta}\right) d\zeta\right\} - 1 \\
&= \frac{1}{\pi}\mathrm{Im}\left\{\int_{\Gamma_1} d(2\log(\zeta-z) - \log\zeta)\right\} - 1 \\
&= \frac{1}{\pi}\Delta_{\Gamma_1}\{2\arg(\zeta-z) - \arg\zeta\} - 1 \\
&= \frac{2}{\pi}\arg\frac{z+1}{z-1} - 2.
\end{aligned}
$$

5122

Determine all continuous functions on $\{z \in \mathbb{C} : 0 < |z| \leq 1\}$ which are harmonic on $\{z : 0 < |z| < 1\}$ and which are identically 0 on $\{z \in \mathbb{C} : |z| = 1\}$.

(*Minnesota*)

Solution.

Suppose $u(z)$ is a continuous function on $\{0 < |z| \leq 1\}$ which is harmonic on $\{0 < |z| < 1\}$ and identically zero on $\{|z| = 1\}$. Let $^*du = -u_y\,dx + u_x\,dy$ and $A = \int_{|z|=r} {}^*du$, where A is a real number not necessarily zero. Denote $v(z) = \int^z {}^*du$, then $v(z)$ is the conjugate harmonic function of $u(z)$, but may be not single-valued. Define

$$f(z) = (u(z) + iv(z)) - \frac{A}{2\pi} \log z,$$

then $f(z)$ is a single-valued analytic function on $\{0 < |z| < 1\}$ and $\operatorname{Re} f(z)$ is identically zero on $\{|z| = 1\}$.

Let $f(z) = \sum_{n=-\infty}^{\infty} a_n z^n$ be the Laurent expansion of $f(z)$ on $\{0 < |z| < 1\}$, then $\overline{\lim}_{n\to\infty} \sqrt[n]{|a_{-n}|} = 0$ and $\overline{\lim}_{n\to\infty} \sqrt[n]{|a_n|} \leq 1$. Define $g(z) = \sum_{n=-\infty}^{\infty} b_n z^n$, satisfying $b_{-n} = -\bar{b}_n$ for $n = 0, 1, 2, \ldots$, and $b_{-n} = a_{-n}$ for $n = 1, 2, \ldots$. Then $g(z)$ is an analytic function on $\{0 < |z| < +\infty\}$. When $|z| = 1$, it follows from $\operatorname{Re} b_0 = 0$ and

$$\operatorname{Re} \sum_{n=-\infty}^{-1} b_n z^n = \operatorname{Re} \sum_{n=1}^{\infty} b_{-n} z^{-n}$$

$$= \operatorname{Re} \sum_{n=1}^{\infty} -\bar{b}_n z^{-n} = -\operatorname{Re} \sum_{n=1}^{\infty} b_n z^n$$

that $\operatorname{Re} g(z) = 0$. Then $f(z) - g(z) = \sum_{n=0}^{\infty} c_n z^n$ is an analytic function in $\{|z| < 1\}$ and $\operatorname{Re}(f(z) - g(z))$ is identically zero on $\{|z| = 1\}$. Consider $F(z) = e^{f(z)-g(z)}$ which is analytic and does not assume zero in $\{|z| < 1\}$, and $|F(z)| = 1$ on $\{|z| = 1\}$, by the maximum and minimum modulus principles, we have $F(z) \equiv e^{i\alpha}$, hence $f(z) = g(z) + i\alpha$.

From the above discussion, we finally obtain

$$u(z) = \operatorname{Re} \sum_{n=-\infty}^{+\infty} b_n z^n + \frac{A}{2\pi} \log |z|,$$

where $b_{-n} = -\bar{b}_n$ and $\overline{\lim}_{n\to\infty} \sqrt[n]{|b_n|} = 0$.

5123

(a) Let $f(z)$ be a holomorphic function in the disc $|z| \leq r$ whose zeros in this disc are given by $a_1, a_2, \ldots, a_n$ counted with multiplicity. Suppose further that $|a_j| < r$ for all $j = 1, 2, \ldots, n$, and $|f(0)| = 1$. Jensen's formula states that

$$\frac{1}{2\pi} \int_0^{2\pi} \log |f(re^{i\theta})| d\theta = \sum_{j=1}^n \log \left(\frac{r}{|a_j|} \right).$$

Prove this.

(b) With the hypotheses and notations of (a), let $n(t)$ be the number of a_j $(j = 1, 2, \ldots, n)$ such that $|a_j| \le t$. Using Jensen's formula, show that

$$\int_0^r n(t) \frac{dt}{t} = \frac{1}{2\pi} \int_0^{2\pi} \log |f(re^{i\theta})| d\theta.$$

(c) For $r < R$ deduce an estimate on $n(r)$ in terms of $\max_{0 \le \theta \le 2\pi} \log |f(Re^{i\theta})|$.

(d) What can be said about the zeros of bounded holomorphic functions in the unit disc?

(*Harvard*)

Solution.

(a) Let

$$F(z) = f(z) \prod_{j=1}^n \frac{r^2 - \bar{a}_j z}{r(z - a_j)},$$

then $F(z)$ is holomorphic and has no zero in the disc $\{|z| \le r\}$, which implies that $\log |F(z)|$ is harmonic in $\{|z| \le r\}$. By the mean value theorem of harmonic functions,

$$\log |F(0)| = \frac{1}{2\pi} \int_0^{2\pi} \log |F(re^{i\theta})| d\theta.$$

Noting that

$$|F(0)| = |f(0)| \prod_{j=1}^n \frac{r}{|a_j|} = \prod_{j=1}^n \frac{r}{|a_j|}$$

and

$$|F(re^{i\theta})| = |f(re^{i\theta})|,$$

we obtain that

$$\sum_{j=1}^{n} \log\left(\frac{r}{|a_j|}\right) = \frac{1}{2\pi} \int_0^{2\pi} \log|f(re^{i\theta})|\,d\theta.$$

(b) It is obvious that $\log\frac{r}{|a_j|} = \int_{|a_j|}^r \frac{dt}{t}$. By the definition of the function $n(t)$ we have

$$\sum_{j=1}^{n} \log\frac{r}{|a_j|} = \sum_{j=1}^{n} \int_{|a_j|}^r \frac{dt}{t} = \int_0^r n(t)\frac{dt}{t},$$

which shows that the identity holds.

(c) Apply the identity in (b), we have

$$\frac{1}{2\pi} \int_0^{2\pi} \log|f(Re^{i\theta})|\,d\theta = \int_0^R n(t)\frac{dt}{t}$$

$$\geq \int_r^R n(t)\frac{dt}{t} \geq n(r)\log\frac{R}{r}.$$

Denote $\max_{0\leq\theta\leq 2\pi} \log|f(Re^{i\theta})|$ by $M(R)$, we obtain

$$n(r) \leq M(R)/\log\frac{R}{r}.$$

(d) Let $f(z)$ be a bounded holomorphic function in $\{z : |z| < 1\}$. We know that $f(z)$ can have countably many zeros. Suppose $z = 0$ is a zero of $f(z)$ of multiplicity $m \geq 0$ with $\frac{f^{(m)}(0)}{m!} = \alpha$, and let the other zeros be ordered by $0 < |a_1| \leq |a_2| \leq \cdots$. Obviously $|a_n| \to 1$. Apply Jensen's formula in (a) to $F(z) = \frac{f(z)}{\alpha z^m}$ with $0 < r < 1$ such that there is no zero of f on $\{|z| = r\}$, we have

$$\frac{1}{2\pi} \int_0^{2\pi} \log|f(re^{i\theta})|\,d\theta = \sum_{|a_j|<r} \log\left(\frac{r}{|a_j|}\right) + \log(|\alpha|r^m).$$

Since $f(z)$ is bounded, we assume

$$\frac{1}{2\pi}\int_0^{2\pi}\log|f(re^{i\theta})|d\theta \le M.$$

For any n, we can choose r such that $r > |a_n|$, and hence

$$\sum_{j=1}^n \log\left(\frac{r}{|a_j|}\right) \le \sum_{|a_j|<r}\log\left(\frac{r}{|a_j|}\right) \le M - \log(|\alpha|r^m).$$

Let $r \to 1$, we obtain

$$\prod_{j=1}^n |a_j| \ge |\alpha|e^{-M} > 0,$$

which implies that the series $\sum_{j=1}^\infty (1 - |a_j|)$ is convergent.

5124

Let $B(z) = \prod_{k=1}^n \frac{a_k-z}{1-\overline{a_k}z}$, where $a_1, a_2, \ldots, a_n$ are distinct points in $D\backslash\{0\}$. Show that

$$B(z) = \prod_{k=1}^n \frac{1}{a_k} + \sum_{k=1}^n \frac{1}{\overline{a_k}\, B'(a_k)(1 - \overline{a_k}z)}.$$

(SUNY, Albany)

Solution.
Let $w = \frac{1}{z}$, then

$$B(z) = \prod_{k=1}^n \frac{a_k - z}{1 - \overline{a_k}z} = \prod_{k=1}^n \frac{a_k\overline{w} - 1}{\overline{w} - \overline{a_k}} = \frac{1}{\overline{B(w)}},$$

and when $z \to \frac{1}{\overline{a_k}}$, we have $w \to a_k$.

By partial fraction expansion,

$$B(z) = \prod_{k=1}^n \frac{a_k - z}{1 - \overline{a_k}z} = c_0 + \sum_{k=1}^n \frac{c_k}{1 - \overline{a_k}z} = c_0 + \sum_{k=1}^n \frac{-\frac{c_k}{\overline{a_k}}}{z - \frac{1}{\overline{a_k}}},$$

where $c_0 = \prod_{k=1}^{n} \frac{1}{a_k}$, and

$$-\frac{c_k}{a_k} = \operatorname{Res}\left\{B(z), \frac{1}{a_k}\right\} = \lim_{z \to \frac{1}{a_k}}\left\{\left(z - \frac{1}{a_k}\right)B(z)\right\}$$

$$= \lim_{w \to a_k}\left\{\left(\frac{1}{\bar{w}} - \frac{1}{a_k}\right)\cdot\frac{1}{\overline{B(w)}}\right\}$$

$$= \lim_{w \to a_k}\overline{\left\{\frac{-(w - a_k)}{a_k w B(w)}\right\}} = -\frac{1}{\overline{a_k^2 B'(a_k)}}.$$

Hence

$$c_k = \frac{1}{\overline{a_k B'(a_k)}}.$$

5125

Does there exist an analytic function in $|z| < 1$ such that $0 < |f(\frac{1}{n})| < e^{-n}$ for $n = 2, 3, 4, \ldots$? Justify your answer.

(*SUNY, Albany*)

Solution.
There exists no analytic function $f(z)$ satisfying the given conditions. We give proof by contradiction. Suppose $f(z) = \sum_{n=0}^{\infty} a_n z^n$, it is obvious that $a_0 = 0$. Assume the first non-zero coefficient among $\{a_1, a_2, a_3, \ldots\}$ is a_k, then

$$f\left(\frac{1}{n}\right) = \frac{a_k}{n^k} + \frac{a_{k+1}}{n^{k+1}} + \cdots = \frac{a_k}{n^k} + o\left(\frac{1}{n^k}\right).$$

It follows from $|f(\frac{1}{n})| < e^{-n}$ that when n is sufficiently large,

$$|a_k| < \frac{2n^k}{e^n} \to 0 \ (n \to \infty),$$

hence $a_n = 0$ for $n = 0, 1, 2, \ldots$. But it violates the fact that $|f(\frac{1}{n})| > 0$. So there exists no such analytic function $f(z)$.

5126

Let $D = \{z \in \mathbb{C} : |z| < 1\}$. Find all holomorphic functions $f : D \to \mathbb{C}$ such that $f\left(\frac{1}{n} + ie^{-n}\right)$ is real for all integers $n \geq 2$.

(*UC, Berkerly*)

Solution.

It is obvious that $f(0)$ is real. Assume $f(z)$ is not a constant, then

$$f(z) = f(0) + a_k x^k + a_{k+1} x^{k+1} + \cdots,$$

where $k \geq 1$. First we have

$$f\left(\frac{1}{n} + ie^{-n}\right) = f(0) + a_k \left(\frac{1}{n} + ie^{-n}\right)^k + a_{k+1} \left(\frac{1}{n} + ie^{-n}\right)^{k+1}$$

$$+ \cdots = f(0) + \frac{a_k}{n^k}\left(1 + O\left(\frac{1}{n}\right)\right).$$

It follows from the realness of $f\left(\frac{1}{n} + ie^{-n}\right)$ and $f(0)$ that a_k must be real. Then we have

$$f\left(\frac{1}{n} + ie^{-n}\right) = f(0) + a_k \left(\frac{1}{n} + ie^{-n}\right)^k + a_{k+1} \left(\frac{1}{n} + ie^{-n}\right)^{k+1}$$

$$+ \cdots = f(0) + \frac{a_k}{n^k} + \frac{a_{k+1}}{n^{k+1}}\left(1 + O\left(\frac{1}{n}\right)\right),$$

and it follows from the realness of $f\left(\frac{1}{n} + ie^{-n}\right)$, $f(0)$ and a_k that a_{k+1} must be real. Continue the above steps, we obtain that for every $k \geq 1$, a_k is real. Hence we have $f(z) = \overline{f(\bar{z})}$, which means that the image of the real axis under the mapping of $f(z)$ lies in the real axis, and the pre-image of the real axis under the mapping of $f(z)$ consists of k curves meeting at the origin with the angle of $\frac{\pi}{k}$. So we cannot claim that $f(z)$ is real when z is real and $f\left(\frac{1}{n} + ie^{-n}\right)$ is real at the same time. The only possibility is that $a_k = 0$ for every positive integer k, which implies that $f(z)$ is a real constant.

Section 2

Geometry of Analytic Functions

5201

Find a one-to-one holomorphic map from the unit disk $\{|z| < 1\}$ onto the slit disk $\{|w| < 1\} - \{[0, 1)\}$.

(*SUNY, Stony Brook*)

Solution.

We construct the map by the following steps:

$$z_1 = \phi_1(z) = i\frac{z+1}{z-1} : \{z : |z| < 1\} \to \{z_1 : \operatorname{Im} z_1 < 0\};$$

$$z_2 = \phi_2(z_1) = \sqrt{z_1^2 - 1} + z_1 \left(\sqrt{z_1^2 - 1}\Big|_{z_1 = -i} = \sqrt{2}i \right) :$$

$$\{z_1 : \operatorname{Im} z_1 < 0\} \to \{z_2 : |z_2| < 1 \text{ and } \operatorname{Im} z_2 > 0\};$$

$$w = \phi_3(z_2) = z_2^2 : \{z_2 : |z_2| < 1 \text{ and } \operatorname{Im} z_2 > 0\} \to$$

$$\{w : |w| < 1\} \backslash \{w : \operatorname{Im} w = 0, 0 \le \operatorname{Re} w < 1\}.$$

Then $w = \phi_3 \circ \phi_2 \circ \phi_1(z) = f(z)$ is a one-to-one holomorphic map from the unit disk $\{|z| < 1\}$ onto the slit disk $\{|w| < 1\} \backslash \{[0, 1)\}$.

5202

(a) Find a function f that conformally maps the region $\{z : |\arg z| < 1\}$ one-to-one onto the region $\{w : |w| < 1\}$. Show that the function you have found satisfies the required conditions.

(b) Is it possible to require that $f(1) = 0$ and $f(2) = \frac{1}{2}$? If yes, give an explicit map; if No, explain why not.

(*Illinois*)

Solution.

(a) $\zeta = f_1(z) = z^{\frac{\pi}{2}} = e^{\frac{\pi}{2} \log z}$ $(\log 1 = 0)$ is a conformal map of $\{z : |\arg z| < 1\}$ onto $\{\zeta : \operatorname{Re} \zeta > 0\}$, and $w = f_2(\zeta) = \frac{\zeta-1}{\zeta+1}$ is a conformal map of $\{\zeta : \operatorname{Re} \zeta > 0\}$ onto $\{w : |w| < 1\}$. Hence

$$w = f(z) = f_2 \circ f_1(z) = \frac{z^{\frac{\pi}{2}} - 1}{z^{\frac{\pi}{2}} + 1}$$

is a conformal map of $\{z : |\arg z| < 1\}$ onto $\{w : |w| < 1\}$ with $f(1) = 0$ and $f(2) = \frac{2^{\frac{\pi}{2}}-1}{2^{\frac{\pi}{2}}+1}$.

(b) Suppose $\tilde{w} = \tilde{f}(z)$ is an arbitrary conformal map of $\{z : |\arg z| < 1\}$ onto $\{\tilde{w} : |\tilde{w}| < 1\}$ with $\tilde{f}(1) = 0$. Then $w = F(\tilde{w}) = f \circ \tilde{f}^{-1}(\tilde{w})$ is a conformal map of $\{\tilde{w} : |\tilde{w}| < 1\}$ onto $\{w : |w| < 1\}$ with $F(0) = 0$, and $\tilde{w} = \tilde{F}(w) = \tilde{f} \circ f^{-1}(w)$ is a conformal map of $\{w : |w| < 1\}$ onto $\{\tilde{w} : |\tilde{w}| < 1\}$ with $\tilde{F}(0) = 0$. By Schwarz Lemma, we have both $|F(\tilde{w})| \leq |\tilde{w}|$ and $|F(\tilde{w})| \leq |w|$, which implies that $|f(z)| = |\tilde{f}(z)|$ for every $z \in \{z : |\arg z| < 1\}$. Since

$$f(2) = \frac{2^{\frac{\pi}{2}} - 1}{2^{\frac{\pi}{2}} + 1},$$

we cannot require that $\tilde{f}(2) = \frac{1}{2}$.

5203

(1) Find one 1-1 onto conformal map f that sends the open quadrant $\{(x, y) : x > 0 \text{ and } y > 0\}$ onto the open lower half disc $\{(x, y) : x^2 + y^2 < 1 \text{ and } y < 0\}$.

(2) Find all such f.

(Toronto)

Solution.

(1) Let $\zeta = \phi_1(z) = z^2$. It is a conformal map of $\{z = x+iy : x > 0 \text{ and } y > 0\}$ onto $\{\zeta = \xi + i\eta : \eta > 0\}$.

Let $w = \phi_2(\zeta) = \sqrt{\zeta^2 - 1} + \zeta$, where $\sqrt{\zeta^2 - 1}\big|_{\zeta=i} = -\sqrt{2}i$. It is a conformal map of $\{\zeta = \xi + i\eta : \eta > 0\}$ onto $\{w = u + iv : u^2 + v^2 < 1 \text{ and } v < 0\}$.

Then $w = \phi_2 \circ \phi_1(z) = \sqrt{z^4 - 1} + z^2$, where $\sqrt{z^4 - 1}\big|_{z=e^{\frac{\pi}{4}i}} = -\sqrt{2}i$ is a required conformal map.

(2) If f is an arbitrary conformal map satisfying the condition of (1), then $\phi_2^{-1} \circ f \circ \phi_1^{-1}(\zeta)$ is a conformal map of the upper half plane onto itself, which can be represented by $\psi(\zeta) = \frac{a\zeta+b}{c\zeta+d}$, where $a, b, c, d \in \mathbb{R}$, $ad - bc > 0$. Hence f can be written as $\phi_2 \circ \psi \circ \phi_1(z)$.

5204

Map the disk $\{|z| < 1\}$ with slits along the segments $[a, 1], [-1, -b]$ $(0 < a < 1, 0 < b < 1)$ conformally on the full disk $\{|w| < 1\}$ by means of a function $w = f(z)$ with $f(0) = 0$, $f'(0) > 0$. Compute $f'(0)$ and the lengths of the arcs corresponding to the slits.

(Harvard)

Solution.

We construct the conformal mapping by the following steps.

(i) $z_1 = \phi_1(z) = z + \frac{1}{z}$: $\{|z| < 1\} \backslash \{[a, 1] \cup [-1, -b]\} \rightarrow \bar{\mathbb{C}} \backslash \{[-b - \frac{1}{b}, a + \frac{1}{a}]\}$. It has the point correspondences $\phi_1(0) = \infty$, $\phi_1(a) = a + \frac{1}{a}$, $\phi_1(-b) = -b - \frac{1}{b}$, $\phi_1(1) = 2$ and $\phi_1(-1) = -2$.

(ii) $z_2 = \phi_2(z_1) = \frac{z_1 + (b + \frac{1}{b})}{-z_1 + (a + \frac{1}{a})}$: $\bar{\mathbb{C}} \backslash \{[-b - \frac{1}{b}, a + \frac{1}{a}]\} \rightarrow \mathbb{C} \backslash [0, +\infty)$. It has the point correspondences

$$\phi_2(\infty) = -1, \quad \phi_2(-2) = \left(\frac{\frac{1}{\sqrt{b}} - \sqrt{b}}{\frac{1}{\sqrt{a}} + \sqrt{a}}\right)^2$$

and

$$\phi_2(2) = \left(\frac{\frac{1}{\sqrt{b}} + \sqrt{b}}{\frac{1}{\sqrt{a}} - \sqrt{a}} \right)^2 ,$$

and it is easy to know that

$$(\phi_2 \circ \phi_1)'(0) = - \left(a + \frac{1}{a} + b + \frac{1}{b} \right) < 0.$$

(iii) $z_3 = \phi_3(z_2) = \sqrt{z_2} : \mathbb{C}\backslash[0, +\infty) \to \{z_3 : \operatorname{Im} z_3 > 0\}$.
It has the point correspondences

$$\phi_3(-1) = i, \quad \phi_3 \left(\left(\frac{\frac{1}{\sqrt{b}} - \sqrt{b}}{\frac{1}{\sqrt{a}} + \sqrt{a}} \right)^2 \right) = \pm \left(\frac{\frac{1}{\sqrt{b}} - \sqrt{b}}{\frac{1}{\sqrt{a}} + \sqrt{a}} \right)$$

and

$$\phi_3 \left(\left(\frac{\frac{1}{\sqrt{b}} + \sqrt{b}}{\frac{1}{\sqrt{a}} - \sqrt{a}} \right)^2 \right) = \pm \left(\frac{\frac{1}{\sqrt{b}} + \sqrt{b}}{\frac{1}{\sqrt{a}} - \sqrt{a}} \right).$$

For the convenience of computation, let

$$A = \frac{\frac{1}{\sqrt{b}} - \sqrt{b}}{\frac{1}{\sqrt{a}} + \sqrt{a}}, \quad B = \frac{\frac{1}{\sqrt{b}} + \sqrt{b}}{\frac{1}{\sqrt{a}} - \sqrt{a}}.$$

We also know that $\phi_3'(-1) = -\frac{i}{2}$.

(iv) $w = \phi_4(z_3) = \frac{z_3 - i}{z_3 + i} : \{z_3 : \operatorname{Im} z_3 > 0\} \to \{w : |w| < 1\}$. It
is obvious that $\phi_4(i) = 0$ and $\phi_4'(i) = -\frac{i}{2}$.

Now we define $w = f(z) = \phi_4 \circ \phi_3 \circ \phi_2 \circ \phi_1(z)$. From the
above discussion, we know that f maps the unit disk with slits
$[-1, -b]$ and $[a, 1]$ conformally onto the unit disk with $f(0) = 0$
and

$$f'(0) = \phi_4'(i) \cdot \phi_3'(-1) \cdot (\phi_2 \circ \phi_1)'(0) = \frac{1}{4} \left(a + \frac{1}{a} + b + \frac{1}{b} \right) > 0.$$

What correspond to the slits are the arc with endpoints $\frac{A-i}{A+i}$ and $\frac{A+i}{A-i}$ containing point $z = -1$ and the arc with endpoints $\frac{B-i}{B+i}$ and $\frac{B+i}{B-i}$ containing point $z = 1$. The lengths of the two arcs are

$$0l_1 = \arg\frac{A-i}{A+i} - \arg\frac{A+i}{A-i} = 4\arctan A = 4\arctan\frac{\frac{1}{\sqrt{b}} - \sqrt{b}}{\frac{1}{\sqrt{a}} + \sqrt{a}},$$

and

$$l_2 = \arg\frac{B+i}{B-i} - \arg\frac{B-i}{B+i} = 4\arctan\frac{1}{B} = 4\arctan\frac{\frac{1}{\sqrt{a}} - \sqrt{a}}{\frac{1}{\sqrt{b}} + \sqrt{b}}.$$

5205

Let $0 < \varepsilon < \pi$, let γ_ε denote the arc $\{e^{it} : \varepsilon \le t \le 2\pi - \varepsilon\}$ and let Ω_ε be the complement of γ_ε in the Riemann sphere. If f is the conformal map of the unit disk onto Ω_ε, $f(0) = 0$, $f'(0) > 0$, describe the part of the unit disc that f maps onto $\{|z| > 1\}$.

(*Stanford*)

Solution.

We are going to find the map f by the following steps:

$$z_1 = \phi_1(z) = e^{i\varepsilon} \cdot \frac{z - e^{-i\varepsilon}}{z - e^{i\varepsilon}} : \{z : |z| < 1\} \to \{z_1 : \operatorname{Im} z_1 < 0\},$$

with $\phi_1(0) = e^{-i\varepsilon}$, $\arg \phi_1'(0) = -\frac{\pi}{2} - \varepsilon$.

$$z_2 = \phi_2(z_1) = \sqrt{z_1} : \{z_1 : \operatorname{Im} z_1 < 0\} \to \{z_2 : \operatorname{Re} z_2 > 0, \operatorname{Im} z_2 < 0\},$$

with $\phi_2(e^{-i\varepsilon}) = e^{-\frac{\varepsilon}{2}i}$, $\arg \phi_2'(e^{-i\varepsilon}) = -\frac{\varepsilon}{2}$.

$$\zeta = \phi_3(z_2) = e^{\frac{\varepsilon}{2}i} \cdot \frac{z_2 - e^{-\frac{\varepsilon}{2}i}}{z - e^{\frac{\varepsilon}{2}i}} : \{z_2 : \operatorname{Re} z_2 > 0, \operatorname{Im} z_2 < 0\} \to D_1$$

(shown in Fig. 5.4), with $\phi_3(e^{-\frac{\varepsilon}{2}i}) = 0$, $\arg \phi_3'(e^{-\frac{\varepsilon}{2}i}) = \frac{\pi}{2} + \frac{\varepsilon}{2}$, where D_1 is a domain bounded by $\{\zeta = e^{i\theta}, \frac{\varepsilon}{2} \le \theta \le 2\pi - \frac{\varepsilon}{2}\}$ and a circular arc l_ε which is orthogonal to $\{|\zeta| = 1\}$ and connects points $e^{\frac{\varepsilon}{2}i}$ and $e^{-\frac{\varepsilon}{2}i}$ in $\{|\zeta| < 1\}$.

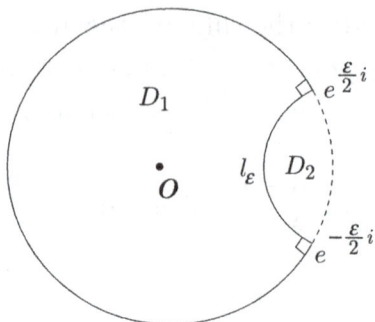

Fig. 5.4.

Let $\Phi(z) = \phi_3 \circ \phi_2 \circ \phi_1(z)$, then Φ maps $\{z : |z| < 1\}$ conformally onto D_1 with $\Phi(0) = 0, \Phi'(0) > 0$. After considering the boundary correspondence, we know that l_ε corresponds to the arc $\{z = e^{it} : |t| < \varepsilon\}$ under the map Φ. Since the symmetric domain of $\{|z| < 1\}$ with respect to arc $\{z = e^{it} : |t| < \varepsilon\}$ is $\{|z| > 1\}$, and the symmetric domain of D_1 with respect to l_ε is $D_2 = \{|\zeta| < 1\}\backslash\overline{D}_1$, by the reflection principle, $\Phi(z)$ can be extended to a conformal map of Ω_ε onto $\{\zeta : |\zeta| < 1\}$. Hence the conformal map f in the problem is nothing but the inverse of Φ, and the domain f maps onto $\{|z| > 1\}$ is D_2, which is bounded by circular arcs l_ε and $\{\zeta = e^{i\theta} : |\theta| \le \frac{\varepsilon}{2}\}$.

5206

Suppose that $w = f(z)$ maps a simply connected region G one-to-one and conformally onto a circular disk D_r with center $w = 0$, radius r, such that $f(a) = 0$ and $|f'(a)| = 1$ for some point $a \in G$.

(1) Prove that the radius $r = r(G, a)$ of D_r is uniquely determined by G and a.

(2) Determine $r(G, a)$ if G is the region between the hyperbola $xy = 1$ $(x > 0, y > 0)$ and the positive axes, and if $a = 1 + \frac{i}{2}$.

(Indiana)

Solution.

(1) Suppose $\zeta = g(z)$ is another conformal map of G onto a circular disk D_{r_1} with center $\zeta = 0$ and radius r_1, such that $g(a) = 0$ and $|g'(a)| = 1$, then $w = F(\zeta) = f \circ g^{-1}(\zeta)$ is a conformal map of $\{\zeta : |\zeta| < r_1\}$ onto $\{w : |w| < r\}$ with $F(0) = 0$ and $|F'(0)| = \left|\frac{f'(a)}{g'(a)}\right| = 1$. Apply Schwarz Lemma to $F(\zeta)$ and we have $|F'(0)| \leq \frac{r}{r_1}$, hence $r_1 \leq r$. For the same reason, apply Schwarz Lemma to $F^{-1}(w)$ and we have $r \leq r_1$, which implies $r_1 = r$. In other words, r is uniquely determined by G and a.

(2) We construct a conformal map of G onto a circular disk D_r in the following steps:

$$z_1 = \phi_1(z) = z^2 : G \to \{z_1 : 0 < \operatorname{Im} z_1 < 2\},$$

with $\phi_1\left(1 + \frac{i}{2}\right) = \frac{3}{4} + i, \left|\phi_1'\left(1 + \frac{i}{2}\right)\right| = \sqrt{5}$.

$$z_2 = \phi_2(z_1) = e^{\frac{\pi}{2} z_1} : \{z_1 : 0 < \operatorname{Im} z_1 < 2\} \to \{z_2 : \operatorname{Im} z_2 > 0\},$$

with $\phi_2\left(\frac{3}{4} + i\right) = ie^{\frac{3\pi}{8}}, \left|\phi_2'\left(\frac{3}{4} + i\right)\right| = \frac{\pi}{2} e^{\frac{3\pi}{8}}$.

$$w = \phi_3(z_2) = \frac{4}{\sqrt{5}\pi} \cdot \frac{z_2 - ie^{\frac{3}{8}\pi}}{z_2 + ie^{\frac{3}{8}\pi}} : \{z_2 : \operatorname{Im} z_2 > 0\}$$

$$\to \left\{w : |w| < \frac{4}{\sqrt{5}\pi}\right\},$$

with $\phi_3(ie^{\frac{3}{8}\pi}) = 0, |\phi_3'(ie^{\frac{3}{8}\pi})| = \frac{2}{\sqrt{5}\pi e^{\frac{3}{8}\pi}}$.

Define $f(z) = \phi_3 \cdot \phi_2 \circ \phi_1(z)$, then $w = f(z) : G \to \{w : |w| < \frac{4}{\sqrt{5}\pi}\}$, with $f(a) = 0, |f'(a)| = 1$. Hence $r(G, a) = \frac{4}{\sqrt{5}\pi}$.

5207

Let $T(z) = \frac{az+b}{cz+d}$ be a Möbius transformation.

(a) Assume that $z_1, z_2 \in \mathbb{C}$ are two distinct fixed points for T, i.e., $T(z_i) = z_i, i = 1, 2$. Show that there exists a constant c such that

$$\frac{T(z) - z_1}{T(z) - z_2} = c \frac{z - z_1}{z - z_2}.$$

(b) Use (a) to find an expression for $T^n(z), n = 1, 2, 3, \ldots$, if

$$T(z) = \frac{1 - 3z}{z - 3}.$$

(Iowa)

Solution.

(a) Let $a \in \mathbb{C}$ be a point different from z_1, z_2. Because the cross ratio is invariant under Möbius transformations, we have

$$(T(z), z_1, T(a), z_2) = (z, z_1, a, z_2),$$

which is

$$\frac{T(z) - z_1}{T(z) - z_2} : \frac{T(a) - z_1}{T(a) - z_2} = \frac{z - z_1}{z - z_2} : \frac{a - z_1}{a - z_2}.$$

Denoting

$$\frac{T(a) - z_1}{T(a) - z_2} : \frac{a - z_1}{a - z_2} = c,$$

we obtain

$$\frac{T(z) - z_1}{T(z) - z_2} = c \frac{z - z_1}{z - z_2}.$$

(b) Since $T^n(z) = T(T^{n-1}(z))$, it is easy to have

$$\frac{T^n(z) - z_1}{T^n(z) - z_2} = c \frac{T^{n-1}(z) - z_1}{T^{n-1}(z) - z_2} = c^2 \frac{T^{n-2}(z) - z_1}{T^{n-2}(z) - z_2} = \cdots = c^n \frac{z - z_1}{z - z_2}.$$

When $T(z) = \frac{1-3z}{z-3}$, by solving the equation $\frac{1-3z}{z-3} = z$, we obtain that $z = \pm 1$ are two fixed points of T. Choose $a = 2$, then $T(a) = 5$, hence $c = \frac{5-1}{5+1} : \frac{2-1}{2+1} = 2$.

It follows from

$$\frac{T^n(z) - 1}{T^n(z) + 1} = 2^n \frac{z - 1}{z + 1}$$

that

$$T^n(z) = \frac{(2^n + 1)z - (2^n - 1)}{(2^n + 1) - (2^n - 1)z}.$$

5208

(a) Justify the statement that "the curves

$$\frac{x^2}{a^2 + \lambda} + \frac{y^2}{b^2 + \lambda} = 1$$

form a family of confocal conics".

(b) Prove that such confocal conics intersect orthogonally, if at all.

(c) Show that the transformation $w = \frac{1}{2}\left(z + \frac{1}{z}\right)$ carries straight lines through the origin and circles centered at the origin into a family of confocal conics.

(Harvard)

Solution.

(a) Without loss of generality, we assume $a > b > 0$. When $-a^2 < \lambda < -b^2$, the curves form a family of hyperbolas, while when $\lambda > -b^2$, the curves form a family of ellipses. Suppose the focuses of the conics are $(\pm c(\lambda), 0)$. When $-a^2 < \lambda < -b^2$,

$$c(\lambda) = \sqrt{(a^2 + \lambda) + [-(b^2 + \lambda)]} = \sqrt{a^2 - b^2}.$$

When $\lambda > -b^2$,

$$c(\lambda) = \sqrt{(a^2 + \lambda) - (b^2 + \lambda)} = \sqrt{a^2 - b^2}.$$

Hence the curves

$$\frac{x^2}{a^2 + \lambda} + \frac{y^2}{b^2 + \lambda} = 1$$

form a family of confocal conics.

(b) Suppose (x_0, y_0) is the intersection point of

$$L_1 : \frac{x^2}{a^2 + \lambda_1} + \frac{y^2}{b^2 + \lambda_1} = 1$$

and

$$L_2 : \frac{x^2}{a^2 + \lambda_2} + \frac{y^2}{b^2 + \lambda_2} = 1,$$

where $\lambda_1 \neq \lambda_2$. It follows from

$$\frac{x_0^2}{a^2 + \lambda_1} + \frac{y_0^2}{b^2 + \lambda_1} = 1$$

and

$$\frac{x_0^2}{a^2 + \lambda_2} + \frac{y_0^2}{b^2 + \lambda_2} = 1$$

that

$$\frac{x_0^2}{(a^2 + \lambda_1)(a^2 + \lambda_2)} + \frac{y_0^2}{(b^2 + \lambda_1)(b^2 + \lambda_2)} = 0.$$

Noting that the tangent vector of L_1 at (x_0, t_0) is $\overrightarrow{\mathcal{T}}_1 = \left(\frac{x_0}{a^2+\lambda_1}, \frac{y_0}{b^2+\lambda_1}\right)$, and the tangent vector of L_2 at (x_0, y_0) is $\overrightarrow{\mathcal{T}}_2 = \left(\frac{x_0}{a^2+\lambda_2}, \frac{y_0}{b^2+\lambda_2}\right)$, we have

$$\overrightarrow{\mathcal{T}}_1 \cdot \overrightarrow{\mathcal{T}}_2 = \frac{x_0^2}{(a^2 + \lambda_1)(a^2 + \lambda_2)} + \frac{y_0^2}{(b^2 + \lambda_1)(b^2 + \lambda_2)} = 0,$$

which implies that the confocal conics intersect orthogonally, if at all.

(c) Let $z = re^{i\theta}$, and

$$w = u + iv = \frac{1}{2}\left(z + \frac{1}{z}\right) = \frac{1}{2}\left(r + \frac{1}{r}\right)\cos\theta + \frac{i}{2}\left(r - \frac{1}{r}\right)\sin\theta.$$

The image of straight lines through the origin is

$$\frac{u^2}{\cos^2\theta} - \frac{v^2}{\sin^2\theta} = 1,$$

which are hyperbolas in w-plane. Because

$$\cos^2\theta + \sin^2\theta = 1,$$

the focuses of the hyperbolas are $(\pm 1, 0)$.

The image of circles centered at the origin is

$$\frac{u^2}{\frac{1}{4}\left(r + \frac{1}{r}\right)^2} + \frac{v^2}{\frac{1}{4}\left(r - \frac{1}{r}\right)^2} = 1,$$

which are ellipses in w-plane. Because $\frac{1}{4}\left(r+\frac{1}{r}\right)^2 - \frac{1}{4}\left(r-\frac{1}{r}\right)^2 = 1$, the focuses of the ellipses are $(\pm 1, 0)$. Hence the transformation

$$w = \frac{1}{2}\left(z + \frac{1}{z}\right)$$

carries straight lines through the origin and circles centered at the origin into a family of confocal conics.

5209

If $f : D(0,1) = \{z : |z| < 1\} \to \mathbb{C}$ is an analytic function which satisfies $f(0) = 0$, and if

$$|\operatorname{Re} f(z)| < 1 \text{ for all } z \in D(0,1),$$

prove that

$$|f'(0)| \leq \frac{4}{\pi}.$$

(*Indiana*)

Solution.

It is easy to know that

$$w = g(\zeta) = \frac{e^{\frac{\pi i}{2}\zeta} - 1}{e^{\frac{\pi i}{2}\zeta} + 1}$$

is a conformal mapping of the domain $\{\zeta : |\operatorname{Re}\zeta| < 1\}$ onto the unit disk $\{w : |w| < 1\}$ with $g(0) = 0$. Hence $w = F(z) = g \circ f(z)$ is analytic in $D(0,1)$ and satisfies $F(0) = 0$ and $|F(z)| < 1$. By Schwarz Lemma, we have $|F'(0)| \leq 1$. Because

$$F'(z) = g'(f(z)) \cdot f'(z) = \frac{\pi i e^{\frac{\pi i}{2}f(z)} \cdot f'(z)}{(e^{\frac{\pi i}{2}f(z)} + 1)^2},$$

it follows from $f(0) = 0$ that

$$|f'(0)| \leq \frac{4}{\pi}.$$

5210

Let $\Omega = \{z \in \mathbb{C}; -1 < \operatorname{Im} z < 1\}$, and let $\mathcal{F}$ be the family of all analytic functions $f : \Omega \to \mathbb{C}$ such that $|f| < 1$ on Ω and $f(0) = 0$. Find

$$\sup_{f \in \mathcal{F}} |f(1)|.$$

(Indiana)

Solution.

It is obvious that

$$\zeta = f_0(z) = \frac{e^{\frac{\pi}{2}z} - 1}{e^{\frac{\pi}{2}z} + 1}$$

is a conformal mapping of Ω onto the unit disk with the origin fixed. For any analytic function $w = f(z) : \Omega \to \mathbb{C}$ such that $|f| < 1$ and $f(0) = 0$, we consider the composite function $w = F(\zeta) = f \circ f_0^{-1}(\zeta)$. $F(\zeta)$ is analytic in the unit disk such that $|F(\zeta)| < 1$ and $F(0) = 0$. By Schwarz Lemma,

$$|F(\zeta)| \leq |\zeta|.$$

Choose $\zeta_0 = \frac{e^{\frac{\pi}{2}} - 1}{e^{\frac{\pi}{2}} + 1}$, we have

$$|F(\zeta_0)| = |f(1)| \leq |\zeta_0| = \frac{e^{\frac{\pi}{2}} - 1}{e^{\frac{\pi}{2}} + 1}.$$

The equality holds if and only if $F(\zeta) = e^{i\theta}\zeta$, which implies

$$\sup_{f \in \mathcal{F}} |f(1)| = \frac{e^{\frac{\pi}{2}} - 1}{e^{\frac{\pi}{2}} + 1},$$

and the supremum is attained by $f(z) = e^{i\theta} f_0(z)$, where θ is a real number.

5211

Let f be an analytic function on $D = \{z; |z| < 1\}$ such that $f(0) = -1$, and suppose that $|1 + f(z)| < 1 + |f(z)|$ whenever $|z| < 1$. Prove that $|f'(0)| \leq 4$.

(*Indiana*)

Solution.

Let $\Omega = \mathbb{C} \backslash \{w = u + iv : u \geq 0 \text{ and } v = 0\}$. It follows from $|1 + f(z)| < 1 + |f(z)|$ that $f(D) \subset \Omega$.

Set $g(w) = \frac{\sqrt{w} - i}{\sqrt{w} + i}$, $(\sqrt{w}|_{w=-1} = i)$. Then $g \circ f(z)$ is an analytic function on D with $g \circ f(0) = 0$ and $|g \circ f(z)| < 1$. By Schwarz Lemma,

$$|(g \circ f)'(0)| \leq 1.$$

Since

$$g'(w) = \frac{i}{\sqrt{w}(\sqrt{w} + i)^2},$$

we have $g'(-1) = -\frac{1}{4}$. From

$$(g \circ f)'(0) = g'(-1)f'(0),$$

we obtain

$$|f'(0)| \leq 4.$$

5212

Let P be the set of holomorphic function f on the open unit disc so that (i) Both the real and imaginary parts of $f(z)$ are positive for $|z| < 1$, (ii) $f(0) = 1 + i$. Let $E = \{f(\frac{1}{2}) : f \in P\}$. Describe E explicitly.

(*Minnesota*)

Solution.

Let $f \in P$ and define

$$\zeta = F(z) = \frac{f^2(z) - 2i}{f^2(z) + 2i}.$$

Then F is a holomorphic function on the unit disc with $F(0) = 0$ and $|F(z)| < 1$. By Schwarz Lemma, we have $|F(z)| \leq |z|$, which implies $|F(\frac{1}{2})| \leq \frac{1}{2}$. It should be noted that when f changes in P, $F(\frac{1}{2})$ can take any value in the disc $\{\zeta : |\zeta| \leq \frac{1}{2}\}$. Because $w = \frac{2i(1+\zeta)}{1-\zeta}$ (that is the inverse of $\zeta = \frac{w-2i}{w+2i}$) is a conformal mapping of $\{\zeta : |\zeta| \leq \frac{1}{2}\}$ onto $\{w : |w - \frac{10}{3}i| \leq \frac{8}{3}\}$, we obtain that the set $\{f^2(\frac{1}{2}) : f \in P\}$ is equal to

$$\left\{ w : \left| w - \frac{10}{3}i \right| \leq \frac{8}{3} \right\}$$

$$= \left\{ w = \rho e^{i\phi} : \left| \phi - \frac{\pi}{2} \right| \leq \arcsin \frac{4}{5}, \rho^2 - \frac{20}{3}\rho \sin \phi + 4 \leq 0 \right\}.$$

Hence

$$E = \left\{ f\left(\frac{1}{2}\right) : f \in P \right\}$$

$$= \left\{ re^{i\theta} : \left| \theta - \frac{\pi}{4} \right| \leq \frac{1}{2}\arcsin \frac{4}{5}, r^4 - \frac{20}{3}r^2 \sin 2\theta + 4 \leq 0 \right\}.$$

If we denote the two roots of $\rho^2 - \frac{20}{3}\rho \sin \phi + 4 = 0$ by $\rho_1(\phi), \rho_2(\phi)$ where $\rho_1(\phi) \leq \rho_2(\phi)$ and $\left| \phi - \frac{\pi}{2} \right| \leq \arcsin \frac{4}{5}$, the set E can also be represented by

$$\left\{ re^{i\theta} : \left| \theta - \frac{\pi}{4} \right| \leq \frac{1}{2} \arcsin \frac{4}{5}, \sqrt{\rho_1(2\theta)} \leq r \leq \sqrt{\rho_1(2\theta)} \right\}.$$

5213

Let

$$\Omega = \left\{ w = u + iv : \frac{u^2}{5^2} + \frac{v^2}{3^2} > 1 \right\}.$$

If $\mathcal{F}$ is the family of all analytic function on Ω such that $|f| \leq 1$ in Ω and $\lim_{w \to \infty} f(w) = 0$, find $\sup_{f \in \mathcal{F}} |f(8)|$. Your answer should be an explicit number, and you should prove your assertion.

(Indiana)

Solution.

Define $w = \phi(z) = 2(\frac{z}{2} + \frac{2}{z})$, it is easy to know that $w = \phi(z)$ is a conformal map of $D = \{z : |z| < 1\}$ onto Ω with $\phi(0) = \infty$ and $\phi(4 - \sqrt{12}) = 8$.

Then $F(z) = f \circ \phi(z) = f(2(\frac{z}{2} + \frac{2}{z}))$ is analytic in D and satisfies $F(0) = 0$ and $|F(z)| \leq 1$. By Schwarz Lemma,

$$|F(z)| \leq |z|.$$

Hence

$$|f(8)| = |F(4 - \sqrt{12})| \leq 4 - \sqrt{12}.$$

This upper bound can be reached if we let $f = \phi^{-1}$ which belongs to family $\mathcal{F}$ and satisfies $\phi^{-1}(8) = 4 - \sqrt{12}$. So we obtain

$$\sup_{f \in \mathcal{F}} |f(8)| = 4 - \sqrt{12}.$$

5214

Let D be the upper-half and let $f \neq id$ be a conformal map of D onto itself such that $f \circ f = id$. Prove that f has a unique fixed point inside D.

(SUNY, Stony Brook)

Solution.

Since f is a conformal map of D onto itself, it can be written as $f(z) = \frac{az+b}{cz+d}$, where $a, b, c, d \in \mathbb{R}$ and $ad - bc > 0$. Then

$$f \circ f(z) = \frac{(a^2 + bc)z + b(a + d)}{c(a + d)z + d^2 + bc}.$$

It follows from $f \circ f = id$ that $b(a + d) = c(a + d) = 0$ and $a^2 + bc = d^2 + bc \neq 0$.

If $a + d \neq 0$, then $b = c = 0$. Hence $ad - bc > 0$ and $a^2 + bc = d^2 + bc$ impies $f = id$, which contradicts the condition $f \neq id$. Thus we have $a + d = 0$ and the inequality $ad - bc > 0$ can be written as $bc + a^2 < 0$.

Now we consider the equation $f(z) = \frac{az+b}{cz+d} = z$, which is equivalent to $cz^2 + (d - a)z - b = 0$. Since $\Delta = (d - a)^2 + 4bc$ is equal to $4bc + 4a^2 < 0$, we know that $f(z) = z$ has two conjugate roots, one in the upper-half plane and the other in the lower-half plane. So f has a unique fixed point inside D.

5215

Let Ω be a convex, open subset of $\mathbb{C}$ and let $f : \Omega \to \mathbb{C}$ be an analytic function satisfying $\operatorname{Re} f'(z) > 0$, $z \in \Omega$. Prove that f is one-to-one in Ω (i.e., f is injective).

(*Indiana*)

Solution.

Let $z_1 \neq z_2$ be two arbitrary points in Ω. $L : z(t) = z_1 + t(z_2 - z_1), t \in [0, 1]$ is the line segment connecting z_1 and z_2. Since Ω is convex, $L \subset \Omega$, we have

$$f(z_2) - f(z_1) = \int_L f'(z)dz = \int_0^1 f'(z(t))(z_2 - z_1)dt.$$

Hence

$$\frac{f(z_2) - f(z_1)}{z_2 - z_1} = \int_0^1 f'(z(t))dt.$$

Since $\operatorname{Re} f'(z) > 0$ for $z \in \Omega$, we know that $\int_0^1 f'(z(t))dt \neq 0$, which implies $f(z_1) \neq f(z_2)$ whenever $z_1 \neq z_2$.

5216

Show that if the polynomial $P(z) = a_n z^n + a_{n-1} z^{n-1} + \cdots + a_1 z + a_0, n > 1$, is one-to-one in the unit disk $|z| < 1$ and $a_1 = 1$, then $|na_n| \leq 1$.

(SUNY, Stony Brook)

Solution.

It follows from the univalence of $P(z)$ in $\{|z| < 1\}$ that $P'(z) = na_n z^{n-1} + (n-1)a_{n-1} z^{n-2} + \cdots + 2a_2 z + a_1 \neq 0$ for all $z \in \{|z| < 1\}$. In other words, the roots of $P'(z)$ are all situated outside the open unit disk. Let $z_1, z_2, \ldots, z_{n-1}$ be the roots of $P'(z)$, then $|z_j| \geq 1$ for $j = 1, 2, \ldots, n - 1$. Because $P'(z)$ can also be written as $na_n(z - z_1)(z - z_2) \cdots (z - z_{n-1})$, by comparing the constant terms, we have

$$(-1)^{n-1} na_n \prod_{j=1}^{n-1} z_j = a_1.$$

Since $a_1 = 1$, we obtain

$$|na_n| = \frac{|a_1|}{\prod_{j=1}^{n-1} |z_j|} \leq 1.$$

5217

Let $P(z)$ be a polynomial on the complex plane, not identically zero; let $H = \{z : \operatorname{Re} z > 0\}$.

(a) If all roots of $P(z)$ lie in H, show that the same is true for the roots of dP/dz.

(b) For any non-vanishing polynomial $P(z)$, use the result in (a) to show that the convex hull of the roots of $P(z)$ contains the roots of dP/dz.

(Courant Inst.)

Solution.

(a) Let $z_1, z_2, \ldots, z_n$ be the zeros of $P(z)$. By assumption,

$$\operatorname{Re} z_j > 0 \quad (j = 1, 2, \ldots, n),$$

and $P(z) = a(z - z_1)(z - z_2) \cdots (z - z_n)$. It follows that

$$(\log P(z))' = \frac{P'(z)}{P(z)} = \frac{1}{z - z_1} + \frac{1}{z - z_2} + \cdots + \frac{1}{z - z_n}.$$

When $z \in \{z : \operatorname{Re} z \leq 0\}$, then $\frac{\pi}{2} < \arg(z - z_j) < \frac{3\pi}{2}$, or equivalently, $\operatorname{Re} \frac{1}{z - z_j} < 0$. Hence $\operatorname{Re} \sum_{j=1}^{n} \frac{1}{z - z_j} < 0$, which shows $\frac{P'(z)}{P(z)}$ cannot be zero on $\{z : \operatorname{Re} z \leq 0\}$.

(b) Let $z_1, z_2, \ldots, z_n$ be the zeros of $P(z)$, and l is a directed straight line passing through two zeros z_k and z_l such that the other zeros are on the right side of l (including on l). Denote the intersectional angle from the positive direction of the imaginary axis to l by θ. When z is on the left side of l, we have $\operatorname{Re}\{e^{-i\theta}(z - z_j)\} < 0$. Hence

$$\operatorname{Re}\left\{e^{i\theta}\frac{P'(z)}{P(z)}\right\} = \operatorname{Re} \sum_{j=1}^{n} \frac{e^{i\theta}}{z - z_j} < 0,$$

which shows that the zeros of $P'(z)$ do not lie on the left side of l. After considering all the directed straight lines passing through two of the zeros of $P(z)$ such that the other zeros are on the right side of the line, we obtain that the zeros of $P'(z)$ lie on the convex hull of the zeros of $P(z)$.

5218

Let $f(z)$ be a Laurent series centered at 0, convergent in $\mathbb{C}\backslash\{0\}$, with residue b at $z = 0$.

(a) Show that there exists ζ on $\{z \in \mathbb{C} : |z| = 1\}$ with

$$|f(\zeta) - \zeta^{-1}| \geq |b - 1|.$$

(b) Characterize those functions with

$$\max_{|\zeta|=1} |f(\zeta) - \zeta^{-1}| = |b - 1|.$$

(Minnesota)

Solution.

(a) Let $f(z) = \sum_{n=-\infty}^{+\infty} b_n z^n$, then

$$b_{-1} = \frac{1}{2\pi i} \int_{|\zeta|=1} f(\zeta) d\zeta = b.$$

Hence

$$b_{-1} = \frac{1}{2\pi i} \int_{|\zeta|=1} (f(\zeta) - \zeta^{-1}) d\zeta.$$

If $|f(\zeta) - \zeta^{-1}| < |b - 1|$ holds for all ζ with $|\zeta| = 1$, then

$$|b - 1| \le \frac{1}{2\pi} \max_{|\zeta|=1} |f(\zeta) - \zeta^{-1}| \cdot \int_{|\zeta|=1} |d\zeta| < |b - 1|,$$

which is a contradiction. Hence there exists ζ with $|\zeta| = 1$ such that

$$|f(\zeta) - \zeta^{-1}| \ge |b - 1|.$$

(b) If $\max_{|\zeta|=1} |f(\zeta) - \zeta^{-1}| = |b - 1|$, it follows from

$$|b - 1| \le \frac{1}{2\pi} \int_{|\zeta|=1} |f(\zeta) - \zeta^{-1}| |d\zeta|$$

that

$$|f(\zeta) - \zeta^{-1}| = |b - 1|$$

holds for all ζ with $|\zeta| = 1$.

Let $f(\zeta) - \zeta^{-1} = (b - 1)e^{i\phi(\theta)}$, where $\zeta = e^{i\theta}$ and $\phi(\theta)$ is a continuous real-valued function. It follows from

$$b - 1 = \frac{1}{2\pi i} \int_{|\zeta|=1} (f(\zeta) - \zeta^{-1}) d\zeta$$

that

$$\frac{1}{2\pi}\int_0^{2\pi} e^{i(\phi(\theta)+\theta)}\,d\theta = 1,$$

which implies that $\phi(\theta) = -\theta$, and hence

$$f(\zeta) - \zeta^{-1} = \frac{b-1}{\zeta}$$

holds on $\{\zeta : |\zeta| = 1\}$. Apply the discreteness of zeros for analytic functions to $f(z) - \frac{b}{z}$, we obtain $f(z) = \frac{b}{z}, z \in \mathbb{C}\backslash\{0\}$.

5219

Assume f is analytic in a neighborhood of $\overline{D}$, f maps D into D, and f maps ∂D into ∂D, where $D = \{z : |z| < 1\}$.
  (a) Show that $\forall z \in \partial D$, $f'(z) \neq 0$.
  (b) Show that $\frac{d}{d\theta}[\arg f(e^{i\theta})] > 0$ for θ in $\mathbb{R}$.
  (c) Assume that $f(0) = f'(0) = 0$ and $f|_{\partial D}$ is a two-to-one map from ∂D onto ∂D. Show that $f(z) \neq 0$ whenever $0 < |z| < 1$.

(*Indiana–Purdue*)

Solution.
  (a) Assume $f'(z_0) = 0$, where $z_0 \in \partial D$. Let $f(z_0) = w_0 \in \partial D$. Then

$$f(z) - w_0 = (z - z_0)^n g(z),$$

where $n \geq 2$,

$$g(z) = b_0 + b_1(z - z_0) + b_2(z - z_0)^2 + \cdots,$$

with $b_0 \neq 0$. Let Γ be an arc in $\overline{D}$ defined by $\Gamma = \{z \in \overline{D} : |z - z_0| = r\}$, and denote by $\Delta_\Gamma \phi(z)$ the change of $\phi(z)$ when z goes along the arc Γ in the counterclockwise sense. It is demanded that r is sufficiently small such that $\Delta_\Gamma \arg(z - z_0) > \frac{3\pi}{4}$ and

$|g(z) - b_0| < \frac{|b_0|}{2}$ when $z \in \Gamma$. It follows from $f(z) - w_0 = (z - z_0)^n g(z)$ $(n \geq 2)$, that

$$\Delta_\Gamma \arg(f(z) - w_0) = n \Delta_\Gamma \arg(z - z_0) + \Delta_\Gamma \arg g(z) > \frac{3\pi}{2} - \frac{\pi}{3} > \pi,$$

which implies that $f(z)$ assumes values outside the disk $\overline{D}$ when $z \in \Gamma$. It is a contradiction to the fact that f maps D into D. Hence $f'(z) \neq 0$ for all $z \in \partial D$.

(b) Let $z = re^{i\theta}$, and $w = f(z) = Re^{i\psi}$. A variation of the Cauchy–Riemann equations for analytic function $w = f(z)$ is

$$r \frac{\partial R}{\partial r} = R \frac{\partial \psi}{\partial \theta}, \quad \frac{\partial R}{\partial \theta} = -rR \frac{\partial \psi}{\partial r}.$$

Since f maps ∂D into ∂D, we know that $\frac{\partial R}{\partial \theta}(e^{i\theta}) = 0$. If $\frac{\partial \psi}{\partial \theta}(e^{i\theta}) = 0$, then at point $e^{i\theta}$, $\frac{\partial R}{\partial r} = \frac{\partial R}{\partial \theta} = \frac{\partial \psi}{\partial r} = \frac{\partial \psi}{\partial \theta} = 0$, which implies that $\frac{\partial f}{\partial z}(e^{i\theta}) = f'(e^{i\theta}) = 0$. But from (a) it is impossible. If $\frac{\partial \psi}{\partial \theta}(e^{i\theta}) < 0$, it follows from $r \frac{\partial R}{\partial r} = R \frac{\partial \psi}{\partial \theta}$ that $\frac{\partial R}{\partial r}(e^{i\theta}) < 0$. Since $R = 1$ when $r = 1$, $\frac{\partial R}{\partial r}(e^{i\theta}) < 0$ implies that $R > 1$ when $r < 1$. This is also impossible. Hence we obtain

$$\frac{\partial \psi}{\partial \theta}(e^{i\theta}) = \frac{d}{d\theta}[\arg f(e^{i\theta})] > 0.$$

(c) Because $f|_{\partial D}$ is a two-to-one map from ∂D onto ∂D, $\frac{1}{2\pi} \Delta_{|z|=1} \arg f(z) = 2$, which implies that $f(z)$ has two zeros (counted by multiplicity) in D. Since $f(0) = f'(0) = 0$, $z = 0$ is a zero of f of multiplicity $m = 2$. Hence $f(z)$ has no zero in $\{0 < |z| < 1\}$.

5220

If $f(z)$ is a univalent (1-1 analytic) function with domain the unit disc such that $f(z) = z + \sum_{n=2}^\infty a_n z^n$, then prove that $g(z) = \sqrt{f(z^2)}$ is an odd analytic univalent function on the unit disc.

(UC, Berkerly)

Solution.

$g(z) = \sqrt{f(z^2)}$ should be understood as $g(z) = z\sqrt{h(z)}$, where $h(z) = \frac{f(z^2)}{z^2}$ is analytic and has no zeros in the unit disk, and $\sqrt{h(z)}$ is defined with the branch $\sqrt{h(0)} = \sqrt{1} = 1$. It is obvious that $g(z)$ is an odd analytic function on the unit disc, so we need only to prove that $g(z)$ is univalent.

Suppose $g(z_1) = g(z_2)$. Squaring both sides of the equality, we have $f(z_1^2) = f(z_2^2)$. Because $f(z)$ is a univalent, we have $z_1^2 = z_2^2$, i.e., $z_1 = \pm z_2$. Since $g(z)$ is an odd function, obviously $z_1 = -z_2$ does not satisfies $g(z_1) = g(z_2)$. So we obtain that $z_1 = z_2$ which implies that $g(z)$ is univalent.

5221

Consider the unit disk D and an analytic function $f : D \to D$. If $f(0) = a \neq 0$, show that f does not have a zero in the disk $\{z : |z| < |a|\}$.

(SUNY, Albany)

Solution.

Assume $f(\lambda) = 0$, Define $g(z) = f\left(\frac{z+\lambda}{1+\lambda z}\right)$, then $g(z) : D \to D$, with $g(0) = 0$. Apply Schwarz Lemma to $g(z) = f\left(\frac{z+\lambda}{1+\lambda z}\right)$ to get $|g(z)| \leq |z|$. Then take $z = -\lambda$, we have $|a| \leq |\lambda|$, which implies that f does not have a zero in the disk $\{z : |z| < |a|\}$.

5222

Let $f(z) = z(1+z)$. Let $V = \{z \in \mathbb{C} : 0 < |z| < \frac{1}{2}, \frac{3}{4}\pi < \arg z < \frac{5}{4}\pi\}$.

(a) Show that if $z \in V$, then $|f(z)| \leq |z|$ and $f(z) \in V$.

(b) Prove that if $z \in V$, then $\lim_{n \to \infty} f^{(n)}(z) = 0$, where $f^{(n)}(z) = f \circ f \circ \cdots \circ f$ is the n-fold composition of f.

(Michigan)

Solution.

(a) First we are going to find the maximum modulus of $\frac{f(z)}{z} = 1 + z$ when $z \in \partial V$:

When $\theta \in \left[\frac{3\pi}{4}, \frac{5\pi}{4}\right]$, we have

$$\left|1 + \frac{1}{2}e^{i\theta}\right| = \sqrt{1 + \frac{1}{4} + \cos\theta} < 1;$$

When $r \in \left[0, \frac{1}{2}\right]$, we have

$$\left|1 + re^{i\frac{3\pi}{4}}\right| = \sqrt{1 + r^2 + 2r\cos\frac{3\pi}{4}} \leq 1,$$

$$\left|1 + re^{i\frac{5\pi}{4}}\right| = \sqrt{1 + r^2 + 2r\cos\frac{5\pi}{4}} \leq 1.$$

Hence we obtain that $|f(z)| \leq |z|$ when $z \in V$.

When $z \in V$ and $\frac{3\pi}{4} < \arg z \leq \pi$, we have $0 \leq \arg(1+z) < \frac{\pi}{4}$, hence we obtain that

$$\frac{3\pi}{4} < \arg f(z) < \frac{5\pi}{4};$$

When $z \in V$ and $\pi \leq \arg z < \frac{5\pi}{4}$, we have $-\frac{\pi}{4} < \arg(1+z) \leq 0$, hence we obtain that

$$\frac{3\pi}{4} < \arg f(z) < \frac{5\pi}{4}.$$

Combine the above results, we obtain that $f(z) \in V$ if $z \in V$.

(b) Now we apply the following **Vitali Theorem:** Let D be a domain, $\{f_n(z)\}$ be a sequence of functions analytic and locally uniformly bounded in D. If there is a series $\{z_n\}_{n=1}^{\infty}$ with $z_n \in D$ which has a limit point belonging to D, and $\{f_n(z)\}$ converges on every point of $\{z_n\}_{n=1}^{\infty}$, then $\{f_n(z)\}$ converges uniformly on compact subsets of D.

For arbitrary $x_0 \in \left(-\frac{1}{2}, 0\right)$, set $x_n = f^{(n)}(x_0)$, then it is easy to prove that $\lim_{n\to\infty} x_n = 0$. By Vitali Theorem, we obtain that $\lim_{n\to\infty} f^{(n)}(z) = 0$, and the convergence is uniform in every compact subsets of V.

Section 3

Complex Integration

5301

Evaluate the integral

$$\int_{|z|=2} e^{e^{\frac{1}{z}}} \, dz.$$

(*Indiana*)

Solution.

Function $e^{e^{\frac{1}{z}}}$ is analytic in $\{z : 0 < |z| < +\infty\}$, and its Laurent expansion around $z = 0$ is:

$$e^{e^{\frac{1}{z}}} = 1 + e^{\frac{1}{z}} + \frac{1}{2!}e^{\frac{2}{z}} + \cdots + \frac{1}{n!}e^{\frac{n}{z}} + \cdots$$

$$= 1 + \left\{ 1 + \frac{1}{z} + \frac{1}{2!} \cdot \frac{1}{z^2} + \cdots \right\}$$

$$+ \frac{1}{2!}\left\{ 1 + \frac{2}{z} + \frac{1}{2!}\left(\frac{2}{z}\right)^2 + \cdots \right\}$$

$$+ \cdots + \frac{1}{n!}\left\{ 1 + \frac{n}{z} + \frac{1}{2!}\left(\frac{n}{z}\right)^2 + \cdots \right\} + \cdots.$$

The coefficient of the term $\frac{1}{z}$ in the above development is

$$1 + 1 + \frac{1}{2!} + \cdots + \frac{1}{(n-1)!} + \cdots = e.$$

By the residue theorem, we obtain

$$\int_{|z|=2} e^{e^{\frac{1}{z}}} dz = 2\pi i \operatorname{Res}\left(e^{e^{\frac{1}{z}}}, 0\right) = 2\pi e i.$$

5302

Evaluate

$$\int_\gamma \frac{dz}{\sin^3 z},$$

where γ is the positively oriented circle $\{|z| = 1\}$.

(Indiana)

Solution.

It is obvious that $\frac{1}{\sin^3 z}$ is analytic in $\{z : 0 < |z| \le 1\}$, and with $z = 0$ as a pole. The Laurent expansion of $\frac{1}{\sin^3 z}$ around $z = 0$ can be obtained as follows:

$$\frac{1}{\sin^3 z} = \frac{1}{\left(z - \frac{1}{3!}z^3 + \frac{1}{5!}z^5 - \cdots\right)^3}$$

$$= \frac{1}{z^3\{1 - (\frac{1}{3!}z^2 - \frac{1}{5!}z^4 + \cdots)\}^3}$$

$$= \frac{1}{z^3}\left\{1 + 3\left(\frac{1}{3!}z^2 - \frac{1}{5!}z^4 + \cdots\right)\right.$$

$$\left. + 6\left(\frac{1}{3!}z^2 - \frac{1}{5!}z^4 + \cdots\right)^2 + \cdots\right\}.$$

Hence the coefficient of the term $\frac{1}{z}$ in the above development is $\frac{1}{2}$. By the residue theorem, we have

$$\int_\gamma \frac{dz}{\sin^3 z} = 2\pi i \operatorname{Res}\left(\frac{1}{\sin^3 z}, 0\right) = \pi i.$$

5303

For what value of a is the function

$$f(z) = \int_1^z \left(\frac{1}{z} + \frac{a}{z^3} \right) \cos z \, dz$$

single-valued?

(*Indiana*)

Solution.

Function $F(z) = (\frac{1}{z} + \frac{a}{z^3}) \cos z$ is analytic in $\{z : 0 < |z| < +\infty\}$, and its Laurent expansion around $z = 0$ is:

$$F(z) = \left(\frac{1}{z} + \frac{a}{z^3} \right) \cos z = \left(\frac{1}{z} + \frac{a}{z^3} \right) \left(1 - \frac{1}{2!}z^2 + \frac{1}{4!}z^4 - \cdots \right)$$

$$= \frac{a}{z^3} + \left(1 - \frac{a}{2} \right) \frac{1}{z} + \left(\frac{a}{24} - \frac{1}{2} \right) z + \cdots .$$

The necessary and sufficient condition for $f(z)$ to be single-valued is that the residue of $F(z)$ at $z = 0$ is zero, i.e., the coefficient of the term $\frac{1}{z}$ in the above development is zero. Hence we obtain $a = 2$.

5304

Define

$$h(z) = \int_0^\infty (1 + zte^{-t})^{-1} e^{-t} \cos(t^2) dt.$$

What is the largest possible P so that $h(z)$ is analytic for $|z| < P$?

(*Indiana–Purdue*)

Solution.

When $z = -e$,

$$h(-e) = \int_0^\infty \frac{\cos(t^2)}{e^t - et} dt.$$

Problems and Solutions in Mathematics

It is easy to see that when $t \to 1$,

$$\frac{\cos(t^2)}{e^t - et} \sim \frac{A}{(t-1)^2},$$

where $A = \frac{2}{e}\cos 1$, which implies that the integral is divergent. Hence P cannot be larger than e.

For any $r < e$, let $|z| \le r$. Consider the integral

$$h(z) = \int_0^\infty \frac{\cos(t^2)}{e^t + zt} dt.$$

It follows from $|e^t + zt| \ge e^t - rt$ and the convergence of the integral

$$\int_0^\infty \frac{|\cos(t^2)|}{e^t - rt} dt$$

that

$$\int_0^\infty \frac{\cos(t^2)}{e^t + zt} dt$$

is uniformly convergent in any compact subset of $\{z : |z| < e\}$. By Weierstrass Theorem, we know that $h(z)$ is analytic in $\{z : |z| < e\}$. Hence the largest possible P is equal to e.

5305

Let $f(z)$ be analytic in $S = \{z \in \mathbb{C}; |z| < 2\}$. Show that

$$\frac{2}{\pi} \int_0^{2\pi} f(e^{it}) \cos^2 \frac{t}{2} dt = 2f(0) + f'(0).$$

(Iowa)

Solution.

It is easy to see that

$$f(0) = \frac{1}{2\pi i} \int_{|z|=1} \frac{f(z)}{z} dz = \frac{1}{2\pi} \int_0^{2\pi} f(e^{it}) dt,$$

$$f'(0) = \frac{1}{2\pi i} \int_{|z|=1} \frac{f(z)}{z^2} dz = \frac{1}{2\pi} \int_0^{2\pi} f(e^{it}) e^{-it} dt.$$

Note that

$$0 = \frac{1}{2\pi i} \int_{|z|=1} f(z)dz = \frac{1}{2\pi} \int_0^{2\pi} f(e^{it})e^{it}\, dt.$$

It follows from the above three equalities that

$$2f(0) + f'(0) = \frac{1}{2\pi} \int_0^{2\pi} f(e^{it})(2 + e^{it} + e^{-it})dt$$

$$= \frac{2}{\pi} \int_0^{2\pi} f(e^{it}) \cos^2 \frac{t}{2} dt.$$

5306

Suppose that the real-valued function u is harmonic in the disk $\{|z| < 2\}$, v is its harmonic conjugate and $u(0) = v(0) = 0$. Show that

$$\int_\gamma u^2(z)v^2(z)\frac{dz}{z} = \frac{1}{6} \int_\gamma (u^4(z) + v^4(z))\frac{dz}{z},$$

where $\gamma(t) = e^{2\pi it}$, $t \in [0,1]$.

(SUNY, Stony Brook)

Solution.

Let $f(z) = u(z) + iv(z)$. Then $f(z)$ is analytic in $\{z : |z| < 2\}$, and we have

$$\int_\gamma f^4(z)\frac{dz}{z} = 2\pi i f^4(0) = 0,$$

$$\int_\gamma \bar{f}^4(z)\frac{dz}{z} = \overline{\left(\int_\gamma f^4(z)\frac{d\bar{z}}{\bar{z}}\right)}$$

$$= \overline{\left(\int_\gamma zf^4(z)d\left(\frac{1}{z}\right)\right)}$$

$$= \overline{\left(-\int_\gamma \frac{f^4(z)}{z}dz\right)} = 0.$$

It follows from

$$u(z) = \frac{f(z) + \overline{f(z)}}{2}$$

and

$$v(z) = \frac{f(z) - \overline{f(z)}}{2i}$$

that

$$\int_\gamma u^2(z)v^2(z)\frac{dz}{z} = -\frac{1}{16}\int_\gamma (f^4(z) + \bar{f}^4(z) - 2|f(z)|^4)\frac{dz}{z}$$

$$= \frac{1}{8}\int_\gamma |f(z)|^4\frac{dz}{z}$$

$$= \frac{1}{8}\int_\gamma (u^4(z) + v^4(z) + 2u^2(z)v^2(z))\frac{dz}{z},$$

which implies that

$$\int_\gamma u^2(z)v^2(z)\frac{dz}{z} = \frac{1}{6}\int_\gamma (u^4(z) + v^4(z))\frac{dz}{z}.$$

5307

Let f be an analytic function on an open set containing $\overline{D(0,1)} = \{z; |z| \le 1\}$.

(a) Prove that

$$\frac{d^n f}{dz^n}(0) = \frac{n!}{\pi}\int_0^{2\pi} e^{-ni\theta}[\text{Re } f(e^{i\theta})]d\theta.$$

(b) If $f(0) = 1$, and if Re $f(z) > 0$ for all points $z \in D(0,1)$, prove that

$$\left|\frac{d^n f}{dz^n}(0)\right| \le 2(n!).$$

(Indiana)

Solution.

(a) Assume that

$$f(z) = \sum_{k=0}^{\infty} a_k z^n,$$

we have

$$\frac{n!}{2\pi} \int_0^{2\pi} \overline{f(e^{i\theta})} e^{-ni\theta} d\theta = \frac{n!}{2\pi} \int_0^{2\pi} \left(\sum_{k=0}^{\infty} \bar{a}_k e^{-ki\theta} \right) e^{-ni\theta} d\theta$$

$$= \frac{n!}{2\pi} \sum_{k=0}^{\infty} \bar{a}_k \left(\int_0^{2\pi} e^{-(n+k)i\theta} d\theta \right) = 0.$$

By Cauchy Integral Formula,

$$\frac{d^n f}{dz^n}(0) = \frac{n!}{2\pi i} \int_{|\zeta|=1} \frac{f(\zeta)}{\zeta^{n+1}} d\zeta = \frac{n!}{2\pi} \int_0^{2\pi} f(e^{i\theta}) e^{-ni\theta} d\theta.$$

Hence

$$\frac{d^n f}{dz^n}(0) = \frac{n!}{2\pi} \int_0^{2\pi} f(e^{i\theta}) e^{-ni\theta} d\theta + \frac{n!}{2\pi} \int_0^{2\pi} \overline{f(e^{i\theta})} e^{-ni\theta} d\theta$$

$$= \frac{n!}{\pi} \int_0^{2\pi} e^{-ni\theta} [\operatorname{Re} f(e^{i\theta})] d\theta.$$

(b) Because $\operatorname{Re} f(z)$ is harmonic on $\overline{D(0,1)}$, by the mean-value formula of harmonic functions,

$$\frac{1}{2\pi} \int_0^{2\pi} \operatorname{Re} f(e^{i\theta}) d\theta = \operatorname{Re} f(0) = 1.$$

Noting that $\operatorname{Re} f(e^{i\theta}) \geq 0$, we have

$$\left| \frac{d^n f}{dz^n}(0) \right| = \left| \frac{n!}{\pi} \int_0^{2\pi} e^{-ni\theta} [\operatorname{Re} f(e^{i\theta})] d\theta \right|$$

$$\leq \frac{n!}{\pi} \int_0^{2\pi} |e^{-ni\theta}| [\operatorname{Re} f(e^{i\theta})] d\theta$$

$$= \frac{n!}{\pi} \int_0^{2\pi} \operatorname{Re} f(e^{i\theta}) d\theta$$

$$= 2(n!).$$

5308

If f is analytic in the unit disk and its derivative satisfies

$$|f'(z)| \leq (1 - |z|)^{-1},$$

show that the coefficients in the expansion

$$f(z) = \sum_{n=0}^{\infty} a_n z^n$$

satisfy $|a_n| < e$ for $n \geq 1$, where e is the base of natural logarithms.

(Stanford)

Solution.

It follows from

$$f(z) = \sum_{n=0}^{\infty} a_n z^n$$

that

$$f'(z) = \sum_{n=1}^{\infty} n a_n z^{n-1},$$

where

$$n a_n = \frac{1}{2\pi i} \int_{|z|=r} \frac{f'(z)}{z^n} dz, \quad (0 < r < 1).$$

It is obvious that

$$|a_1| = |f'(0)| \leq 1 < e.$$

For $n > 1$, we choose $r = 1 - \frac{1}{n}$,

$$|a_n| = \frac{1}{2\pi n} \left| \int_{|z|=1-\frac{1}{n}} \frac{f'(z)}{z^n} dz \right|$$

$$\leq \frac{1}{2\pi n} \cdot \frac{(\frac{1}{n})^{-1}}{(1-\frac{1}{n})^n} \cdot 2\pi \left(1 - \frac{1}{n}\right)$$

$$= \left(1 + \frac{1}{n-1}\right)^{n-1} < e.$$

5309

Let $f = u + iv$ be an entire function.

(a) Show that if $u^2(z) \geq v^2(z)$ for all $z \in \mathbb{C}$, then f must be a constant.

(b) Show that if $|f(z)| \leq A + B|z|^h$ for all $z \in \mathbb{C}$ with some positive numbers A, B, h, then $f(z)$ is a polynomial of degree bounded by h.

(Stanford)

Solution.

(a) Let

$$F(z) = e^{-f^2(z)} = e^{-(u^2(z)-v^2(z))-2iu(z)v(z)}.$$

Then $F(z)$ is an entire function with

$$|F(z)| = e^{-(u^2(z)-v^2(z))} \leq 1.$$

By Liouville's Theorem, $F(z)$ must be a constant, which implies that $f(z)$ is a constant,

(b) Let

$$f(z) = \sum_{n=0}^{\infty} a_n z^n.$$

Then

$$a_n = \frac{1}{2\pi i} \int_{|z|=R} \frac{f(z)}{z^{n+1}} dz.$$

For any integer $n > h$,

$$|a_n| \leq \frac{1}{2\pi} \int_{|z|=R} \left| \frac{f(z)}{z^{n+1}} \right| \cdot |dz| = \frac{1}{2\pi R^n} \int_0^{2\pi} |f(Re^{i\theta})| d\theta$$

$$\leq \frac{A + BR^h}{R^n}.$$

Letting $R \to +\infty$, we obtain that $a_n = 0$, which implies that $f(z)$ is a polynomial of degree bounded by h.

5310

Let f be an entire function that satisfies $|\mathrm{Re}\{f(z)\}| \leq |z|^n$ for all z, where n is a positive integer. Show that f is a polynomial of degree at most n.

(*Indiana*)

Solution.

Let R be an arbitrary positive number. Then it follows from Schwarz's Theorem that when $|z| < R$,

$$f(z) = \frac{1}{2\pi i} \int_{|\zeta|=R} \mathrm{Re}\{f(\zeta)\} \cdot \frac{\zeta + z}{\zeta - z} \cdot \frac{d\zeta}{i\zeta} + i\,\mathrm{Im}\{f(0)\}.$$

Especially when $|z| = \frac{R}{2}$,

$$|f(z)| \leq \frac{1}{2\pi} \cdot 3R^n \cdot 2\pi + |\mathrm{Im}\{f(0)\}| = 3R^n + |\mathrm{Im}\{f(0)\}|,$$

which implies that there exist constants A, B such that

$$|f(z)| \leq A|z|^n + B$$

holds for all $z \in \mathbb{C}$.

Let

$$f(z) = \sum_{k=0}^{\infty} a_k z^k,$$

where

$$a_k = \frac{1}{2\pi i} \int_{|z|=r} \frac{f(z)}{z^{k+1}} dz.$$

Hence when $k > n$,

$$|a_k| \le \frac{1}{2\pi} \int_{|z|=r} \frac{|f(z)|}{|z|^{k+1}} |dz| \le \frac{Ar^n + B}{r^k} \to 0 \quad (r \to +\infty),$$

which shows that $f(z)$ is a polynomial of degree at most n.

5311

Compute the double integral

$$\iint_D \cos z \, dx \, dy$$

where D is the disk given by $\{z = x + iy \in \mathbb{C} : x^2 + y^2 < 1\}$.

(*Iowa*)

Solution 1.

For any positive integer n, we have

$$\iint_D z^n dx \, dy = \left(\int_0^1 r^{n+1} dr \right) \left(\int_0^{2\pi} e^{in\theta} d\theta \right) = 0.$$

Hence

$$\iint_D \cos z \, dx \, dy = \iint_D \left(\sum_{n=0}^{\infty} \frac{(-1)^n}{(2n)!} z^{2n} \right) dx \, dy$$

$$= \sum_{n=0}^{\infty} \left(\iint_D \frac{(-1)^n}{(2n)!} z^{2n} dx \, dy \right) = \iint_D 1 \, dx \, dy = \pi.$$

Solution 2.

First we have the following complex forms of Green's formula:

$$\iint_D w_z \, dx \, dy = \iint_D \frac{1}{2}(w_x - i w_y) dx \, dy$$

$$= -\frac{1}{2i} \int_{\partial D} w(dx - i \, dy) = -\frac{1}{2i} \int_{\partial D} w \, d\bar{z},$$

$$\iint_D w_{\bar{z}} \, dx \, dy = \iint_D \frac{1}{2}(w_x + i \, w_y) dx \, dy$$

$$= \frac{1}{2i} \int_{\partial D} w(dx + i \, dy) = \frac{1}{2i} \int_{\partial D} w \, dz.$$

The problem can be solved directly by either one of the above two forms:

$$\iint_D \cos z \, dx \, dy = \frac{1}{2i} \int_{|z|=1} \bar{z} \cos z \, dz = \frac{1}{2i} \int_{|z|=1} \frac{\cos z}{z} dz = \pi;$$

or

$$\iint_D \cos z \, dx \, dy = -\frac{1}{2i} \int_{|z|=1} \sin z \, d\bar{z} = -\frac{1}{2i} \int_{|z|=1} \sin z \, d\left(\frac{1}{z}\right)$$

$$= \frac{1}{2i} \int_{|z|=1} \frac{\sin z}{z^2} dz = \pi.$$

5312

Let

$$f(z) = \sum_{n=0}^{\infty} a_n z^n$$

be analytic in $D = \{|z| < 1\}$ and assume that the integral

$$A = \iint_D |f'(z)|^2 dx \, dy$$

is finite.

(a) Express A in terms of the coefficients a_n.

(b) Prove that

$$|f(z) - f(0)| \leq \sqrt{\frac{A}{\pi} \log \frac{1}{1 - |z|^2}}$$

for $z \in D$.

(Indiana)

Solution.

(a) By

$$f'(z) = \sum_{n=1}^{\infty} n a_n z^{n-1},$$

we have

$$A = \iint_D |f'(z)|^2 dx\,dy = \int_0^1 r\,dr \int_0^{2\pi} (f'(re^{i\theta}))\overline{(f'(re^{i\theta}))}d\theta$$

$$= \int_0^1 r\,dr \int_0^{2\pi} \left(\sum_{n=1}^{\infty} n a_n r^{n-1} e^{i(n-1)\theta} \right)$$

$$\times \left(\sum_{n=1}^{\infty} n \bar{a}_n r^{n-1} e^{-i(n-1)\theta} \right) d\theta.$$

Noting that

$$\int_0^{2\pi} e^{ik\theta} \cdot e^{-il\theta} d\theta = \begin{cases} 0, & k \neq l, \\ 2\pi, & k = l, \end{cases}$$

we obtain that

$$A = \iint_D |f'(z)|^2 dx\,dy = \int_0^1 r\,dr \int_0^{2\pi} \sum_{n=1}^{\infty} n^2 |a_n|^2 r^{2n-2}\,d\theta$$

$$= 2\pi \int_0^1 \sum_{n=1}^{\infty} n^2 |a_n|^2 r^{2n-1}\,dr = \pi \sum_{n=1}^{\infty} n |a_n|^2.$$

(b) By Cauchy's inequality, we have

$$|f(z) - f(0)| = \left| \sum_{n=1}^{\infty} a_n z^n \right| = \left| \sum_{n=1}^{\infty} \left(\sqrt{n} a_n \cdot \frac{1}{\sqrt{n}} z^n \right) \right|$$

$$\leq \sqrt{\sum_{n=1}^{\infty} n|a_n|^2 \cdot \sum_{n=1}^{\infty} \frac{1}{n}|z|^{2n}} = \sqrt{\frac{A}{\pi} \log \frac{1}{1-|z|^2}}.$$

5313

Let f be analytic in $\{0 < |z| < 1\}$ and in L^2 with respect to planar Lebesque measure. Is 0 a removable singularity? Proof or counterexample.

(Stanford)

Solution.

The answer to the problem is Yes.

Let the Laurent expansion of f in $\{z : 0 < |z| < 1\}$ be

$$f(z) = \sum_{n=-\infty}^{\infty} a_n z^n,$$

where

$$a_n = \frac{1}{2\pi i} \int_{|z|=r<1} \frac{f(z)}{z^{n+1}} dz, \quad (n = 0, \pm 1, \pm 2, \ldots).$$

From

$$|a_n| \leq \frac{1}{2\pi} \int_0^{2\pi} \frac{|f(re^{i\theta})|}{r^n} d\theta,$$

we have

$$|a_n|^2 r^{2n+1} \leq \left(\frac{1}{2\pi} \int_0^{2\pi} |f(re^{i\theta})| \sqrt{r}\, d\theta \right)^2$$

$$\leq \frac{1}{2\pi} \int_0^{2\pi} |f(re^{i\theta})|^2 r\, d\theta.$$

Let $\varepsilon < 1$ be a small positive number, and then

$$\int_\varepsilon^1 |a_n|^2 r^{2n+1}\, dr \le \frac{1}{2\pi} \int_0^{2\pi} \int_\varepsilon^1 |f(re^{i\theta})|^2 r\, dr\, d\theta$$

$$< \frac{1}{2\pi} \iint_{0<|z|<1} |f(z)|^2 dx\, dy.$$

Then a_n must be zero when $n \le -1$. Otherwise, let $\varepsilon \to +0$, the left side of the above inequality will tend to infinity, while the right side of the inequality is finite, which leads to a contradiction. Hence

$$f(z) = \sum_{n=0}^{\infty} a_n z^n,$$

which shows that $z = 0$ is a removable singularity of f.

5314

Evaluate the integral

$$\int_{|z|=\rho} \frac{|dz|}{|z-a|^2}, \quad |a| \ne \rho.$$

(Indiana)

Solution.

Let $z = \rho e^{i\theta}$, $a = r e^{i\phi}$.

$$\int_{|z|=\rho} \frac{dz}{|z-a|^2} = \int_0^{2\pi} \frac{\rho\, d\theta}{\rho^2 + r^2 - \rho r(e^{i(\theta-\phi)} + e^{i(\phi-\theta)})}$$

$$= \int_0^{2\pi} \frac{\rho\, d\theta}{\rho^2 + r^2 - \rho r(e^{i\theta} + e^{-i\theta})}$$

$$= \int_{|z|=\rho} \frac{\rho\, dz/(iz)}{\rho^2 + r^2 - rz - \rho^2 r/z}$$

$$= \int_{|z|=\rho} \frac{\rho i\, dz}{rz^2 - (\rho^2 + r^2)z + \rho^2 r}$$

When $r < \rho$,

$$\int_{|z|=\rho} \frac{|dz|}{|z-a|^2} = \int_{|z|=\rho} \frac{\rho i \, dz}{r(z - \frac{\rho^2}{r})(z-r)}$$

$$= 2\pi i \cdot \frac{\rho i}{r} \mathrm{Res}\left(\frac{1}{(z - \frac{\rho^2}{r})(z-r)}, r\right)$$

$$= \frac{2\pi\rho}{\rho^2 - r^2}.$$

When $r > \rho$,

$$\int_{|z|=\rho} \frac{|dz|}{|z-a|^2} = \int_{|z|=\rho} \frac{\rho i \, dz}{r(z - \frac{\rho^2}{r})(z-r)}$$

$$= 2\pi i \cdot \frac{\rho i}{r} \mathrm{Res}\left(\frac{1}{(z - \frac{\rho^2}{r})(z-r)}, \frac{\rho^2}{r}\right)$$

$$= \frac{2\pi\rho}{r^2 - \rho^2}.$$

5315

Evaluate

$$\int_0^{\frac{\pi}{2}} \frac{dx}{a + \sin^2 x}, \quad |a| > 1,$$

by the method of residues.

(Columbia)

Solution.
Denote

$$I(a) = \int_0^{\frac{\pi}{2}} \frac{dx}{a + \sin^2 x}.$$

It is obvious that $I(a)$ is an analytic function in $\{a : |a| > 1\}$. Then we have

$$I(a) = \int_0^{\frac{\pi}{2}} \frac{dx}{a + \sin^2 x} = \int_0^{\frac{\pi}{2}} \frac{2\,dx}{2a + 1 - \cos 2x}$$

$$= \int_0^{\pi} \frac{dx}{2a + 1 - \cos x} = \frac{1}{2} \int_{-\pi}^{\pi} \frac{dx}{2a + 1 - \cos x}.$$

Let $z = e^{ix}$, then

$$dx = \frac{dz}{iz},$$

$$\cos x = \frac{z + z^{-1}}{2},$$

and

$$I(a) = \int_{|z|=1} \frac{i\,dz}{z^2 - 2(2a + 1)z + 1}.$$

Denote the two roots of $z^2 - 2(2a + 1)z + 1 = 0$ by z_1 and z_2. Since $z_1 \cdot z_2 = 1$, we may assume that $|z_1| > 1$, $|z_2| < 1$. By the residue theorem we have

$$I(a) = \int_{|z|=1} \frac{i\,dz}{(z - z_1)(z - z_2)} = \frac{2\pi}{z_1 - z_2}$$

$$= \frac{2\pi}{\sqrt{(z_1 + z_2)^2 - 4z_1 z_2}} = \frac{\pi}{2\sqrt{a(a + 1)}}.$$

It should be noted that $\dfrac{\pi}{2\sqrt{a(a+1)}}$ is also analytic in $\{a : |a| > 1\}$, and the branch of $\sqrt{a(a + 1)}$ should be chosen by arg $\sqrt{a(a + 1)}|_{a>1} = 0$.

5316

Consider the function

$$g(z, \theta) = \frac{1}{1 + z \sin \theta}.$$

(a) Use the residue theorem to find an explicit formula for

$$f(z) = \int_0^{2\pi} g(z, \theta) d\theta$$

when $|z| < 1$.

(b) Integrate the Taylor expansion

$$g(z, \theta) = \sum_{n=0}^{\infty} g_n(\theta) z^n$$

term by term to find the coefficients in the Taylor expansion

$$f(z) = \sum_{n=0}^{\infty} f_n z^n.$$

(c) Verify directly that (a) and (b) agree when $|z| < 1$.

(*Courant Inst.*)

Solution.

(a) Let $\zeta = e^{i\theta}$. Then

$$\sin \theta = \frac{e^{i\theta} - e^{-i\theta}}{2i} = \frac{\zeta^2 - 1}{2i\zeta},$$

and

$$f(z) = \int_0^{2\pi} g(z, \theta) d\theta = \int_{|\zeta|=1} \frac{2i\zeta}{z\zeta^2 + 2i\zeta - z} \cdot \frac{d\zeta}{i\zeta}$$

$$= \int_{|\zeta|=1} \frac{2d\zeta}{z(\zeta - \zeta_1)(\zeta - \zeta_2)},$$

where $\zeta_1 = \frac{i}{z}(\sqrt{1 - z^2} - 1)$, $\zeta_2 = \frac{i}{2}(-\sqrt{1 - z^2} - 1)$, and the single-valued branch of $\sqrt{1 - z^2}$ in $\{|z| < 1\}$ is defined by $\sqrt{1 - z^2}|_{z=0} = 1$. Because $|\zeta_1 \cdot \zeta_2| = 1$, we know that $\zeta_1 \in \{|\zeta| < 1\}$ and $\zeta_2 \in \{|\zeta| > 1\}$. Hence

$$f(z) = 2\pi i \cdot \frac{2}{z} \cdot \frac{1}{\zeta_1 - \zeta_2} = \frac{2\pi}{\sqrt{1 - z^2}}.$$

(b) It follows from $|\sin\theta| \leq 1$ and $|z| < 1$ that

$$g(z,\theta) = \sum_{k=0}^{\infty}(-1)^k \sin^k\theta \cdot z^k, \quad (|z| < 1).$$

Since the series converges uniformly for all $\theta \in [0, 2\pi]$, the integration with respect to θ can be taken term by term, and

$$f(z) = \int_0^{2\pi}\left(\sum_{k=0}^{\infty}(-1)^k \sin^k\theta \cdot z^k\right)d\theta = \sum_{k=0}^{\infty}a_k z^k,$$

where

$$a_k = \int_0^{2\pi}(-1)^k \sin^k\theta\, d\theta.$$

It is easy to obtain that $a_{2n-1} = 0$ and

$$a_{2n} = 4\int_0^{\frac{\pi}{2}}\sin^{2n}\theta\, d\theta = \frac{(2n-1)!!}{(2n)!!}\cdot 2\pi.$$

(c) In order to verify that (a) and (b) agree when $|z| < 1$, we develop the function $f(z)$ in (a) into a power series:

$$f(z) = \frac{2\pi}{\sqrt{1-z^2}} = 2\pi(1-z^2)^{-\frac{1}{2}} = 2\pi\sum_{n=0}^{\infty}(-1)^n C_{-\frac{1}{2}}^n z^{2n}.$$

Since

$$(-1)^n C_{-\frac{1}{2}}^n = (-1)^n\frac{(-\frac{1}{2})(-\frac{3}{2})\cdots(-\frac{2n-1}{2})}{n!} = \frac{(2n-1)!!}{(2n)!!},$$

we know that the results in (a) and (b) agree when $|z| < 1$.

5317

If a is real, show that

$$\lim_{R\to\infty}\int_{-R}^{R}e^{-(x+ia)^2}\,dx$$

582   *Problems and Solutions in Mathematics*

exists and is independent of a.

(UC, Irvine)

Solution.

First we have

$$\left| e^{-(x+ia)^2} \right| \le e^{a^2} \cdot e^{-x^2}.$$

It follows from the existence of

$$\lim_{R \to \infty} \int_{-R}^{R} e^{-x^2} dx$$

that

$$\lim_{R \to \infty} \int_{-R}^{R} e^{-(x+ia)^2} dx$$

exists.

Define $f(z) = e^{-z^2}$ and choose the contour of integration $\Gamma = \Gamma_1 \cup \Gamma_2 \cup \Gamma_3 \cup \Gamma_4$ as shown in Fig. 5.5.

As $f(z)$ is analytic inside Γ, by Cauchy integral theorem,

$$\int_{\Gamma} f(z)dz = \int_{\Gamma_1} f(z)dz + \int_{\Gamma_2} f(z)dz + \int_{\Gamma_3} f(z)dz + \int_{\Gamma_4} f(z)dz$$

$$= \int_{-R}^{R} e^{-x^2} dx + ie^{-R^2} \int_0^a e^{y^2 - 2Ryi} dy$$

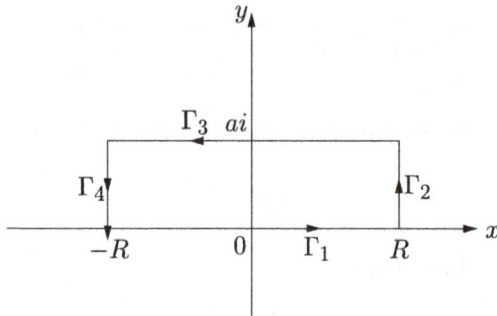

Fig. 5.5.

$$-\int_{-R}^{R} e^{-(x+ia)^2} dx - ie^{-R^2} \int_{0}^{a} e^{y^2+2Ryi} dy$$
$$= 0.$$

Letting $R \to \infty$, it follows from the facts that $e^{-R^2} \to 0 \ (R \to \infty)$ and

$$\left| \int_{0}^{a} e^{y^2 \pm 2Ryi} dy \right| \leq \int_{0}^{a} e^{y^2} dy$$

that

$$\lim_{R \to \infty} \int_{-R}^{R} e^{-(x+ia)^2} dx = \lim_{R \to \infty} \int_{-R}^{R} e^{-x^2} dx = \sqrt{\pi}.$$

5318

Let $n \geq 2$ be an integer. Compute

$$\int_{0}^{\infty} \frac{1}{1+x^n} dx.$$

(*Iowa*)

Solution.

Let $f(z) = \frac{1}{1+z^n}$, and select the integral contour Γ as shown in Fig. 5.6. $f(z)$ has one simple pole $z = e^{\frac{\pi}{n}i}$ inside Γ. By the

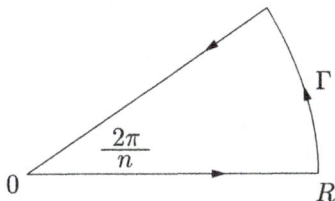

Fig. 5.6.

residue theorem, we have

$$\int_\Gamma f(z)dz = 2\pi i \operatorname{Res}(f, e^{\frac{\pi}{n}i}).$$

$$\int_\Gamma f(z)dz = \int_0^R \frac{dx}{1+x^n} + \int_0^{\frac{2\pi}{n}} i\, Re^{i\theta} f(Re^{i\theta})d\theta + \int_R^0 \frac{e^{\frac{2\pi}{n}i}dx}{1+x^n}$$

$$= (1 - e^{\frac{2\pi}{n}i}) \int_0^R \frac{dx}{1+x^n} + \int_0^{\frac{2\pi}{n}} i\, Re^{i\theta} f(Re^{i\theta})d\theta.$$

It is obvious that

$$\lim_{R\to\infty} \int_0^{\frac{2\pi}{n}} i\, Re^{i\theta} f(Re^{i\theta})d\theta = 0,$$

and

$$\operatorname{Res}(f, e^{\frac{\pi}{n}i}) = \left. \frac{1}{(1+z^n)'} \right|_{z=e^{\frac{\pi}{n}i}} = \frac{1}{ne^{\frac{n-1}{n}\pi i}}.$$

Letting $R \to \infty$, we obtain

$$\int_0^\infty \frac{dz}{1+x^n} = \frac{2\pi i}{ne^{\frac{n-1}{n}\pi i}(1 - e^{\frac{2\pi}{n}i})} = \frac{2\pi i}{n(e^{\frac{\pi}{n}i} - e^{-\frac{\pi}{n}i})} = \frac{\pi}{n\sin\frac{\pi}{n}}.$$

5319

Evaluate

$$\int_0^\infty \cos(x^2)dx$$

with full justification.

(Minnesota)

Solution.
Define

$$f(z) = e^{-z^2},$$

and choose the contour of integration $\Gamma = \sum_{j=1}^3 \Gamma_j$ as shown in Fig. 5.7. Because $f(z) = e^{-z^2}$ is analytic on Γ and inside Γ, by

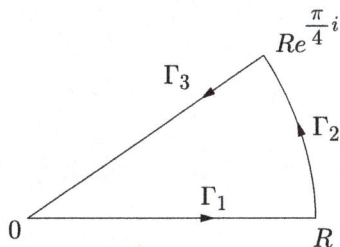

Fig. 5.7.

Cauchy integral theorem, we have

$$\int_\Gamma f(z)dz = \sum_{j=1}^{3} \int_{\Gamma_j} f(z)dz = 0.$$

For the integral of $f(z)$ on Γ_2, we make a change of variable by $w = z^2$, then

$$\int_{\Gamma_2} f(z)dz = \int_{\gamma_2} e^{-w} \cdot \frac{dw}{2w^{\frac{1}{2}}},$$

where

$$\gamma_2 = \left\{ w : |w| = R^2, \ 0 \leq \arg w \leq \frac{\pi}{2} \right\}.$$

By Jordan's Lemma, we have

$$\lim_{R\to\infty} \int_{\Gamma_2} f(z)dz = 0.$$

For the integral of $f(z)$ on Γ_3, we have

$$\int_{\Gamma_3} f(z)dz = -\int_0^R e^{-x^2 i} e^{\frac{\pi}{4}i} dx$$

$$= -\int_0^R \frac{\sqrt{2}}{2}(\cos x^2 + \sin x^2)dx$$

$$- i\int_0^R \frac{\sqrt{2}}{2}(\cos x^2 - \sin x^2)dx.$$

It is well known that

$$\int_{\Gamma_1} f(z)dz = \int_0^R e^{-x^2}dx \to \frac{\sqrt{\pi}}{2}$$

when $R \to \infty$. Hence we obtain by letting $R \to \infty$ that

$$\int_0^\infty (\cos x^2 + \sin x^2)dx + i \int_0^\infty (\cos x^2 - \sin x^2)dx = \frac{\sqrt{2\pi}}{2},$$

which implies

$$\int_0^\infty \cos x^2\, dx = \int_0^\infty \sin x^2\, dx = \frac{\sqrt{2\pi}}{4}.$$

5320

Evaluate

$$\int_0^\infty \frac{\sin^2 x}{x^2}dx.$$

(*Iowa*)

Solution.
Define

$$f(z) = \frac{1 - e^{2iz}}{z^2},$$

and select the integral contour Γ as shown in Fig. 5.8. Because $f(z)$ is analytic inside Γ, by Cauchy integral theorem,

$$\int_\Gamma f(z)dz = 0,$$

where

$$\int_\Gamma f(z)dz = \int_\varepsilon^R \frac{1 - e^{2ix}}{x^2}dx + \int_0^\pi i\, Re^{i\theta} f(Re^{i\theta})d\theta$$

$$+ \int_{-R}^{-\varepsilon} \frac{1 - e^{2ix}}{x^2}dx + \int_\pi^0 i\varepsilon e^{i\theta} f(\varepsilon e^{i\theta})d\theta$$

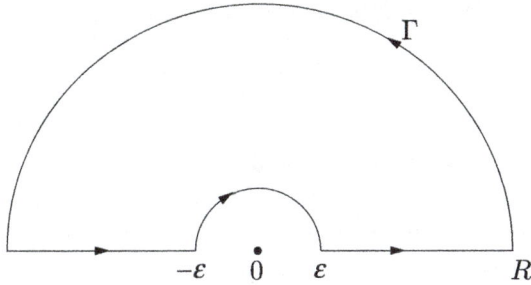

Fig. 5.8.

$$= \int_{\varepsilon}^{R} \frac{2 - e^{2ix} - e^{-2ix}}{x^2}\,dx + \int_{0}^{\pi} i\,Re^{i\theta}f(Re^{i\theta})d\theta$$

$$+ \int_{\pi}^{0} i\varepsilon e^{i\theta}f(\varepsilon e^{i\theta})d\theta$$

$$= \int_{\varepsilon}^{R} \frac{4\sin^2 x}{x^2}\,dx + \int_{0}^{\pi} i\,Re^{i\theta}f(Re^{i\theta})d\theta$$

$$+ \int_{\pi}^{0} i\varepsilon e^{i\theta}f(\varepsilon e^{i\theta})d\theta.$$

It is easy to see that

$$\lim_{R\to\infty} \int_{0}^{\pi} i\,Re^{i\theta}f(Re^{i\theta})d\theta = 0$$

and

$$\lim_{\varepsilon\to 0} \int_{\pi}^{0} i\varepsilon e^{i\theta}f(\varepsilon e^{i\theta})d\theta = -\pi i\,\mathrm{Res}(f,0).$$

Since the Laurent expansion of f about $z = 0$ is

$$f(z) = \sum_{n=-1}^{\infty} a_n z^n,$$

where $a_{-1} = -2i$, we know that $\mathrm{Res}(f,0) = -2i$.

Letting $\varepsilon \to 0$ and $R \to \infty$, we obtain

$$\int_0^\infty \frac{\sin^2 x}{x^2}\,dx = \frac{\pi}{2}.$$

5321

Let $f(z)$ be holomorphic in the unit disk $|z| \le 1$. Prove that

$$\int_0^1 f(x)\,dx = \frac{1}{2\pi i} \int_{|z|=1} f(z)\log z\,dz,$$

where respective integration goes along the straight line from 0 to 1 and along the positively oriented unit circle starting from the point $z = 1$. The branch of log is chosen to be real for positive z.

(SUNY, Stony Brook)

Solution.

Let the contour of integration Γ be shown as in Fig. 5.9, and the single-valued branch of $\log z$ be chosen by $\arg z|_{z=-1} = \pi$. Since $f(z)\log z$ is holomorphic inside the contour Γ, by Cauchy integral theorem,

$$\int_\Gamma f(z)\log z\,dz = 0,$$

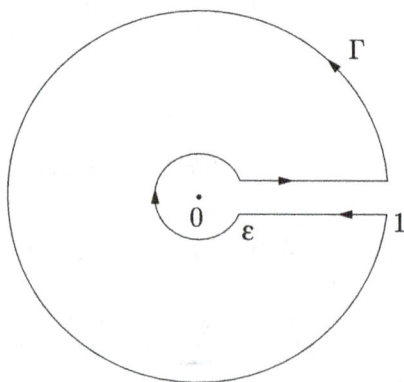

Fig. 5.9.

where

$$\int_\Gamma f(z)\log z\,dz = \int_\varepsilon^1 f(x)\log x\,dx + \int_{|z|=1} f(z)\log z\,dz$$

$$+ \int_1^\varepsilon f(x)(\log x + 2\pi i)dx$$

$$+ \int_{2\pi}^0 f(\varepsilon e^{i\theta})\log(\varepsilon e^{i\theta})i\varepsilon e^{i\theta}d\theta$$

$$= -2\pi i \int_\varepsilon^1 f(x)dx + \int_{|z|=1} f(z)\log z\,dz$$

$$- \int_0^{2\pi} f(\varepsilon e^{i\theta})\log(\varepsilon e^{i\theta})i\varepsilon e^{i\theta}d\theta.$$

It is easy to see that

$$\lim_{\varepsilon\to 0}\int_0^{2\pi} f(\varepsilon e^{i\theta})\log(\varepsilon e^{i\theta})i\varepsilon e^{i\theta}d\theta = 0.$$

Letting $\varepsilon \to 0$, we obtain

$$\int_0^1 f(x)dx = \frac{1}{2\pi i}\int_{|z|=1} f(z)\log z\,dz,$$

where the integration contour $|z| = 1$ has starting point and end point $z = 1$, and the value of $\log z$ at the starting point $z = 1$ is defined as 0.

5322

Find the value of

$$\int_0^{2\pi} \log|a + be^{i\phi}|d\phi$$

where a and b are complex constants, not both equal to zero.

(*Harvard*)

Solution.

First we assume $|a| > |b|$, and then the multi-valued analytic function $\log(a + bz)$ has single-valued branch on $\{z : |z| \le 1\}$. Take $e^{i\phi} = z$, then $d\phi = \frac{dz}{iz}$, and

$$\int_0^{2\pi} \log|a + be^{i\phi}|d\phi = \mathrm{Re}\left\{\int_0^{2\pi} \log(a + be^{i\phi})d\phi\right\}$$

$$= \mathrm{Re}\left\{\int_{|z|=1} \frac{\log(a + bz)}{iz}dz\right\}$$

$$= \mathrm{Re}\{2\pi \log a\} = 2\pi \log|a|.$$

When $|a| < |b|$, we have

$$\int_0^{2\pi} \log|a + be^{i\phi}|d\phi = \int_0^{2\pi} \log|\bar{b} + \bar{a}e^{i\phi}|d\phi$$

$$= 2\pi \log|\bar{b}| = 2\pi \log|b|.$$

In the case $|a| = |b|$, let $b = ae^{i\alpha}$. Then

$$\int_0^{2\pi} \log|a + be^{i\phi}|d\phi = \int_0^{2\pi} (\log|a| + \log|1 + e^{i(\phi+\alpha)}|)d\phi$$

$$= 2\pi \log|a| + \int_{-\pi}^{\pi} \log|1 + e^{i\phi}|d\phi.$$

In order to evaluate the integral

$$\int_{-\pi}^{\pi} \log|1 + e^{i\phi}|d\phi,$$

we define

$$f(z) = \frac{\log(1 + z)}{z},$$

where the single-valued branch is defined by $\log(1 + z)|_{z=0} = 0$. Choose a contour of integration $\Gamma = \Gamma_\varepsilon \cup \gamma_\varepsilon$ as shown in

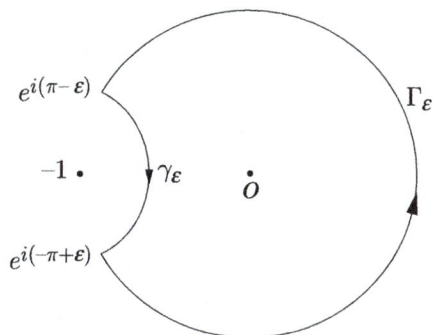

Fig. 5.10.

Fig. 5.10. Since $f(z)$ is analytic on Γ and inside Γ, by Cauchy integral theorem, $\int_\Gamma f(z)dz = 0$. Because

$$\left| \int_{\gamma_\varepsilon} f(z)dz \right| \le \frac{\log \frac{1}{\varepsilon} + \frac{\pi}{2}}{1 - \varepsilon} \cdot \pi\varepsilon \to 0 \quad (\varepsilon \to 0),$$

we have

$$\int_{-\pi}^{\pi} \log|1 + e^{i\phi}|d\phi = \text{Re} \int_{-\pi}^{\pi} \log(1 + e^{i\phi})d\phi$$

$$= \lim_{\varepsilon \to 0} \text{Re} \left\{ \int_{\Gamma_\varepsilon} \log(1 + z)\frac{dz}{iz} \right\}$$

$$= \lim_{\varepsilon \to 0} \text{Re} \left\{ \frac{1}{i} \int_\Gamma f(z)dz \right\} = 0.$$

Hence we obtain

$$\int_0^{2\pi} \log|a + be^{i\phi}|d\phi = 2\pi \max\{\log|a|, \log|b|\}.$$

5323

Evaluate

$$\int_0^\infty \frac{\log x}{(1 + x)^3}dx.$$

(Iowa)

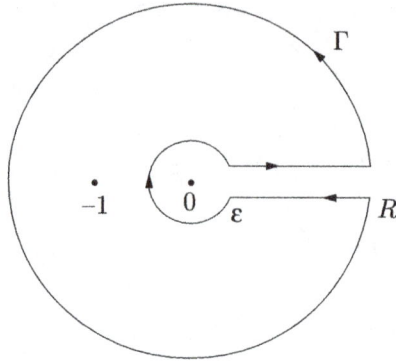

Fig. 5.11.

Solution.

Let

$$f(z) = \frac{\log^2 z}{(1+z)^3},$$

and select the integral path Γ as shown in Fig. 5.11. The single-valued branch of $\log z$ is chosen by $\arg z|_{z=-1} = \pi$. By the residue theorem, we have

$$\int_\Gamma f(z)dz = 2\pi i\, \mathrm{Res}(f, -1),$$

where

$$\int_\Gamma f(z)dz = \int_\varepsilon^R \frac{\log^2 x}{(1+x)^3}dx + \int_0^{2\pi} i\, Re^{i\theta} f(Re^{i\theta})d\theta$$

$$+ \int_R^\varepsilon \frac{(\log x + 2\pi i)^2}{(1+x)^3}dx + \int_{2\pi}^0 i\varepsilon e^{i\theta} f(\varepsilon e^{i\theta})d\theta$$

$$= \int_\varepsilon^R \frac{-4\pi i \log x + 4\pi^2}{(1+x)^3}dx + \int_0^{2\pi} i\, Re^{i\theta} f(Re^{i\theta})d\theta$$

$$+ \int_{2\pi}^0 i\varepsilon e^{i\theta} f(\varepsilon e^{i\theta})d\theta.$$

It is obvious that

$$\lim_{R \to \infty} \int_0^{2\pi} i\, Re^{i\theta} f(Re^{i\theta}) d\theta = 0$$

and

$$\lim_{\varepsilon \to 0} \int_{2\pi}^0 i\varepsilon e^{i\theta} f(\varepsilon e^{i\theta}) d\theta = 0.$$

In order to find $\text{Res}(f, -1)$, we consider the Laurent expansion of f about $z = -1$:

$$f(z) = \frac{\log^2[(z+1) - 1]}{(z+1)^3} = \frac{(\pi i + \log[1 - (z+1)])^2}{(z+1)^3}$$

$$= \frac{(\pi i - (z+1) - \frac{1}{2}(z+1)^2 - \cdots)^2}{(z+1)^3}$$

$$= \sum_{n=-3}^{\infty} a_n(z+1)^n,$$

where $a_{-1} = 1 - \pi i$. Hence

$$2\pi i\, \text{Res}(f, -1) = 2\pi i + 2\pi^2.$$

As $\varepsilon \to 0$ and $R \to \infty$, it turns out that

$$\int_0^\infty \frac{-4\pi i \log x + 4\pi^2}{(1+x)^3} dx = 2\pi i + 2\pi^2.$$

Comparing the imaginary parts on the two sides of the above identity, we obtain

$$\int_0^\infty \frac{\log x}{(1+x)^3} dx = -\frac{1}{2}.$$

5324

Evaluate the following integrals:

(a) $\int_{-i\infty}^{+i\infty} \frac{dz}{(z^2-4)\log(z+1)}$ (the integration is over the imaginary axis),

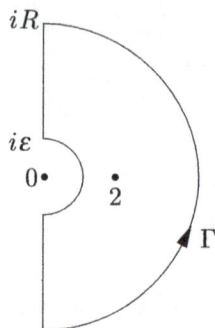

Fig. 5.12.

(b) $\int_0^\infty \frac{x^\alpha}{x^3+1} dx$ for α in the range $-1 < \alpha < 2$.

(Courant Inst.)

Solution.

(a) Define

$$f(z) = \frac{1}{(z^2-4)\log(z+1)}.$$

The single-valued branch for $\log(z+1)$ is chosen by $\log(z+1)|_{z=0} = 0$, and the contour Γ of integration is shown in Fig. 5.12. As $f(z)$ is analytic on and inside Γ except a simple pole at $z=2$, we have

$$\int_\Gamma f(z)dz = 2\pi i \operatorname{Res}(f,2),$$

where

$$\int_\Gamma f(z)dz = \int_{-\frac{\pi}{2}}^{\frac{\pi}{2}} f(Re^{i\theta})i\,Re^{i\theta}\,d\theta - \int_{-\frac{\pi}{2}}^{\frac{\pi}{2}} f(\varepsilon e^{i\theta})i\varepsilon e^{i\theta}\,d\theta$$

$$- \int_{-iR}^{-i\varepsilon} f(z)dz - \int_{i\varepsilon}^{iR} f(z)dz,$$

and

$$\operatorname{Res}(f,2) = \lim_{z\to2} \frac{(z-2)}{(z^2-4)\log(z+1)} = \frac{1}{4\log 3}.$$

Because

$$\lim_{R\to\infty}\int_{-\frac{\pi}{2}}^{\frac{\pi}{2}} f(Re^{i\theta})iRe^{i\theta}\,d\theta = 0$$

and

$$\lim_{\varepsilon\to 0}\int_{-\frac{\pi}{2}}^{\frac{\pi}{2}} f(\varepsilon e^{i\theta})i\varepsilon e^{i\theta}\,d\theta = \pi i\,\mathrm{Res}(f,0) = -\frac{\pi i}{4},$$

by letting $\varepsilon \to 0$ and $R \to \infty$, we obtain

$$\int_{-i\infty}^{+i\infty} \frac{dz}{(z^2-4)\log(z+1)} = \frac{\pi i}{4}\left(1 - \frac{2}{\log 3}\right).$$

(b) Define

$$f(z) = \frac{z^\alpha}{z^3+1}.$$

The single-valued branch for z^α is chosen by $\arg z|_{z=x>0} = 0$, and the contour Γ of integration is shown in Fig. 5.13. As $f(z)$ is analytic on and inside Γ except a simple pole at $z = e^{\frac{\pi}{3}i}$, we have

$$\int_\Gamma f(z)dz = 2\pi i\,\mathrm{Res}(f, e^{\frac{\pi}{3}i}),$$

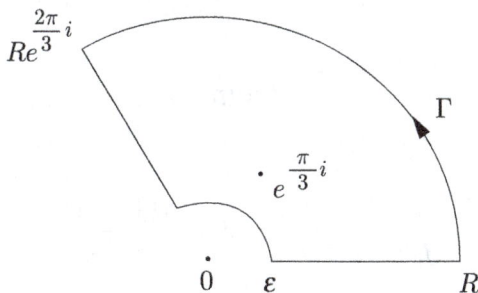

Fig. 5.13.

where

$$\int_{\Gamma} f(z)dz = \int_{\varepsilon}^{R} f(x)dx + \int_{0}^{\frac{2\pi}{3}} f(Re^{i\theta})i\,Re^{i\theta}\,d\theta$$

$$- \int_{\varepsilon}^{R} e^{\frac{2\pi\alpha}{3}i}f(x) \cdot e^{\frac{2\pi}{3}i}\,dx - \int_{0}^{\frac{2\pi}{3}} f(\varepsilon e^{i\theta})i\varepsilon e^{i\theta}\,d\theta,$$

and

$$\text{Res}(f, e^{\frac{\pi}{3}i}) = \lim_{z \to e^{\frac{\pi}{3}i}} (z - e^{\frac{\pi}{3}i})f(z)$$

$$= \frac{e^{\frac{\pi\alpha}{3}i}}{3e^{\frac{2\pi}{3}i}}$$

$$= \frac{1}{3e^{\frac{\pi}{3}(2-\alpha)i}}.$$

Because

$$\lim_{R \to \infty} \int_{0}^{\frac{2\pi}{3}} f(Re^{i\theta})iRe^{i\theta}\,d\theta = 0$$

when $\alpha < 2$ and

$$\lim_{\varepsilon \to 0} \int_{0}^{\frac{2\pi}{3}} f(\varepsilon e^{i\theta})i\varepsilon e^{i\theta}\,d\theta = 0$$

when $\alpha > -1$, by letting $\varepsilon \to 0$ and $R \to \infty$, we obtain

$$\int_{0}^{\infty} \frac{x^{\alpha}}{x^3 + 1}dx = \frac{\pi}{3\sin(\frac{2-\alpha}{3}\pi)}.$$

5325

Show that

$$\int_{0}^{\infty} \frac{x^{\alpha}}{(1+x^2)^2}dx = \frac{\pi(1-\alpha)}{4\cos(\frac{\pi\alpha}{2})},$$

for $-1 < \alpha < 3$, $\alpha \neq 1$. What happens if $\alpha = 1$?

(*Harvard*)

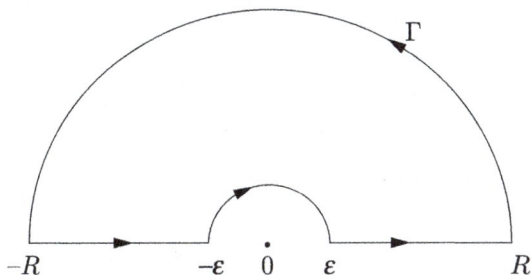

Fig. 5.14.

Solution.

Let

$$f(z) = \frac{z^\alpha}{(1+z^2)^2},$$

where $(\arg z^\alpha)_{z=x>0} = 0$, and select the integral path Γ as shown in Fig. 5.14. By the residue theorem, we have

$$\int_\Gamma f(z)dz = 2\pi i \, \text{Res}(f(z), i),$$

where

$$\begin{aligned}
\int_\Gamma f(z)dz &= \int_\varepsilon^R \frac{x^\alpha}{(1+x^2)^2}dx + \int_0^\pi i\,Re^{i\theta}f(Re^{i\theta})d\theta \\
&\quad + \int_{-R}^{-\varepsilon} \frac{(-x)^\alpha e^{i\pi\alpha}}{(1+x^2)^2}dx + \int_\pi^0 i\varepsilon e^{i\theta}f(\varepsilon e^{i\theta})d\theta \\
&= (1+e^{i\pi\alpha})\int_\varepsilon^R \frac{x^\alpha}{(1+x^2)^2}dx + \int_0^\pi i\,Re^{i\theta}f(Re^{i\theta})d\theta \\
&\quad + \int_\pi^0 i\varepsilon e^{i\theta}f(\varepsilon e^{i\theta})d\theta,
\end{aligned}$$

and

$$\text{Res}(f(z), i) = \lim_{z\to i}\left[\frac{z^\alpha}{(z+i)^2}\right]' = \frac{1-\alpha}{4i}e^{i\frac{\pi\alpha}{2}}.$$

It follows from $\alpha < 3$ that

$$\lim_{R\to\infty}\int_0^\pi i\,Re^{i\theta}f(Re^{i\theta})d\theta = 0,$$

and from $\alpha > -1$ that

$$\lim_{\varepsilon\to 0}\int_\pi^0 i\varepsilon e^{i\theta}f(\varepsilon e^{i\theta})d\theta = 0.$$

Letting $\varepsilon \to 0$ and $R \to \infty$, we obtain

$$(1+e^{i\pi\alpha})\int_0^\infty \frac{x^\alpha}{(1+x^2)^2}dx = \frac{\pi(1-\alpha)}{2}e^{i\frac{\pi\alpha}{2}}.$$

When $\alpha \neq 1$,

$$\int_0^\infty \frac{x^\alpha}{(1+x^2)^2}dx = \frac{\pi(1-\alpha)}{4\cdot\left(\frac{e^{i\frac{\pi\alpha}{2}}+e^{-i\frac{\pi\alpha}{2}}}{2}\right)} = \frac{\pi(1-\alpha)}{4\cos\left(\frac{\pi\alpha}{2}\right)}.$$

when $\alpha = 1$.

$$\int_0^\infty \frac{x}{(1+x^2)^2}dx = \lim_{\alpha\to 1}\frac{\pi(1-\alpha)}{4\cos\left(\frac{\pi\alpha}{2}\right)} = \frac{1}{2}.$$

5326

(a) Prove that

$$\int_0^\infty e^{ix}x^{-\alpha}\,dx$$

converges if $0 < \alpha < 1$.

(b) Use complex integration to show that

$$\int_0^\infty x^{-\alpha}\cos x\,dx = \sin\frac{\pi\alpha}{2}\cdot\Gamma(-\alpha+1).$$

(*Harvard*)

Solution.

(a)

$$\int_0^\infty e^{ix} x^{-\alpha}\, dx = \left(\int_0^1 x^{-\alpha} \cos x\, dx + \int_1^\infty x^{-\alpha} \cos x\, dx \right)$$
$$+\, i \int_0^\infty x^{-\alpha} \sin x\, dx.$$

It follows from $\alpha < 1$ that

$$\int_0^1 x^{-\alpha} \cos x\, dx$$

is convergent. It is also obvious that

$$\left| \int_1^A \cos x\, dx \right| \le 2, \quad \left| \int_0^A \sin x\, dx \right| \le 2,$$

$x^{-\alpha}$ is monotonic decreasing and

$$\lim_{x \to +\infty} x^{-\alpha} = 0 \quad \text{for } \alpha > 0.$$

By Dirichlet's criterion, we know that $\int_1^\infty x^{-\alpha} \cos x\, dx$ and $\int_0^\infty x^{-\alpha} \sin x\, dx$ are also convergent. Hence $\int_0^\infty e^{ix} x^{-\alpha}\, dx$ is convergent when $0 < \alpha < 1$.

(b) Let $f(z) = z^{-\alpha} e^{-z}$, and the contour of integration Γ is chosen as shown in Fig. 5.15. The single-valued branch of $f(z)$ on Γ is defined by $z^{-\alpha}|_{z=x>0} > 0$.

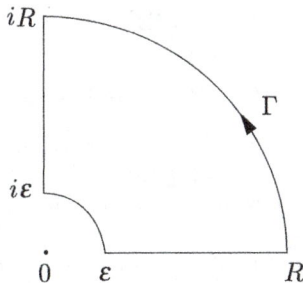

Fig. 5.15.

By Cauchy integral theorem,

$$\int_\Gamma f(z)dz = \int_\varepsilon^R x^{-\alpha}e^{-x}\,dx + \int_0^{\frac{\pi}{2}} i\,Re^{i\theta}f(Re^{i\theta})d\theta$$

$$+ \int_R^\varepsilon x^{-\alpha}e^{-\frac{\pi\alpha}{2}i}e^{-ix}i\,dx + \int_{\frac{\pi}{2}}^0 i\varepsilon e^{i\theta}f(\varepsilon e^{i\theta})d\theta = 0.$$

It follows from $\alpha < 1$ that

$$\lim_{\varepsilon\to0}\int_{\frac{\pi}{2}}^0 i\varepsilon e^{i\theta}f(\varepsilon e^{i\theta})d\theta = 0,$$

and from $\alpha > 0$ and Jordan's Lemma that

$$\lim_{R\to\infty}\int_0^{\frac{\pi}{2}} i\,Re^{i\theta}f(Re^{i\theta})d\theta = 0.$$

Letting $\varepsilon \to 0$ and $R \to \infty$, we have

$$\Gamma(-\alpha+1) = \int_0^\infty x^{-\alpha}e^{-x}\,dx = ie^{-\frac{\pi\alpha}{2}i}\int_0^\infty x^{-\alpha}e^{-ix}\,dx.$$

Multiplying both sides by $e^{\frac{\pi\alpha}{2}i}$, and comparing the imaginary parts, we obtain

$$\int_0^\infty x^{-\alpha}\cos x\,dx = \sin\frac{\pi\alpha}{2}\Gamma(-\alpha+1).$$

5327

Use a change of contour to show that

$$\int_0^\infty \frac{\cos(\alpha x)}{x+\beta}dx = \int_0^\infty \frac{te^{-\alpha\beta t}}{t^2+1}dt,$$

provided that α and β are positive. Define the left side as a limit of proper integral and show that the limit exists.

(Courant Inst.)

Solution.

Since

$$\left| \int_0^A \cos(\alpha x) dx \right| \le \frac{2}{\alpha},$$

$\frac{1}{x+\beta}$ is monotonic with respect to x and

$$\lim_{x \to +\infty} \frac{1}{x+\beta} = 0,$$

the convergence of the integral

$$\int_0^\infty \frac{\cos(\alpha x)}{x+\beta} dx$$

follows from Dirichlet's criterion.

Define

$$f(z) = \frac{e^{-\alpha z}}{z + \beta i},$$

and choose the contour of integration $\Gamma = \Gamma_1 \cup \Gamma_2 \cup \Gamma_3$ as shown in Fig. 5.16.

By Cauchy integral theorem, we have

$$\int_\Gamma f(z)dz = \int_0^R \frac{e^{-\alpha x}}{x + \beta i} dx + \int_{\Gamma_2} f(z)dz - \int_0^R \frac{e^{-\alpha x i}}{x + \beta} dx$$

$$= \int_0^R \frac{e^{-\alpha x}(x - \beta i)}{x^2 + \beta^2} dx + \int_{\Gamma_2} f(z)dz - \int_0^R \frac{e^{-\alpha x i}}{x + \beta} dx$$

$$= 0.$$

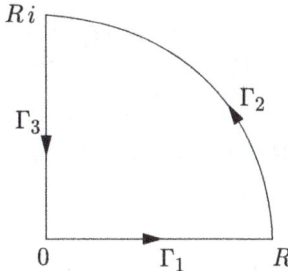

Fig. 5.16.

It follows from Jordan's Lemma that

$$\lim_{R\to\infty}\int_{\Gamma_2} f(z)dz = 0.$$

Letting $R \to \infty$ and considering the real part in the above identity, we obtain

$$\int_0^\infty \frac{\cos(\alpha x)}{x+\beta}dx = \int_0^\infty \frac{xe^{-\alpha x}}{x^2+\beta^2}dx = \int_0^\infty \frac{te^{-\alpha\beta t}}{t^2+1}dt.$$

5328

(a) Let c be the unit circle in the complex plane, and let f be a continuous $\mathbb{C}$-valued function on c. Show that

$$F_f(z) = \int_c \frac{f(\zeta)}{\zeta-z}d\zeta$$

is a holomorphic function of z in the interior of the unit disk.

(b) Find a continuous f on c which is not identically zero, but so that the associated function F is identically zero.

(Minnesota)

Solution.

(a) Let z_0 be an arbitrary point in the unit disk. Then $1 - |z_0| = \rho > 0$. Choosing $\delta > 0$ such that $\delta < \rho$, we prove that

$$F_f(z) = \int_c \frac{f(\zeta)}{\zeta-z}d\zeta$$

has a power series expansion in $\{|z - z_0| \le \delta\}$.

It is clear that

$$\left|\frac{z-z_0}{\zeta-z_0}\right| \le \frac{\delta}{\rho} < 1$$

when $|z - z_0| \le \delta$ and $\zeta \in c$. We can also assume $|f(\zeta)| \le M$ because f is continuous on c. Thus

$$\frac{f(\zeta)}{\zeta - z} = \frac{f(\zeta)}{(\zeta - z_0) - (z - z_0)} = \frac{f(\zeta)}{\zeta - z_0} \cdot \frac{1}{1 - \frac{z - z_0}{\zeta - z_0}}$$

$$= \frac{f(\zeta)}{\zeta - z_0} \sum_{n=0}^{\infty} \left(\frac{z - z_0}{\zeta - z_0}\right)^n$$

$$= \sum_{n=0}^{\infty} \frac{f(\zeta)}{\zeta - z_0} \cdot \left(\frac{z - z_0}{\zeta - z_0}\right)^n.$$

As

$$\left| \frac{f(\zeta)}{\zeta - z_0} \left(\frac{z - z_0}{\zeta - z_0}\right)^n \right| \le \frac{M}{\rho} \cdot \left(\frac{\delta}{\rho}\right)^n,$$

and $\sum_{n=0}^{\infty} \frac{M}{\rho} \left(\frac{\delta}{\rho}\right)^n$ is convergent, the series $\sum_{n=0}^{\infty} \frac{f(\zeta)}{\zeta - z_0} \left(\frac{z - z_0}{\zeta - z_0}\right)^n$ converges uniformly for all $\zeta \in c$. Hence termwise integration is permissible, and we obtain

$$F_f(z) = \int_c \frac{f(\zeta)}{\zeta - z} d\zeta = \sum_{n=0}^{\infty} a_n (z - z_0)^n,$$

where $|z - z_0| \le \delta$ and

$$a_n = \int_c \frac{f(\zeta)}{(\zeta - z_0)^{n+1}} d\zeta.$$

Since z_0 is arbitrarily chosen in the unit disk, $F_f(z)$ is holomorphic in $\{|z| < 1\}$.

(b) Take $f(\zeta) = \frac{1}{\zeta}$ $(|\zeta| = 1)$. Then

$$F_f(z) = \int_c \frac{1}{\zeta(\zeta - z)} d\zeta = \int_c \frac{1}{z} \left(\frac{1}{\zeta - z} - \frac{1}{\zeta} \right) d\zeta$$

$$= \frac{1}{z}(2\pi i - 2\pi i) = 0.$$

In fact, $f(\zeta)$ can be taken as $\frac{1}{(\zeta - z_0)^n}$ for any positive integer n and fixed $z_0 \in \{z : |z| < 1\}$. When $\zeta \in c$,

$$\int_c \frac{\overline{f(\zeta)}}{\zeta - z} d\zeta = \int_c \frac{1}{(\frac{1}{\zeta} - \bar{z})(\frac{1}{\zeta} - \bar{z}_0)^n} d\left(\frac{1}{\zeta} \right)$$

$$= \int_c \frac{-\zeta^{n-1}}{(1 - \bar{z}\zeta)(1 - \bar{z}_0\zeta)^n} d\zeta = 0.$$

5329

Let $[a, b]$ be a finite interval in $\mathbb{R}$ and define, for z in $D = \mathbb{C} - [a, b]$,

$$f(z) = \int_a^b \frac{dt}{t - z}.$$

Show that $f(z)$ is analytic in D. Given c, $a < c < b$, calculate the limit of $f(z)$ as z tends to c from the upper half plane and as z tends to c from the lower half plane.

$$\textit{(UC, Irvine)}$$

Solution.

For any $z_0 \in D$, choose $\delta > 0$ sufficiently small such that $\{z : |z - z_0| \leq \delta\} \cap \{z = x + iy : y = 0, a \leq x \leq b\} = \emptyset$. When $|z - z_0| < \delta$, $a \leq t \leq b$, we have

$$\frac{1}{t - z} = \frac{1}{(t - z_0) - (z - z_0)} = \frac{1}{t - z_0} \cdot \frac{1}{1 - \frac{z - z_0}{t - z_0}}$$

$$= \sum_{n=0}^{\infty} \frac{(z - z_0)^n}{(t - z_0)^{n+1}},$$

and the series converges uniformly for t with $a \le t \le b$. Hence

$$f(z) = \int_a^b \frac{dt}{t-z} = \int_a^b \left(\sum_{n=0}^\infty \frac{(z-z_0)^n}{(t-z_0)^{n+1}} \right) dt$$

$$= \sum_{n=0}^\infty \left(\int_a^b \frac{dt}{(t-z_0)^{n+1}} \right) (z-z_0)^n$$

holds for $z \in \{z : |z-z_0| < \delta\}$, which implies $f(z)$ is analytic in $\{z : |z-z_0| < \delta\}$. Since z_0 is an arbitrary point in D, we obtain that $f(z)$ is analytic in D.

For $z \in D$, $f(z)$ can also be represented explicitly by

$$f(z) = \int_a^b \frac{dt}{t-z} = \int_a^b d\log(t-z) = \log \frac{z-b}{z-a},$$

where the single-valued branch is defined by $\arg(\frac{z-b}{z-a})|_{z=x_0>b} = 0$. Let Γ_1 and Γ_2 be two continuous curves connecting $z = x_0 > b$ and $z = c$ in the upper half plane and the lower half plane respectively. Then the limit of $f(z)$ as z tends to c from the upper half plane is

$$\log \left| \frac{c-b}{c-a} \right| + i\Delta_{\Gamma_1} \arg \frac{z-b}{z-a} = \log \left| \frac{c-b}{c-a} \right| + \pi i,$$

while the limit of $f(z)$ as z tends to c from the lower half plane is

$$\log \left| \frac{c-b}{c-a} \right| + i\Delta_{\Gamma_2} \arg \frac{z-b}{z-a} = \log \left| \frac{c-b}{c-a} \right| - \pi i.$$

5330

For each $z \in U = \{z : \operatorname{Im} z > 0\}$ define

$$g(z) = \frac{1}{2\pi i} \int_{-1}^1 \frac{\sin^2 t}{t-z} dt.$$

Determine which points $a \in \mathbb{R}$ have the following property: there exist $\varepsilon > 0$ and an analytic function f on $D(a, \varepsilon)$ such that $f(z) = g(z)$ for all $z \in U \cap D(a, \varepsilon)$.

(*Indiana*)

Solution.

Let Γ be the half unit circle in the lower half plane whose direction is defined from point $z = -1$ to point $z = 1$, and define a function

$$f(z) = \frac{1}{2\pi i} \int_\Gamma \frac{\sin^2 t}{t - z} dt.$$

It follows from the Cauchy integral theorem that when $z \in U$, $f(z) \equiv g(z)$. With a similar reason as in problem 5328, $f(z)$ is analytic in the complement of Γ. Hence we obtain that for any $a \in \mathbb{R}$, $a \neq \pm 1$, there exists $\varepsilon > 0$ ($\varepsilon < \min\{|a - 1|, |a + 1|\}$) such that $f(z)$ is analytic in $D(a, \varepsilon) = \{z : |z - a| < \varepsilon\}$ and $f(z) = g(z)$ for all $z \in U \cap D(a, \varepsilon)$.

When $a = \pm 1$, such an $f(z)$ does not exist. The reason is as follows: As

$$\frac{\sin^2 t}{t - z} = \frac{\sin^2 t - \sin^2 z}{t - z} + \frac{\sin^2 z}{t - z},$$

where $\frac{\sin^2 t - \sin^2 z}{t - z}$ is an analytic function of two variables for $(t, z) \in \mathbb{C} \times \mathbb{C}$, we know that

$$h(z) = \frac{1}{2\pi i} \int_{-1}^{1} \frac{\sin^2 t - \sin^2 z}{t - z} dt$$

is analytic for $z \in \mathbb{C}$. But

$$\frac{1}{2\pi i} \int_{-1}^{1} \frac{\sin^2 z}{t - z} dt = \frac{\sin^2 z}{2\pi i} \int_{-1}^{1} d\log(t - z) = \frac{\sin^2 z}{2\pi i} \log \frac{z - 1}{z + 1},$$

which has branch points $z = \pm 1$, hence $g(z)$ cannot be analytically continued to $D(\pm 1, \varepsilon)$.

5331

Prove that there exists a constant C such that for every polynomial P,

$$\left| \int_{-\frac{1}{2}}^{\frac{1}{2}} P(x)dx \right| \leq C \int_{|z|=1} |P(z)| \cdot |dz|.$$

(SUNY, Albany)

Solution.

It follows from $P(x) = \frac{1}{2\pi i} \int_{|z|=1} \frac{P(z)}{z-x} dz$ that

$$\int_{-\frac{1}{2}}^{\frac{1}{2}} P(x)dx = \frac{1}{2\pi i} \int_{|z|=1} P(z) \left(\int_{-\frac{1}{2}}^{\frac{1}{2}} \frac{dx}{z-x} \right) dz,$$

$$\left| \int_{-\frac{1}{2}}^{\frac{1}{2}} P(x)dx \right| \leq \frac{1}{2\pi} \int_{|z|=1} |P(z)| \cdot \left| \ln \frac{z + \frac{1}{2}}{z - \frac{1}{2}} \right| \cdot |dz|.$$

Since

$$\ln \frac{z + \frac{1}{2}}{z - \frac{1}{2}} = \ln \left| \frac{z + \frac{1}{2}}{z - \frac{1}{2}} \right| + i \arg \left(\frac{z + \frac{1}{2}}{z - \frac{1}{2}} \right),$$

$$\frac{1}{3} \leq \left| \frac{z + \frac{1}{2}}{z - \frac{1}{2}} \right| \leq 3, \quad \left| \arg \left(\frac{z + \frac{1}{2}}{z - \frac{1}{2}} \right) \right| \leq 2 \arctan \frac{1}{2},$$

we obtain that

$$\left| \int_{-\frac{1}{2}}^{\frac{1}{2}} P(x)dx \right| \leq C \int_{|z|=1} |P(z)| \cdot |dz|$$

where $C = \sqrt{\ln^2 3 + 4 \arctan^2 \frac{1}{2}}$.

5332

Let $p(z) = z^n + a_{n-1} z^{n-1} + \cdots + a_1 z + a_0$ be a complex polynomial. Show that there must be at least one point with $|z| = 1$ and $|p(z)| \geq 1$.

(UC, Los Angeles)

Solution.

It follows from $\frac{1}{2\pi i}\int_{|z|=1}\frac{p(z)}{z^{n+1}}dz = 1$ that

$$1 \le \frac{1}{2\pi}\int_{|z|=1}|p(z)|\cdot|dz|.$$

If $\max_{|z|=1}|p(z)| < 1$, then we have

$$\frac{1}{2\pi}\int_{|z|=1}|p(z)|\cdot|dz| < 1,$$

which is a contradiction.

5333

Let f be analytic and absolutely integrable with respect to the Lebesgue area measure $dA = dxdy$ in the unit disk Δ. Prove that for $a \in \Delta$

$$\iint_\Delta f(z)\frac{\bar{z}}{(1-a\bar{z})^3}dA(z) = \frac{1}{2}f'(a).$$

(*SUNY, Albany*)

Solution.

By Cauchy Integral Formula, we have

$$f'(a) = \frac{1}{2\pi i}\int_{\partial\Delta}\frac{f(z)}{(z-a)^2}dz = \frac{1}{2\pi i}\int_{\partial\Delta}\frac{\bar{z}^2 f(z)}{(1-a\bar{z})^2}dz.$$

Apply Green Formula $\int_{\partial\Omega}g(z)dz = 2i\iint_\Omega\frac{\partial g}{\partial\bar{z}}dxdy$ to the above integral, we obtain

$$f'(a) = \frac{1}{2\pi i}\int_{\partial\Delta}\frac{\bar{z}^2 f(z)}{(1-a\bar{z})^2}dz$$

$$= \frac{1}{\pi}\iint_\Delta f(z)\frac{\partial}{\partial\bar{z}}\left(\frac{\bar{z}^2}{(1-a\bar{z})^2}\right)dxdy$$

$$= \frac{2}{\pi}\iint_\Delta f(z)\frac{\bar{z}}{(1-a\bar{z})^3}dxdy.$$

5334

If f is analytic in the unit disk Δ, continuous in $\bar{\Delta}$ and maps Δ into itself, prove that for every point $a \in \Delta$,

$$|f^{(3)}(a)| \leq \frac{6(1 + |a|^2)}{(1 - |a|^2)^3}.$$

(SUNY, Albany)

Solution.

By Cauchy Integral Formula, we have

$$f^{(3)}(a) = \frac{3!}{2\pi i} \int_{|z|=1} \frac{f(z)}{(z-a)^4} dz.$$

Hence

$$|f^{(3)}(a)| \leq \frac{3!}{2\pi} \int_{|z|=1} \frac{|dz|}{|z-a|^4} = \frac{3!}{2\pi} \int_{|z|=1} \frac{dz/iz}{(z-a)^2(\bar{z}-\bar{a})^2}$$

$$= \frac{3!}{2\pi i} \int_{|z|=1} \frac{z\,dz}{(z-a)^2(1-\bar{a}z)^2}$$

$$= 6\left(\frac{z}{(1-\bar{a}z)^2}\right)'_{z=a} = \frac{6(1+|a|^2)}{(1-|a|^2)^3}.$$

5335

Let f be holomorphic in a neighborhood of $\bar{U} = \{z : |z| \leq 1\}$. Let $f(z) = \sum_{k=0}^{\infty} a_k z^k$ and $s_n(z) = \sum_{k=0}^{n} a_k z^k$. Show that, among all polynomials P of degree n or less, the integral $\frac{1}{2\pi}\int_0^{2\pi} |f(e^{i\theta}) - p(e^{i\theta})|^2 d\theta$ attains its minimum for $p = s_n$.

(SUNY, Albany)

Solution.

First we have $\int_0^{2\pi} e^{ik\theta}\, d\theta = \begin{cases} 0, & k \neq 0, \\ 2\pi, & k = 0. \end{cases}$ Then it follows that

$$\int_0^{2\pi} \left| \sum_{k=0}^{\infty} c_k e^{ik\theta} \right|^2 d\theta = \int_0^{2\pi} \left(\sum_{k=0}^{\infty} c_k e^{ik\theta} \right) \left(\sum_{k=0}^{\infty} \overline{c_k} e^{-ik\theta} \right) d\theta$$

$$= \int_0^{2\pi} \sum_{k=0}^{\infty} |c_k|^2 d\theta = 2\pi \sum_{k=0}^{\infty} |c_k|^2.$$

Assume $p(z) = \sum_{k=0}^{n} b_k z^k$, then we have

$$\frac{1}{2\pi} \int_0^{2\pi} |f(e^{i\theta}) - p(e^{i\theta})|^2 d\theta$$

$$= \int_0^{2\pi} \left| \sum_{k=0}^{n} (a_k - b_k) e^{ik\theta} + \sum_{k=n+1}^{\infty} a_k e^{ik\theta} \right|^2 d\theta$$

$$= 2\pi \left(\sum_{k=0}^{n} |a_k - b_k|^2 + \sum_{k=n+1}^{\infty} |a_k|^2 \right) \geq 2\pi \sum_{k=n+1}^{\infty} |a_k|^2.$$

The equality holds if and only if when $b_k = a_k$ for $k = 0, 1, 2, \ldots, n$, i.e., $\frac{1}{2\pi} \int_0^{2\pi} |f(e^{i\theta}) - p(e^{i\theta})|^2 d\theta$ attains its minimum for $p = s_n$.

5336

Suppose f is analytic in the unit disk D. For $z \in D$ and $0 < r < 1 - |z|$, let $B(z, r) = \{w \in D : |w - z| < r\}$. Show that $|f(z)|^\pi \leq \frac{1}{\pi r^2} \iint_{B(z,r)} |f(w)|^\pi dA(w)$, where $dA(w)$ is area measure on D.

(*SUNY, Albany*)

Solution.

It follows from the Cauchy Integral Formula $f(z) = \frac{1}{2\pi i} \int_{|w-z|=\rho} \frac{f(w)}{w-z} dw$, where $0 < \rho < r$, that

$$|f(z)| \leq \frac{1}{2\pi} \int_0^{2\pi} |f(z + \rho e^{i\theta})| d\theta.$$

Hence we have

$$|f(z)| \int_0^r \rho d\rho \leq \frac{1}{2\pi} \int_0^r \int_0^{2\pi} |f(z + \rho e^{i\theta})| \rho d\rho d\theta,$$

$$|f(z)| \leq \frac{1}{\pi r^2} \iint_{B(z,r)} |f(w)| dA(w).$$

Let $p > 1$, $q > 1$, $\frac{1}{p} + \frac{1}{q} = 1$, and apply the Hölder Inequality to the above inequality.

$$|f(z)| \leq \frac{1}{\pi r^2} \iint_{B(z,r)} |f(w)| dA(w)$$

$$\leq \frac{1}{\pi r^2} \left(\iint_{B(z,r)} |f(w)|^p dA(w) \right)^{\frac{1}{p}} \left(\iint_{B(z,r)} 1^q dA(w) \right)^{\frac{1}{q}}$$

$$= \left(\frac{1}{\pi r^2} \right)^{\frac{1}{p}} \left(\iint_{B(z,r)} |f(w)|^p dA(w) \right)^{\frac{1}{p}}.$$

Take $p = \pi$, we obtain

$$|f(z)|^{\pi} \leq \frac{1}{\pi r^2} \iint_{B(z,r)} |f(w)|^{\pi} dA(w).$$

5337

Let f be a smooth bounded function in the unit disk Δ and $F(z)$ be given by

$$F(z) = -\frac{1}{\pi} \iint_{\Delta} \frac{f(w) dA(w)}{w - z}.$$

Prove that $F(z)$ is analytic outside of $\bar{\Delta}$ and for every $z \in \Delta$,

$$\frac{\partial F}{\partial \bar{z}} = f(z).$$

(SUNY, Albany)

Solution.

Let $z_0 \in \mathbb{C}\backslash\bar{\Delta}$ and $\inf_{w \in \Delta} |w - z_0| = r > 0$. Then for $z \in \{z : |z - z_0| < \frac{r}{2}\}$ and $w \in \Delta$,

$$\frac{1}{w-z} = \frac{1}{(w-z_0)-(z-z_0)} = \frac{1}{(w-z_0)(1-\frac{z-z_0}{w-z_0})}$$

$$= \sum_{n=0}^{\infty} \frac{(z-z_0)^n}{(w-z_0)^{n+1}},$$

and the convergence of the series is uniform for $z \in \{z : |z-z_0| < \frac{r}{2}\}$ and $w \in \Delta$, hence we have

$$F(z) = -\frac{1}{\pi} \iint_{\Delta} \frac{f(w)dA(w)}{w-z} = \sum_{n=0}^{\infty} a_n(z-z_0)^n,$$

where $a_n = -\frac{1}{\pi} \iint_{\Delta} \frac{f(w)}{(w-z_0)^{n+1}} dA(w)$, which implies that $F(z)$ is analytic at $z_0 \in \mathbb{C}\backslash\bar{\Delta}$.

If $z_0 = \infty$, for $w \in \Delta$ and z in the neighborhood of ∞,

$$\frac{1}{w-z} = -\frac{1}{z}\cdot\frac{1}{1-\frac{w}{z}} = -\sum_{n=0}^{\infty}\frac{w^n}{z^{n+1}},$$

and the convergence of the series is uniform for $w \in \Delta$ and z in the neighborhood of ∞, hence we have

$$F(z) = -\frac{1}{\pi} \iint_{\Delta} \frac{f(w)dA(w)}{w-z} = \sum_{n=0}^{\infty}\frac{b_{n+1}}{z^{n+1}},$$

where $b_{n+1} = -\frac{1}{\pi} \iint_{\Delta} f(w)w^n dA(w)$, which implies that $F(z)$ is also analytic at point $z_0 = \infty$, so $F(z)$ is analytic outside of $\bar{\Delta}$.

Let $r \in (0,1)$, $\Delta_r = \{z : |z| < r\}$. Now we extend f to a smooth function $g \in C_0^1$ with compact support in the complex plane such that $g(z) = f(z)$ when $z \in \Delta_r$. Define $G(z) = -\frac{1}{\pi} \iint_{\mathbb{C}} \frac{g(w)dA(w)}{w-z}$. Then

$$F(z) = -\frac{1}{\pi} \iint_{\Delta_r} \frac{f(w)dA(w)}{w-z} - \frac{1}{\pi} \iint_{\Delta \backslash \Delta_r} \frac{f(w)dA(w)}{w-z}$$

$$G(z) = -\frac{1}{\pi} \iint_{\Delta_r} \frac{g(w)dA(w)}{w-z} - \frac{1}{\pi} \iint_{\mathbb{C} \backslash \Delta_r} \frac{g(w)dA(w)}{w-z},$$

where $-\frac{1}{\pi} \iint_{\Delta \backslash \Delta_r} \frac{f(w)dA(w)}{w-z}$ and $-\frac{1}{\pi} \iint_{\mathbb{C} \backslash \Delta_r} \frac{g(w)dA(w)}{w-z}$ are analytic in Δ_r, so for $z \in \Delta_r$, we have $\frac{\partial F}{\partial \bar{z}}(z) = \frac{\partial G}{\partial \bar{z}}(z)$.

$$G(z) = -\frac{1}{\pi} \iint_{\mathbb{C}} \frac{g(w)dA(w)}{w-z} = -\frac{1}{\pi} \iint_{\mathbb{C}} \frac{g(\varsigma + z)dA(\varsigma)}{\varsigma},$$

$$\frac{\partial G}{\partial \bar{z}} = -\frac{1}{\pi} \iint_{\mathbb{C}} \frac{g_{\bar{z}}(\varsigma + z)dA(\varsigma)}{\varsigma} = -\frac{1}{\pi} \iint_{\mathbb{C}} \frac{g_{\bar{w}}(w)dA(w)}{w-z}.$$

Apply Green Formula $\int_{\partial \Omega} \varphi(z)dz = 2i \iint_{\Omega} \frac{\partial \varphi}{\partial \bar{z}} dxdy$, we obtain for $z \in \Delta_r$,

$$\frac{\partial F}{\partial \bar{z}} = \frac{\partial G}{\partial \bar{z}} = -\frac{1}{\pi} \iint_{\mathbb{C}} \frac{g_{\bar{w}}(w)dA(w)}{w-z}$$

$$= \lim_{\varepsilon \to 0} \left\{ -\frac{1}{\pi} \iint_{|w-z|>\varepsilon} \frac{g_{\bar{w}}(w)dA(w)}{w-z} \right\}.$$

$$= \lim_{\varepsilon \to 0} \left\{ \frac{1}{2\pi i} \int_{|w-z|=\varepsilon} \frac{g(w)}{w-z}dw \right\} = g(z) = f(z).$$

Since $r \in (0,1)$ is arbitrary, we know that $\frac{\partial F}{\partial \bar{z}} = f(z)$ holds for every $z \in \Delta$.

Section 4

The Maximum Modulus and Argument Principles

5401

Let $a \in \mathbb{C}, |a| \leq 1$, and consider the polynomial

$$P(z) = \frac{a}{2} + (1 - |a|^2)z - \frac{\bar{a}}{2}z^2.$$

Show that $|P(z)| \leq 1$ whenever $|z| \leq 1$.

(*Indiana*)

Solution.

$$P(z) = \frac{a}{2} + (1 - |a|^2)z - \frac{\bar{a}}{2}z^2$$

$$= z\left[(1 - |a|^2) + \frac{1}{2}\left(\frac{a}{z} - \bar{a}z\right)\right].$$

When $|z| = 1$,

$$\mathrm{Re}\left(\frac{a}{z} - \bar{a}z\right) = \mathrm{Re}\left[\frac{a}{z} - \overline{(\bar{a}z)}\right] = \mathrm{Re}\left[\frac{a}{z} - \frac{a}{z}\right] = 0,$$

$$\left|\mathrm{Im}\left(\frac{a}{z} - \bar{a}z\right)\right| \leq 2|a|.$$

Hence when $|z| = 1$,

$$|P(z)|^2 = (1 - |a|^2)^2 + \left(\mathrm{Im} \left[\frac{1}{2} \left(\frac{a}{z} - \bar{a}z \right) \right] \right)^2$$
$$\leq (1 - 2|a|^2 + |a|^4) + |a|^2 = 1 - |a|^2 + |a|^4 \leq 1.$$

By the maximum modulus principle, $|P(z)| \leq 1$ whenever $|z| \leq 1$.

5402

Let f be holomorphic in the unit disk $\{|z| < 1\}$, continuous in $\{|z| \leq 1\}$ and $|f(z)| = 1$ whenever $|z| = 1$. Prove that f is a rational function.

(*SUNY, Stony Brook*)

Solution.

If $f(z)$ has infinite many zeros, by the isolatedness of the zeros of holomorphic functions, the zeros must have limit points on the boundary of the unit disk. But it will violate the fact that f is continuous in $\{|z| \leq 1\}$ and $|f(z)| = 1$ whenever $|z| = 1$. Hence f has only finite zeros in the unit disk. Denote all these zeros by $z_1, z_2, \ldots, z_n$, multiple zeros being repeated, and define

$$F(z) = f(z) \bigg/ \prod_{k=1}^{n} \left(\frac{z - z_k}{1 - \bar{z}_k z} \right).$$

Then $F(z)$ is holomorphic in $\{|z| < 1\}$, continuous in $\{|z| \leq 1\}$ and $|F(z)| = 1$ when $|z| = 1$. By the maximum modulus principle, $|F(z)| \leq 1$ in $\{|z| \leq 1\}$. Since $F(z)$ has no zero in $\{|z| \leq 1\}$, $\frac{1}{F(z)}$ is also holomorphic in $\{|z| < 1\}$, continuous in $\{|z| \leq 1\}$ and $\left| \frac{1}{F(z)} \right| = 1$ when $|z| = 1$. Application of the maximum modulus principle to $\frac{1}{F(z)}$ yields $|F(z)| \geq 1$ in $\{|z| \leq 1\}$. Hence $|F(z)| = 1$ holds in $\{|z| \leq 1\}$, which implies $F(z) = e^{i\alpha}$ with α a real number. So we obtain

$$f(z) = e^{i\alpha} \prod_{k=1}^{n} \left(\frac{z - z_k}{1 - \bar{z}_k z} \right).$$

5403

Let f be a continuous function on $\overline{U} = \{z : |z| \leq 1\}$ such that f is analytic in U. If $f = 1$ on the half-circle $\gamma = \{e^{i\theta} : 0 \leq \theta \leq \pi\}$, prove that $f = 1$ everywhere in $\overline{U}$.

(*Indiana*)

Solution.
Define $F(z) = (f(z) - 1)(f(-z) - 1)$, then $F(z)$ is also continuous on $\overline{U}$ and analytic in U. When $z \in \partial U$, we have either $f(z) - 1 = 0$ or $f(-z) - 1 = 0$. Hence $F(z) = 0$ holds for all $z \in \overline{U}$, which implies either $f(z) - 1 \equiv 0$ or $f(-z) - 1 \equiv 0$. Since $f(z) - 1 \equiv 0$ is equivalent to $f(-z) - 1 \equiv 0$, we obtain $f(z) \equiv 1$ for all $z \in \overline{U}$.

Remark. The condition that "$f = 1$ on the half-circle γ" can be weakened to that "$f = 1$ on an arc $\gamma = \{e^{i\theta} : 0 \leq \theta \leq \frac{\pi}{n}\}$, where n is a natural number". In this case, the proof is the same except that $F(z)$ is defined by

$$F(z) = (f(z) - 1)(f(ze^{\frac{\pi}{n}i}) - 1)(f(ze^{\frac{2\pi}{n}i}) - 1) \cdots (f(ze^{\frac{2n-1}{n}\pi i}) - 1).$$

5404

Let S denote the sector in the complex plane given by $S = \{z : -\frac{\pi}{4} < \arg z < \frac{\pi}{4}\}$. Let $\overline{S}$ denote the closure of S. Let f be a continuous complex function on $\overline{S}$ which is holomorphic in S. Suppose further

(1) $|f(z)| \leq 1$ for all z in the boundary of S;
(2) $|f(x + iy)| \leq e^{\sqrt{x}}$ for all $x + iy \in S$.

Prove that $|f(z)| \leq 1$ for all $z \in S$.

(*SUNY, Stony Brook*)

Solution.

Let $F(z) = e^{-\varepsilon z} f(z)$, where $\varepsilon > 0$ is an arbitrary fixed number. Then $F(z)$ is also continuous on $\overline{S}$ and analytic in S. When z is on the boundary of S, $|F(z)| = e^{-\varepsilon x}|f(x)| \leq 1$. When $|z| \to +\infty$ $(-\frac{\pi}{4} < \arg z < \frac{\pi}{4})$, $|F(z)| \leq e^{-\varepsilon x} \cdot e^{\sqrt{x}} \to 0$. By the maximum modulus principle, we have $|F(z)| \leq 1$ for all $z \in S$, which implies $|f(z)| \leq |e^{\varepsilon z}| = e^{\varepsilon x}$ for all $z \in S$. Because $\varepsilon > 0$ can be arbitrarily chosen, letting $\varepsilon \to 0$, we obtain $f|(z)| \leq 1$ for all $z \in S$.

5405

Let K be a compact, connected subset of $\mathbb{C}$ containing more than one point and let f be a one-to-one conformal map $\bar{\mathbb{C}} \backslash K$ onto $\Delta = \{z : |z| < 1\}$ with $f(\infty) = 0$. If p is a polynomial of degree n for which $|p(z)| \leq 1$ for $z \in K$, prove that

$$|p(z)| \leq |f(z)|^{-n} \quad \text{for } z \in \mathbb{C} \backslash K.$$

(*Indiana*)

Solution.

Because f is a one-to-one conformal map of $\bar{\mathbb{C}} \backslash K$ onto Δ with $f(\infty) = 0$, it has a simple zero at $z = \infty$. Since p is a polynomial of degree n, it has a pole of order n at $z = \infty$. Hence the function $F(z) = p(z) f^n(z)$ is analytic in $\bar{\mathbb{C}} \backslash K$ which contains point $z = \infty$. As $f(z)$ maps $\bar{\mathbb{C}} \backslash K$ onto $\Delta = \{z : |z| < 1\}$, we have $\lim_{z \to K} |f(z)| = 1$. Together with $|p(z)| \leq 1$ for $z \in K$, we know that the limit of $|F(z)|$ when z tends to K cannot be larger than 1. Apply the maximum modulus principle to $F(z)$ on $\bar{\mathbb{C}} \backslash K$, we obtain $|F(z)| \leq 1$ for $z \in \bar{\mathbb{C}} \backslash K$, which implies $|p(z)| \leq |f(z)|^{-n}$ for all $z \in \mathbb{C} \backslash K$.

5406

Suppose f and g (non-constant functions) are analytic in a region G and continuous on the closure $\overline{G}$ of the region. Assume that $\overline{G}$ is compact. Prove that $|f| + |g|$ achieves its maximum value on the boundary of G.

(*Iowa*)

Solution.

Assume that $|f| + |g|$ achieves its maximum value c ($c > 0$) at $z_0 \in \overline{G}$, we prove that if $z_0 \in G$, then f and g must be constants.

Let

$$|f(z_0)| = f(z_0)e^{i\phi_1}, \quad |g(z_0)| = g(z_0)e^{i\phi_2}.$$

Then for fixed ϕ_1 and ϕ_2,

$$F(z) = f(z)e^{i\phi_1} + g(z)e^{i\phi_2}$$

is analytic in G and continuous on $\overline{G}$. It follows from

$$|F(z)| \le |f(z)| + |g(z)| \le c,$$
$$F(z_0) = f(z_0)e^{i\phi_1} + g(z_0)e^{i\phi_2} = |f(z_0)| + |g(z_0)| = c$$

and $z_0 \in G$ that

$$F(z) = f(z)e^{i\phi_1} + g(z)e^{i\phi_2}$$

must be the constant c.

Without loss of generality, we assume that f is not a constant, and try to lead to a contradiction. Since the image of an open set $\{z : |z - z_0| < \delta\} \subset G$ under f is an open set which contains point $f(z_0)$, $f(z)$ assumes all the values $f(z) = f(z_0) + \varepsilon e^{i\phi}$ for small $\varepsilon > 0$ and $0 \le \phi < 2\pi$ in $\{z : |z - z_0| < \delta\}$. Then when $\phi + \phi_1 \ne 0, \pi$, we have

$$
\begin{aligned}
|f(z)| + |g(z)| &= |f(z)| + |c - f(z)e^{i\phi_1}| \\
&= |f(z_0) + \varepsilon e^{i\phi}| + |c - f(z_0)e^{i\phi_1} - \varepsilon e^{i(\phi+\phi_1)}| \\
&= |\varepsilon e^{i(\phi+\phi_1)} + f(z_0)e^{i\phi_1}| + |\varepsilon e^{i(\phi+\phi_1)} - g(z_0)e^{i\phi_2}| \\
&> f(z_0)e^{i\phi_1} + g(z_0)e^{i\phi_2} = c,
\end{aligned}
$$

which contradicts that z_0 is a maximum value point of $|f| + |g|$. Hence f must be a constant, which also implies g is a constant too.

5407

Suppose $f(z)$ is an entire function with

$$|f(z)| \leq \frac{1}{|\text{Re}\, z|}, \quad \text{all } z.$$

Show that $f(z)$ is identically 0.

(Iowa)

Solution.

For any $R > 0$, consider function

$$g(z) = (z - Ri)(z + Ri)f(z).$$

When $|z| = R$, and $\text{Im}\, z \geq 0$, denote by θ the angle between the line perpendicular to the imaginary axis and the line passing through z and Ri. Then $0 \leq \theta \leq \frac{\pi}{4}$, and

$$\left| \frac{z - Ri}{\text{Re}\, z} \right| = \sec \theta \leq \sqrt{2}.$$

When $|z| = R$, and $\text{Im}\, z < 0$, denote by θ the angle between the line perpendicular to the imaginary axis and the line passing through z and $-Ri$. Then $0 \leq \theta < \frac{\pi}{4}$, and

$$\left| \frac{z + Ri}{\text{Re}\, z} \right| = \sec \theta < \sqrt{2}.$$

It follows from the above discussion that when $|z| = R$,

$$|g(z)| = |(z - Ri)(z + Ri)f(z)| \leq \left| \frac{(z - Ri)(z + Ri)}{\text{Re}\, z} \right| \leq 2\sqrt{2}R.$$

By the maximum modulus principle, when $|z| < R$,

$$|f(z)| = \left| \frac{g(z)}{(z - Ri)(z + Ri)} \right| \leq \frac{2\sqrt{2}R}{R^2 - |z|^2}.$$

Now fixing z, and letting $R \to +\infty$, we obtain $f(z) = 0$. Since R can be arbitrarily large, we have $f(z) = 0$ for all $z \in \mathbb{C}$.

5408

Suppose f is analytic on $\{z; 0 < |z| < 1\}$ and

$$|f(z)| \le \log \frac{1}{|z|}.$$

Show that $f \equiv 0$.

(Indiana)

Solution.

Denote the Laurent expansion of f on $\{z; 0 < |z| < 1\}$ by

$$f(z) = \sum_{n=-\infty}^{\infty} a_n z^n,$$

where

$$a_n = \frac{1}{2\pi i} \int_{|z|=r<1} \frac{f(z)}{z^{n+1}} dz.$$

It follows that

$$|a_n| \le \frac{1}{2\pi} \int_{|z|=r} \left| \frac{f(z)}{z^{n+1}} \right| \cdot |dz| \le \frac{1}{r^n} \log \frac{1}{r}.$$

When $n < 0$, letting $r \to 0$, we have

$$a_n = 0 \quad (n = -1, -2, \ldots),$$

which implies $z = 0$ is a removable singularity of f. In other words, f can be extended to an analytic function of the unit disk.

Since $\log \frac{1}{|z|} = 0$ when $|z| = 1$. By the maximum modulus principle, we obtain

$$f \equiv 0.$$

5409

Let f be an analytic function on $D = \{z : |z| < 1\}$, $f(D) \subseteq D$ and $f(0) = 0$.

(a) Prove that $|f(z) + f(-z)| \le 2|z|^2$ for all z in D and if equality occurs for some non-zero z in D, then $f(z) = e^{i\alpha} z^2$.

(b) Prove that

$$\left| \int_{-1}^{1} f(x)dx \right| \le \frac{2}{3}.$$

(*Indiana*)

Solution.

(a) Let $F(z) = f(z) + f(-z)$, then $F(0) = 0$,

$$F'(0) = \lim_{z \to 0} \frac{F(z)}{z} = \lim_{z \to 0} \left(\frac{f(z)}{z} - \frac{f(-z)}{-z} \right) = 0.$$

Hence $\frac{F(z)}{z^2}$ is analytic in D, and when z tends to ∂D, the limit of $\left| \frac{F(z)}{z^2} \right|$ cannot be larger than 2. By the maximum modulus principle, $|f(z) + f(-z)| \le 2|z|^2$ holds for all $z \in D$.

If equality occurs for some non-zero z in D, we have

$$f(z) + f(-z) = 2e^{i\alpha} z^2,$$

where α is a real constant.

Let

$$f(z) = \sum_{n=1}^{\infty} a_n z^n,$$

it follows from

$$f(z) + f(-z) = 2e^{i\alpha} z^2$$

that

$$a_2 = e^{i\alpha}, \quad a_4 = a_6 = \cdots = 0.$$

Because $|f(z)| < 1$ for $z \in D$, we have

$$\lim_{r \to 1} \frac{1}{2\pi} \int_0^{2\pi} |f(re^{i\theta})|^2 d\theta = \sum_{n=1}^{\infty} |a_n|^2 \leq 1.$$

Since $a_2 = e^{i\alpha}$, the other coefficients must be zero, which implies $f(z) = e^{i\alpha} z^2$.

(b)

$$\left| \int_{-1}^{1} f(x) dx \right| = \left| \int_{-1}^{0} f(x) dx + \int_{0}^{1} f(x) dx \right|$$

$$= \left| \int_{0}^{1} (f(x) + f(-x)) dx \right| \leq \int_{0}^{1} 2x^2 \, dx = \frac{2}{3}.$$

5410

If f is analytic and $|f(z)| < 1$ on $\{z : |z| \leq 1\}$, prove that $f(z)$ has a fixed point.

(*Rutgers*)

Solution.

Let $F(z) = f(z) - z$ and $G(z) = -z$.
When $|z| = 1$,

$$|F(z) - G(z)| = |f(z)| < 1 = |G(z)|.$$

By Rouché's Theorem, $F(z)$ and $G(z)$ have the same number of zeros in $\{z : |z| < 1\}$. Since $G(z)$ has only one simple zero in $\{z : |z| < 1\}$, we conclude that $f(z) - z$ has one zero in $\{z : |z| < 1\}$, which implies that $f(z)$ has a fixed point in $\{z : |z| < 1\}$.

5411

Let $f(z) = z + e^{-z}, \lambda > 1$. Prove or disprove: $f(z)$ takes the value λ exactly once in the right half-plane. If the answer is yes, is the point necessarily real? Justify.

(*Iowa*)

Solution.

Let R be a sufficiently large real number such that $R > 2\lambda$. Take a closed curve Γ on the right half-plane, where

$$\Gamma = \{z = x + iy : x = 0, -R \le y \le R\}$$
$$\cup \left\{z : |z| = R, -\frac{\pi}{2} \le \arg z \le \frac{\pi}{2}\right\}.$$

Define

$$F(z) = \lambda - z - e^{-z}$$

and

$$G(z) = \lambda - z.$$

When $z \in \Gamma$,

$$|F(z) - G(z)| = |e^{-z}| \le 1 < |G(z)|.$$

Since $G(z)$ has exactly one zero inside Γ, it follows from Rouché's Theorem that $F(z)$ has exactly one zero inside Γ. Because R can be arbitrarily large, $F(z)$ has exactly one zero in the right half-plane. Hence $f(z)$ takes value λ exactly once in the right half-plane.

Take $z = x \ge 0$. We have

$$F(x) = \lambda - x - e^{-x},$$

which is a real-valued function of real variable x. Since $F(x)$ is continuous and $F(0) > 0$,

$$\lim_{x \to +\infty} F(x) = -\infty,$$

there must exist x_0, $0 < x_0 < +\infty$, such that $F(x_0) = 0$. In other words, the point z in the right half-plane such that $f(z) = \lambda$ is necessarily real.

5412

Suppose f is analytic in a region which contains the closed unit disc $\{z : |z| \le 1\}$. Assume f is non-zero on the unit circle

$\{z : |z| = 1\}$. Let C denote the unit circle traversed in the counterclockwise sense. Suppose that

$$(1) \quad \frac{1}{2\pi i} \int_C \frac{f'(z)}{f(z)} dz = 2,$$

$$(2) \quad \frac{1}{2\pi i} \int_C z \frac{f'(z)}{f(z)} dz = 0,$$

and

$$(3) \quad \frac{1}{2\pi i} \int_C z^2 \frac{f'(z)}{f(z)} dz = \frac{1}{2}.$$

Find the location of the zeros of f in the open unit disc $\{z : |z| < 1\}$.

(*Iowa*)

Solution.

Assume $z_1, z_2, \ldots, z_n$ are the zeros of $f(z)$ in $\{z : |z| < 1\}$, multiple zeros being repeated. Then

$$f(z) = g(z) \prod_{j=1}^{n} (z - z_j),$$

where $g(z)$ is analytic and has no zero in $\{z : |z| \leq 1\}$. We have

$$\frac{1}{2\pi i} \int_C \frac{f'(z)}{f(z)} dz = \frac{1}{2\pi i} \int_C d \log f(z)$$

$$= \frac{1}{2\pi i} \int_C \left(\frac{g'(z)}{g(z)} + \sum_{j=1}^{n} \frac{1}{z - z_j} \right) dz$$

$$= n.$$

It follows from (1) that $n = 2$, i.e., $f(z)$ has two zeros in the unit disk. Then for $f(z) = (z - z_1)(z - z_2)g(z)$,

$$\frac{1}{2\pi i} \int_C z \frac{f'(z)}{f(z)} dz = \frac{1}{2\pi i} \int_C \left(\frac{z}{z - z_1} + \frac{z}{z - z_2} + \frac{zg'(z)}{g(z)} \right) dz$$

$$= z_1 + z_2 = 0$$

and

$$\frac{1}{2\pi i}\int_C z^2 \frac{f'(z)}{f(z)}dz = \frac{1}{2\pi i}\int_C \left(\frac{z^2}{z-z_1} + \frac{z^2}{z-z_2} + \frac{z^2 g'(z)}{g(z)}\right)dz$$

$$= z_1^2 + z_2^2 = \frac{1}{2},$$

which show that $z_{1,2} = \pm\frac{1}{2}$. Hence $z = \pm\frac{1}{2}$ are the only zeros of $f(z)$ in the unit disc.

5413

(a) How many roots does this equation

$$z^4 + z + 5 = 0$$

have in the first quadrant?

(b) How many of them have argument between $\frac{\pi}{4}$ and $\frac{\pi}{2}$?

$$(Indiana\text{-}Purdue)$$

Solution.

(a) Let R be sufficiently large such that when $|z| = R$,

$$|z^4 + 5| > |z|.$$

Set

$$f(z) = z^4 + z + 5$$

and

$$g(z) = z^4 + 5.$$

Choose a closed curve

$$\Gamma = \{z = x + iy; 0 \le x \le R, y = 0\}$$
$$\cup \left\{z : |z| = R, 0 \le \arg z \le \frac{\pi}{2}\right\}$$
$$\cup \{z = x + iy : x = 0, 0 \le y \le R\}.$$

It is obvious that

$$|f(z) - g(z)| < |g(z)|$$

holds when $z \in \Gamma$. By Rouché's Theorem, the numbers of the zeros of f and g inside Γ are equal. Since g has only one zero inside Γ, f has also one zero inside Γ. Noting that R can be arbitrarily large, we know that

$$z^4 + z + 5 = 0$$

has one root in the first quadrant.

(b) Let R be sufficiently large such that when $|z| = R$, $\frac{z+5}{z^4}$ is approximately zero. Set

$$f(z) = z^4 + z + 5$$

and

$$\Gamma_1 = \{z = x + iy : x = 0, 0 \le y \le R\},$$
$$\Gamma_2 = \{z = re^{\frac{\pi}{4}i} : 0 \le r \le R\},$$

and

$$\Gamma_3 = \left\{z : |z| = R, \frac{\pi}{4} \le \arg z \le \frac{\pi}{2}\right\}.$$

It is easy to see that $\mathrm{Im} f(z) > 0$ when

$$z \in (\Gamma_1 \cup \Gamma_2) \setminus \{z = 0\},$$

$f(0) = 5$, and

$$f(Ri) \in \{w : 0 < \arg w < \varepsilon\}, \quad f(Re^{\frac{\pi}{4}i}) \in \{w : \pi - \varepsilon < \arg w < \pi\}$$

where $\varepsilon > 0$ is very small. We also know that

$$\Delta_{\Gamma_3} \arg f(z) = \Delta_{\Gamma_3} \arg z^4 + \Delta_{\Gamma_3} \arg \left(1 + \frac{z+5}{z^4}\right),$$

where $\Delta_{\Gamma_3} \arg f(z)$ denotes the change of $\arg f(z)$ when z goes continuously from $Re^{\frac{\pi}{4}i}$ to Ri along Γ_3. It is obvious that

$$\Delta_{\Gamma_3} \arg z^4 = \pi,$$

while

$$\Delta_{\Gamma_3} \arg \left(1 + \frac{z+5}{z^4}\right)$$

is very small. Let $\Gamma = \Gamma_1 \cup \Gamma_2 \cup \Gamma_3$ is taken once counterclockwise, it follows from the above discussion that

$$\Delta_{\Gamma} \arg f(z) = 2\pi.$$

By the argument principle, the number of the roots of $f(z) = 0$ inside Γ is equal to

$$\frac{1}{2\pi i}\int_{\Gamma}\frac{f'(z)}{f(z)}dz = \frac{1}{2\pi i}\Delta_{\Gamma}\log f(z) = \frac{1}{2\pi}\Delta_{\Gamma}\arg f(z) = 1.$$

Hence

$$f(z) = z^4 + z + 5 = 0$$

has exactly one root in the domain

$$\left\{z : \frac{\pi}{4} < \arg z < \frac{\pi}{2}\right\}.$$

5414

Prove that the equation $\sin z = z$ has infinitely many solutions in $\mathbb{C}$.

(*Indiana*)

Solution.
Let

$$f(z) = \sin z - z$$

and $z = x + iy$, then $f(z)$ can be written as

$$f(z) = \frac{e^{iz} - e^{-iz}}{2i} - z = \frac{i}{2}(e^{y-xi} - e^{-y+xi}) - (x + iy).$$

For any fixed natural number n, choose a positive number $t \gg \log n$ and a closed contour $\Gamma = \Gamma_1 \cup \Gamma_2 \cup \Gamma_3 \cup \Gamma_4$ in the counterclockwise sense, where

$$\Gamma_1 = \{z = x + iy : 2n\pi \le x \le 2(n+1)\pi, y = 0\},$$
$$\Gamma_2 = \{z = x + iy : x = 2(n+1)\pi, 0 \le y \le t\},$$
$$\Gamma_3 = \{z = x + iy : 2n\pi \le x \le 2(n+1)\pi, y = t\},$$

and

$$\Gamma_4 = \{z = x + iy : x = 2n\pi, 0 \le y \le t\}.$$

Then we consider the image of Γ under $w = f(z)$:

$$f(\Gamma_1) = \{w = u + iv : -2(n+1)\pi \le u \le -2n\pi, v = 0\}$$

with the direction from the right to the left;

$$f(\Gamma_2) = \left\{ w = u + iv : u = -2(n+1)\pi, 0 \le v \le \frac{1}{2}(e^t - e^{-t}) - t \right\}$$

with the direction upwards; $f(\Gamma_3)$ lies in the annulus

$$\left\{ w : \frac{1}{2}e^t - \left(\frac{1}{2}e^{-t} + t + 2(n+1)\pi \right) \le |w| \right.$$
$$\left. \le \frac{1}{2}e^t + \left(\frac{1}{2}e^{-t} + t + 2(n+1)\pi \right) \right\}$$

starting from

$$w = -2(n+1)\pi + i \left(\frac{1}{2}e^t - \frac{1}{2}e^{-t} - t \right)$$

and ending at

$$w = -2n\pi + i \left(\frac{1}{2}e^t - \frac{1}{2}e^{-t} - t \right)$$

in the counterclockwise sense;

$$f(\Gamma_4) = \left\{ w = u + iv : u = -2n\pi, 0 \le v \le \frac{1}{2}(e^t - e^{-t}) - t \right\}$$

with the direction downwards.

Hence the winding number of $f(\Gamma)$ around $w = 0$ is 1. By the argument principle, $f(z) = \sin z - z$ has one zero inside the contour Γ. Since n is arbitrarily chosen, we conclude that $\sin z = z$ has infinitely many solutions in $\mathbb{C}$.

Remark. This problem can also be proved by Hadamard's Theorem.

Assume that

$$f(z) = \sin z - z$$

has only finite zeros in $\mathbb{C}$, and denote all the zeros by $z_1, z_2, \ldots, z_n$, multiple zeros being repeated. By Hadamard's Theorem, $f(z)$ can be written as

$$f(z) = e^{g(z)} p(z),$$

where

$$p(z) = \prod_{k=1}^{n} (z - z_k)$$

and $g(z)$ is a polynomial.

It is obvious that $f(z)$ is an entire function of order $\lambda = 1$, where

$$\lambda = \varlimsup_{r \to \infty} \frac{\log \log\{\max_{|z|=r} |f(z)|\}}{\log r},$$

which implies that $g(z)$ must be a polynomial of degree 1. Hence we have

$$\sin z - z = e^{az+b} p(z).$$

Let $z = x + iy$ and x be fixed. By letting $y \to +\infty$ and $y \to -\infty$ respectively, and comparing the increasing order on both sides, we obtain that $\operatorname{Im} a < 0$ in the former case and that $\operatorname{Im} a > 0$ in the latter case. This contradiction implies that $\sin z = z$ has infinite many solutions in $\mathbb{C}$.

5415

(a) Let f be a non-constant analytic function in the annulus $\{1 < |z| < 2\}$ and suppose that $|f| = 5$ on the boundary. Show that f has at least two zeros.

(b) If f is meromorphic in the annulus, is the statement in part (a) still true?

(*Stanford*)

Solution.

(a) Let $D = \{z : 1 < |z| < 2\}$ and $\partial D = \Gamma_1 \cup \Gamma_2$, where $\Gamma_1 = \{z : |z| = 2\}$ is in the counterclockwise sense, and $\Gamma_2 = \{z : |z| = 1\}$ is in the clockwise sense. Because f is non-constant analytic in D and $|f| = 5$ when $z \in \partial D$, we know that both $f(\Gamma_1)$ and $f(\Gamma_2)$ must be $\{w : |w| = 5\}$ in the counterclockwise sense. Hence $\frac{1}{2\pi}\Delta_{\Gamma_1}\arg f(z) \geq 1$ and $\frac{1}{2\pi}\Delta_{\Gamma_2}\arg f(z) \geq 1$. In other words,

$$\frac{1}{2\pi}\Delta_{\partial D}\arg f(z) \geq 2,$$

which shows by the argument principle that f has at least two zeros in D.

(b) If f is meromorphic in D, the statement in (a) is not true. It might occur that $f(\Gamma_1)$ and $f(\Gamma_2)$ are two subarcs of $\{w : |w| = 5\}$, or both $f(\Gamma_1)$ and $f(\Gamma_2)$ are $\{w : |w| = 5\}$ in the clockwise sense. In the latter case, f has no zero in D. The following is a counterexample. Let $g(\zeta)$ be a conformal map of

$$\left\{\zeta = \xi + i\eta : \frac{\xi^2}{a^2} + \frac{\eta^2}{b^2} < 1, \text{ where } a = \frac{1}{2}\left(\sqrt{2} + \frac{1}{\sqrt{2}}\right),\right.$$
$$\left. b = \frac{1}{2}\left(\sqrt{2} - \frac{1}{\sqrt{2}}\right)\right\}$$

onto $\{w : |w| > 5\}$ with the normalization $g(0) = \infty$, $g'(0) > 0$. Then

$$f(z) = g\left(\frac{1}{2}\left(\frac{z}{\sqrt{2}} + \frac{\sqrt{2}}{z}\right)\right)$$

is a non-constant meromorphic function in D with $|f| = 5$ when $z \in \partial D$. But f has no zero in D.

5416

Let n be a positive integer, and let P be a polynomial of exact degree $2n$:

$$P(z) = a_0 + a_1 z + a_2 z^2 + \cdots + a_{2n} z^{2n},$$

where each $a_j \in \mathbb{C}$, and $a_{2n} \neq 0$. Suppose that there is no real number x such that $P(x) = 0$, and suppose that

$$\lim_{r \to \infty} \int_{-r}^r \frac{P'(x)}{P(x)} dx = 0.$$

Prove that P has exactly n roots (counted with multiplicity) in the open upper half plane $\{z \in \mathbb{C} : \text{Im } z > 0\}$.

(Indiana)

Solution.

Let $r > 0$ be sufficiently large such that when $|z| = r$,

$$|a_{2n} z^{2n}| > |a_0 + a_1 z + \cdots + a_{2n-1} z^{2n-1}|.$$

Take a closed contour $\Gamma = \Gamma_1 \cup \Gamma_2$ in the counterclockwise sense, where

$$\Gamma_1 = \{z = re^{i\theta} : 0 \leq \theta \leq \pi\}$$

and

$$\Gamma_2 = \{z = x + iy : -r \leq x \leq r, y = 0\}.$$

Then the number of zeros of $P(z)$ inside Γ is equal to

$$\frac{1}{2\pi i}\int_\Gamma \frac{P'(z)}{P(z)}dz = \frac{1}{2\pi i}\int_{\Gamma_1}\frac{P'(z)}{P(z)}dz + \frac{1}{2\pi i}\int_{\Gamma_2}\frac{P'(z)}{P(z)}dz.$$

It is already known that

$$\lim_{r\to\infty}\int_{\Gamma_2}\frac{P'(z)}{P(z)}dz = \lim_{r\to\infty}\int_{-r}^{r}\frac{P'(x)}{P(x)}dx = 0.$$

We also have

$$\frac{1}{2\pi i}\int_{\Gamma_1}\frac{P'(z)}{P(z)}dz = \frac{1}{2\pi i}\int_{\Gamma_1} d\log P(z)$$
$$= \frac{1}{2\pi}\Delta_{\Gamma_1}\arg P(z) = \frac{1}{2\pi}\Delta_{\Gamma_1}\arg(a_{2n}z^{2n})$$
$$+ \frac{1}{2\pi}\Delta_{\Gamma_1}\arg\left(1 + \frac{a_0 + a_1 z + \cdots + a_{2n-1}z^{2n-1}}{a_{2n}z^{2n}}\right).$$

Note that

$$\frac{1}{2\pi}\Delta_{\Gamma_1}\arg(a_{2n}z^{2n}) = n$$

and

$$\lim_{r\to\infty}\frac{1}{2\pi}\Delta_{\Gamma_1}\arg\left(1 + \frac{a_0 + a_1 z + \cdots + a_{2n-1}z^{2n-1}}{a_{2n}z^{2n}}\right) = 0,$$

we obtain that P has exactly n roots (counted with multiplicity) in the open upper half plane.

5417

Consider the function

$$f(z) = 1 + \frac{1}{z} + \frac{1}{2!}\frac{1}{z^2} + \cdots + \frac{1}{n!}\frac{1}{z^n}.$$

(a) What does the integral

$$\frac{1}{2\pi i}\int_{|z|=r}\frac{f'(z)}{f(z)}dz$$

count?

(b) What is the value of the integral for large n and fixed r?

(c) What does this tell you about the zeros of $f(z)$ for large n?

(Courant Inst.)

Solution.

(a) Let

$$F(\zeta) = f\left(\frac{1}{\zeta}\right) = 1 + \zeta + \frac{1}{2!}\zeta^2 + \frac{1}{3!}\zeta^3 + \cdots + \frac{1}{n!}\zeta^n.$$

From

$$\frac{1}{2\pi i}\int_{|z|=r}\frac{f'(z)}{f(z)}dz = -\frac{1}{2\pi i}\int_{|\zeta|=\frac{1}{r}}\frac{f'\left(\frac{1}{\zeta}\right)}{f\left(\frac{1}{\zeta}\right)}\cdot\frac{d\zeta}{-\zeta^2}$$

$$= -\frac{1}{2\pi i}\int_{|\zeta|=\frac{1}{r}}\frac{F'(\zeta)}{F(\zeta)}d\zeta,$$

we know that the negative of

$$\frac{1}{2\pi i}\int_{|z|=r}\frac{f'(z)}{f(z)}dz$$

represents the number of zeros of $F(\zeta)$ in $\{|\zeta| < \frac{1}{r}\}$, which is just the number of zeros of $f(z)$ in $\{|z| > r\}$.

(b) When $n \to \infty$, $F(\zeta)$ converges to e^ζ uniformly in any compact subset of $\mathbb{C}$. Let

$$\min_{|\zeta|=\frac{1}{r}}|e^\zeta| = m,$$

then $m > 0$.

When n is sufficiently large,

$$|F(\zeta) - e^\zeta| < m \le |e^\zeta|$$

for $|\zeta| = \frac{1}{r}$, which implies the numbers of zeros for $F(\zeta)$ and e^ζ in $\{|\zeta| < \frac{1}{r}\}$ are equal. Since e^ζ has no zero in $\mathbb{C}$, we obtain

$$\frac{1}{2\pi i}\int_{|z|=r}\frac{f'(z)}{f(z)}dz = 0$$

for fixed r and large n.

(c) From the above discussion, we conclude that for any fixed $r > 0$, when n is sufficiently large, there is no zero of $f(z)$ in $\{|z| > r\}$. In other words, all the n zeros of $f(z)$ are in $\{|z| \leq r\}$.

5418

(a) Suppose that $f(z)$ is analytic in the closed disk $|z| \leq R$, and that there is a unique, simple solution z_1 of the equation $f(z) = w$ in $\{|z| < R\}$. Show that this solution is given by the formula

$$z_1 = \frac{1}{2\pi i} \int_{|z|=R} \frac{zf'(z)}{f(z) - w} dz.$$

(b) Show that, if the integer n is sufficiently large, the equation

$$z = 1 + \left(\frac{z}{2}\right)^n$$

has exactly one solution with $|z| < 2$.

(c) If z_1 is the solution in (b), show that

$$\lim_{n \to \infty} (z_1 - 1)^{\frac{1}{n}} = \frac{1}{2}.$$

(*Courant Inst.*)

Solution.
  (a) Let

$$f(z) - w = (z - z_1)Q(z),$$

where $Q(z)$ is analytic and has no zero in $\{|z| < R\}$. Then

$$\frac{f'(z)}{f(z) - w} = [\log(f(z) - w)]' = [\log(z - z_1) + \log Q(z)]'$$

$$= \frac{1}{z - z_1} + \frac{Q'(z)}{Q(z)}.$$

Hence

$$\frac{1}{2\pi i} \int_{|z|=R} \frac{zf'(z)}{f(z)-w} dz$$

$$= \frac{1}{2\pi i} \int_{|z|=R} \frac{z}{z-z_1} dz + \frac{1}{2\pi i} \int_{|z|=R} \frac{zQ'(z)}{Q(z)} dz$$

$$= \frac{1}{2\pi i} \int_{|z|=R} \frac{z}{z-z_1} dz = z_1.$$

(b) Let

$$f_n(z) = z - 1 - \left(\frac{z}{2}\right)^n, \quad g(z) = z - 1,$$

and

$$\Gamma_\varepsilon = \{|z| = 2 - \varepsilon\}.$$

For fixed large n, we choose $\varepsilon > 0$ sufficiently small such that when $z \in \Gamma_\varepsilon$,

$$|f_n(z) - g(z)| = \left|\frac{z}{2}\right|^n = \left(1 - \frac{\varepsilon}{2}\right)^n < 1 - \varepsilon \le |g(z)|.$$

Hence $f_n(z)$ and $g(z)$ have the same number of zeros in $\{|z| < 2 - \varepsilon\}$, and the number is 1. Since ε can be arbitrarily small, the equation $z = 1 + (\frac{z}{2})^n$ has exactly one solution (denoted by $z_1^{(n)}$) in $\{|z| < 2\}$.

(c) $f_n(x)$ is a continuous real-valued function for $1 \le x \le \frac{3}{2}$. When n is sufficiently large, we have $f_n(1) < 0$ and $f_n(\frac{3}{2}) > 0$.

Hence we have $z_1^{(n)} \in (1, \frac{3}{2})$. It follows from

$$|z_1^{(n)} - 1| = \left|\frac{z_1^{(n)}}{2}\right|^n \le \left(\frac{3}{4}\right)^n \to 0$$

that

$$\lim_{n \to \infty} z_1^{(n)} = 1,$$

which implies

$$\lim_{n\to\infty} \left(z_1^{(n)} - 1\right)^{\frac{1}{n}} = \lim_{n\to\infty} \frac{z_1^{(n)}}{2} = \frac{1}{2}.$$

5419

Let

$$\Omega = D(0,1) \backslash \left\{ \frac{1}{2}, -\frac{1}{2} \right\}.$$

Find all analytic functions $f : \Omega \to \Omega$ with the following property: if γ is any cycle in Ω which is not homologous to zero (mod Ω), then $f * \gamma$ is not homologous to zero (mod Ω).

(Indiana)

Solution.

Since f is analytic in Ω and bounded by $|f(z)| < 1$, the points $z = \pm\frac{1}{2}$ must be the removable singularities of f. Let

$$\gamma_1 = \left\{ \left| z - \frac{1}{2} \right| = \varepsilon \right\}, \quad \gamma_2 = \left\{ \left| z + \frac{1}{2} \right| = \varepsilon \right\},$$

where $\varepsilon > 0$ is small, and the directions of γ_1 and γ_2 are both in the counterclockwise sense. Since γ_1, γ_2 are not homologous to zero (mod Ω), $f * \gamma_1$ and $f * \gamma_2$ are also not homologous to zero (mod Ω). As ε tends to zero, $f(\gamma_1)$ and $f(\gamma_2)$ will tend to either $w = \frac{1}{2}$ or $w = -\frac{1}{2}$, because otherwise, $f * \gamma_1$ or $f * \gamma_2$ will be homologous to zero (mod Ω). Hence we obtain

$$f\left(\pm\frac{1}{2}\right) = \pm\frac{1}{2}.$$

Now we claim that the case that $f\left(\frac{1}{2}\right) = f\left(-\frac{1}{2}\right)$ will not happen. If, for example,

$$f\left(\frac{1}{2}\right) = f\left(-\frac{1}{2}\right) = \frac{1}{2},$$

we assume that $z = \frac{1}{2}$ is a zero of $f(z) - \frac{1}{2}$ of order n and $z = -\frac{1}{2}$ is a zero of $f(z) - \frac{1}{2}$ of order m, then

$$f * (m\gamma_1 - n\gamma_2)$$

is homologous to zero (mod Ω), while $m\gamma_1 - n\gamma_2$ not homologous to zero (mod Ω), which is a contradiction. Thus we obtain either

$$f\left(\frac{1}{2}\right) = \frac{1}{2}, \quad f\left(-\frac{1}{2}\right) = -\frac{1}{2}$$

or

$$f\left(\frac{1}{2}\right) = -\frac{1}{2}, \quad f\left(-\frac{1}{2}\right) = \frac{1}{2}.$$

In the case of

$$f\left(\frac{1}{2}\right) = \frac{1}{2}, \quad f\left(-\frac{1}{2}\right) = -\frac{1}{2},$$

we consider the function

$$F(z) = \frac{f(z) - \frac{1}{2}}{1 - \frac{1}{2}f(z)} : \frac{z - \frac{1}{2}}{1 - \frac{1}{2}z}$$

which is analytic in $D(0,1)$ and satisfies $|F(z)| \leq 1$. It follows from $F\left(-\frac{1}{2}\right) = 1$ that $F(z) \equiv 1$, which implies that $f(z) = z$.

In the case of

$$f\left(\frac{1}{2}\right) = -\frac{1}{2}, \quad f\left(-\frac{1}{2}\right) = \frac{1}{2},$$

we consider the function

$$G(z) = \frac{f(z) + \frac{1}{2}}{1 + \frac{1}{2}f(z)} : \frac{z - \frac{1}{2}}{1 - \frac{1}{2}z}$$

which is also analytic in $D(0,1)$ and satisfies $|G(z)| \leq 1$. It follows from $G\left(-\frac{1}{2}\right) = -1$ that $G(z) \equiv -1$, which implies that $f(z) = -z$. Thus we conclude that the functions which satisfy the requirements of the problem are $f(z) = z$ and $f(z) = -z$.

5420

Let $f(z)$ be a holomorphic function on $\{|z| < 1\}$ and continuous up to $\{|z| \le 1\}$. Let M be the supremum of $|f|$ on $\{|z| \le 1\}$. Let L be the intersection of $\{|z| \le 1\}$ and $\{\operatorname{Re} z = \frac{1}{2}\}$. Let m be the supremum of $|f|$ on L. Show that

$$|f(0)|^3 \le mM^2$$

(Harvard)

Solution.

Consider the triangle T with three vertexes $z_1 = \frac{1+i\sqrt{3}}{2}$, $z_2 = -1$ and $z_3 = \frac{1-i\sqrt{3}}{2}$. Define $F(z) = f(z)f\left(ze^{i\frac{2\pi}{3}}\right)f\left(ze^{i\frac{4\pi}{3}}\right)$, then $F(z)$ is holomorphic in the triangle T, and for $\varsigma \in \partial T$, $|F(\varsigma)| \le mM^2$. By the maximum modulus principle, we obtain that

$$|F(0)| = |f(0)|^3 \le mM^2.$$

5421

If $f : D \to \mathbb{C}$, where $D = \{z \in \mathbb{C} : |z| < 1\}$, is a bounded analytic function, show that

$$\sup_{z \in D}\{(1 - |z|^2)|f'(z)|\} \le \sup_{z \in D}\{|f(z)|\}.$$

(SUNY, Albany)

Solution.

Assume $\sup_{z \in D}\{|f(z)|\} = M$, define $g(z) = \frac{f(z)}{M}$, then $g(z) : D \to D$.

Apply Schwarz Lemma to $g(z)$, we have

$$\frac{|f'(z)|}{M} = |g'(z)| \le \frac{1 - |g(z)|^2}{1 - |z|^2} \le \frac{1}{1 - |z|^2}.$$

Hence for every $z \in D$, we have $(1 - |z|^2)|f'(z)| \le M$, which implies that

$$\sup_{z \in D}\{(1 - |z|^2)|f'(z)|\} \le \sup_{z \in D}\{|f(z)|\}.$$

5422

Let f be a bounded analytic function in the right half-plane. Prove that if $f(n) = 0$ for $n = 1, 2, 3, \ldots$, then $f \equiv 0$.

(*SUNY, Albany*)

Solution.

Let $L = \{w : \operatorname{Re} w > 0\}$, $\Delta = \{z : |z| < 1\}$, then $w = \varphi(z) = \frac{1+z}{1-z}$ is a conformal mapping of Δ onto L. Define $F = f \circ \varphi$, then $F(z)$ is a bounded analytic function defined in Δ, and has zeros $a_n = \frac{n-1}{n+1} = 1 - \frac{2}{n+1}$ for $n = 1, 2, 3, \ldots$. Assume $|F(z)| \leq M$. If $F(z)$ is not identically zero, then there is a positive integer j such that $F(z) = c_j z^j + c_{j+1} z^{j+1} + \cdots$ with $c_j \neq 0$.

For any positive integer $n \geq 2$, define $F_n(z) = \dfrac{F(z)}{z^j \prod_{k=2}^{n} \frac{a_k - z}{1 - a_k z}}$, which is also analytic in Δ. Because on $\partial\Delta$, $|F_n(z)| = |F(z)|$, by the maximum modulus principle, $|F_n(z)| \leq M$. Take $z = 0$, we have $|c_j| \leq M |\prod_{k=2}^{n} a_k|$. Let $n \to \infty$, since the infinite product $\prod_{k=2}^{\infty} a_k$ diverges to zero, we obtain that $c_j = 0$, which is a contradiction. Hence $F \equiv 0$ which implies $f \equiv 0$.

5423

Let f and g be analytic on a domain Ω and continuous on $\Omega \cup \partial\Omega$. If $|f(z) + g(z)| < |f(z)| + |g(z)|$ on $\partial\Omega$, prove that f and g have the same number of zeroes in Ω, counting multiplicity.

(*SUNY, Albany*)

Solution.

It is obvious that on $\partial\Omega$ $f(z) \neq 0$ and $g(z) \neq 0$. Define $F(z) = \frac{f(z)}{g(z)}$, then when $z \in \partial\Omega$, $|F(z) + 1| < |F(z)| + 1$ which implies that $F(z)$ does not take positive real number when $z \in \partial\Omega$.

Hence

$$\int_{\partial\Omega} d \arg F(z) = \Delta_{\partial\Omega} \arg F(z) = 0,$$

which shows that $F(z)$ has the same number of zeroes and poles in Ω, counting multiplicity. So f and g have the same number of zeroes in Ω, counting multiplicity.

Section 5

Series and Normal Families

5501

Let

$$\sum_{n=0}^{\infty} a_n z^n$$

have a radius of convergence r and let the function $f(z)$ to which it converges have exactly one singular point z_0, on $|z| = r$, which is a simple pole. Prove that

$$\lim_{n \to \infty} a_n / a_{n+1} = z_0.$$

(*Indiana*)

Solution.

Assume that the residue of $f(z)$ at z_0 is A, and define

$$F(z) = f(z) - \frac{A}{z - z_0}.$$

Then $F(z)$ is analytic on $\{z : |z| \le r\}$. In other words, the Taylor expansion of $F(z)$ at $z = 0$ has a radius of convergence larger than r. Hence the power series

$$F(z) = \sum_{n=0}^{\infty} a_n z^n - \frac{A}{z - z_0}$$

$$= \sum_{n=0}^{\infty} a_n z^n + \sum_{n=0}^{\infty} A \frac{z^n}{z_0^{n+1}}$$

$$= \sum_{n=0}^{\infty} \left(a_n + \frac{A}{z_0^{n+1}} \right) z^n$$

is convergent at $z = z_0$, which implies

$$\lim_{n \to \infty} \left(a_n + \frac{A}{z_0^{n+1}} \right) z_0^n = 0.$$

It follows that

$$\lim_{n \to \infty} a_n z_0^n = -\frac{A}{z_0} \neq 0$$

and

$$\lim_{n \to \infty} a_{n+1} z_0^{n+1} = -\frac{A}{z_0} \neq 0,$$

and we obtain

$$\lim_{n \to \infty} \frac{a_n}{a_{n+1}} = z_0.$$

5502

(1) Show that the series

$$- \sum_{n \geq 1} \alpha^n / n$$

is convergent for $1 \neq \alpha \in \mathbb{C}$ with $|\alpha| = 1$.

(2) Show that this series converges to $\log(1 - \alpha)$ for such α.

(*Minnesota*)

Solution.

(1) Let $\alpha = e^{it}$, $t \in (0, 2\pi)$, then

$$-\sum_{n\geq 1} \frac{\alpha^n}{n} = -\sum_{n\geq 1} \frac{\cos nt + i \sin nt}{n}.$$

For $t \in (0, 2\pi)$ we have

$$\left| \sum_{k=1}^{n} \cos kt \right| = \left| \frac{\sin \frac{t}{2} - \sin \frac{2n+1}{2}t}{2 \sin \frac{t}{2}} \right| \leq \frac{1}{\sin \frac{t}{2}},$$

and

$$\left| \sum_{k=1}^{n} \sin kt \right| = \left| \frac{\cos \frac{t}{2} - \cos \frac{2n+1}{2}t}{2 \sin \frac{t}{2}} \right| \leq \frac{1}{\sin \frac{t}{2}}.$$

Because $\frac{1}{n}$ tends to zero monotonically, by Dirichlet's criterion we know that both $\sum_{n\geq 1} \frac{\cos nt}{n}$ and $\sum_{n\geq 1} \frac{\sin nt}{n}$ converge, which shows that $-\sum_{n\geq 1} \frac{\alpha^n}{n}$ is convergent for $1 \neq \alpha \in \mathbb{C}$ with $|\alpha| = 1$.

(2) Let

$$f(z) = -\sum_{n\geq 1} \frac{z^n}{n} \quad (|z| < 1).$$

Differentiating term by term, we have

$$f'(z) = -\sum_{n\geq 1} z^{n-1} = \frac{-1}{1-z}.$$

Integrating both sides on the above identity, we obtain $f(z) = \log(1 - z)$, for $|z| < 1$.

Let $\alpha = e^{it}$, $z = re^{it}$ where $0 < r < 1$, $0 < t < 2\pi$. It follows from Abel's limit theorem that

$$-\sum_{n\geq 1} \frac{\alpha^n}{n} = -\sum_{n\geq 1} \frac{e^{int}}{n} = \lim_{r\to 1^-} -\sum_{n\geq 1} \frac{(re^{it})^n}{n}$$

$$= \lim_{r\to 1^-} \log(1 - re^{it}) = \log(1 - e^{it}) = \log(1 - \alpha).$$

5503

Consider a power series

$$\sum_{n=1}^{\infty} \frac{1}{n} z^{n!}.$$

Show that the series converges to a holomorphic function on the open unit disk centered at origin. Prove that the boundary of the disk is the natural boundary of the function.

(Columbia)

Solution.

First of all, we prove the following proposition: If the radius of convergence of

$$f(z) = \sum_{n=0}^{\infty} a_n z^n$$

is equal to 1 and $a_n \geq 0$ for all n, then $z = 1$ is a singular point of $f(z)$. Assume the proposition is false, i.e., $z = 1$ is a regular point of f, then for fixed $x \in (0, 1)$ there exists a small real number $\delta > 0$ such that the power series expansion of f at point x is convergent at $z = 1 + \delta$. Suppose the series is

$$\sum_{k=0}^{\infty} b_k (z - x)^k,$$

where

$$b_k = \frac{f^{(k)}(x)}{k!} = \frac{1}{k!} \sum_{n=k}^{\infty} n(n-1)\cdots(n-k+1)a_n x^{n-k}.$$

Thus

$$\sum_{k=0}^{\infty} b_k (z-x)^k = \sum_{k=0}^{\infty} \sum_{n=k}^{\infty} \frac{n(n-1)\cdots(n-k+1)}{k!} a_n (z-x)^k x^{n-k}$$

is convergent at $z = 1 + \delta$. Noting that when $z = 1 + \delta$ the right side in the above identity is a convergent double series with positive terms, and hence the order of summation can be changed, we assert that when $z = 1 + \delta$,

$$\sum_{k=0}^{\infty} b_k (z-x)^k = \sum_{k=0}^{\infty} \sum_{n=k}^{\infty} \frac{n(n-1)\cdots(n-k+1)}{k!} a_n (z-x)^k x^{n-k}$$

$$= \sum_{n=0}^{\infty} a_n \sum_{k=0}^{n} \frac{n(n-1)\cdots(n-k+1)}{k!} (z-x)^k x^{n-k}$$

$$= \sum_{n=0}^{\infty} a_n z^n,$$

which contradicts the statement that the radius of convergence of

$$\sum_{n=0}^{\infty} a_n z^n$$

is equal to 1.

Now we return to the power series

$$F(z) = \sum_{n=1}^{\infty} \frac{1}{n} z^{n!}.$$

It follows from

$$\varlimsup_{n \to \infty} \sqrt[n!]{\frac{1}{n}} = 1$$

that the radius of convergence of

$$\sum_{n=1}^{\infty} \frac{1}{n} z^{n!}$$

is equal to 1. By the above proposition, $z = 1$ is a singular point of $F(z)$. For any natural numbers p and q,

$$F(z e^{\frac{2q}{p} \pi i}) = \sum_{n=1}^{p-1} \frac{1}{n} (z e^{\frac{2q}{p} \pi i})^{n!} + \sum_{n=p}^{\infty} \frac{1}{n} z^{n!}.$$

Since $z = 1$ is a singular point of

$$\sum_{n=p}^{\infty} \frac{1}{n} z^{n!},$$

it is also a singular point of $F\left(ze^{\frac{2q}{p}\pi i}\right)$. In other words, $z = e^{\frac{2q}{p}\pi i}$ is a singular point of $F(z)$. Since the set $\left\{e^{\frac{2q}{p}\pi i} : p, q = 1, 2, \dots\right\}$ is dense on $\{|z| = 1\}$, we conclude that the unit circle $\{|z| = 1\}$ is the natural boundary of $F(z)$.

Remark. By the above discussion, the boundary of the unit disk is also the natural boundary of the function $\sum_{n=1}^{\infty} \frac{1}{n^2} z^{n!}$ although the series is absolutely and uniformly convergent on the closure of the unit disk.

5504

Suppose f is analytic in $U = \{|z| < 1\}$ with $f(0) = 0$ and $|f(z)| < 1$ for all $z \in U$. If the sequence $\{f_n\}$ is defined by composition

$$f_n(z) = \underbrace{f(f(\cdots f(z))\cdots)}_{n}$$

and

$$f_n(z) \to g(z)$$

for all $z \in U$, prove that either $g(z) = 0$ or $g(z) = z$.

(Indiana–Purdue)

Solution.
By Schwarz Lemma, it follows from $f(0) = 0$ and $|f(z)| < 1$ that $|f(z)| \le |z|$ for all $z \in U$, and if $|f(z)| = |z|$ for some $z \ne 0$, then $f(z) = e^{i\alpha}z$ where α is a real number.

In the case when $f(z) = e^{i\alpha}z$. $f_n(z) = e^{in\alpha}z$. Since $f_n(z)$ is convergent, we obtain $\alpha = 0$, which implies that $f(z) = z$ and $g(z) = z$.

In other cases, we have

$$\left|\frac{f(z)}{z}\right| < 1$$

for all $z \in U$. Let $0 < r < 1$. Then

$$\max_{|z|\leq r}\left|\frac{f(z)}{z}\right| = \lambda < 1.$$

For all $z \in \{|z| \leq r\}$, we have

$$|f(z)| \leq \lambda|z|,$$
$$|f_2(z)| = |f(f(z))| \leq \lambda|f(z)| \leq \lambda^2|z|,$$
$$\cdots$$
$$|f_n(z)| = |f(f_{n-1}(z))| \leq \lambda|f_{n-1}(z)| \leq \lambda^n|z|,$$
$$\cdots$$

Hence $f_n(z)$ converges to zero uniformly in $\{|z| \leq r\}$. Since $0 < r < 1$ is arbitrarily chosen, we obtain $g(z) = 0$ for all $z \in U$.

5505

Let $\{f_n\}_{n=1}^{\infty}$ be a sequence of analytic functions in a domain D which converges uniformaly on compact subsets of D to a function f on D.

(a) Prove that if $f_n(z) \neq 0$ for all $n \geq 1$ and $z \in D$, then either f is identically zero in D or $f(z) \neq 0$ for all $z \in D$.

(b) If each f_n is one-to-one on D, show that f is either constant or one-to-one on D.

(UC, Irvine)

Solution.

(a) First of all, we know from Weierstrass' Theorem that f is analytic on D. Suppose f is not identically zero, but has a zero point $z_0 \in D$. Since the zeros of a non-zero analytic function are isolated, there exists $r > 0$, such that $f(z) \neq 0$ when

$$z \in \{z : 0 < |z - z_0| \leq r\} \subset D.$$

Let m be the minimum value of $|f(z)|$ on

$$\{z : |z - z_0| = r\}.$$

Then $m > 0$. As $\{f_n\}$ converges to $f(z)$ uniformly on compact subsets of D, we know that for sufficiently large n,

$$|f_n(z) - f(z)| < m \leq |f(z)|$$

holds on $\{z : |z - z_0| = r\}$. It follows from Rouché's Theorem that f_n and f have the same number of zeros in $\{z : |z-z_0| < r\}$. Since z_0 is a zero of f, f_n must have a zero in $\{z : |z - z_0| < r\}$, which is a contradiction to the assumption that $f_n(z) \neq 0$ for all $z \in D$.

(b) Suppose f is not a constant, and is not one-to-one on D. Then there exist $z_1, z_2 \in D$ ($z_1 \neq z_2$), such that $f(z_1) = f(z_2)$ (denote it by a). Choose $r > 0$ sufficiently small, such that

$$\{z : |z - z_1| \leq r\} \cap \{z : |z - z_2| \leq r\} = \emptyset,$$

$$\{z : |z - z_1| \leq r\} \cup \{z : |z - z_2| \leq r\} \subset D,$$

and $f(z) - a \neq 0$ in $\{z : 0 < |z-z_1| \leq r\} \cup \{z : 0 < |z-z_2| \leq r\}$. Let m be the minimum value of $|f(z) - a|$ on $\{z : |z - z_1| = r \text{ or } |z - z_2| = r\}$. Then $m > 0$. With the same reason as in (a), when n is sufficiently large,

$$|(f_n(z) - a) - (f(z) - a)| = |f_n(z) - f(z)| < m \leq |f(z) - a|$$

holds on $\{z : |z - z_1| = r \text{ or } |z - z_2| = r\}$. It follows from Rouché's Theorem that $f_n(z) - a$ and $f(z) - a$ have the same number of zeros in $\{z : |z - z_1| < r\}$ and $\{z : |z - z_2| < r\}$ respectively. In other words, there exists

$$z_1' \in \{z : |z - z_1| < r\}$$

and

$$z_2' \in \{z : |z - z_2| < r\},$$

such that $f_n(z_1') - a = 0$ and $f_n(z_2') - a = 0$, which implies $f_n(z_1') = f_n(z_2')$ $(z_1' \neq z_2')$. This is a contradiction to the assumption that f_n is one-to-one on D.

5506

Let $D \subset \mathbb{C}$ be a bounded domain, and let $\{f_n\}$ be a sequence of analytic automorphisms of D such that

$$\lim_{n \to \infty} f_n(a) = b \in \partial D$$

for some point $a \in D$. Prove that

$$\lim_{n \to \infty} f_n(z) = b$$

for every $z \in D$.

(Indiana)

Solution.
Take $a_0 \in D$, $a_0 \neq a$. If $\{f_n(a_0)\}$ does not converge to b, there exists a subsequence of $\{f_n(a_0)\}$ converging to $b_0 \neq b$. Without loss of generality, we assume

$$\lim_{n \to \infty} f_n(a) = b \in \partial D,$$

$$\lim_{n \to \infty} f_n(a_0) = b_0 \neq b.$$

Since $\{f_n(z)\}$ is a normal family, there is a subsequence $\{f_{n_k}(z)\}$ converging uniformly on compact subsets of D to $f(z)$. Because $f(a) \neq f(a_0)$, $f(z)$ is a non-constant analytic function of D.

Let r be sufficiently small such that $f(z) - b$ has no zero in $\{z : 0 < |z - a| \leq r\} \subset D$, then $m = \min\{|f(z) - b| : |z - a| = r\} > 0$. Since $\{f_{n_k}\}$ converges uniformly to f on $\{z : |z - a| = r\}$, when k is sufficiently large,

$$|f_{n_k}(z) - f(z)| = |(f_{n_k}(z) - b) - (f(z) - b)| < m \leq |f(z) - b|$$

on $\{z : |z-a| = r\}$. By Rouché's Theorem, $f_{n_k}(z)-b$ has zero(s) in $\{z : |z - a| < r\}$, which is a contradiction to the fact that f_{n_k} does not assume the value $b \in \partial D$ in D because f_{n_k} is an automorphism of D.

5507

Which of the following families are normal, and which is compact? Justify your answers.

(a) $\mathcal{F} = \{f : f$ is analytic in D, $f(0) = 0$, diam $f(D) \leq 2\}$

(b) $\mathcal{G} = \{g : g$ is analytic in D, $g(0) = 1$, Re$\{g\} > 0$, diam $g(D) \geq 1\}$.

Here the diameter of a set S is diam $S = \sup\{|z - \zeta| : z, \zeta \in S\}$.

(*Indiana*)

Solution.

(a) For any $f \in \mathcal{F}$, it follows from $f(0) = 0$ and diam $f(D) \leq 2$ that $|f(z)| \leq 2$, which shows that $\mathcal{F}$ is normal.

Let $\{f_n\}$ be a sequence of functions in $\mathcal{F}$. Then there exists a subsequence $\{f_{n_k}\}$ converging uniformly in compact subsets of D to $f(z)$, which obviously satisfies the conditions that $f(z)$ is analytic in D and $f(0) = 0$. For any two fixed points $z, \zeta \in D$, we have

$$|f_{n_k}(z) - f_{n_k}(\zeta)| \leq 2$$

because diam $f_{n_k}(D) \leq 2$. We choose a compact subset $K \subset D$ such that $z, \zeta \in K$. It follows from the uniform convergence of $\{f_{n_k}\}$ on K that

$$|f(z) - f(\zeta)| \leq 2.$$

Since $z, \zeta \in D$ can be arbitrarily chosen, we obtain diam $f(D) \leq 2$, hence $f(z) \in \mathcal{F}$, which shows that $\mathcal{F}$ is also compact.

(b) Let $\{g_n\}$ be any sequence of functions in $\mathcal{G}$. Then for $G_n(z) = e^{-g_n(z)}$, we have $|G_n(z)| < 1$. Hence there exists a subsequence $\{G_{n_k}\}$ converging uniformly in compact subsets of D to a function $G(z)$ which is either a constant or a non-constant analytic function in D. If $G(z)$ is a constant, then the constant is e^{-1} because

$$G(0) = \lim_{n \to \infty} e^{-g_n(0)} = e^{-1};$$

if $G(z)$ is non-constant analytic, since $G_n(z) \neq 0$ for all $z \in D$, by Hurwitz's Theorem, we have $G(z) \neq 0$ for all $z \in D$. Hence we can define an analytic function $g(z) = -\log G(z)$, where the single-valued branch is chosen by $g(0) = -\log G(0) = 1$, and we conclude that

$$g_{n_k}(z) = -\log G_{n_k}(z)$$

converges uniformly in compact subsets of D to $g(z)$, which shows that family $\mathcal{G}$ is normal. But family $\mathcal{G}$ is not compact. First we can choose a sequence of functions $g_n(z)$ in $\mathcal{G}$ as follows: $g_n(z)$ is a conformal mapping of D onto

$$\Omega_n = \left\{ w : |w - 1| < \frac{1}{4} \right\} \cup \{w : |w - 3| < 1\}$$

$$\cup \left\{ w : |\operatorname{Im} w| < \frac{1}{n}, 1 < \operatorname{Re} w < 3 \right\}$$

satisfying $g_n(0) = 1$, $g_n'(0) > 0$. By the Riemann Mapping Theorem, such a mapping g_n exists and is unique, and it is obvious that g_n satisfies all the conditions required by the family $\mathcal{G}$. Because the domain sequence $\{\Omega_n\}$ converges to $\Omega = \{w : |w - 1| < \frac{1}{4}\}$ which is called the kernel of $\{\Omega_n\}$ with respect to $w = 1$, by Caratheodory's Theorem, $\{g_n(z)\}$ converges uniformly in compact subsets of D to $g(z)$ which is a conformal mapping of D onto Ω. Since diam $g(D) = \frac{1}{2}$, $g(z)$ does not belong to the family $\mathcal{G}$, which shows that $\mathcal{G}$ is not compact.

5508

Suppose that $1 \leq p < \infty$ and $c \geq 0$ is a real number. Let $\mathcal{F}$ be the set of all analytic functions f on $\{|z| < 1\}$ such that

$$\sup_{0 \leq r < 1} \int_0^{2\pi} |f(re^{i\theta})|^p d\theta \leq c.$$

Show that $\mathcal{F}$ is a normal family.

(*Illinois*)

Solution.

It suffices to prove that the functions in $\mathcal{F}$ are uniformly bounded on every compact set of $\{|z| < 1\}$. We prove the assertion by contradiction. If it is not the case, then there exist $z_n \in D$, $f_n \in \mathcal{F}$ such that $z_n \to z_0 \in D$ and $f_n(z_n) \to \infty$.

Let $1 - |z_0| = 3r$. Then when n is sufficiently large, $|z_n - z_0| < r$. By Cauchy integral formula,

$$f_n(z_n) = \frac{1}{2\pi i} \int_{|\zeta - z_0| = \rho} \frac{f_n(\zeta)}{\zeta - z_n} d\zeta, \quad (2r < \rho < 3r).$$

Hence

$$|f_n(z_n)| \leq \frac{3}{2\pi} \int_0^{2\pi} |f_n(z_0 + \rho e^{i\theta})| d\theta$$

$$\leq \frac{3}{2\pi} \left(\int_0^{2\pi} |f_n(z_0 + \rho e^{i\theta})|^p d\theta \right)^{\frac{1}{p}} \cdot \left(\int_0^{2\pi} d\theta \right)^{\frac{1}{q}}$$

$$= \frac{3}{(2\pi)^{\frac{1}{q}}} \left(\int_0^{2\pi} |f_n(z_0 + \rho e^{i\theta})|^p d\theta \right)^{\frac{1}{q}}$$

where

$$\frac{1}{p} + \frac{1}{q} = 1.$$

Then

$$\frac{2\pi}{3^p}|f_n(z_n)|^p \int_{2r}^{3r} \rho \, d\rho \le \int_{2r}^{3r}\int_0^{2\pi} |f_n(z_0 + \rho e^{i\theta})|^p \rho \, d\rho \, d\theta$$

$$\le \int_0^1 \int_0^{2\pi} |f(\rho e^{i\theta})|^p \rho \, d\rho \, d\theta \le \frac{c}{2}.$$

As $n \to \infty$, the left side of the above inequality tends to infinity, while the right side of the inequality is a constant. The contradiction implies that $\mathcal{F}$ is a normal family.

5509

(a) Let f be holomorphic for $|z| < R$ and satisfy $f(0) = 0$, $f'(0) \ne 0$, $f(z) \ne 0$ for $0 < |z| < r \le R$. Let C be the circle $|z| = \rho$ where $\rho < r$. Show that

$$g(w) = \frac{1}{2\pi i} \int_C \frac{t f'(t) dt}{f(t) - w}$$

define a holomorphic function of w for

$$|w| < m = \min_\theta |f(\rho e^{i\theta})|,$$

and that $z = g(w)$ is the unique solution of

$$f(z) = w$$

that tends to zero with w.

(b) Find the Taylor's expansion of $g(w)$, and apply this to find the explicit series expansion of the root of the equation

$$z^3 + 3z - w = 0$$

that tends to zero with w.

(*Harvard*)

Solution.

(a) It follows from

$$|w| < m = \min_\theta |f(\rho e^{i\theta})|$$

that when $t \in C$,

$$\frac{1}{f(t) - w} = \frac{1}{f(t)(1 - \frac{w}{f(t)})} = \sum_{n=0}^{\infty} \frac{w^n}{f(t)^{n+1}}.$$

Hence

$$g(w) = \frac{1}{2\pi i} \int_C \frac{t f'(t) dt}{f(t) - w} = \sum_{n=0}^{\infty} \left(\frac{1}{2\pi i} \int_C \frac{t f'(t)}{f(t)^{n+1}} dt \right) w^n,$$

which implies that $g(w)$ is holomorphic in $\{w : |w| < m\}$.

Let Γ be the image of C under f where C is the circle $\{z : |z| = \rho\}$ taken once counterclockwise. Because

$$|w| < m = \min_{\theta} |f(\rho e^{i\theta})|,$$

the winding number

$$n(\Gamma, 0) = n(\Gamma, w),$$

which shows that $f(z)$ and $f(z) - w$ have the same number of zeros in $\{z : |z| < \rho\}$. Since $z = 0$ is the only simple zero of f in $\{z : |z| < \rho\}$, we know that $f(z) = w$ has a unique solution in $\{z : |z| < \rho\}$. Denote the unique solution by z_1, then

$$f(t) - w = (t - z_1) Q(t)$$

where $Q(t)$ is analytic and has no zero in $\{t : |t| < \rho\}$, and

$$\frac{f'(t)}{f(t) - w} = [\log(f(t) - w)]' = [\log(t - z_1) + \log Q(t)]'$$

$$= \frac{1}{t - z_1} + \frac{Q'(t)}{Q(t)}.$$

Hence

$$g(w) = \frac{1}{2\pi i} \int_C \frac{t f'(t) dt}{f(t) - w}$$

$$= \frac{1}{2\pi i} \int_C \frac{t}{t - z_1} dt + \frac{1}{2\pi i} \int_C \frac{t Q'(t)}{Q(t)} dt$$

$$= z_1,$$

which shows that $g(w)$ is just the unique solution of $f(z) = w$. As the constant term in the Taylor expansion of $g(w)$ is

$$\frac{1}{2\pi i} \int_C \frac{tf'(t)}{f(t)} dt$$

which is obviously zero, we assert that the unique solution $g(w)$ tends to zero together with w.

(b) Let

$$f(z) = z^3 + 3z,$$

then

$$g(w) = \frac{1}{2\pi i} \int_C \frac{tf'(t)dt}{f(t) - w} = \sum_{n=1}^{\infty} a_n w^n,$$

where

$$a_n = \frac{1}{2\pi i} \int_C \frac{tf'(t)}{f(t)^{n+1}} dt = \frac{1}{2\pi i} \int_C \frac{t^2+1}{3^n t^n} \left(1 + \frac{t^2}{3}\right)^{-n-1} dt.$$

After some computation, we obtain $a_{2k} = 0$ and

$$a_{2k-1} = \frac{1}{2\pi i} \int_C \frac{1}{3^{2k-1} t^{2k-1}} \left[t^2 C_{-2k}^{k-2} \left(\frac{t^2}{3}\right)^{k-2} \right.$$

$$\left. + C_{-2k}^{k-1} \left(\frac{t^2}{3}\right)^{k-1} \right] dt = \frac{1}{3^{3k-2}} (3 C_{-2k}^{k-2} + C_{-2k}^{k-1}).$$

5510

Find an explicit formula for a meromorphic function f whose only singularities are simple poles at $-1, -2, -3, \ldots$ with residue n at $z = -n$. Prove in detail that your function has all the required properties.

(*Illinois*)

Solution.

By Mittag–Leffler's Theorem, we construct

$$f(z) = \sum_{n=1}^{\infty} \left(\frac{n}{z+n} - 1 + \frac{z}{n} \right) = \sum_{n=1}^{\infty} \frac{z^2}{n(n+z)}.$$

For any natural number N, when $|z| \leq N$, $n \geq 2N$,

$$\left| \frac{z^2}{n(n+z)} \right| \leq \frac{2N^2}{n^2}.$$

Hence

$$\sum_{n=2N}^{\infty} \frac{z^2}{n(n+z)}$$

converges uniformly in $\{|z| \leq N\}$ to a function which is analytic in $\{|z| < N\}$. In addition,

$$\sum_{n=1}^{2N-1} \frac{z^2}{n(n+z)}$$

is a meromorphic function whose only singularities in $\{|z| < N\}$ are simple poles at $z = -1, -2, \ldots, -N+1$ with residue n at $z = -n$. So $f(z)$ is analytic in $\{|z| < N\} \backslash \{-1, -2, \ldots, -N+1\}$, and $z = -1, -2, \ldots, -N+1$ are its simple poles with residue n at $z = -n$.

Because N can be chosen arbitrarily large, it is obvious that $f(z)$ has all the required properties of the problem.

5511

(a) Does there exist a sequence of polynomials $\{P_n\}$ such that $P_n(z) \to \frac{1}{z^2}$ uniformly on the annulus $1 < |z| < 2$? If Yes, give an explicit formula for the P_n; if No, explain why not.

(b) Does there exist an entire function g whose zero-set is $\{\sqrt{n}(1+i) : n = 0, 1, 2, 3, \ldots\}$? If Yes, give an explicit formula for g; if No, explain why not.

(*Illinois*)

Solution.

(a) No. If there exists a sequence of polynomials $\{P_n\}$ such that $P_n(z) \to \frac{1}{z^2}$ uniformly on $\{1 < |z| < 2\}$, then for any $\varepsilon \in (0, \frac{1}{4})$, there exists $N > 0$ such that when $n > N$, $|P_n(z) - \frac{1}{z^2}| < \varepsilon$ holds for all $z \in \{1 < |z| < 2\}$. Multiply both sides by $|z|^2$, we have

$$|z^2 P_n(z) - 1| < \varepsilon |z|^2 < 4\varepsilon < 1 \quad \text{for} \quad z \in \{1 < |z| < 2\}.$$

Because $z^2 P_n(z) - 1$ is an analytic function in $\{|z| < 2\}$, it follows from the maximum modulus principle that

$$|z^2 P_n(z) - 1| < 1$$

holds for all $z \in \{|z| < 2\}$. The contradiction follows by taking $z = 0$ in the inequality.

(b) Yes. The function g can be chosen as

$$g(z) = z \prod_{n=1}^{\infty} \left(1 - \frac{z}{a_n}\right) e^{\frac{z}{a_n} + \frac{1}{2}\left(\frac{z}{a_n}\right)^2},$$

where $a_n = \sqrt{n}(1 + i)$.

For any $R > 0$, let $|z| \le R$ and choose $N > R^2$. Then when $n \ge N$,

$$\log\left[\left(1 - \frac{z}{a_n}\right) e^{\frac{z}{a_n} + \frac{1}{2}\left(\frac{z}{a_n}\right)^2}\right]$$

$$= \log\left(1 - \frac{z}{a_n}\right) + \frac{z}{a_n} + \frac{1}{2}\left(\frac{z}{a_n}\right)^2$$

$$= -\frac{1}{3}\left(\frac{z}{a_n}\right)^3 - \cdots - \frac{1}{m}\left(\frac{z}{a_n}\right)^m - \cdots.$$

It is easy to see that

$$\left|\log\left[\left(1 - \frac{z}{a_n}\right) e^{\frac{z}{a_n} + \frac{1}{2}\left(\frac{z}{a_n}\right)^2}\right]\right| < \frac{R^3}{n^{3/2}}.$$

Because $\sum_{n=N}^{\infty} \frac{R^3}{n^{3/2}}$ converges, we know that

$$\sum_{n=N}^{\infty} \log\left[\left(1 - \frac{z}{a_n}\right) e^{\frac{z}{a_n} + \frac{1}{2}\left(\frac{z}{a_n}\right)^2}\right]$$

is analytic in $\{z : |z| < R\}$, which implies

$$\prod_{n=N}^{\infty} \left(1 - \frac{z}{a_n}\right) e^{\frac{z}{a_n} + \frac{1}{2}\left(\frac{z}{a_n}\right)^2}$$

is analytic in $\{z : |z| < R\}$. Hence $g(z)$ is analytic in $\{z : |z| < R\}$, and its zeros in $\{z : |z| < R\}$ are $0, a_1, a_2, \ldots, a_k$ ($\frac{R^2}{2} - 1 \le k < \frac{R^2}{2}$). Since R can be arbitrarily large, we see that $g(z)$ is an entire function with the required zero-set.

5512

State whether the following statement is True or False, and prove your assertion.

For each positive integer n there exists an entire function f_n such that

$$\max_{1 \le |z| \le 2} |\operatorname{Re} f_n(z) - \log|z|| < \frac{1}{n}.$$

(Indiana)

Solution.

False.

We prove the assertion by contradiction.

If for each positive number n there exists an entire function f_n such that

$$\max_{1 \le |z| \le 2} |\operatorname{Re} f_n(z) - \log|z|| < \frac{1}{n},$$

then for $1 \le |z| \le 2$, we have

$$-1 \le \operatorname{Re} f_n(z) \le 1 + \log 2.$$

Define $F_n(z) = e^{f_n(z)}$. Then $F_n(z)$ are entire functions with no zeros, and

$$\frac{1}{e} \le |F_n(z)| = e^{\operatorname{Re} f_n(z)} \le 2e$$

for $1 \le |z| \le 2$. By the maximum modulus principle, $|F_n(z)| \le 2e$ for $|z| \le 2$. Hence $\{F_n(z)\}$ is a normal family in $\{z : |z| < 2\}$, and there exists a subsequence $\{F_{n_k}(z)\}$ converging locally uniformly to an analytic function $F(z)$ in $\{z : |z| < 2\}$. Since $|F_n(z)| \ge \frac{1}{e}$ for $1 \le |z| \le 2$, $F(z)$ cannot be identically zero, and by Hurwitz's Theorem $F(z)$ has no zero in $\{z : |z| < 2\}$. But we have for $1 \le |z| < 2$,

$$|F(z)| = \lim_{k \to \infty} |F_{n_k}(z)| = \lim_{k \to \infty} e^{\operatorname{Re} f_{n_k}(z)} = e^{\log|z|} = |z|,$$

which implies that $F(z) = \alpha z$ with $|\alpha| = 1$ in $\{z : |z| < 2\}$. This is a contradiction to the fact that $F(z)$ has no zero in $\{z : |z| < 2\}$.

5513

Let $G = D\backslash(-1, 0]$, where $D = \{z : |z| < 1\}$.

(a) Give a single-valued definition for z^i in G.

(b) Why should there exist a sequence of polynomials P_n such that

$$\lim_{n \to \infty} P_n(z) = z^i$$

for all z in G?

(c) Can the polynomials be chosen so that there exists a constant M with $|P_n(z)| \le M$ for all $z \in G$ and all n? Justify your answer.

(*Indiana*)

Solution.

(a) z^i is defined by $e^{i \log z}$. In domain G, single-valued branch of $\log z$ can be chosen. For example, a single-valued branch of z^i in G can be defined by $\arg z|_{0<z<1} = 0$.

(b) Choose

$$K_n = \left\{ z : \frac{1}{n} \le |z| \le 1 - \frac{1}{n}, \quad -\pi + \frac{1}{n} \le \arg z \le \pi - \frac{1}{n} \right\}$$

where $n > 2$. Then $K_n \subset K_{n+1}$, and

$$\lim_{n \to \infty} K_n = G.$$

Because the complement of K_n is connected and contains $z = \infty$, we know by Runge's Theorem that there exists a sequence of polynomials which converges uniformly on K_n to z^i. In other words, we can find a polynomial $P_n(z)$ such that

$$|P_n(z) - z^i| < \frac{1}{n}$$

for all $z \in K_n$. Hence $\{P_n(z); n = 1, 2, \ldots\}$ converges to z^i uniformly on compact subsets of G.

(c) No. If there exists M with $|P_n(z)| \le M$ for all $z \in G$ and all n, then because $P_n(z)$ are continuous on D, $|P_n(z)| \le M$ for all $z \in D$ and all n. It follows that $\{P_n(z)\}$ is a normal family in D, and there exists a subsequence $P_{n_k}(z)$ which converges uniformly on compact subsets of D to an analytic function $f(z)$ in D. Since $P_n(z)$ converges to z^i in G, hence $z^i = f(z)$ for $z \in G$, which implies that z^i can be extended to a single-valued analytic function in D. It is obvious impossible, so the contradiction is obtained.

5514

(a) Prove that

$$\frac{\pi^2}{\sin^2 \pi z} = \sum_{n=-\infty}^{\infty} \frac{1}{(z-n)^2}.$$

(b) Use this to show that

$$\pi \cot \pi z = \frac{1}{z} + \sum_{n=1}^{\infty} \frac{2z}{z^2 - n^2}.$$

Justify your steps.

(c) Develop $\pi \cot \pi z$ in a Laurent series about the origin directly and by use of (b), with enough terms to find the values of $\sum_{n=1}^{\infty} \frac{1}{n^2}$ and $\sum_{n=1}^{\infty} \frac{1}{n^4}$.

(Harvard)

Solution.

(a) Let

$$f(z) = \frac{\pi^2}{\sin^2 \pi z}.$$

The singular part of f at $z = n$ $(n = 0, \pm 1, \pm 2, \ldots)$ is $\frac{1}{(z-n)^2}$. Now we consider the series

$$\sum_{n=-\infty}^{\infty} \frac{1}{(z-n)^2}.$$

For any natural number N and $|z| \leq N$,

$$\left| \frac{1}{(z-n)^2} \right| \leq \frac{4}{n^2}$$

holds for $n \geq 2N$ and $n \leq -2N$. It follows from the convergence of $\sum_{n=2N}^{\infty} \frac{4}{n^2}$ and $\sum_{-\infty}^{n=-2N} \frac{4}{n^2}$ that $\sum_{n=-\infty}^{\infty} \frac{1}{(z-n)^2}$ is analytic in

$$\{|z| < N\} \backslash \{z = 0, \pm 1, \ldots, \pm(N-1)\}.$$

Because N can be arbitrarily large, we obtain the result that

$$\sum_{n=-\infty}^{\infty} \frac{1}{(z-n)^2}$$

is a meromorphic function which has the same singularities as $f(z)$.

Let

$$g(z) = f(z) - \sum_{n=-\infty}^{\infty} \frac{1}{(z-n)^2}.$$

Then $g(z)$ is an entire function.

As $f(z)$ and

$$\sum_{n=-\infty}^{\infty} \frac{1}{(z-n)^2}$$

are both periodic functions with period equal to 1, we restrict z in the strip

$$\{z : 0 < \operatorname{Re} z \leq 1\}.$$

It is obvious that

$$\lim_{\operatorname{Im} z \to \pm\infty} f(z) = 0.$$

As the convergence of

$$\sum_{n=-\infty}^{\infty} \frac{1}{(z-n)^2}$$

is uniform for

$$|\operatorname{Im} z| \geq 1,$$

the limit of the series for $\operatorname{Im} z \to \pm\infty$ can be obtained by taking the limit in each term and the limit is also zero. Hence $g(z)$ is a bounded entire function, which implies that $g(z)$ is a constant. It is obvious that the constant must be zero. Thus we obtain the identity

$$\frac{\pi^2}{\sin^2 \pi z} = \sum_{n=-\infty}^{\infty} \frac{1}{(z-n)^2}. \tag{1}$$

(b) Let

$$F(z) = \pi \cot \pi z.$$

The singular part of F at $z = n$ $(n = 0, \pm 1, \pm 2, \ldots)$ is $\frac{1}{z-n}$. Now we consider the series

$$\frac{1}{z} + \sum_{n=1}^{\infty} \left(\frac{1}{z-n} + \frac{1}{z+n} \right) = \frac{1}{z} + \sum_{n=1}^{\infty} \frac{2z}{z^2 - n^2}.$$

With similar discussion to that in (a), we know that

$$\frac{1}{z} + \sum_{n=1}^{\infty} \frac{2z}{z^2 - n^2}$$

is a meromorphic function which has the same singularities as $F(z)$.

Let

$$F(z) = \frac{1}{z} + \sum_{n=1}^{\infty} \frac{2z}{z^2 - n^2} + G(z).$$

Then $G(z)$ is an entire function. Differentiating both sides of the above identity, we obtain

$$\frac{\pi^2}{\sin^2 \pi z} = \frac{1}{z^2} + \sum_{n=1}^{\infty} \left[\frac{1}{(z - n)^2} + \frac{1}{(z + n)^2} \right] - G'(z)$$

Comparing this identity with (1), we have $G'(z) = 0$ which implies that $G = c$ (c is a constant). For

$$F(z) = \frac{1}{z} + \sum_{n=1}^{\infty} \frac{2z}{z^2 - n^2} + c,$$

it follows from the fact that $F(z)$ and

$$\frac{1}{z} + \sum_{n=1}^{\infty} \frac{2z}{z^2 - n^2}$$

are both odd functions that $c = 0$. Hence we obtain

$$\pi \cot \pi z = \frac{1}{z} + \sum_{n=1}^{\infty} \frac{2z}{z^2 - n^2}. \qquad (2)$$

(c) The Laurent expansion of $\pi \cot \pi z$ about the origin is

$$\pi \cot \pi z = \frac{1}{z} - \frac{\pi^2}{3} z - \frac{\pi^4}{45} z^3 - \cdots . \qquad (3)$$

It follows from (2) and (3) that around the origin,

$$\sum_{n=1}^{\infty} \frac{1}{z^2 - n^2} = -\frac{\pi^2}{6} - \frac{\pi^4}{90}z^2 - \cdots . \tag{4}$$

Take $z = 0$, we obtain

$$\sum_{n=1}^{\infty} \frac{1}{n^2} = \frac{\pi^2}{6}.$$

After differentiating (4) on both sides, we can also obtain

$$\sum_{n=1}^{\infty} \frac{1}{n^4} = \frac{\pi^4}{90}.$$

5515

Let $z_1, \ldots, z_n$ be distinct complex numbers. Let f and g be polynomials, f of degree $\leq n - 2$ and

$$g(z) = (z - z_1) \cdots (z - z_n).$$

(a) Show that

$$\sum_{j=1}^{n} \frac{f(z_j)}{g'(z_j)} = 0.$$

(b) Show that there exists a polynomial of degree $\leq n - 2$ with $f(z_j) = a_j$ if and only if

$$\sum_{j=1}^{n} \frac{a_j}{g'(z_j)} = 0.$$

(c) Given a sequence of complex numbers $z_1, z_2, \ldots$ such that $|z_n| \to \infty$, does there exist an entire function f with $f(z_j) = a_j$? Can you write this function down?

(Harvard)

Solution.

(a) Take R sufficiently large such that

$$z_1, z_2, \ldots, z_n \in \{|z| < R\}.$$

Because

$$g(z) = (z - z_1)(z - z_2) \cdots (z - z_n)$$

is of degree n, while $f(z)$ is of degree $\leq n - 2$,

$$\int_{|z|=R} \frac{f(z)}{g(z)} dz = \lim_{R \to \infty} \int_{|z|=R} \frac{f(z)}{g(z)} dz = 0.$$

Since

$$\int_{|z|=R} \frac{f(z)}{g(z)} dz = 2\pi i \sum_{j=1}^{n} \operatorname{Res}\left(\frac{f(z)}{g(z)}, z_j\right) = 2\pi i \sum_{j=1}^{n} \frac{f(z_j)}{g'(z_j)},$$

we obtain

$$\sum_{j=1}^{n} \frac{f(z_j)}{g'(z_j)} = 0.$$

(b) If $f(z)$ is a polynomial of degree $\leq n - 2$ with $f(z_j) = a_j$ $(j = 1, 2, \ldots, n)$, then by (a), we have

$$\sum_{j=1}^{n} \frac{f(z_j)}{g'(z_j)} = \sum_{j=1}^{n} \frac{a_j}{g'(z_j)} = 0.$$

If $a_1, a_2, \ldots, a_n$ are n complex numbers such that

$$\sum_{j=1}^{n} \frac{a_j}{g'(z_j)} = 0,$$

we construct the function $f(z)$ by

$$f(z) = \sum_{j=1}^{n} \frac{a_j}{g'(z_j)} \cdot \frac{g(z)}{(z - z_j)}.$$

For each j,

$$\frac{g(z)}{z - z_j} = (z - z_1) \cdots (z - z_{j-1})(z - z_{j+1}) \cdots (z - z_n)$$

is a polynomial of degree $n - 1$, and the coefficient of z^{n-1} is 1. Since

$$\sum_{j=1}^{n} \frac{a_j}{g'(z_j)} = 0,$$

the coefficient of z^{n-1} of $f(z)$ is zero. In other words, $f(z)$ is a polynomial of degree $\leq n - 2$.

Because

$$\lim_{z \to z_j} \frac{a_j}{g'(z_j)} \cdot \frac{g(z)}{(z - z_j)} = a_j,$$

while for $k \neq j$,

$$\frac{a_k}{g'(z_k)} \cdot \frac{g(z)}{z - z_k}\bigg|_{z=z_j} = 0,$$

$f(z)$ satisfies the condition $f(z_j) = a_j$ $(j = 1, 2, \ldots, n)$.

(c) For the given sequence $z_1, z_2, \ldots$, such that $|z_n| \to \infty$, by the Weierstrass Theorem about the canonical product of entire functions, we can construct an entire function $g(z)$ with simple zeros $z_1, z_2, \ldots$. Then we define

$$f(z) = \sum_{n=1}^{\infty} u_n(z) = \sum_{n=1}^{\infty} e^{\gamma_n(z-z_n)} \frac{g(z)}{z - z_n} \cdot \frac{a_n}{g'(z_n)},$$

where γ_n is chosen such that when $|z| \leq \frac{|z_n|}{2}$,

$$|u_n(z)| = \left| e^{\gamma_n(z-z_n)} \frac{g(z)}{z - z_n} \cdot \frac{a_n}{g'(z_n)} \right| \leq \frac{1}{n^2}.$$

Because $|z_n| \to \infty$, for any $R > 0$, there exists $N > 0$ such that $|z_n| > 2R$ when $n \geq N$. Hence

$$|u_n(z)| \leq \frac{1}{n^2}$$

holds for all $|z| \leq R$ when $n \geq N$. In other words, $\sum_{n=1}^{\infty} u_n(z)$ converges uniformly for all $|z| \leq R$, so that $f(z)$ is analytic in $\{|z| < R\}$. Since R can be arbitrarily large, $f(z)$ is an entire function.

It is easy to see that

$$\lim_{z \to z_n} u_n(z) = \lim_{z \to z_n} e^{\gamma_n(z - z_n)} \frac{g(z)}{z - z_n} \cdot \frac{a_n}{g'(z_n)} = a_n,$$

while for $k \neq n$,

$$u_k(z_n) = 0,$$

which implies that $f(z)$ is an entire function satisfying the required condition.

5516

For $a_n = 1 - \frac{1}{n^2}$, let $f(z) = \prod_{n=1}^{\infty} \frac{a_n - z}{1 - a_n z}$.

(1) Show that f defines a holomorphic function on $D = \{z : |z| < 1\}$.

(2) Prove that f does not have an analytic continuation to any larger disk $\{z : |z| < r\}$ where $r > 1$.

(UC, Los Angeles)

Solution.

(1) In order to prove that f defines a non-zero holomorphic function on $D = \{z : |z| < 1\}$, we prove that the infinite product $\prod_{n=1}^{\infty} \frac{a_n - z}{1 - a_n z}$ converges uniformly in compact subset of $D = \{z : |z| < 1\}$. It follows from $\frac{a_n - z}{1 - a_n z} = 1 - \frac{1}{n^2} \cdot \frac{1+z}{1 - a_n z}$ that we need only to prove that $\sum_{n=1}^{\infty} \frac{1}{n^2} \cdot \left| \frac{1+z}{1 - a_n z} \right|$ converges uniformly in $D_r = \{z : |z| \leq r < 1\}$.

Let $w_n = \frac{1+z}{1 - a_n z}$, then $z = \frac{w_n - 1}{a_n w_n + 1}$, it follows from $|z| \leq r < 1$ that $|w_n| < \frac{1+r}{1-r}$, which implies $\sum_{n=1}^{\infty} \frac{1}{n^2} \cdot \left| \frac{1+z}{1 - a_n z} \right|$ converges uniformly in $D_r = \{z : |z| \leq r < 1\}$, hence f defines a non-zero holomorphic function on $D = \{z : |z| < 1\}$.

(2) $f(z)$ is a non-zero holomorphic function on $D = \{z : |z| < 1\}$, which has zeros $\{a_n\}$ and $\lim_{n \to \infty} a_n = 1$. Hence $f(z)$ is not holomorphic at point $z = 1$, so $f(z)$ does not have an analytic continuation to any larger disk $\{z : |z| < r\}$ where $r > 1$.

5517

Let f be holomorphic in a neighborhood of 0 and suppose that the series $\sum_{k=0}^{\infty} f^{(k)}(0)$ converges. Show that f can be extended to an entire function.

(*SUNY, Albany*)

Solution.

We need only to prove that the radius of convergence of the power series $\sum_{k=0}^{\infty} \frac{f^{(k)}(0)}{k!} z^k$ is infinity. For this purpose, set $z = x \in (0, +\infty)$ is an arbitrary positive real number. Now we prove that $\sum_{k=0}^{\infty} \frac{f^{(k)}(0)}{k!} x^k$ is convergent.

Set $a_k = \frac{x^k}{k!}$, $b_k = f^{(k)}(0)$. $\{a_k\}$ is a real number sequence which converges to zero, and $\{a_k\}$ is monotonically decreasing from the k-th term, where $k > x$. It is obvious that sequence $\{a_k\}$ is bounded, i.e., there exists $M > 0$ such that $a_k \leq M$ for every positive integer k.

Since $\sum_{k=0}^{\infty} b_k$ is convergent, so for arbitrary $\varepsilon > 0$, there exists $N > x$, such that when $m > n > N$, it holds that $|\sum_{k=n+1}^{m} b_k| < \varepsilon$.

Because $\{a_k\}$ is monotonically decreasing when $k > x$, we can apply Abel Lemma to obtain that

$$\left| \sum_{k=n+1}^{m} a_k b_k \right| < \varepsilon(|a_{n+1}| + 2|a_m|) \leq 3M\varepsilon,$$

which implies that $\sum_{k=0}^{\infty} \frac{f^{(k)}(0)}{k!} x^k$ is convergent.

5518

Let F denote the family of functions $f(z) = \sum_{n=0}^{\infty} a_n z^n$ which are holomorphic in the unit disk D and satisfy $|a_n| \leq n$ for $n = 0, 1, 2, \ldots$. Prove that any sequence in F has a subsequence which converges uniformly on compact subset of D.

(*SUNY, Albany*)

Solution.

We need only to prove that the family F is uniformly bounded on compact subset of D. For every $f \in F$ and $|z| \leq r < 1$,

$$|f(z)| = \left| \sum_{n=0}^{\infty} a_n z^n \right| \leq \sum_{n=1}^{\infty} n r^n = \frac{r}{(1-r)^2},$$

which implies that F is uniformly bounded on compact subset of D.

5519

Let $f(z)$ be a bounded analytic function on the open right half plane $\{z = x + iy : x > 0\}$ such that $f(z) \to 0$ as z tends to 0 along the positive real axis. Suppose $0 < \theta < \frac{\pi}{2}$. Prove that $f(z) \to 0$ as $z \to 0$ uniformly in the sector $|\arg z| \leq |\theta|$.

(*UC, Los Angeles*)

Solution.

Define $f_n(z) = f(\frac{z}{n})$, then the functions $f_n(z)$ are uniformly bounded in the right half plane. It follows from the fact that $f(z) \to 0$ as z tends to 0 along the positive real axis that for every point $z = x \in (r, R)$, where $0 < r < R < +\infty$, $\{f_n(z)\}$ converges to zero. By Vitali Theorem, $\{f_n(z)\}$ converges to an analytic function $F(z)$ uniformly in each compact subset of the right half plane, and the limit function $F(z)$ is identically zero since it equals to zero on the interval (r, R).

Now take the compact subset as $S = \{z : |\arg z| \leq \theta, 0 < r \leq |z| \leq R\}$, where $0 < \theta < \frac{\pi}{2}$ and $R > 2r$. $\{f_n(z)\}$ converges to zero uniformly in S, i.e., for arbitrary $\varepsilon > 0$, there exists a positive integer N, such that when $n > N$, $|f_n(z)| < \varepsilon$ holds for all $z \in S$.

Take $\delta = \frac{R}{N+1}$, when $|\arg z| \leq \theta$ and $|z| < \delta$, there is $n > N$ and $z' \in S$ such that $z' = nz$. Hence $|f(z)| = |f_n(z')| < \varepsilon$, which implies that $f(z) \to 0$ as $z \to 0$ uniformly in the sector $|\arg z| \leq |\theta|$.

5520

Let D be a domain (connected open set) in $\mathbb{C}$ and let $\{u_n\}$ be a sequence of harmonic functions $u_n : D \to (0, +\infty)$. Show that if $u_n(z_0) \to 0$ for some $z_0 \in D$, then $u_n \to 0$ uniformly on compact subset of D.

(UC, Los Angeles)

Solution.

Let $f_n(z) = u_n(z) + iv_n(z)$ be analytic on $O(z_0, \rho_0) = \{z : |z - z_0| < \rho_0\} \subset D$, and define $F_n(z) = e^{-f_n(z)}$, then $|F_n(z)| \leq 1$ which implies that $\{F_n\}$ is a normal family. So there exists a subsequence for any subsequence of $\{F_n\}$ (assume it still to be $\{F_{n_k}\}$) which converges uniformly in compact subset of $O(z_0, \rho_0) \subset D$ to an analytic function $F(z)$ with $|F(z_0)| = 1$. So $F(z)$ attains its maximum modulus in an inner point of $O(z_0, \rho_0)$. By the maximum modulus principle, $F(z) \equiv e^{i\theta}$ is a constant function. So we deduce that $\{F_n\}$ converges to $F(z) \equiv e^{i\theta}$ uniformly in a neighborhood of point z_0, which implies that $\{u_n(z)\}$ converges to zero uniformly in a neighborhood of point z_0.

For an arbitrary point $z' \in D$, let $\gamma : [0, 1] \to D$ with $\gamma(0) = z_0, \gamma(1) = z'$ be a continuous path in D which connects z_0 and z', then each F_n can be analytically continued from z_0

to z' along γ, so $\{F_n\}$ is also a normal family in a simply connected domain which contains the path γ. Since $\{F_n\}$ converges to $F(z) \equiv e^{i\theta}$ uniformly in a neighborhood of point z_0, we know that $\{F_n\}$ converges to $F(z) \equiv e^{i\theta}$ uniformly in a neighborhood of point z', which implies that $\{u_n(z)\}$ converges to zero uniformly in a neighborhood of point z'. Since $z' \in D$ is arbitrary, we proved that $u_n \to 0$ locally uniformly in D, which is equivalent to $u_n \to 0$ uniformly on compact subset of D.

PART 6

Partial Differential Equations

Section 1

General Theory

6101

(a) Let $A = (a_{ij})_{i,j=1}^n$ be a real matrix. Show that

$$\int_{|x|<1} (x, Ax)dx = \frac{\omega_n}{n(n+2)} \text{ trace } A.$$

Here $(\cdot, \cdot)$ is the dot product of vectors in $\mathbb{R}^n$ and ω_n is the area of the unit sphere in $\mathbb{R}^n$.

(b) Show that $u \in C_0^2(\mathbb{R}^n)$ implies

$$\int_{\mathbb{R}^n} (\Delta u)^2 dx = \sum_{i,j=1}^n \int_{\mathbb{R}^n} |D_{ij}u|^2 dx.$$

(Iowa)

Solution.

(a) It is not difficult to verify that

$$\int_{|x|<1} a_{ij} x_i x_j \, dx = 0, \quad \forall i \neq j$$

and

$$\int_{|x|<1} x_1^2 \, dx = \cdots = \int_{|x|<1} x_n^2 \, dx.$$

677

Therefore,

$$\int_{|x|<1} (x, Ax)dx = \int_{|x|<1} \sum_{i=1}^{n} a_{ii} x_i^2 \, dx$$

$$= \frac{1}{n} \operatorname{trace} A \int_{|x|<1} |x|^2 dx$$

$$= \frac{\omega_n}{n(n+2)} \operatorname{trace} A.$$

(b) First let $u \in C_0^\infty(\mathbb{R}^n)$. By applying Green's formula, we get immediately

$$\int_{\mathbb{R}^n} (\Delta u)^2 dx = \sum_{i,j=1}^{n} \int_{\mathbb{R}^n} D_{ii}u \cdot D_{jj}u \, dx$$

$$= \sum_{i,j=1}^{n} \int_{\mathbb{R}^n} |D_{ij}u|^2 dx.$$

As $C_0^\infty(\mathbb{R}^n)$ is dense in $H_0^2(\mathbb{R}^n)$, the conclusion is true for $u \in C_0^2(\mathbb{R}^n)$.

6102

Let T be a distribution on $\mathbb{R}$ and suppose that $T' = 0$ on $\mathbb{R}$. Show that $T = \operatorname{const}$; i.e., show that there is a number a such that

$$T(\phi) = \int_{\mathbb{R}} a\phi \, dx \quad \text{for all } \phi \in C_0^\infty(\mathbb{R}).$$

(Indiana)

Solution.
$T' = 0$ if and only if

$$T(\phi') = 0, \quad \forall \phi \in C_0^\infty(\mathbb{R}).$$

It is easy to verify that a function ϕ in $C_0^\infty(\mathbb{R})$ is the derivative of a function in $C_0^\infty(\mathbb{R})$ if and only if

$$\int_\mathbb{R} \phi \, dx = 0.$$

Take a function $\rho \in C_0^\infty(\mathbb{R})$ such that

$$\int_\mathbb{R} \rho \, dx = 1.$$

Then it is easy to verify that for any $\phi \in C_0^\infty(\mathbb{R})$

$$\psi = \phi - \int_\mathbb{R} \phi \, dx \rho$$

satisfies the condition

$$\int_\mathbb{R} \psi \, dx = 0.$$

Hence $T(\psi) = 0$ and

$$T(\phi) = \int_\mathbb{R} a\phi \, dx, \quad \forall \phi \in C_0^\infty(\mathbb{R})$$

where $a = T(\rho)$.

6103

Let $u \in H^s(\mathbb{R}^n)$ with $s > n/2$. Show that $\lim_{|x|\to\infty} u(x) = 0$.
(*Cincinnati*)

Solution.
We show first that $\hat{u} \in L^1(\mathbb{R}^n)$ if $u \in H^s(\mathbb{R}^n)$ with $s > n/2$. In fact, $u \in H^s(\mathbb{R}^n)$ if and only if $(1+|\xi|^2)^{s/2}\hat{u} \in L^2(\mathbb{R}^n)$. And it is clear that $(1+|\xi|^2)^{-s/2} \in L^2(\mathbb{R}^n)$ if $s > n/2$. Therefore, we have $\hat{u} \in L^1(\mathbb{R}^n)$.

Then we get immediately

$$u(x) = \frac{1}{(2\pi)^n} \int_{\mathbb{R}^n} e^{ix\cdot\xi}\hat{u}(\xi)d\xi \to 0$$

as $|x| \to \infty$.

6104

Let u be defined for $f \in \mathcal{D}((0,1))$ by

$$\langle u, f \rangle = \sum_{n=0}^{\infty} f^{(n)}\left(\frac{1}{n}\right).$$

Determine whether u is a distribution on $(0, 1)$, and support your answer.

(*Indiana*)

Solution.

It is clear that u, defined above, is a linear functional in $\mathcal{D}((0,1))$. Let $\{f_k\}$ be a sequence in $\mathcal{D}((0,1))$ such that

$$f_k \to 0 \text{ in } \mathcal{D}((0,1)), \quad \text{as } k \to \infty.$$

Then we have a compact interval $[a, b] \subset (0, 1)$ such that

$$\text{supp } f_k \subset [a, b], \quad \forall k$$

and

$$f_k^{(n)} \to 0 \text{ uniformly on } [a, b] \text{ as } k \to \infty,$$

for any non-negative integer n.

Let N be a positive integer such that

$$\frac{1}{N} < a.$$

Then we have

$$\langle u, f_k \rangle = \sum_{n=0}^{N-1} f_k^{(n)}\left(\frac{1}{n}\right) \to 0, \quad \text{as } k \to \infty.$$

Therefore, u is continuous in $\mathcal{D}((0,1))$, and is a distribution on $(0, 1)$.

6105

Let

$$B = \{(x, y) \mid x^2 + y^2 < 1\}.$$

For which $p \geq 1$ does the function

$$u(x, y) = \frac{x}{\sqrt{x^2 + y^2}}$$

belong to $W^{1,p}(B)$?

(*Iowa*)

Solution.

Let

$$u_x(x, y) = \frac{1}{\sqrt{x^2 + y^2}} - \frac{x^2}{(x^2 + y^2)^{3/2}},$$

$$u_y(x, y) = -\frac{xy}{(x^2 + y^2)^{3/2}}.$$

It is clear that $u_x, u_y \in L^1(B)$.

We show that

$$\frac{\partial}{\partial x} u = u_x \quad \text{and} \quad \frac{\partial}{\partial y} u = u_y$$

in the sense of distributions, that is

$$\int_B u\phi_x \, dx \, dy = -\int_B u_x\phi \, dx \, dy$$

and

$$\int_B u\phi_y \, dx \, dy = -\int_B u_y\phi \, dx \, dy$$

for all $\phi \in C_0^\infty(B)$. By B_ε we denote the ball centred at the origin with the radius $\varepsilon < 1$. Let $\Omega_\varepsilon = B\backslash\bar{B}_\varepsilon$ and $S_\varepsilon = \partial B_\varepsilon$. Then we have

$$\int_{\Omega_\varepsilon} u\phi_x \, dxdy = -\int_{\Omega_\varepsilon} u_x\phi \, dxdy + \int_{S_\varepsilon} u\phi \cos(\vec{n}, x) ds$$

for any $\phi \in C_0^\infty(B)$. Letting $\varepsilon \to 0$ in the above inequality, we get $\frac{\partial}{\partial x} u = u_x$. Similarly, $\frac{\partial}{\partial y} u = u_y$.

It is not difficult to verify that

$$u, u_x, u_y \in L^p(B)$$

for $1 \leq p < 2$. Therefore, $u \in W^{1,p}(B)$ for $1 \leq p < 2$. And we can verify that

$$u_y \notin L^2(B),$$

which means that $u \notin W^{1,p}(B)$ for $p = 2$.

6106

Let Σ be a smooth hypersurface dividing $\mathbb{R}^n$ into two disjoint open regions Ω_1 and Ω_2. Denote by $\nu = \nu(x)$ the unit normal to Σ pointing into Ω_2. Suppose $v : \mathbb{R}^n \to \mathbb{R}^n$ solves the equation

$$\text{div } v(x) = b(x) \quad \text{for } x \in \mathbb{R}^n - \Sigma \qquad (*)$$

in the classical (C^1) sense where b is a continuous function. Assume furthermore that v lies in both $C^0(\bar{\Omega}_1)$ and $C^0(\bar{\Omega}_2)$ (but not necessarily $C^0(\mathbb{R}^n)$!!) with

$$v^+(x_0) \equiv \lim_{\substack{x \to x_0 \\ x \in \Omega_1}} v(x), \quad v^-(x_0) \equiv \lim_{\substack{x \to x_0 \\ x \in \Omega_2}} v(x)$$

for each $x_0 \in \Sigma$. Derive a necessary and sufficient condition in terms of v^+, v^- and ν for v to be a distribution solution of $(*)$ on all of $\mathbb{R}^n$.

(Indiana)

Solution.

v is a distribution solution of $(*)$ in $\mathbb{R}^n$ if and only if

$$-\int_{\mathbb{R}^n} v \cdot \nabla \phi \, dx = \int_{\mathbb{R}^n} b\phi \, dx, \quad \forall \phi \in C_0^\infty(\mathbb{R}^n). \qquad (1)$$

By Green's formula, we have

$$-\int_{\mathbb{R}^n} v \cdot \nabla \phi \, dx = -\int_{\Omega_1} v \cdot \nabla \phi \, dx - \int_{\Omega_2} v \cdot \nabla \phi \, dx$$

$$= -\int_{\Sigma} \phi(v^+ - v^-) \cdot \nu \, ds + \int_{\mathbb{R}^n} \text{div } v \cdot \phi \, dx.$$

Hence the equation (1) is equivalent to

$$\int_{\Sigma} \phi(v^+ - v^-) \cdot \nu \, ds = 0, \quad \forall \phi \in C_0^\infty(\mathbb{R}^n).$$

Therefore v is a distribution solution to $(*)$ on all of $\mathbb{R}^n$ if and only if

$$(v^+ - v^-) \cdot \nu = 0 \quad \text{on } \Sigma.$$

6107

Recall the definition of the curl operator in two dimensions acting on a vector field $U(x,y) = (u(x,y), v(x,y))$:

$$\nabla \times U \equiv u_y - v_x.$$

(a) Give a definition of what it means for a (not necessarily continuous) vector field U to have curl zero in the sense of distributions.

(b) Suppose U takes the form

$$U(x,y) \equiv \begin{cases} U_1, & (x,y) \in \Omega_1, \\ U_2, & (x,y) \in \Omega_2, \end{cases}$$

where Ω_1 and Ω_2 are open sets such that $\bar{\Omega}_1 \cup \bar{\Omega}_2 = \mathbb{R}^2$, $\Gamma \equiv \partial\Omega_1 \cap \partial\Omega_2$ is a smooth curve, and U_1 and U_2 are (different) constant vectors. If U has curl zero in $\mathbb{R}^2$ in the sense of distributions, describe as completely as possible the curve Γ.

(*Indiana*)

Solution.

(a) Let $u, v \in L_{\text{loc}}(\mathbb{R}^2)$. Then

$$\text{curl}\, U = 0$$

in the sense of distributions is defined as

$$\int_{\mathbb{R}^2} (-u\phi_y + v\phi_x)\,dx\,dy = 0, \quad \forall \phi \in C_0^\infty(\mathbb{R}^2).$$

(b) Let $U_1 = (u_1, v_1)$, $U_2 = (u_2, v_2)$, where u_1, v_1, u_2 and v_2 are constants. Then curl $U = 0$ if and only if

$$\int_{\Omega_1} (-u_1\phi_y + v_1\phi_x)dx\,dy + \int_{\Omega_2} (-u_2\phi_y + v_2\phi_x)dx\,dy = 0,$$

$$\forall \phi \in C_0^\infty(\mathbb{R}^2).$$

Integrating by parts, we have

$$\int_{\Gamma} (u_1 - u_2)\phi\,dx + (v_1 - v_2)\phi\,dy = 0, \quad \forall \phi \in C_0^\infty(\mathbb{R}^2).$$

Therefore, Γ is a straight line defined by

$$(u_1 - u_2)dx + (v_1 - v_2)dy = 0.$$

6108

Let p be a polynomial in n variables, and assume that the set $\{x : p(ix) = 0\}$ is bounded. Prove that there exists a tempered distribution E on $\mathbb{R}^n$ such that $p(D)E - \delta \in S$. Here S is the Schwartz class of rapidly decreasing functions, and δ is the Dirac distribution defined by $\delta \cdot \phi = \phi(0)$.

(*Indiana*)

Solution.
As the set $\{\xi : p(i\xi) = 0\}$ is bounded, there exists a positive constant R such that

$$\{\xi : p(i\xi) = 0\} \subset \{\xi : |\xi| < R\}.$$

Construct a function $\chi_R \in C_c^\infty(\mathbb{R}^n)$ such that

$$\chi_R(\xi) = 1, \quad \forall \xi \in \{\xi : |\xi| < R\}.$$

Set

$$\hat{E}(\xi) = \begin{cases} 0, & |\xi| < R, \\ (1 - \chi_R(\xi))\frac{1}{p(i\xi)}, & |\xi| \geq R. \end{cases}$$

It is clear that $\hat{E} \in S'(\mathbb{R}^n) \cap C^\infty(\mathbb{R}^n)$, where $S'(\mathbb{R}^n)$ denotes the space of tempered distributions on $\mathbb{R}^n$, and that

$$p(i\xi)\hat{E}(\xi) = \begin{cases} 0, & |\xi| < R, \\ (1 - \chi_R(\xi)), & |\xi| \geq R. \end{cases}$$

It is easy to see that

$$p(i\xi)\hat{E}(\xi) - 1 \in C_0^\infty(\mathbb{R}^n).$$

Therefore, we get

$$p(D)E - \delta \in S,$$

where $E = F^{-1}(\hat{E}) \in S'(\mathbb{R}^n)$.

6109

Let $P(D)$ be an m-th order linear constant coefficient partial differential operator on $\mathbb{R}^n$; i.e.,

$$P(D) = \sum_{|\alpha| \leq m} C_\alpha D^\alpha$$

where α denotes a multi-index $\alpha = (\alpha_1, \alpha_2, \ldots, \alpha_n)$ and the C_α are constants. Suppose

$$P(i\xi) \equiv \sum_{|\alpha| \leq m} C_\alpha(i\xi)^\alpha \neq 0 \quad \text{for all } \xi \in \mathbb{R}^n - \{0\}.$$

If $u \in S'$ satisfies $P(D)u = 0$, prove that u must be a polynomial. (Here S' denotes the space of tempered distributions on $\mathbb{R}^n$.)

(*Indiana*)

Solution.

Let $\hat{u}(\xi)$ be the Fourier transform of u. Transforming the equation, we get

$$P(i\xi)\hat{u}(\xi) = 0.$$

From the transformed equation, we can see that

$$\text{supp } \hat{u} = \{0\}.$$

Therefore, $\hat{u}(\xi)$ must be finite linear combinations of the derivatives of the Dirac delta function $\delta(\xi)$ at $\{0\}$, i.e.,

$$\hat{u}(\xi) = \sum_{|\alpha| \le N} a_\alpha \delta^{(\alpha)}(\xi),$$

where $\alpha = (\alpha_1, \alpha_2, \ldots, \alpha_n)$ is a multi-index and a_α are constants. Hence

$$u(x) = \sum_{|\alpha| \le N} a_\alpha F^{-1}(\delta^{(\alpha)}) = \frac{1}{(2\pi)^n} \sum_{|\alpha| \le N} a_\alpha (-ix)^\alpha$$

is a polynomial.

6110

(a) Show that

$$\lim_{t \to 0^+} \frac{1}{\sqrt{4\pi t}} e^{-x^2/4t} = \delta(x)$$

in $\mathcal{D}'$.

(b) Show how to define $\frac{1}{x}$ as a distribution (i.e., define the Cauchy Principle Value and show that it is in $\mathcal{D}'$).

(c) Calculate rigorously the derivative of $\log |x|$ and express the result in terms of your answer to (b).

(d) Calculate the Fourier transform of the distribution $\frac{1}{x}$.

(*Indiana*)

Solution.

(a) It is well-known that

$$\int_{-\infty}^{\infty} \frac{1}{\sqrt{4\pi t}} e^{-x^2/4t} dx = 1 \quad \forall t > 0.$$

For any $\phi \in \mathcal{D}$, we have

$$\int_{-\infty}^{\infty} \frac{1}{\sqrt{4\pi t}} e^{-x^2/4t} \phi(x) dx = \frac{1}{\sqrt{\pi}} \int_{-\infty}^{\infty} e^{-y^2} \phi(2\sqrt{t}y) dy.$$

Then it is not difficult to verify that

$$\lim_{t \to 0^+} \left| \int_{-\infty}^{\infty} \frac{1}{\sqrt{4\pi t}} e^{-x^2/4t} \phi(x) dx - \phi(0) \right|$$

$$= \lim_{t \to 0^+} \frac{1}{\sqrt{\pi}} \left| \int_{-\infty}^{\infty} e^{-y^2} (\phi(2\sqrt{t}y) - \phi(0)) dy \right| = 0.$$

Hence

$$\lim_{t \to 0^+} \frac{1}{\sqrt{4\pi t}} e^{-x^2/4t} = \delta(x).$$

(b) We define the distribution $\frac{1}{x}$ by its Cauchy Principle Value $V_p \frac{1}{x}$ as follows:

$$\left\langle V_p \frac{1}{x}, \phi \right\rangle = \lim_{\varepsilon \to 0^+} \left(\int_{-\infty}^{-\varepsilon} \frac{\phi(x)}{x} dx + \int_{\varepsilon}^{\infty} \frac{\phi(x)}{x} dx \right) \quad \forall \phi \in \mathcal{D}.$$

The linearity of the functional $V_p \frac{1}{x}$ is obvious. Further, suppose that $\phi_n \to 0$ in $\mathcal{D}$. Let $\operatorname{supp} \phi_n \subset (-a, a)$ $\forall n$. Then we have

$$\left\langle V_p \frac{1}{x}, \phi_n \right\rangle = \lim_{\varepsilon \to 0^+} \int_{\varepsilon}^{a} \frac{\phi_n(x) - \phi_n(-x)}{x} dx \to 0, \quad \text{as } n \to \infty.$$

Therefore, $V_p \frac{1}{x}$ is a distribution in $\mathcal{D}'$.

(c) For any $\phi \in \mathcal{D}$, we have

$$\langle \log |x|, \phi'(x) \rangle = \int_{-\infty}^{\infty} \log |x| \phi'(x) dx$$

$$= \lim_{\varepsilon \to 0^+} \left(\int_{-\infty}^{-\varepsilon} + \int_{\varepsilon}^{\infty} \right) \log |x| \phi'(x) dx$$

$$= \lim_{\varepsilon \to 0^+} \left((\phi(-\varepsilon) - \phi(\varepsilon)) \log \varepsilon - \left(\int_{-\infty}^{-\varepsilon} + \int_{\varepsilon}^{\infty} \right) \frac{1}{x} \phi(x) dx \right)$$

$$= -\left\langle V_p \frac{1}{x}, \phi \right\rangle.$$

Hence $\frac{d}{dx} \log |x| = V_p \frac{1}{x}$.

(d) By $F(f)$ we denote the Fourier transform of a tempered distribution f. For any $\phi \in \mathcal{S}$, we have

$$\left\langle F\left(V_p\frac{1}{x}\right), \phi\right\rangle = \left\langle V_p\frac{1}{x}, F(\phi)\right\rangle$$

$$= \lim_{\varepsilon\to 0+}\left(\int_{-\infty}^{-\varepsilon} + \int_{\varepsilon}^{\infty}\right)\frac{1}{\xi}\int_{-\infty}^{\infty} e^{-ix\xi}\phi(x)dxd\xi$$

$$= -2i\lim_{\varepsilon\to 0+}\int_{\varepsilon}^{\infty}\frac{1}{\xi}\int_{-\infty}^{\infty} \sin(x\xi)\phi(x)dxd\xi$$

$$= -2i\lim_{M\to\infty}\int_{-\infty}^{\infty}\phi(x)\int_{0}^{M}\frac{1}{\xi}\sin(x\xi)d\xi dx$$

$$= -i\pi\int_{-\infty}^{\infty}(\operatorname{sign} x)\phi(x)dx.$$

Here, we have applied the integral equality

$$\frac{1}{\pi}\int_{-\infty}^{\infty}\frac{1}{\xi}\sin(x\xi)d\xi = 1 \quad \text{for } x > 0.$$

Therefore, we obtain

$$F\left(V_p\frac{1}{x}\right) = -i\pi\operatorname{sign} x.$$

6111

(a) Let

$$Lu(x) = \sum_{|\alpha|\le m} a_\alpha D^\alpha u(x) \quad x \in \mathbb{R}^n$$

where the $a'_\alpha s$ are constants, and α is a multi-index. What is a fundamental solution of L? When is it unique?

(b) If I_+ denotes the Heaviside step function.

$$I_+(\xi) = \begin{cases} 1, & \xi \ge 0, \\ 0, & \xi < 0. \end{cases}$$

Show that

$$F(x,t) = \frac{1}{2}I_+(t)I_+(t - |x|)$$

is a fundamental solution of the wave operator $W(u) = u_{tt} - u_{xx}$ on $\mathbb{R}^2$.

(c) Express the solution of the Cauchy problem

$$u_{tt} - u_{xx} = \phi(x,t), \quad -\infty < x < \infty, \quad t > 0$$
$$u(x,0) = f(x), \quad u_t(x,0) = g(x), \quad -\infty < x < \infty$$

(f, g, ϕ sufficiently nice) in terms of the above fundamental solution. Simplify your result to obtain D'Alembert's solution of this problem.

(*Indiana*)

Solution.

(a) A distribution $E \in D'(\mathbb{R}^n)$ is called a fundamental solution of the differential operator L if

$$LE(x) = \delta(x),$$

where $\delta(x)$ is the Dirac function.

Let L^* be the formal adjoint of L defined by

$$L^* v(x) = \sum_{|\alpha| \leq m} (-1)^{|\alpha|} D^\alpha(a_\alpha v(x)).$$

If $L^* S$ is dense in S, where S is the space of rapidly decreasing functions, then the operator L has at most one fundamental solution in S'.

(b) We need only to show that

$$\int_{\mathbb{R}^2} F(x,t) \left(\frac{\partial^2 \phi}{\partial t^2} - \frac{\partial^2 \phi}{\partial x^2} \right) dx dt = \phi(0,0) \quad \forall \phi \in C_0^\infty(\mathbb{R}^2).$$

In fact

$$2 \int_{\mathbb{R}^2} F(x,t)(\phi_{tt} - \phi_{xx}) dx dt = \int_{|x|<t} (\phi_{tt} - \phi_{xx}) dx dt$$

$$= -\int_0^\infty dt \int_{-t}^t \phi_{xx}(x,t) dx + \int_{-\infty}^0 dx \int_{-x}^\infty \phi_{tt}(x,t) dt$$

$$+ \int_0^\infty dx \int_x^\infty \phi_{tt}(x,t) dt$$

$$= \int_0^\infty (\phi_x(-t,t) - \phi_x(t,t))dt - \int_0^\infty (\phi_t(-x,x) + \phi_t(x,x))dx$$

$$= -\int_0^\infty (\phi_t(t,t) + \phi_x(t,t))dt - \int_0^\infty (\phi_t(-t,t) - \phi_x(-t,t))dt$$

$$= 2\phi(0,0).$$

(c) The solution of the Cauchy problem

$$u_{tt} - u_{xx} = \phi(x,t), \quad x \in \mathbb{R}, t > 0$$
$$u(x,0) = u_t(x,0) = 0$$

is given by

$$u(x,t) = \int_0^t d\tau \int_{-\infty}^\infty \phi(x-\xi,t-\tau)F(\xi,\tau)d\xi$$

$$= \frac{1}{2}\int_0^t d\tau \int_{|\xi|<r} \phi(x-\xi,t-\tau)d\xi$$

$$= \frac{1}{2}\int_0^t d\tau \int_{x-(t-\tau)}^{x+(t-\tau)} \phi(\xi,\tau)d\xi.$$

The solution of the Cauchy problem

$$u_{tt} - u_{xx} = 0, \quad x \in \mathbb{R}, t > 0$$
$$u(x,0) = 0, \quad u_t(x,0) = g(x)$$

is given by

$$u(x,t) = \int_{-\infty}^\infty g(x-\xi)F(\xi,t)d\xi$$

$$= \frac{1}{2}\int_{-t}^t g(x-\xi)d\xi$$

$$= \frac{1}{2}\int_{x-t}^{x+t} g(\xi)d\xi.$$

Therefore, the solution of the original Cauchy problem can be obtained as

$$u(x,t) = \frac{1}{2}\frac{\partial}{\partial t}\int_{x-t}^{x+t} f(\xi)d\xi + \frac{1}{2}\int_{x-t}^{x+t} g(\xi)d(\xi)$$
$$+ \frac{1}{2}\int_0^t d\tau \int_{x-(t-\tau)}^{x+(t-\tau)} \phi(\xi,\tau)d\xi.$$

This is just the D'Alembert's formula when $\phi \equiv 0$.

6112

Let u be the solution of the initial problem

$$u_{tt} = -u_{x_j x_j x_k x_k}, \quad x \in \mathbb{R}^3, t > 0,$$
$$u(x,0) = 0, \quad u_t(x,0) = \phi(x),$$

where summation is understood in the pde, and ϕ is a C^∞ function on $\mathbb{R}^3$ with compact support. Assume that there is a constant C such that

$$\|u(\cdot,t)\|_{L^2(\mathbb{R}^3)} \leq Ct^{1/4}/\log t,$$

and prove that

$$\int \phi(x)dx = 0.$$

(Indiana)

Solution.

By applying Fourier transform to the problem, we have

$$\hat{u}_{tt} + |\xi|^4 \hat{u} = 0,$$
$$\hat{u}(\xi,0) = 0, \quad \hat{u}_t(\xi,0) = \hat{\phi}(\xi),$$

where $\hat{u}$ and $\hat{\phi}$ denote the Fourier transforms of u and ϕ with respect to x, respectively.

By solving the above transformed problem, it yields that

$$\hat{u}(\xi,t) = \frac{\hat{\phi}(\xi)}{|\xi|^2} \sin(|\xi|^2 t).$$

Problems and Solutions in Mathematics

Therefore, we have

$$\int_{\mathbb{R}^3} |\hat{u}(\xi,t)|^2 d\xi = \int_{\mathbb{R}^3} \frac{|\hat{\phi}(\xi)|^2}{|\xi|^4} \sin^2(|\xi|^2 t) d\xi.$$

Set $\eta = t^{\frac{1}{2}}\xi$ in the integral in the right side of the above equality. It is reduced to

$$\int_{\mathbb{R}^3} |\hat{u}(\xi,t)|^2 d\xi = t^{\frac{1}{2}} \int_{\mathbb{R}^3} |\hat{\phi}(t^{-\frac{1}{2}}\eta)|^2 \frac{\sin^2(|\eta|^2)}{|\eta|^4} d\eta.$$

This implies that

$$\hat{\phi}(0) = \int_{\mathbb{R}^3} \phi(x) dx = 0.$$

In fact, if $|\hat{\phi}(0)| = 2a \neq 0$, then there exists a positive constant r such that

$$|\hat{\phi}(\xi)| \geq a, \quad \forall \xi \in \mathbb{R}^3, |\xi| < r.$$

And when $t > \frac{1}{r^2}$, we have

$$\int_{\mathbb{R}^3} |\hat{u}(\xi,t)|^2 d\xi \geq t^{\frac{1}{2}} \int_{|\eta|\leq 1} |\hat{\phi}(t^{-\frac{1}{2}}\eta)|^2 \frac{\sin^2(|\eta|^2)}{|\eta|^4} d\eta$$

$$\geq a^2 t^{\frac{1}{2}} \int_{|\eta|\leq 1} \frac{\sin^2(|\eta|^2)}{|\eta|^4} d\eta.$$

This contradicts the condition satisfied by the solution u, given in the problem.

6113

Given $\phi \in S$ (rapidly decreasing functions) over $\mathbb{R}^1$, consider the solution u of the Schrödinger equation

$$iu_t + u_{xx} = 0 \quad (x \in \mathbb{R}^1, \ t > 0),$$
$$u(x,0) = \phi(x).$$

Show that

$$\lim_{t \to \infty} \int_{|x|<t} |u(x,t)|^2 dx = \int_{|\xi|<\frac{1}{2}} |\hat{\phi}(\xi)|^2 d\xi.$$

Here $\hat{\phi}$ denotes the Fourier Transform of ϕ.

(Indiana)

Solution.

It can be verified that the Poisson's formula for the Cauchy problem of the heat equation applies to the Schrödinger equation. So we have

$$u(x,t) = \frac{1}{2\sqrt{\pi t}} e^{-\frac{\pi}{4}i} \int_{\mathbb{R}^1} \phi(y) e^{i\frac{(x-y)^2}{4t}} dy.$$

And then

$$\int_{|x|<t} |u(x,t)|^2 dx = \frac{1}{4\pi t} \int_{|x|<t} \left| \int_{\mathbb{R}^1} \phi(y) e^{i\frac{(x-y)^2}{4t}} dy \right|^2 dx.$$

Set $x = 2\xi t$, we get

$$\int_{|x|<t} |u(x,t)|^2 dx = \frac{1}{2\pi} \int_{|\xi|<\frac{1}{2}} \left| \int_{\mathbb{R}^1} \phi(y) e^{i\frac{(2\xi t - y)^2}{4t}} dy \right|^2 d\xi$$

$$= \frac{1}{2\pi} \int_{|\xi|<\frac{1}{2}} \left| \int_{\mathbb{R}^1} \phi(y) e^{-i\xi y + i\frac{y^2}{4t}} dy \right|^2 d\xi$$

$$\to \frac{1}{2\pi} \int_{|\xi|<\frac{1}{2}} \left| \int_{\mathbb{R}^1} \phi(y) e^{-i\xi y} dy \right|^2 d\xi$$

$$= \int_{|\xi|<\frac{1}{2}} |\hat{\phi}(\xi)|^2 d\xi, \quad \text{as } t \to \infty.$$

Here the Fourier transform of ϕ is defined by

$$\hat{\phi}(\xi) = \frac{1}{\sqrt{2\pi}} \int_{\mathbb{R}^1} \phi(y) e^{-i\xi y} dy.$$

6114

Suppose u is a smooth solution of the following problem

$$u_{xxt} + u_{xx} - u^3 = 0, \quad \text{in} \quad [0,1] \times [0, \infty),$$

$$u(0, t) = u(1, t) = 0 \quad \text{for } t \in [0, \infty)$$

with initial data $u(x, 0) = x(x - 1)$. Derive a differential inequality for $w := \int_0^1 |u_x(x, t)|^2 dx$, and show that $u(x, t)$ uniformly tends to zero as $t \to \infty$.

(UC, Los Angeles)

Solution.

Multiplying both sides of the equation by u and integrating the resulting equation in $[0, 1]$, we have

$$\int_0^1 (u_{xxt}u + u_{xx}u - u^4)dx = 0.$$

Integrating by parts and noting the boundary conditions, we obtain

$$\frac{1}{2}\frac{d}{dt}w(t) + w(t)^2 = -\int_0^1 u^4\, dx$$

and the differential inequality

$$\frac{d}{dt}\left(-\frac{1}{w(t)}\right) + 2 \le 0.$$

Solving above differential inequality and noting that

$$w(0) = \int_0^1 (2x - 1)^2 dx = \frac{1}{3},$$

we get

$$w(t) \le \frac{w(0)}{1 + 2w(0)t} = \frac{1}{2t + 3}.$$

It is not difficult to verify that

$$|u(x, t)| \le \int_0^1 |u_x(x, t)|^2 dx$$

(in fact, it is 1-dimensional form of Poincare's inequality), from which we get immediately that $u(x,t)$ uniformly tends to zero as $t \to \infty$.

6115

Assume that $g \in L^2(\mathbb{R})$ and consider the Cauchy problem

$$u_t + u_x = \varepsilon u_{xxx}, u(0,x) = g(x). \quad (CP)_\varepsilon$$

(i) For every ε, write a formula for the solution $u^\varepsilon(t,x)$ using Fourier transform. Show that

$$\|u^\varepsilon(t,\cdot)\|_{L^2(\mathbb{R})} = \|g\|_{L^2(\mathbb{R})} \quad \forall t \geq 0.$$

(ii) Prove that, as $\varepsilon \to 0$, the solutions u^ε of $(CP)_\varepsilon$ converge to the function $v(t,x) = g(x-t)$. More precisely, show that $\|u^\varepsilon(t) - v(t)\|_{L^2(\mathbb{R})} \to 0$, for every fixed time t.

(Penn State)

Solution.

(i) By applying the Fourier transform to the problem, we have

$$\tilde{u}_t + i(\xi + \varepsilon\xi^3)\tilde{u} = 0, \quad \tilde{u}(0,\xi) = \tilde{g}(\xi),$$

where $\tilde{u}$ and $\tilde{g}$ denote the Fourier transforms of u and g with respect to x, respectively.

By solving the transformed problem, it yields that

$$\tilde{u}^\varepsilon(t,\xi) = \tilde{g}(\xi)e^{-i(\xi+\varepsilon\xi^3)t}.$$

Applying the Parserval equality, we get from the above expression that

$$\|u^\varepsilon(t,\cdot)\|_{L^2(\mathbb{R})} = \|g\|_{L^2(\mathbb{R})}.$$

(ii) Note that

$$\tilde{v}(t,\xi) = \int_{\mathbb{R}} e^{-ix\xi} g(x-t)dx = e^{-it\xi}\tilde{g}(\xi).$$

Then for any fixed t, we have

$$\|\tilde{u}^\varepsilon(t,\cdot) - \tilde{v}(t,\cdot)\|_{L^2(\mathbb{R})}^2 = \int_{\mathbb{R}} |\tilde{g}(\xi)(e^{-i(\xi+\varepsilon\xi^3)t} - e^{-it\xi})|^2 d\xi$$

$$= \int_{\mathbb{R}} |\tilde{g}(\xi)|^2 |e^{-i\varepsilon\xi^3 t} - 1|^2 d\xi.$$

As $g \in L^2(\mathbb{R})$, using Lebesgue dominated convergence theorem and then Parserval equality, we get immediately that

$$\|u^\varepsilon(t,\cdot) - v(t,\cdot)\|_{L^2(\mathbb{R})} \to 0, \quad \text{as } \varepsilon \to 0.$$

Section 2

Elliptic Equations

6201

Let $u \not\equiv 0$ satisfy $u \in C^2(\mathbb{R}^n)$, $\Delta u = 0$ on $\mathbb{R}^n$. Show that

$$\int_{\mathbb{R}^n} u^2 dx$$

does not exist.

(Indiana)

Solution.

There exists a point $x^0 \in \mathbb{R}^n$ such that $u(x^0) \neq 0$. Applying the mean-value property for the harmonic function u, we have

$$u(x^0) = \frac{1}{\omega_n \rho^{n-1}} \int_{|x-x^0|=\rho} u(x)\, dS_x$$

where ω_n is the surface area of the unit sphere in $\mathbb{R}^n$. Schwarz's inequality gives

$$u^2(x^0) \leq \left(\frac{1}{\omega_n \rho^{n-1}} \right)^2 \int_{|x-x^0|=\rho} u^2(x) dS_x \int_{|x-x^0|=\rho} dS_x$$

$$= \frac{1}{\omega_n \rho^{n-1}} \int_{|x-x^0|=\rho} u^2(x) dS_x,$$

that is

$$\int_{|x-x^0|=\rho} u^2(x)dS_x \geq \omega_n \rho^{n-1} u^2(x^0).$$

Then we have

$$\int_{\mathbb{R}^n} u^2(x)dx \geq \int_{|x-x^0|\leq r} u^2(x)dx$$

$$= \int_0^r \left(\int_{|x-x^0|=\rho} u^2(x)dS_x \right) d\rho$$

$$\geq \omega_n u^2(x^0) \int_0^r \rho^{n-1} d\rho$$

$$= \frac{\omega_n}{n} u^2(x^0) r^n,$$

for any $r > 0$. The conclusion follows as $r \to \infty$.

6202

Let $u \in C^0(\Omega)$ be weakly harmonic in an open set Ω, i.e., the relation

$$\int_\Omega u\Delta\phi\, dx = 0$$

holds for all $\phi \in C_0^2(\Omega)$. Show that u is then harmonic in Ω.

(*Iowa*)

Solution.

Set

$$\rho(x) = \begin{cases} Ce^{\frac{1}{|x|^2-1}}, & |x| < 1 \\ 0, & |x| \geq 1, \end{cases}$$

where C is a constant such that

$$\int_{\mathbb{R}^n} \rho(x)dx = 1.$$

For $x \in \Omega$, set

$$\rho_\varepsilon(y) = \varepsilon^{-n} \rho\left(\frac{x-y}{\varepsilon}\right).$$

Then it is easy to see that

$$\rho_\varepsilon(y) \in C_0^\infty(\Omega),$$

provided $\varepsilon < \operatorname{dist}(x, \partial\Omega)$. Therefore, we have

$$\int_\Omega u(y) \Delta_y \rho_\varepsilon(y) = 0,$$

i.e.,

$$0 = \varepsilon^{-n} \int_\Omega u(y) \Delta_x \rho\left(\frac{x-y}{\varepsilon}\right) dy = \Delta_x u_\varepsilon(x),$$

where

$$u_\varepsilon(x) = \varepsilon^{-n} \int_\Omega u(y) \rho\left(\frac{x-y}{\varepsilon}\right) dy.$$

This implies that for any compact set $K \subset \Omega$, u_ϵ is harmonic in K if $\varepsilon < \operatorname{dist}(K, \partial\Omega)$.

It is well-known that

$$u_\varepsilon \to u \text{ uniformly in } K, \quad \text{as } \varepsilon \to 0.$$

Therefore u is harmonic in Ω.

6203

Let $f(x, y)$ be a locally bounded function, harmonic in $x \in \mathbb{R}^2$ and continuous in $y \in \mathbb{R}^2$. Show that f is continuous in (x, y), i.e., as a function on $\mathbb{R}^4$.

(*Indiana*)

Solution.

Let $(x^0, y^0) \in \mathbb{R}^4$ and $R > 0$ be given. Then there exists a constant $M > 0$ such that

$$|f(x, y)| \leq M, \quad \forall (x, y) \in \mathbb{R}^4, \quad |x - x^0|^2 + |y - y^0|^2 < 2R^2.$$

Applying the mean-value theorem to the harmonic function $\frac{\partial f}{\partial x_1}$, we have

$$\frac{\partial f}{\partial x_1}(\xi, y) = \frac{4}{\pi R^2} \int_{|x-\xi|<\frac{R}{2}} \frac{\partial f}{\partial x_1}(x, y)dx,$$

$$\forall (\xi, y) \in \mathbb{R}^4, |\xi - x^0| < \frac{R}{2}, |y - y^0| < R.$$

By Green's formula, we get

$$\left| \frac{\partial f}{\partial x_1}(\xi, y) \right| = \frac{4}{\pi R^2} \left| \int_{|x-\xi|=\frac{R}{2}} f(x, y) \cos(n, x_1) dS_x \right|$$

$$\leq \frac{4M}{R}, \quad \forall (\xi, y) \in \mathbb{R}^4, |\xi - x^0| < \frac{R}{2}, |y - y^0| < R.$$

For $\frac{\partial f}{\partial x_2}$, we have the same estimation.

These estimations imply that $f(x, y)$ is continuous at $x = x^0$ uniformly with respect to y near y^0. Then the conclusion that $f(x, y)$ is continuous at $(x^0, y^0) \in \mathbb{R}^4$ follows immediately.

6204

Suppose u is a function defined on $\mathbb{R}^n$ such that

(i) u is bounded and continuous on the half-space $x_n \geq 0$ of $\mathbb{R}^n$,

(ii) $u = 0$ on $x_n = 0$,

(iii) u is harmonic in $x_n > 0$.

Prove that $u \equiv 0$ on $x_n \geq 0$.

(*Indiana*)

Solution.

Redefine the function u in the half-space $x_n < 0$ as an odd function of x_n:

$$u(x', x_n) = -u(x', -x_n), \quad \forall x_n < 0$$

where $x' = (x_1, \ldots, x_{n-1})$. By using the uniqueness theorem of the Dirichlet problem and the Poisson's formula of the Dirichlet problem in a ball, we can verify that the redefined function u is harmonic in any ball centred at the origin in $\mathbb{R}^n$. Therefore, u is bounded and harmonic in $\mathbb{R}^n$. Thanks to the Liouville's Theorem, we get $u \equiv 0$ immediately.

6205

Let $u(x)$ be C^2 on the half-space $\mathbb{R}^n_+ = \{x : x_n > 0\}$ and continuous on the boundary $\partial \mathbb{R}^n_+ = \{x : x_n = 0\}$, and let $g(x)$ be a compactly supported, C^1 function on $\partial \mathbb{R}^n_+ = \{x : x_n = 0\}$. If u, bounded, satisfies

$$\begin{cases} \Delta u = 0, & x_n > 0, \\ u = g, & x_n = 0, \end{cases} \tag{3}$$

show that its "tangential" derivatives $\frac{\partial u}{\partial x_j}$, $j = 1, \ldots, n-1$, are bounded in magnitude by $\max |\nabla g|$ on all of $\mathbb{R}^n_+ = \{x : x_n > 0\}$. (Warning. You must justify any assumptions of regularity you make on the solution u.)

(*Indiana*)

Solution.

By the uniqueness of the bounded solution to the Dirichlet problem (3), we can show that the unique bounded solution to the problem (3) is given by

$$u(x', x_n) = \frac{2x_n}{\omega_n} \int_{\mathbb{R}^{n-1}} \frac{g(\xi')d\xi'}{(x_n^2 + |x' - \xi'|^2)^{n/2}},$$

where $x' = (x_1, \ldots, x_{n-1})$, $\xi' = (\xi_1, \ldots, \xi_{n-1})$ and ω_n is the surface measure of the unit sphere in $\mathbb{R}^n$.

It is not difficult to verify from the above formula that

$$\frac{\partial}{\partial x_j} u(x', x_n) = \frac{2x_n}{\omega_n} \int_{\mathbb{R}^{n-1}} \frac{\frac{\partial}{\partial \xi_j} g(\xi')d\xi'}{(x_n^2 + |x' - \xi'|^2)^{n/2}},$$

$$j = 1, \ldots, n-1$$

and

$$1 = \frac{2x_n}{\omega_n} \int_{\mathbb{R}^{n-1}} \frac{d\xi'}{(x_n^2 + |x' - \xi'|^2)^{n/2}}.$$

Therefore, we have

$$\left| \frac{\partial}{\partial x_j} u(x', x_n) \right| \le \max_{\xi' \in \mathbb{R}^{n-1}} \left| \frac{\partial}{\partial \xi_j} g(\xi') \right|, \quad j = 1, \ldots, n-1,$$

which are just we want to show.

6206

Show that the problem

$$\Delta u = -1 \qquad \text{for } |x| < 1, |y| < 1,$$

$$u = 0 \qquad \text{for } |x| = 1,$$

$$\frac{\partial u}{\partial x} - \frac{\partial u}{\partial y} = 0 \quad \text{for } |y| = 1$$

has at most one solution. (Here $u = u(x, y)$ is a function of two variables x and y, and is a classical solution.)

(*Cincinnati*)

Solution.

We need only prove that the problem

$$\Delta u = 0 \qquad \text{for } |x| < 1, |y| < 1,$$

$$u = 0 \qquad \text{for } |x| = 1,$$

$$\frac{\partial u}{\partial x} - \frac{\partial u}{\partial y} = 0 \quad \text{for } |y| = 1$$

has the unique solution $u \equiv 0$. It is easy to see that

$$\int_{|x|<1, |y|<1} u \, \Delta u \, dx dy = 0.$$

Integrating the above integral by parts, we have

$$0 = \int_{|x|<1,|y|<1} u\,\Delta u\,dxdy$$

$$= \int_{-1}^{1} \left(u\frac{\partial u}{\partial x}(1,y) - u\frac{\partial u}{\partial x}(-1,y) \right) dy$$

$$+ \int_{-1}^{1} \left(u\frac{\partial u}{\partial y}(x,1) - u\frac{\partial u}{\partial y}(x,-1) \right) dx$$

$$- \int_{|x|<1,|y|<1} \left(\left(\frac{\partial u}{\partial x}\right)^2 + \left(\frac{\partial u}{\partial y}\right)^2 \right) dxdy$$

$$= \int_{-1}^{1} \left(u\frac{\partial u}{\partial x}(x,1) - u\frac{\partial u}{\partial x}(x,-1) \right) dx$$

$$- \int_{|x|<1,|y|<1} \left(\left(\frac{\partial u}{\partial x}\right)^2 + \left(\frac{\partial u}{\partial y}\right)^2 \right) dxdy$$

$$= \frac{1}{2}(u^2(1,1) - u^2(-1,1) - u^2(1,-1) + u^2(-1,-1))$$

$$- \int_{|x|<1,|y|<1} \left(\left(\frac{\partial u}{\partial x}\right)^2 + \left(\frac{\partial u}{\partial y}\right)^2 \right) dxdy$$

$$= - \int_{|x|<1,|y|<1} \left(\left(\frac{\partial u}{\partial x}\right)^2 + \left(\frac{\partial u}{\partial y}\right)^2 \right) dxdy.$$

Then we get $u \equiv 0$ immediately.

6207

Let Ω be a bounded normal domain in $\mathbb{R}^3$, and suppose α is such that a non-trivial solution exists to

$$\begin{cases} -\nabla^2 u = \alpha^2 u & \text{in } \Omega, \\ u = 0 & \text{on } \partial\Omega. \end{cases} \tag{$*$}$$

(a) Show that α is a real number, and that u can be chosen to be real-valued.

(b) Suppose $\alpha_1 \neq \alpha_2$ are such that non-trivial real solutions u_1 and u_2 satisfy $(*)$. Show that

$$\int_\Omega u_1(\vec{r})u_2(\vec{r})dv = 0.$$

(Indiana–Purdue)

Solution.

(a) Multiplying the equation in the problem $(*)$ by the complex conjugate of the solution u, and integrating the resulting equation in Ω, we obtain

$$\int_\Omega |\nabla u|^2 dv = \alpha^2 \int_\Omega |u|^2 dv,$$

which implies that α is a real number. Then it is clear that u can be chosen to be real-valued.

(b) Suppose that u_1 and u_2 are chosen to be real-valued. Multiplying the equation satisfied by u_1

$$-\nabla^2 u_1 = \alpha_1^2 u_1 \quad \text{in } \Omega$$

and the equation satisfied by u_2

$$-\nabla^2 u_2 = \alpha_2^2 u_2 \quad \text{in } \Omega$$

by u_2 and u_1, respectively, and integrating the resulting equations in Ω, we can obtain

$$\int_\Omega \nabla u_1 \cdot \nabla u_2 \, dv = \alpha_1^2 \int_\Omega u_1 u_2 \, dv$$

and

$$\int_\Omega \nabla u_1 \cdot \nabla u_2 \, dv = \alpha_2^2 \int_\Omega u_1 u_2 \, dv.$$

From the above integral equalities, we get immediately

$$(\alpha_1^2 - \alpha_2^2) \int_\Omega u_1 u_2 \, dv = 0,$$

which implies the conclusion of the problem.

6208

Let $\Omega \subset \mathbb{R}^n$ be an open set. Let u be a continuous function on Ω. Prove the equivalence of these two notions of "subharmonic".

(i) For every $x \in \Omega$ and $r > 0$ such that the ball $B(x,r)$ centered at x with radius r satisfies $B(x,r) \subset\subset \Omega$ one has

$$u(x) \le \frac{1}{\omega_n r^{n-1}} \int_{\partial B(x,r)} u(y) dS_y.$$

(Here ω_n denotes the surface area of the unit sphere in $\mathbb{R}^n$ and $B(x,r) \subset\subset \Omega$ means that the closure of B is contained within Ω.)

(ii) For every ball $B \subset\subset \Omega$ and every harmonic function $h \in C^0(\bar{B})$ such that $u \le h$ on ∂B, one has $u \le h$ in B.

(Indiana)

Solution.

Let u be subharmonic in the sense of (i). By a standard argument, we can show that u cannot take its maximum in Ω unless u is a constant.

Let h be the function given in (ii). Then $v = u - h$ is also subharmonic in the sense of (i). Hence that $v \le 0$ on ∂B implies that $v \le 0$ in B.

Conversely, let u be subharmonic in the sense of (ii). We define the function h as follows:

$$\begin{cases} \Delta h = 0 & \text{in } B(x,r), \\ h = u & \text{on } \partial B(x,r). \end{cases}$$

From the definition of subharmonicity in the sense of (ii), and the mean-value theorem, we have

$$u(x) \leq h(x) = \frac{1}{w_n r^{n-1}} \int_{\partial B(x,r)} u(y) dS_y.$$

This completes the proof.

6209

Let $\Omega \subseteq \mathbb{R}^n$ be an open set, and let u be harmonic in Ω and continuous in $\bar{\Omega}$.

(a) Show that $\frac{\partial u}{\partial x_i}$ is harmonic in Ω for each i.

(b) Let $B \subset \Omega$, where B is the open ball centered at x of radius r. Show that

$$|\nabla u(x)| \leq \frac{n}{r} \sup\{|u(y)| : |x - y| = r\}.$$

(c) Show that, if Ω is bounded and $x \in \Omega$,

$$|\nabla u(x)| \leq \frac{n}{d(x)} \sup\{|u(y)| : y \in \Omega\}$$

where $d(x) = \mathrm{dist}(x, \partial\Omega)$.

(*Indiana*)

Solution.

(a) The conclusion is clear.

(b) Applying the mean-value theorem for the harmonic function $\frac{\partial u}{\partial x_i}$ in the ball B, we have

$$\frac{\partial u}{\partial x_i}(x) = \frac{n}{w_n r^n} \int_B u_{x_i}(y) dy,$$

where w_n is the surface area of the unit sphere in $\mathbb{R}^n$, hence w_n/n is the volume of the unit ball.

By using the Green's formula, the above equality can be rewritten as

$$\frac{\partial u}{\partial x_i}(x) = \frac{n}{w_n r^n} \int_{\partial B} u(y) \cos(\vec{n}, x_i) dS.$$

And therefore, we have

$$\left| \frac{\partial u}{\partial x_i}(x) \right| \leq \frac{n}{\omega_n r^n} \int_{\partial B} |u(y)| dS$$

$$\leq \frac{n}{r} \sup_{\partial B} |u(y)|.$$

(c) The conclusion can be obtained from the result of the part (b) immediately.

6210

Let $\Omega \subseteq \mathbb{R}^3$ be bounded and open with smooth boundary. Let $u \in C^1(\bar{\Omega}) \cap C^2(\Omega)$ solve

$$\Delta u + k^2 u = 0 \quad \text{in } \Omega \ (k > 0).$$

Derive an appropriate mean-value property for the solution.

(*Indiana*)

Solution.

First we look for the fundamental solutions of the equation. Let $r = |x|$. Then the spherically symmetric solutions depending only on r satisfy

$$\frac{d^2}{dr^2}(ru) + k^2 ru = 0.$$

For this ordinary differential equation, there are two linearly independent solutions:

$$u = \frac{1}{r}\cos(kr), \ \frac{1}{r}\sin(kr).$$

Let u be a solution to the equation $\Delta u + k^2 u = 0$. By the fundamental solution $\frac{1}{4\pi r}\cos kr$, it is not difficult to verify that

$$u(x^0) = -\frac{1}{4\pi}\int_{\partial\Omega}\left(u(x)\frac{\partial}{\partial n}\left(\frac{1}{r_{x^0 x}}\cos(kr_{x^0 x}) \right) \right.$$

$$\left. -\frac{1}{r_{x^0 x}}\cos(kr_{x^0 x})\frac{\partial}{\partial n}u(x) \right) dS_x, \quad \forall x^0 \in \Omega,$$

where $r_{x^0x} = |x - x^0|$.

Let $R < \text{dist}(x^0, \partial\Omega)$ be a fixed positive constant. By $B(x^0, R)$ we denote the ball centered at x^0 with radius R. Applying the above integral formula in the domain $B(x^0, R)$, we get

$$u(x^0) = -\frac{1}{4\pi} \int_{\partial B(x^0, R)} \left(u(x) \frac{\partial}{\partial r_{x^0x}} \left(\frac{1}{r_{x^0x}} \cos(kr_{x^0x}) \right) \right.$$

$$\left. - \frac{1}{r_{x^0x}} \cos(kr_{x^0x}) \frac{\partial}{\partial r_{x^0x}} u(x) \right) dS_x. \tag{1}$$

Now we look for a function $g(x, x^0)$, which satisfies

$$(\Delta_x + k^2)g(x, x^0) = 0, \quad x \in B(x^0, R),$$

$$g(x, x^0) = \frac{1}{4\pi r_{x^0x}} \cos(kr_{x^0x}), \quad x \in \partial B(x^0, R).$$

It is easy to see that the function

$$g(x, x^0) = \cot(kR) \frac{\sin(kr_{x^0x})}{r_{x^0x}}$$

satisfies the above conditions. For this function, it is easy to verify that

$$0 = -\frac{1}{4\pi} \int_{\partial B(x^0, R)} \left(u(x) \frac{\partial}{\partial r_{x^0x}} g(x, x^0) \right.$$

$$\left. - g(x, x^0) \frac{\partial}{\partial r_{x^0x}} u(x) \right) dS_x. \tag{2}$$

Subtracting the equation (2) from the equation (1), we obtain

$$u(x^0) = -\frac{1}{4\pi} \int_{\partial B(x^0, R)} u(x) \frac{\partial}{\partial r_{x^0x}} \left(\frac{1}{r_{x^0x}} \cos(kr_{x^0x}) \right.$$

$$\left. - \cot(kR) \frac{1}{r_{x^0x}} \sin(kr_{x^0x}) \right) dS_x$$

$$= \frac{k}{4\pi R \sin(kR)} \int_{\partial B(x^0, R)} u(x) dS_x.$$

This is a mean-value theorem for equation $\Delta u + k^2 u = 0$.

6211

Let $u \in C^2(\mathbb{R}^2)$ satisfy

$$\Delta u = u + 1. \tag{1}$$

Prove that its spherical mean, defined by

$$v(x) = \frac{1}{2\pi|x|} \int_{\partial B(0,|x|)} u(y) dS_y, \quad x \neq 0$$

and $v(0) = u(0)$, also satisfies (1), for $x \neq 0$. Here, $B(0,r)$ denotes the ball of radius r, dS_y the element of arc length.

(Indiana)

Solution.

Set $r = |x|$. Then we have

$$v(r) = \frac{1}{2\pi} \int_{\partial B(0,1)} u(r\omega) \, dS_\omega$$

and

$$\frac{\partial}{\partial r} v(r) = \frac{1}{2\pi} \int_{\partial B(0,1)} \sum_{i=1}^{2} u_{x_i}(r\omega) \omega_i \, dS_\omega$$

$$= \frac{1}{2\pi r} \int_{\partial B(0,r)} \sum_{i=1}^{2} u_{y_i}(y) \omega_i \, dS_y.$$

Applying Green's formula, we get

$$\frac{\partial}{\partial r} v(r) = \frac{1}{2\pi r} \int_{B(0,r)} \Delta u(y) dy$$

$$= \frac{1}{2\pi r} \int_0^r \left(\int_{\partial B(0,\rho)} (u(y) + 1) dS_y \right) d\rho.$$

Then it follows that

$$\frac{\partial}{\partial r}\left(r\frac{\partial}{\partial r}v(r)\right) = \frac{1}{2\pi}\int_{\partial B(0,r)}(u(y)+1)dS_y$$
$$= r(v(r)+1).$$

Therefore

$$\Delta v(r) \equiv \frac{1}{r}\frac{\partial}{\partial r}\left(r\frac{\partial}{\partial r}v(r)\right) = v(r)+1.$$

The proof is completed.

6212

Let $B = \{(x,y) \mid x^2 + y^2 < 1\}$. Prove that the problem

$$\Delta u = \frac{1}{\sqrt{x^2 + y^2}} \quad \text{in } B, \quad u = 0 \text{ on } \partial B$$

has exactly one weak solution u in $H_0^1(B)$.

(Iowa)

Solution.
Let

$$v = \frac{x}{\sqrt{x^2+y^2}}, \quad w = \frac{y}{\sqrt{x^2+y^2}} \quad \text{and} \quad f = \frac{1}{\sqrt{x^2+y^2}}.$$

It is easy to see that $v, w \in L^2(B)$ and

$$f = \frac{\partial v}{\partial x} + \frac{\partial w}{\partial y}$$

in the sense of distributions (see the solution to the problem 6105). This implies that $f \in H^{-1}(B)$. Hence the problem admits a unique weak solution $u \in H_0^1(B)$.

6213

(a) Let Ω be a bounded domain in $\mathbb{R}^n$ with smooth boundary $\partial\Omega$, and $f \in L^2(\Omega)$, and $g \in L^2(\partial\Omega)$ be given. Consider the problem

$$\int_\Omega (\operatorname{grad} u) \cdot (\operatorname{grad} v) + \lambda\, uv\, dx = \int_\Omega fv\, dx + \int_{\partial\Omega} gv\, d\sigma$$

$$\forall v \in W^{1,2}(\Omega) \qquad (1)$$

where $\lambda > 0$. Show that the problem (1) has a unique solution $u \in W^{1,2}(\Omega)$.

(b) Will the conclusion of part (a) be valid if $\lambda = 0$? Explain.

(c) Suppose that f, g and $\partial\Omega$ are sufficiently smooth, write problem (1) in the classical form.

(Cincinnati)

Solution.

(a) Define the inner product in $W^{1,2}(\Omega)$ by

$$(u, v)_1 = \int_\Omega [(\operatorname{grad} u) \cdot (\operatorname{grad} v) + \lambda uv]dx, \quad \forall u, v \in W^{1,2}(\Omega).$$

And denote the corresponding norm in $W^{1,2}(\Omega)$ by $\|\cdot\|_1$. Then by applying Cauchy inequality and the trace theorem in Sobolev's spaces, we get

$$\left|\int_\Omega fv\, dx + \int_{\partial\Omega} gv\, d\sigma\right| \leq \|f\|_{L^2(\Omega)}\|v\|_{L^2(\Omega)} + \|g\|_{L^2(\partial\Omega)}\|v\|_{L^2(\partial\Omega)}$$

$$\leq C\|v\|_1,$$

where C is a constant. Therefore,

$$F(v) = \int_\Omega fv\, dx + \int_{\partial\Omega} gv\, d\sigma$$

is a bounded linear functional in $W^{1,2}(\Omega)$. Thanks to the Riesz Theorem, there exists a function $u \in W^{1,2}(\Omega)$ such that

$$F(v) = (u, v)_1, \quad \forall v \in W^{1,2}(\Omega).$$

Then u is a solution to the problem (1). The uniqueness is clear.

(b) By the classical form of the problem, it is easy to see that the conclusion of the part (a) is false if $\lambda = 0$ even for smooth f and g.

(c) Suppose that f, g and $\partial\Omega$ are sufficiently smooth. Integrating by parts, we have

$$\int_\Omega (\operatorname{grad} u) \cdot (\operatorname{grad} v)dx = \int_{\partial\Omega} \frac{\partial u}{\partial n} v\, d\sigma - \int_\Omega (\Delta u)v\, dx.$$

Therefore, the classical form of the problem (1) is

$$-\Delta u + \lambda u = f \quad \text{in } \Omega,$$

$$\frac{\partial u}{\partial n} = g \quad \text{on } \partial\Omega.$$

6214

Let Ω be a unit ball in $\mathbb{R}^n$ centered at zero. Consider the following problem.

Find $u \in H_0^1(\Omega)$ such that

$$\int_\Omega \nabla u \cdot \nabla\phi\, dx = \int_{\Omega\cap\{x_n=0\}} \phi\, dx', \quad \forall\phi \in H_0^1(\Omega),$$

where $\mathbb{R}^n \ni x = (x_1, \ldots, x_{n-1}, x_n) = (x', x_n)$. Prove the existence of a unique solution of that problem.

Denote $u^+ = u|_{\Omega\cap\{x_n>0\}}$ and $u^- = u|_{\Omega\cap\{x_n<0\}}$, and assume that $u^\pm$ are regular enough. Show that $\Delta u^\pm = 0$ and that

$$\frac{\partial u^-}{\partial x_n} - \frac{\partial u^+}{\partial x_n} = 1 \quad \text{on } \Omega\cap\{x_n=0\}.$$

(Cincinnati)

Solution.

Define the scalar product in $H_0^1(\Omega)$ by

$$(u, v)_1 = \int_\Omega \nabla u \cdot \nabla v\, dx.$$

By the Cauchy inequality and the trace theory in Sobolev spaces, we can get

$$\left| \int_{\Omega \cap \{x_n = 0\}} \phi \, dx' \right| \leq \left(\int_{\Omega \cap \{x_n = 0\}} dx' \right)^{1/2} \left(\int_{\Omega \cap \{x_n = 0\}} |\phi|^2 dx' \right)^{1/2}$$

$$\leq C \|\phi\|_1,$$

where $\| \cdot \|_1$ is the norm in $H_0^1(\Omega)$. According to the Riesz Theorem, there exists $u \in H_0^1(\Omega)$ such that

$$(u, \phi)_1 = \int_{\Omega \cap \{x_n = 0\}} \phi \, dx', \quad \forall \phi \in H_0^1(\Omega).$$

This proves the existence of the solution to the problem.

The uniqueness of the solution to the problem is clear.

Assume that $u^{\pm}$ are regular enough. Write the equation in the following form

$$\int_{\Omega_+} \nabla u^+ \cdot \nabla \phi \, dx + \int_{\Omega_-} \nabla u^- \cdot \nabla \phi \, dx = \int_{\Omega \cap \{x_n = 0\}} \phi \, dx',$$

where $\Omega_+ = \Omega \cap \{x_n > 0\}$ and $\Omega_- = \Omega \cap \{x_n < 0\}$. Integrating the above equation by parts, we have

$$- \int_{\Omega_+} \Delta u^+ \cdot \phi \, dx - \int_{\Omega_-} \Delta u^- \cdot \phi \, dx$$

$$+ \int_{\Omega \cap \{x_n = 0\}} \left(\frac{\partial u^-}{\partial x_n} - \frac{\partial u^+}{\partial x_n} \right) \phi \, dx'$$

$$= \int_{\Omega \cap \{x_n = 0\}} \phi \, dx', \quad \forall \phi \in H_0^1(\Omega).$$

From the above equation, we get immediately that

$$\Delta u^{\pm} = 0 \quad \text{in } \Omega_{\pm},$$

$$\frac{\partial u^-}{\partial x_n} - \frac{\partial u^+}{\partial x_n} = 1 \quad \text{on } \Omega \cap \{x_n = 0\}.$$

The proof is completed.

6215

(a) Let $u \in W_0^{1,2}(\Omega)$ satisfy

$$\int_\Omega \nabla u \cdot \nabla \phi \, dx \geq 0 \quad \forall \phi \in W_0^{1,2}(\Omega), \phi \geq 0. \qquad (I)$$

Show that $u \geq 0$ a.e. in Ω.

(b) Let $u \in W^{1,2}(\Omega)$ satisfy (I) above, show that

$$\inf_\Omega u \geq \inf_{\partial\Omega} u \quad (\text{ess inf})$$

(*Cincinnati*)

Solution.

(a) This is a corollary of the conclusion in (b) of this problem.

(b) If $\inf_{\partial\Omega} u = -\infty$, the proposition is true.

Assume that $\inf_{\partial\Omega} u = l > -\infty$. Let

$$\phi(x) = \max\{l - u, 0\}.$$

Then it can be verified that $\phi \in W_0^{1,2}(\Omega)$ and

$$\nabla\phi = \begin{cases} -\nabla u, & u < l \\ 0, & u \geq l. \end{cases}$$

Now we have

$$\int_\Omega \nabla u \cdot \nabla\phi \, dx = -\int_\Omega |\nabla\phi|^2 dx \leq 0.$$

From the above inequality and (I), it holds that

$$\int_\Omega |\nabla\phi|^2 dx = 0.$$

By the Poincaré inequality with the above estimate, we obtain

$$\int_\Omega |\phi|^2 dx = 0.$$

Hence

$$\phi = 0 \quad \text{a.e. in } \Omega$$

This implies that $u \geq l$ a.e. in Ω.

6216

Consider functions $u_n, v_n, w \in W^{2,2}(\Omega) \cap W_0^{1,2}(\Omega)$, such that

$$-\Delta v_n + u_n = w,$$

$$-\Delta u_n - nv_n = 0,$$

in Ω. Prove that $u_n \to w$ strongly in $L^2(\Omega)$, as $n \to +\infty$.

(Cincinnati)

Solution.

Multiplying the first equation and the second equation by u_n and v_n respectively, and then integrating the resulting equations in Ω, we get

$$-\int_\Omega \Delta v_n \cdot u_n \, dx + \|u_n\|_{L^2}^2 = \int_\Omega w u_n \, dx,$$

$$-\int_\Omega \Delta u_n \cdot v_n \, dx - n\|v_n\|_{L^2}^2 = 0.$$

By using Green's formula and noting that $u_n, v_n \in W_0^{1,2}(\Omega)$, both equations above can be written as

$$\int_\Omega \nabla v_n \cdot \nabla u_n \, dx + \|u_n\|_{L^2}^2 = \int_\Omega w u_n \, dx,$$

$$\int_\Omega \nabla u_n \cdot \nabla v_n \, dx - n\|v_n\|_{L^2}^2 = 0.$$

Then we have

$$\|u_n\|_{L^2}^2 + n\|v_n\|_{L^2}^2 = \int_\Omega w u_n \, dx$$

$$\leq \frac{1}{2}\|w\|_{L^2}^2 + \frac{1}{2}\|u_n\|_{L^2}^2.$$

Therefore, we obtain

$$\{u_n\} \text{ is bounded in } L^2(\Omega) \tag{1}$$

and

$$v_n \to 0 \text{ strongly in } L^2(\Omega). \tag{2}$$

Multiplying the first equation by v_n and integrating it on Ω, we have

$$\|\nabla v_n\|_{L^2}^2 + \int_\Omega u_n v_n \, dx = \int_\Omega w v_n \, dx.$$

By noting (1) and (2), it follows from above equality that

$$v_n \to 0 \text{ strongly in } W_0^{1,2}(\Omega). \tag{3}$$

Multiplying the first equation and the second equation by Δu_n and Δv_n respectively, and then integrating the resulting equations, we can get

$$-\int_\Omega \Delta v_n \Delta u_n \, dx - \|\nabla u_n\|_{L^2}^2 = \int_\Omega w \Delta u_n \, dx,$$

$$-\int_\Omega \Delta u_n \Delta v_n \, dx + n\|\nabla v_n\|_{L^2}^2 = 0.$$

Hence

$$\|\nabla u_n\|_{L^2}^2 + n\|\nabla v_n\|_{L^2}^2 = \int_\Omega \nabla w \cdot \nabla u_n \, dx.$$

This implies that

$$\{u_n\} \text{ is bounded in } W_0^{1,2}(\Omega). \tag{4}$$

Multiply the first equation by Δv_n. By similar consideration, we can get finally

$$-\|\Delta v_n\|_{L^2}^2 - \int_\Omega \nabla u_n \cdot \nabla v_n \, dx = \int_\Omega \nabla w \cdot \nabla v_n \, dx.$$

By noting (3) and (4), it follows from the above equality that

$$\Delta v_n \to 0 \text{ strongly in } L^2(\Omega).$$

This gives us the conclusion of the problem.

6217

Assume $\Omega \subset \mathbb{R}^n$ is a bounded open set with smooth boundary and let $f \in C_c^\infty(\Omega)$. Suppose $u_k \in C^\infty(\Omega)$ satisfies $\Delta u_k + u_k = f$ and has the property that for some positive number M,

$$\int_\Omega u_k(x)^2 dx \le M$$

for $k = 1, 2, \ldots$. Prove that there exist a function $u \in C^\infty(\Omega)$ satisfying $\Delta u + u = f$ and a subsequence $\{u_{k_j}\}$ such that u_{k_j} converges uniformly to u on each compact subset of Ω.

(*Indiana*)

Solution.

As $\{u_k\}$ is bounded in $L^2(\Omega)$, there exists a subsequence $\{u_{k_j}\}$ such that u_{k_j} converges to u weakly in $L^2(\Omega)$, and therefore in $\mathcal{D}'(\Omega)$. Then $\Delta u + u = f$ in the sense of the distribution. By the regularity theorem of the elliptic equations, we have $u \in C^\infty(\Omega)$.

Now we show the second conclusion. Set $v_k = u_k - u_1$. It is clear that v_k satisfies

$$\Delta v_k + v_k = 0 \quad \text{in } \Omega$$

and

$$\int_\Omega |v_k(x)|^2 dx \le C, \quad \forall k = 1, 2, \ldots.$$

Hereafter, by C we denote various constants.

Let Ω_1 be a subdomain of Ω such that $\bar{\Omega}_1 \subset\subset \Omega$ and $R = \text{dist}(\partial\Omega, \Omega_1)$ fixed. For any $x \in \bar{\Omega}_1$, by $B(x, r)$ we denote the ball with the radius r ($< R$) centered at x. By the mean-value theorem for the equation $\Delta u + u = 0$ (for the case $n = 3$, see the solution to the problem 6210), we have

$$h(r)v_k(x) = \int_{\partial B(x,r)} v_k(y)dS_y,$$

where $h(r)$ is a continuous function of r. Integrating the above equality from 0 to R with respect to r, we find that

$$v_k(x) = \left(\int_0^R h(r)dr \right)^{-1} \int_{B(x,R)} v_k(y)dy.$$

Here R can be chosen so small that $\int_0^R h(r)dr \neq 0$. Therefore, it holds that

$$|v_k(x)| \leq C, \quad \forall x \in \bar{\Omega}, \quad k = 1, 2, \ldots. \tag{1}$$

Set $v_k = w_k + z_k$, where w_k and z_k are defined by

$$\Delta w_k = -v_k \quad \text{in } \Omega_1,$$

$$w_k = 0 \quad \text{on } \partial\Omega_1$$

and

$$\Delta z_k = 0 \quad \text{in } \Omega_1,$$

$$z_k = v_k \quad \text{on } \partial\Omega_1,$$

respectively. By applying the solution formula of the Dirichlet's problem for the Poisson's equation, we can get from the estimation (1)

$$|\nabla w_k(x)| \leq C, \quad \forall x \in \bar{\Omega}_1, k = 1, 2, \ldots.$$

It is clear that

$$\max_{\bar{\Omega}_1} |z_k(x)| \leq C, \quad \forall k = 1, 2, \ldots.$$

And by applying the mean-value property for the harmonic solutions, we can obtain from the above estimation

$$|\nabla z_k(x)| \leq C \text{ in any compact subset of } \Omega_1.$$

Then by using the Ascoli–Arzela Theorem, we prove the second conclusion of the problem immediately.

6218

Let Ω be a bounded Lipschits domain in $\mathbb{R}^n$, and

$$I(v) = \frac{1}{2} \int_\Omega |\nabla v|^2 dx,$$

$$\mathcal{A} = \{v \in H_0^1(\Omega) : h_1 \le v \le h_2 \quad \text{a.e. in } \Omega\},$$

where $h_1, h_2 : \Omega \to \mathbb{R}$ are given smooth functions.

(a) Show that if $\mathcal{A} \ne \emptyset$, there exists $u \in \mathcal{A}$ satisfying

$$I(u) = \min_{v \in \mathcal{A}} I(v).$$

(b) Show that

$$\int_\Omega \nabla u \nabla (v - u) dx \ge 0 \quad \forall v \in \mathcal{A}.$$

(Iowa)

Solution.

(a) Let $\{u_n\}$ be the minimizing sequence of $I(v)$ in $\mathcal{A}$, i.e.,

$$\lim_{n \to \infty} I(u_n) = \inf_{v \in \mathcal{A}} I(v).$$

Define the norm of v in $H_0^1(\Omega)$ as

$$\|v\|_1 = \left(\int_\Omega |\nabla v|^2 dx \right)^{1/2}.$$

Then we get the boundedness of $\{\|u_n\|_1\}$. Therefore, there exists a subsequence of $\{u_n\}$, for simplicity we still denote it by $\{u_n\}$, such that

$$u_n \rightharpoonup u \quad \text{in } H_0^1(\Omega)$$

and

$$u_n \to u \quad \text{in } L^2(\Omega).$$

And there exists a subsequence of $\{u_n\}$, denoted still by $\{u_n\}$, such that

$$u_n \rightarrow u \quad \text{a.e. in } \Omega.$$

Therefore, we get $u \in \mathcal{A}$ and that

$$\|u\|_1 \leq \liminf_{n\to\infty} \|u_n\|_1 = \lim_{n\to\infty} 2I(u_n)$$

by the weak lower semicontinuity of $\|v\|_1$. This implies that

$$I(u) = \min_{v\in\mathcal{A}} I(v).$$

The conclusion of (a) is proved.

(b) It is easy to verify that

$$u + t(v - u) \in \mathcal{A}, \quad \forall v \in \mathcal{A}, 0 \leq t \leq 1.$$

Let

$$F(t) = \int_\Omega \nabla(u + t(v - u)) \cdot \nabla(u + t(v - u))dx.$$

Then $F(t)$ has the minimum at $t = 0$. Therefore we have

$$F'(0) \geq 0.$$

This implies

$$\int_\Omega \nabla u \cdot \nabla(v - u)dx \geq 0.$$

The proof is completed.

6219

Let $\Omega \subset \mathbb{R}^n$, $n > 1$, be a bounded domain with smooth boundary. Let $u \in C^1(\bar{\Omega})$ be harmonic in Ω.

(a) Prove that

$$\max_{\Omega} |\nabla u| = \max_{\partial\Omega} |\nabla u|.$$

(Note. In both parts (a) and (b) of this question, you may cite, without proof, any standard maximum principles you are using.)

(b) Suppose there exist functions $\Phi_1, \Phi_2 \in C^2(\Omega) \cap C^1(\bar{\Omega})$ such that

$$\Delta\Phi_1 \leq 0, \quad \Delta\Phi_2 \geq 0, \quad \Phi_1 = u = \Phi_2 \quad \text{on } \partial\Omega.$$

Prove that

$$\max_{\Omega} |\nabla u| \leq \max \left(\max_{\partial\Omega} |\nabla\Phi_1|, \max_{\partial\Omega} |\nabla\Phi_2| \right).$$

(*Indiana*)

Solution.

(a) The conclusion is an immediate corollary of the following inequality:

$$\Delta(|\nabla u|^2) = 2 \sum_{i,j=1}^{n} u_{x_i x_j}^2 \geq 0.$$

(b) For the function $\Phi_2 - u$, we have

$$\Delta(\Phi_2 - u) \geq 0 \quad \text{in } \Omega$$

and

$$\Phi_2 - u = 0 \quad \text{on } \partial\Omega.$$

From the maximum principle, it holds that

$$\sup_{\Omega} (\Phi_2 - u) = \max_{\partial\Omega} (\Phi_2 - u) = 0,$$

i.e.,

$$\Phi_2(x) \leq u(x), \quad \forall x \in \Omega.$$

Similarly, we have

$$u(x) \leq \Phi_1(x), \quad \forall x \in \Omega.$$

Therefore, it holds that

$$\frac{\Phi_2(x) - \Phi_2(x^0)}{|x - x^0|} \leq \frac{u(x) - u(x^0)}{|x - x^0|}$$

$$\leq \frac{\Phi_1(x) - \Phi_1(x^0)}{|x - x^0|}, \quad \forall x \in \Omega, x^0 \in \partial\Omega.$$

This inequality implies the conclusion of the part (b).

6220

Let Ω be an open subset of $\mathbb{R}^n$. Suppose $u \in C^2(\bar{\Omega})$ is a solution of the equation $\Delta u = u^3$ with the property that $|\nabla u(x)| \leq 1$ for each $x \in \partial\Omega$. Prove that $|\nabla u(x)| \leq 1$ for all $x \in \Omega$.

(*Indiana*)

Solution.

Set

$$w = |\nabla u|^2.$$

It is easy to see that $w \in C^2(\Omega) \cap C^1(\bar{\Omega})$ and

$$\Delta w = 2 \sum_{i,j=1}^{n} u_{x_i x_j}^2 + 6u^2 |\nabla u|^2 \geq 0.$$

Therefore, w takes its maximum on $\bar{\Omega}$ at some point on $\partial\Omega$. Hence $w \leq 1$ on $\partial\Omega$ implies that $w \leq 1$ in Ω.

6221

Consider (a nonlinear) equation

$$-\Delta u = \lambda u^2(1 - u) \quad \text{in } \Omega,$$

$$u = 0 \quad \text{on } \partial\Omega, \tag{1}$$

where $\lambda > 0$ is a parameter, and $\partial\Omega$ sufficiently smooth.

(a) Prove that for any solution u of (1), $0 \leq u \leq 1$.

(b) Show that if $\lambda > 0$ is sufficiently small, the only solution of (1) is $u \equiv 0$.

(*Cincinnati*)

Solution.

(a) If $u \leq 1$ is false, then there exists $x^0 \in \Omega$ such that u takes maximum at x^0 and

$$u(x^0) > 1.$$

For the maximum point x^0, it is clear that

$$\Delta u(x^0) \leq 0,$$

i.e.,

$$-\Delta u(x^0) \geq 0.$$

This contradicts the assumption that

$$\lambda u^2(1 - u)(x^0) < 0.$$

Hence $u \leq 1$.

As $\lambda u^2(1 - u) \geq 0$, u can not take minimum in Ω unless $u \equiv 0$. Hence $u \geq 0$.

(b) Multiplying the equation by u and integrating the resulting equality on Ω, we can get

$$\int_\Omega |\nabla u|^2 dx = \lambda \int_\Omega u^3(1 - u)dx.$$

By using the conclusion of (a), it holds that

$$\int_\Omega |\nabla u|^2 dx \leq \lambda \int_\Omega |u|^2 dx.$$

Applying Poincaré inequality, we obtain from the above inequality

$$\int_\Omega |\nabla u|^2 dx \leq \lambda C \int_\Omega |\nabla u|^2 dx,$$

where C is a constant. If $\lambda < C^{-1}$, we have $u \equiv 0$ immediately.

6222

Let Ω be a bounded normal domain in $\mathbb{R}^n$ with smooth boundary $\partial\Omega$. Let $\vec{n}$ be the exterior unit normal on $\partial\Omega$. Show that the only solution of the boundary value problem for the biharmonic equation:

$$\Delta(\Delta u) = 0 \quad \text{in } \Omega$$

$$u = 0 \quad \text{on } \partial\Omega,$$

$$\frac{\partial u}{\partial n} = 0 \quad \text{on } \partial\Omega,$$

is the trivial one $u \equiv 0$. Assume $u \in C^4(\bar\Omega)$.

(Indiana–Purdue)

Solution.

Integrating by parts and taking the boundary conditions into account, we have

$$0 = \int_\Omega \Delta(\Delta u) u\, dx$$

$$= \int_{\partial\Omega} \frac{\partial}{\partial n}(\Delta u) u\, dS - \int_\Omega \nabla(\Delta u) \cdot \nabla u\, dx$$

$$= \int_{\partial\Omega} \left(\frac{\partial}{\partial n}(\Delta u) u - \Delta u \frac{\partial u}{\partial n} \right) dS + \int_\Omega |\Delta u|^2 dx$$

$$= \int_\Omega |\Delta u|^2 dx.$$

Therefore, u is harmonic in Ω. Taking into account $u = 0$ on $\partial\Omega$, we get $u \equiv 0$ immediately.

6223

let Ω be a bounded domain in $\mathbb{R}^n$ with smooth boundary $\partial\Omega$ and assume $u \in C^2(\bar\Omega)$ satisfies the following

$$\Delta u = f \quad \text{in} \quad \Omega, \quad u = \phi \quad \text{on} \quad \partial\Omega,$$

where $f \in C^0(\bar{\Omega})$ and $\phi \in C^0(\partial\bar{\Omega})$. Let $B_R(O)$ be the smallest ball centered at the origin O such that $\Omega \subset B_R(O)$. Show that

$$\sup_{\Omega} |u| \leq \sup_{\partial\Omega} |\phi| + \frac{R^2}{2n} \cdot \sup_{\Omega} |f|.$$

Also give an example to demonstrate that the estimate is sharp, i.e., the equality sign can be attained in your example.

(Indiana)

Solution.

Let $a = \sup_{\partial\Omega} |\phi|$, $b = \sup_{\Omega} |f|$. Consider the following auxiliary function

$$w(x) = \left(a + \frac{b}{2n}(R^2 - |x|^2) \right) \pm u(x).$$

It is easy to verify that

$$\Delta w = -b \pm f \leq 0 \quad \text{in} \quad \Omega$$

and $w \geq 0$ on $\partial\Omega$. By the Maximum Principle, we have

$$w \geq 0 \quad \text{in} \quad \bar{\Omega}$$

and then

$$|u(x)| \leq a + \frac{b}{2n}(R^2 - |x|^2), \forall x \in \bar{\Omega}.$$

Therefore

$$\sup_{\Omega} |u| \leq a + \frac{b}{2n} R^2.$$

In fact, for Laplacian equation, i.e., $f \equiv 0$, the equality sign in the estimate always holds. We can also give a one-dimensional example with f not vanishing identically: $\Omega = (-1.1)$, $u = 1 - x^2$. In this example, $\phi = 0$, $f = -2$, $\sup_{\Omega} |u| = 1$, and $R = 1$, the equality sign in the estimate holds.

6224

Inside unit disc $\Omega = \{x \in \mathbb{R}^2, |x| \leq 1\}$, construct two solutions u_1, u_2 of the Hamilton–Jacobi equation

$$|\nabla u|^2 - 1 = 0 \quad \text{for a.e. } x \in \Omega$$
$$u = 0 \quad \text{for } |x| = 1,$$

with $u_1 \geq 0$ while $u_2 \leq 0$. Prove that the negative solution u_2 can not be obtained as a limit of parabolic (elliptic) approximation. Indeed, if u^ϵ denotes a solution to

$$|\nabla u^\epsilon|^2 - 1 = \epsilon \Delta u^\epsilon \quad \text{for } x \in \Omega$$
$$u^\epsilon = 0 \quad \text{for } |x| = 1,$$

then u^ϵ cannot converge to u_2 as $\epsilon \to +0$.

(Penn State)

Solution.

Setting $x_1 = r \cos\theta$ and $x_2 = r \sin\theta$, we have

$$u_{x_1} = u_r \cos\theta - u_\theta \sin\theta$$

and

$$u_{x_2} = u_r \sin\theta + u_\theta \cos\theta.$$

Then the equation can be rewritten as

$$|u_r|^2 + |u_\theta|^2 = 1.$$

Taking the symmetry of the domain into account, we try to find solutions independent of θ, which satisfy the equation

$$|u_r|^2 = 1.$$

So we get two solutions to the problem

$$u_1 = 1 - r, \quad u_2 = r - 1.$$

Now we consider the elliptic boundary value problem

$$|\nabla u^\epsilon|^2 - 1 = \epsilon \Delta u^\epsilon \quad \text{for } x \in \Omega$$

$$u^\epsilon = 0 \quad \text{for } x \in \partial\Omega.$$

We conclude that

$$u^\epsilon \geq 0 \quad \forall x \in \Omega.$$

In fact, if it is false, u^ϵ must have its minimum at a interior point x^0 of Ω and

$$\nabla u^\epsilon(x^0) = 0.$$

Therefore from the equation, we have

$$\epsilon \Delta u^\epsilon(x^0) = -1 < 0.$$

This is impossible for the point x^0, at which u^ϵ attains the minimum.

As u^ϵ is always not negative, we conclude immediately that u^ϵ cannot converge to u_2 as $\epsilon \to +0$.

Section 3

Parabolic Equations

6301

Consider the Cauchy problem for the Heat Operator in $\mathbb{R}^1$:

$$u_t = u_{xx} \quad (-\infty < x < \infty, t > 0)$$
$$u(x, 0) = f(x) \quad (-\infty < x < \infty),$$

where f is bounded, continuous, and satisfies

$$\int_{-\infty}^{\infty} |f(x)|^2 dx < \infty.$$

Show that there exists a constant C such that

$$|u(x, t)| \leq \frac{C}{t^{\frac{1}{4}}}$$

for all $-\infty < x < \infty, t > 0$.

(*Indiana*)

Solution.

The solution to the problem is given by the Poisson's formula

$$u(x, t) = \frac{1}{2\sqrt{\pi t}} \int_{-\infty}^{\infty} f(y) e^{-\frac{(x-y)^2}{4t}} dy.$$

By the Cauchy inequality, we have

$$|u(x,t)| \leq \frac{1}{2\sqrt{\pi t}} \left(\int_{-\infty}^{\infty} |f(y)|^2 dy \right)^{\frac{1}{2}} \left(\int_{-\infty}^{\infty} e^{-\frac{(x-y)^2}{2t}} dy \right)^{\frac{1}{2}}.$$

Set $y = x + \sqrt{2t}\eta$ in the second integral of the above inequality. Then we get immediately

$$|u(x,t)| \leq \frac{C}{t^{\frac{1}{4}}}.$$

6302

Let $\Omega \subset \mathbb{R}^n$ be an open set with smooth boundary and suppose $u \in C^{\infty}(\Omega \times [0,\infty))$ is a solution of the equation

$$u_t - \Delta u = f$$

with $u = 0$ on $\partial\Omega \times [0,\infty)$. Assume that

$$\lim_{t \to \infty} \int_{\Omega} f(x,t)^2 dx = 0.$$

Let

$$\psi(t) = \int_{\Omega} u(x,t)^2 dx.$$

Prove that $\psi(t) \to 0$ as $t \to +\infty$.

(*Indiana*)

Solution.
Multiplying the equation by u and then integrating the resulting equation on Ω, we can get

$$\frac{d}{dt} \left(\frac{1}{2} \|u(\cdot,t)\|_{L^2(\Omega)}^2 \right) + \|\nabla u(\cdot,t)\|_{L^2(\Omega)}^2 = \int_{\Omega} fu \, dx.$$

The Poincaré inequality gives

$$\|u(\cdot,t)\|_{L^2(\Omega)}^2 \leq C\|\nabla u(\cdot,t)\|_{L^2(\Omega)}^2,$$

where $C > 0$ is a constant. For any $\varepsilon > 0$, it holds that

$$\left| \int_{\Omega} fu \, dx \right| \leq \frac{1}{2} \left(\varepsilon\psi(t) + \frac{1}{\varepsilon}\|f(\cdot,t)\|_{L^2(\Omega)}^2 \right).$$

Taking $\varepsilon = 1/C$, we obtain

$$\frac{d}{dt}\psi(t) + \frac{1}{C}\psi(t) \leq C\|f(\cdot,t)\|_{L^2(\Omega)}^2.$$

Solving this differential inequality, we have

$$\psi(t) \leq e^{-\frac{1}{C}t}\psi(0) + C\int_0^t e^{-\frac{1}{C}(t-\tau)}\|f(\cdot,\tau)\|_{L^2(\Omega)}^2 d\tau.$$

It is not difficult to verify that

$$\int_0^t e^{-\frac{1}{C}(t-\tau)}\|f(\cdot,\tau)\|_{L^2(\Omega)}^2 d\tau \to 0, \quad \text{as } t \to +\infty$$

if $\|f(\cdot,t)\|_{L^2(\Omega)}^2 \to 0$ as $t \to +\infty$. This completes the proof.

6303

Consider the Cauchy problem

$$u_t = u_{xx} \quad (-\infty < x < \infty, t > 0),$$
$$u(x,0) = f(x),$$

where f is bounded and continuous.

(a) Using the Fourier Transform, construct explicitly a fundamental solution for this problem, and write down the solution u.

(b) State a maximum principle for this problem. Why is the Tychanov non-uniqueness example possible?

(c) Suppose in addition that $f \in L^1(\mathbb{R})$. Show that

$$\lim_{t\to\infty} u(x,t) = 0.$$

(d) Show how to solve the problem

$$u_t = u_{xx} \quad (x > 0, t > 0)$$
$$u(x,0) = f(x) \quad (x > 0)$$
$$u(0,t) = 0 \quad (t > 0)$$

by using (a) and an appropriate extension of f.

(Indiana)

Solution.

(a) Let $\hat{u}(\xi, t)$ be the Fourier transform of $u(x, t)$ with respect to x. Transforming the following Cauchy problem with the initial data $\delta(x)$:

$$u_t = u_{xx} \quad (-\infty < x < \infty, t > 0)$$
$$u(x, 0) = \delta(x),$$

we get

$$\hat{u}_t + \xi^2 \hat{u} = 0,$$
$$\hat{u}(\xi, 0) = 1.$$

The solution to the transformed problem is given by

$$\hat{u}(\xi, t) = e^{-\xi^2 t}.$$

Then the inverse Fourier transform of $\hat{u}$ gives the fundamental solution to the problem

$$u(x, t) = \frac{1}{2\pi} \int_{\mathbb{R}^1} e^{ix\xi - \xi^2 t} d\xi = \frac{1}{2\sqrt{\pi t}} e^{-\frac{x^2}{4t}}.$$

From this fundamental solution, we can get the explicit representation for the solution $u(x, t)$ to the general Cauchy problem with the initial data $f(x)$

$$u(x, t) = \frac{1}{2\sqrt{\pi t}} \int_{\mathbb{R}^1} \phi(y) e^{-\frac{(x-y)^2}{4t}} dy.$$

(b) The maximum principle can be stated as follows: Suppose that $u(x, t)$ is bounded and continuous on $\mathbb{R} \times [0, T]$, and satisfies the heat equation in $\mathbb{R} \times (0, T]$ with the initial data $f(x)$. Then it holds that

$$\inf_{x \in \mathbb{R}} f(x) \leq u(x, t) \leq \sup_{x \in \mathbb{R}} f(x) \quad \text{for } (x, t) \in \mathbb{R} \times (0, T].$$

The Techanov non-uniqueness example is possible for some unbounded solutions.

(c) The conclusion is clear.

(d) Defined an odd function $\phi(x)$ as follows:

$$\phi(x) = \begin{cases} f(x), & x > 0, \\ -f(-x), & x < 0. \end{cases}$$

Then by solving the following Cauchy problem:

$$u_t = u_{xx} \quad (-\infty < x < \infty, t > 0),$$
$$u(x,0) = \phi(x),$$

we get the solution to the initial-boundary problem

$$u(x,t) = \frac{1}{2\sqrt{\pi t}} \int_0^\infty f(y) \left(e^{-\frac{(x-y)^2}{4t}} - e^{-\frac{(x+y)^2}{4t}} \right) dy.$$

6304

Consider the problem

$$u_t = u_{xx} \quad (x \in \mathbb{R}, t > 0)$$
$$u(x,0) = u_0(x) \in L^2(\mathbb{R}).$$

Find a bound on $\int_{-\infty}^\infty |u_x(x,t)|^2 dx$ in $t > 0$.

(*Indiana*)

Solution.

From the Poisson's formula for the Cauchy problem of the heat equation, we can find that

$$u_x(x,t) = -\frac{1}{4\sqrt{\pi}t^{\frac{3}{2}}} \int_{-\infty}^\infty u_0(y)(x-y)e^{-\frac{(x-y)^2}{4t}} dy.$$

Then by using the Cauchy inequality, we have

$$|u_x(x,t)|^2 \leq \frac{1}{16\pi t^3} \int_{-\infty}^\infty |u_0(y)|^2 e^{-\frac{(x-y)^2}{4t}} dy$$
$$\cdot \int_{-\infty}^\infty (x-y)^2 e^{-\frac{(x-y)^2}{4t}} dy.$$

And therefore, it holds that

$$\int_{-\infty}^{\infty} |u_x(x,t)|^2 dx \le \frac{1}{16\pi t^3} \|u_0\|_{L^2(\mathbb{R})}^2 \int_{-\infty}^{\infty} e^{-\frac{y^2}{4t}} dy \cdot \int_{-\infty}^{\infty} y^2 e^{-\frac{y^2}{4t}} dy$$

$$= \frac{1}{2t} \|u_0\|_{L^2(\mathbb{R})}^2.$$

We can also get the result by the method of energy integrals.

Multiplying the heat equation by u and tu_t respectively, and integrating the resulting equations in $\mathbb{R} \times (0,T)$, we can obtain

$$\frac{1}{2} \int_{-\infty}^{\infty} |u(x,T)|^2 dx + \int_0^T \int_{-\infty}^{\infty} |u_x(x,t)|^2 dx dt$$

$$= \frac{1}{2} \int_{-\infty}^{\infty} |u_0(x)|^2 dx$$

and

$$\int_0^T \int_{-\infty}^{\infty} t|u_t(x,t)|^2 dx dt = \int_0^T \int_{-\infty}^{\infty} tu_{xx}(x,t)u_t(x,t)dx dt.$$

Adding the double of the second equality to the first one, we have

$$\frac{1}{2} \int_{-\infty}^{\infty} |u(x,T)|^2 dx + \int_0^T \int_{-\infty}^{\infty} |u_x(x,t)|^2 dx dt$$

$$+ 2 \int_0^T \int_{-\infty}^{\infty} t|u_t(x,t)|^2 dx dt$$

$$= \frac{1}{2} \int_{-\infty}^{\infty} |u_0(x)|^2 dx + 2 \int_0^T \int_{-\infty}^{\infty} tu_{xx}(x,t)u_t(x,t)dx dt.$$

It can be calculated that

$$2 \int_0^T \int_{-\infty}^{\infty} tu_{xx}(x,t)u_t(x,t)dx dt$$

$$= -2 \int_0^T \int_{-\infty}^{\infty} tu_x(x,t)u_{xt}(x,t)dx dt$$

$$= -\int_0^T \frac{d}{dt}\left(\int_{-\infty}^\infty |u_x(x,t)|^2 dxt\right) dt$$

$$+ \int_0^T \int_{-\infty}^\infty |u_x(x,t)|^2 dxdt$$

$$= -T\int_{-\infty}^\infty |u_x(x,T)|^2 dx + \int_0^T \int_{-\infty}^\infty |u_x(x,t)|^2 dxdt.$$

Here we have applied the assumption

$$\lim_{\varepsilon \to 0} \varepsilon \int_{-\infty}^\infty |u_x(x,\varepsilon)|^2 dx = 0,$$

otherwise an approach to the function $u_0(x)$ by a sequence of functions in $C_0^\infty(\mathbb{R}^n)$ can be used.

From the above two equalities, we have

$$\frac{1}{2}\int_{-\infty}^\infty |u(x,T)|^2 dx + T\int_{-\infty}^\infty |u_x(x,T)|^2 dx$$

$$+ 2\int_0^T \int_{-\infty}^\infty t|u_t(x,t)|^2 dxdt$$

$$= \frac{1}{2}\int_{-\infty}^\infty |u_0(x)|^2 dx.$$

Therefore, we get an estimation

$$\int_{-\infty}^\infty |u_x(x,t)|^2 dx \le \frac{1}{2t}\int_{-\infty}^\infty |u_0(x)|^2 dx, \quad \forall t > 0.$$

6305

For the indicated domain Ω (interior of the parabola) let Ω_T denote the points (x,t) with $x \in \Omega$ and $t < T$. Let B_1 denote the open line-segment forming the top of $\partial\Omega_T$, and let $B_2 = \partial\Omega_T \backslash B_1$. Let $u(x,t) \in C^0(\overline{\Omega}_T)$ with $u_{xx}, u_t \in C^0(\Omega_T \cup B_1)$ be a solution of

$$u_{xx} - u_t + a(x,t)u = f(x,t)$$

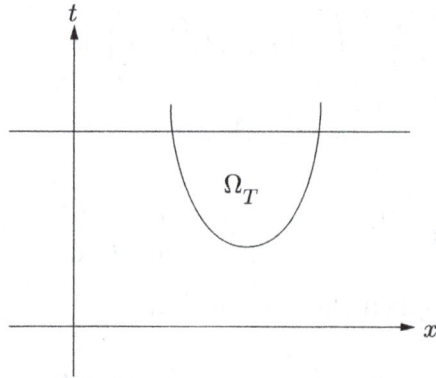

Fig. 6.1.

with $a, f \in C^0(\overline{\Omega})$. Show that if $a < 0$, $f \leq 0$ in Ω_T, then every non-constant solution u assumes its negative minimum (if one exists) on B_2. What can you say if $a < 0$, $f \geq 0$ in Ω_T?

(*Iowa*)

Solution.

Suppose that u assumes its negative minimum at $(x^0, t^0) \in \Omega_T \cup B_1$. Then we have

$$u(x^0, t^0) < 0, \quad u_t(x^0, t^0) \leq 0$$

and

$$u_{xx}(x^0, t^0) \geq 0.$$

Hence

$$u_{xx}(x^0, t^0) - u_t(x^0, t^0) + a(x^0, t^0)u(x^0, t^0) > 0.$$

This contradicts $f(x^0, t^0) \leq 0$. (Here $a < 0$ should be assumed in $\Omega_T \cup B_1$.)

If $a < 0, f \geq 0$ in Ω_T, we can say that every non-constant solution assumes its positive maximum (if there exists one) on B_2.

6306

Consider the boundary-value problem

$$u_{xx} + u_t = 0 \qquad (0 \le x \le \pi \text{ and } t > 0),$$

B.C. $\quad u(0, t) = 0,$

$\qquad u_x(\pi, t) = 0,$

I.C. $\quad u(x, 0) = f(x).$

Use the sequence $\{f_n(x) = \frac{2}{n+1} \sin\left(\frac{n+1}{2}x\right)\}$ and an appropriate space of continuous functions to decide whether this problem is well-posed.

(Indiana–Purdue)

Solution.

When n is even integer, $f_n(x)$ is the eigenfunction of the corresponding two-point boundary value problem. By setting $u = f_n(x)T_n(t)$, we can find the solution u_n to the problem with the initial data $f(x) = f_n(x)$

$$u_n(x, t) = \frac{2}{n+1} \sin\left(\frac{n+1}{2}x\right) e^{\left(\frac{n+1}{2}\right)^2 t}.$$

Let

$$\|f\| = \max_{0 \le x \le \pi} |f(x)|$$

for all $f \in C[0, \pi]$. It is clear that

$$\|f_n\| \to 0 \quad \text{as } n \to \infty.$$

But for any fixed $t > 0$

$$\|u_n(\cdot, t)\| \to \infty \quad \text{as } n \to \infty.$$

This implies that the problem is not well-posed in the space of continuous functions with the norm defined above.

6307

In Probability Theory one encounters the Ornstein–Uhlenbeck process, in which the particles of Brownian motion are subjected to an elastic force. One is required to solve

$$u_t = u_{xx} - xu_x \quad (x \in \mathbb{R}, t > 0)$$
$$u(x,0) = u_0(x).$$

Assume $u_0 \in \mathcal{S}$ and find u. (You should be able to find u explicitly; if not, leave your result in the form of a one-dimensional integral.)

(Indiana)

Solution.

By $\hat{u}(\xi, t)$ we denote the Fourier transform of u with respect to x. It is easy to see that

$$F(xu_x) = i\frac{\partial}{\partial \xi}F(u_x) = -\frac{\partial}{\partial \xi}(\xi\hat{u}) = -\hat{u} - \xi\frac{\partial \hat{u}}{\partial \xi}.$$

Then the transformed equation and initial condition are in the following form

$$\frac{\partial \hat{u}}{\partial t} - \xi\frac{\partial \hat{u}}{\partial \xi} = (1 - \xi^2)\hat{u} \quad \xi \in \mathbb{R}, \quad t > 0,$$
$$\hat{u}(\xi, 0) = \hat{u}_0(\xi).$$

Solving the above Cauchy problem for the first order linear partial differential equation, we can obtain

$$\hat{u}(\xi, t) = e^{t - \frac{1}{2}(e^{2t} - 1)\xi^2}\hat{u}_0(e^t\xi).$$

It is not difficult to verify that

$$F^{-1}(\hat{u}_0(e^t\xi)) = e^{-t}u_0(e^{-t}x)$$

and

$$F^{-1}(e^{-\frac{1}{2}(e^{2t} - 1)\xi^2}) = \frac{1}{\sqrt{2\pi(e^{2t} - 1)}}e^{-\frac{x^2}{2(e^{2t} - 1)}}.$$

Therefore, we get the solution to the problem

$$u(x,t) = \frac{1}{\sqrt{2\pi(e^{2t}-1)}} \int_{\mathbb{R}^1} u_0(e^{-t}y)e^{-\frac{(x-y)^2}{2(e^{2t}-1)}} \, dy.$$

Remark. From the above solution formula, we can find a convenient transformation of variables to the problem. In fact, set

$$\tau = \frac{1}{2}(1 - e^{-2t}), \quad \xi = e^{-t}x.$$

Then the problem is transformed into the following

$$\frac{\partial u}{\partial \tau} = \frac{\partial^2 u}{\partial \xi^2} \quad (\xi \in \mathbb{R}, \tau > 0)$$

$$\tau = 0: \quad u = u_0(\xi).$$

The above solution formula can be obtained easily from the Poisson's formula for the Cauchy problem of the heat equation.

6308

Let L be the usual heat operator

$$L = \frac{\partial}{\partial t} - \left(\frac{\partial^2}{\partial x_1^2} + \cdots + \frac{\partial^2}{\partial x_n^2} \right),$$

let $\phi \in D(\mathbb{R}^n)$, and let h be a fixed point of $\mathbb{R}^n$. Consider the problem

$$(Lu)(x,t) = u(x-h,t), \quad x \in \mathbb{R}^n, t > 0,$$
$$u(x,0) = \phi(x), \quad x \in \mathbb{R}^n$$

for an unknown function $u(x,t)$.

(a) Derive an explicit representation for a solution $u(x,t)$ of the above problem in terms of the initial data. (You should prove that your u is well-defined, but you need not prove that it represents a smooth solution.)

(b) Assume that ϕ is the Fourier transform of a function which vanishes in the ball of radius $R \geq 0$ centered at the origin. Show then that the solution u satisfies

$$\|u(\cdot, t)\|_{L^2(\mathbb{R}^n)} \leq e^{(1-R^2)t}\|\phi\|_{L^2(\mathbb{R}^n)}.$$

(Indiana)

Solution.

(a) Let $\hat{u}(\xi, t)$ be the Fourier transform of u with respect to x. Transforming the above problem we obtain

$$\begin{cases} \hat{u}_t + (|\xi|^2 - e^{-ih\cdot\xi})\hat{u} = 0, & \xi \in \mathbb{R}^n, t > 0, \\ \hat{u}(\xi, 0) = \hat{\phi}(\xi), & x \in \mathbb{R}^n \end{cases}$$

which admids the following solution

$$\hat{u}(\xi, t) = \hat{\phi}(\xi)e^{-(|\xi|^2 - \exp(-ih\cdot\xi))t}.$$

The inverse Fourier transform of $\hat{u}$ gives us the explicit representation for the solution $u(x, t)$ to the problem

$$u(x, t) = \frac{1}{(2\pi)^n}\int_{\mathbb{R}^n} \hat{\phi}(\xi)e^{ix\cdot\xi - (|\xi|^2 - \exp(-ih\cdot\xi))t}\, d\xi.$$

It is clear that $e^{ix\cdot\xi - (|\xi|^2 - \exp(-ih\cdot\xi))t} \in L^1(\mathbb{R}^n)$ and $\hat{\phi}(\xi) \in \mathcal{S}$. Therefore, the above representation is well defined.

(b) By the Plancherel theorem, we have

$$\|u(\cdot, t)\|_{L^2(\mathbb{R}^n)}^2 = (2\pi)^{-n}\|\hat{u}(\cdot, t)\|_{L^2(\mathbb{R}^n)}^2$$

$$= (2\pi)^{-n}\int_{\mathbb{R}^n} |\hat{\phi}(\xi)|^2 e^{2(\cos(h\cdot\xi) - |\xi|^2)t}\, d\xi$$

$$= (2\pi)^{-n}\int_{|\xi| \geq R} |\hat{\phi}(\xi)|^2 e^{2(\cos(h\cdot\xi) - |\xi|^2)t}\, d\xi$$

$$\leq (2\pi)^{-n}e^{2(1-R^2)t}\int_{\mathbb{R}^n} |\hat{\phi}(\xi)|^2 d\xi$$

$$= e^{2(1-R^2)t}\|\phi\|_{L^2(\mathbb{R}^n)}^2.$$

This is just the conclusion of the problem.

6309

Let $\Omega \subset \mathbb{R}^n$ be a bounded domain and assume $u(x,t) \geq 0$ is a function with $u \in C^2(\overline{\Omega} \times [0,\infty))$, which solves the equation

$$u_t - \Delta u = -u^4$$

(heat conduction with heat loss due to radiation) with the boundary condition $u|_{\partial\Omega} = 0$. Prove that we can find a constant C such that

$$E(1) = \int_\Omega u^2(x,1)dx \leq C$$

regardless of the initial value $u(x,0)$.

(*Iowa*)

Solution.
Multiplying both sides of the equation by u and integrating on Ω, we have

$$\int_\Omega u_t u \, dx - \int_\Omega \Delta u \cdot u \, dx = -\int_\Omega u^5 \, dx.$$

By using the Green's formula and noting the boundary condition, we get from above equality

$$\frac{1}{2}\frac{d}{dt}E(t) + \int_\Omega |\nabla u|^2 dx = -\int_\Omega u^5 \, dx, \tag{1}$$

where $E(t) = \int_\Omega u^2 \, dx$.

The Hölder inequality gives

$$E(t) \leq \left(\int_\Omega u^5 dx\right)^{\frac{2}{5}} \left(\int_\Omega dx\right)^{\frac{3}{5}}$$

$$= |\Omega|^{\frac{3}{5}} \left(\int_\Omega u^5 dx\right)^{\frac{2}{5}},$$

i.e.,

$$\int_\Omega u^5 \, dx \geq |\Omega|^{-\frac{3}{2}} E^{\frac{5}{2}}(t).$$

Using above estimation, we get from (1)

$$\frac{1}{2}\frac{d}{dt}E(t) \le -|\Omega|^{-\frac{3}{2}}E^{\frac{5}{2}}(t).$$

The above inequality can be written as

$$\frac{d}{dt}E^{-\frac{3}{2}}(t) \ge 3|\Omega|^{-\frac{3}{2}}.$$

Integrating the above inequality from 0 to t, we have

$$E^{-\frac{3}{2}}(t) \ge E^{-\frac{3}{2}}(0) + 3|\Omega|^{-\frac{3}{2}}t.$$

It follows that

$$E(1) \le \left(\frac{1}{3}\right)^{\frac{2}{3}}|\Omega|.$$

This completes the proof.

6310

Let

$$u \in C^2(B(0,r) \times (0,T]) \cap C(\overline{B(0,r)} \times [0,T])$$

satisfy

$$\begin{cases} u_t + u^3 + \sin(u) = \Delta_x u, & B(0,r) \times (0,T], \\ u = h, & \partial B(0,r) \times (0,T], \\ u = g, & \overline{B(0,r)} \times \{0\} \end{cases}$$

with g, h continuous and $g, h < 1$. Prove that max $u < 1$ on $\overline{B(0,r)} \times [0,T]$.

(Indiana)

Solution.
If the conclusion of the problem is false, then there exists $(x^0, t^0) \in B(0,r) \times (0,T]$ such that u takes the maximum at (x^0, t^0) and

$$u(x^0, t^0) \ge 1.$$

It is clear that

$$u_t(x^0, t^0) \geq 0 \quad \text{and} \quad \Delta_x(x^0, t^0) \leq 0.$$

Therefore, we must have

$$(u_t - \Delta_x u)(x^0, t^0) \geq 0. \tag{1}$$

But from the equation we get

$$(u_t - \Delta_x u)(x^0, t^0) = -(u^3 + \sin(u))(x^0, t^0)$$
$$= \begin{cases} -1 - \sin 1 < 0, & \text{if } u(x^0, t^0) = 1, \\ -u^3(x^0, t^0) - \sin u(x^0, t^0) < 0, & \text{if } u(x^0, t^0) > 1. \end{cases}$$

This contradicts (1).

6311

Let $u \in C^2(\Omega \times (0, T)) \cap C^0(\overline{\Omega} \times [0, T])$ be a solution to the problem

$$\begin{cases} u_t = \Delta u - u^3 & x \in \Omega, t > 0, \\ u = 0 & x \in \partial\Omega, t > 0, \\ u(x, 0) = f(x) & x \in \Omega, \end{cases}$$

where Ω is a smooth bounded domain in $\mathbb{R}^n$.

(a) Prove that

$$\|u\|_{L^2(\Omega)}(t) \leq \|f\|_{L^2(\Omega)} \quad \text{for all } t \in (0, T].$$

(b) Prove that

$$\|u\|_{L^\infty(\Omega)}(t) \leq \|f\|_{L^\infty(\Omega)} \quad \text{for all } t \in (0, T].$$

(c) Prove that the solution to this boundary value problem is unique.

(d) Show that for $f \in L^4(\Omega) \cap H^1(\Omega)$, the following bound also holds:

$$\|u\|_{L^4(\Omega)}^4(t) + \|u\|_{H^1(\Omega)}^2(t) \leq \|f\|_{L^4(\Omega)}^4 + \|f\|_{H^1(\Omega)}^2.$$

(*Indiana*)

Solution.

(a) Multiplying the equation by u, and integrating the resulting equation on Ω, we get

$$\int_\Omega u_t u \, dx = \int_\Omega \Delta u \cdot u \, dx - \int_\Omega u^4 \, dx.$$

By applying Green's formula, we obtain from above integral equality

$$\frac{d}{dt} \frac{1}{2} \int_\Omega u^2 \, dx = -\int_\Omega |\nabla u|^2 \, dx - \int_\Omega u^4 \, dx \le 0.$$

Then we get the conclusion immediately.

(b) First we show that u cannot take the positive maximum in $\Omega \times (0, T]$. In fact, if u take the positive maximum at $(x^0, t^0) \in \Omega \times (0, T]$, then we must have

$$u_t(x^0, t^0) \ge 0 \quad \text{and} \quad \Delta u(x^0, t^0) \le 0.$$

Hence

$$(u_t - \Delta u)(x^0, t^0) \ge 0.$$

This contradicts the inequality

$$-u^3(x^0, t^0) < 0.$$

In a similar way, we can show that u cannot take the negative minimum in $\Omega \times (0, T]$.

Therefore, we have

$$\|u\|_{L^\infty(\Omega)}(t) \le \|f\|_{L^\infty(\Omega)}, \quad \forall t \in (0, T].$$

(c) Let u_1 and u_2 be both solutions to the boundary value problem, and $w = u_1 - u_2$. The w solves the following problem:

$$\begin{cases} w_t = \Delta w - (u_1^2 + u_1 u_2 + u_2^2)w, & x \in \Omega, t > 0, \\ w = 0, & x \in \partial\Omega, t > 0, \\ w(x, 0) = 0, & x \in \Omega. \end{cases}$$

By a similar deduction as done in (a), we can get

$$\|w\|_{L^2(\Omega)}(t) \leq 0, \quad \forall t \in (0, T]$$

for above problem.

This implies that $u_1 = u_2$, and the uniqueness of the solution is proved.

(d) Multiplying the equation by u_t and u^3 respectively, and integrating the resulting equations, we get

$$\int_\Omega u_t^2 \, dx = \int_\Omega \Delta u \cdot u_t \, dx - \int_\Omega u^3 u_t \, dx$$

and

$$\int_\Omega u_t u^3 = \int_\Omega \Delta u \cdot u^3 \, dx - \int_\Omega u^6 \, dx.$$

By applying Green's formula to the first terms in the right side of the both equalities above, we can obtain

$$\frac{d}{dt} \left(\frac{1}{4} \int_\Omega u^4 \, dx \right) + \frac{d}{dt} \left(\frac{1}{2} \int_\Omega |\nabla u|^2 \, dx \right) = - \int_\Omega u_t^2 \, dx \leq 0$$

and

$$\frac{d}{dt} \left(\frac{1}{4} \int_\Omega u^4 \, dx \right) = - \int_\Omega 3u^2 |\nabla u|^2 \, dx - \int_\Omega u^6 \, dx \leq 0.$$

Adding the both inequalities above, we have

$$\frac{d}{dt} \left(\int_\Omega u^4 dx + \int_\Omega |\nabla u|^2 dx \right) \leq 0.$$

This differential inequality, combining with (a) if necessary, implies the conclusion of (d).

6312

For each $p \in \mathbb{R}^n$, let $A(p)$ be a real positive definite $n \times n$ matrix with entries $a_{ij}(p)$. Assume $A \in C^1(\mathbb{R}^n)$. Let $\Omega \subset \mathbb{R}^n$ be a bounded open set. Prove (from first principles) the following comparison principle:

Let u and v be C^2 solutions to the inequalities

$$u_t \leq a_{ij}(\nabla u)u_{x_i x_j} \quad \text{for } (x,t) \in \Omega \times (0, \infty)$$

and

$$v_t \geq a_{ij}(\nabla v)v_{x_i x_j} \quad \text{for } (x,t) \in \Omega \times (0, \infty),$$

respectively. Suppose, furthermore, that u and v are continuous on $\overline{\Omega} \times [0, \infty)$ and satisfy the condition $u \leq v$ on the parabolic boundary

$$(\Omega \times \{0\}) \cup (\partial\Omega \times [0, \infty)).$$

Then $u \leq v$ on $\overline{\Omega} \times [0, \infty)$.

$$(Indiana)$$

Solution.

Set $w = u - v$ and

$$b_k(x,t) = \sum_{i,j} \int_0^1 \frac{\partial a_{ij}}{\partial u_{x_k}}(\nabla v + \theta(\nabla u - \nabla v))d\theta v_{x_i x_j}.$$

Then from the given differential inequalities, we can obtain a differential inequality for w as follow:

$$\sum_{i,j=1}^n a_{ij}(\nabla u)w_{x_i x_j} + \sum_{k=1}^n b_k(x,t)w_{x_k} - w_t \geq 0.$$

By applying the maximum principle to the above differential inequality, we find

$$w \leq 0 \quad \text{on } \overline{\Omega} \times [0, \infty).$$

This is the result we need.

6313

For $\Omega \subset \mathbb{R}^n$ bounded and open, let $u \in C^2(\overline{\Omega} \times [0, \infty))$. Assume $a_{ij}(x, t, z, p)$ and $b_i(x, t, z)$ are continuous functions of their arguments and assume that

$$\sum_{i,j=1}^n a_{ij}(x, t, z, p)\xi_i\xi_j \geq \lambda(x, t, z, p)|\xi|^2 > 0$$

$$\text{for every } \xi \in \mathbb{R}^n - \{0\},$$

where λ is a positive function. Suppose u satisfies the inequality

$$u_t - \sum_{i,j=1}^n a_{ij}(x, t, u, \nabla u)u_{x_i x_j}$$

$$- \sum_{i=1}^n b_i(x, t, u)u_{x_i} \geq 0 \quad \text{for } x \in \Omega, t > 0.$$

Prove that for any $T > 0$

$$\min_{\overline{\Omega} \times [0,T]} u(x, t) = \min_Q u(x, t)$$

where $Q = (\Omega \times \{0\}) \cup (\partial\Omega \times [0, T])$.

(Indiana)

Solution.

First we prove that if $w \in C^2(\overline{\Omega} \times [0, \infty))$ satisfies

$$w_t - \sum_{i,j=1}^n a_{ij}(x, t, u, \nabla u)w_{x_i x_j}$$

$$- \sum_{i=1}^n b_i(t, x, u)w_{x_i} > 0, \quad \text{at } (x, t) \in \Omega \times (0, T], \qquad (1)$$

then (x, t) cannot be the minimum point of w in $\Omega \times (0, T]$. In fact, at the minimum point (x, t), it must hold that

$$w_t \leq 0, \quad w_{xi} = 0$$

and

$$\sum_{i,j=1}^{n} a_{ij}(x,t,u,\nabla u)w_{x_i x_j} \geq 0,$$

which contradicts (1).

Now suppose that u takes the minimum at $(x^0, t^0) \in \Omega \times (0, T]$ and

$$u(x^0, t^0) < \min_{Q} u.$$

Take $\alpha > 0$ large enough so that

$$\alpha a_{11}(x,t,u,\nabla u) + b_1(x,t,u,\nabla u) > 0, \quad \text{for } (x,t) \in \overline{\Omega} \times [0,T].$$

Then

$$w = u - \varepsilon e^{\alpha x_1}$$

takes its minimum at some point in $\Omega \times (0, T]$, if $\varepsilon > 0$ is sufficiently small. However, for the function w we have

$$w_t - \sum_{i,j=1}^{n} a_{ij}(x,t,u,\nabla u)w_{x_i x_j} - \sum_{i=1}^{n} b_i(x,t,u)w_{x_i}$$
$$\geq \varepsilon\alpha(a_{11}(x,t,u,\nabla u)\alpha + b_1(x,t,u))e^{\alpha x_1}$$
$$> 0, \quad \text{for all } (x,t) \in \Omega \times (0,T].$$

This is in contradiction with w having minimum point in $\Omega \times (0, T]$.

6314

Let Ω be a bounded open set of $\mathbb{R}^n$ with smooth boundary $\partial\Omega$. Let $u \in C^2(\Omega \times (0, \infty))$, with u continuous on $\Omega \times [0, \infty)$, be a solution of the initial-boundary problem

$$u_t - \Delta u = \sin(u) \quad \text{for } x \in \Omega, t > 0$$
$$u(x,t) = 0 \quad \text{for } x \in \partial\Omega$$
$$u(x,0) = f(x) \quad \text{for } x \in \Omega.$$

Show that if $f(x) \leq 1$, then $u(x,t) \leq e^t$ for all $x \in \Omega, t > 0$.

(*Maryland*)

Solution.

Set

$$u = e^t v.$$

Then the function v satisfies

$$v_t - \Delta v + v = e^{-t} \sin(e^t v).$$

We rewrite this equation in the following form

$$v_t - \Delta v + \left(1 - \frac{\sin(e^t v)}{e^t v} \right) v = 0.$$

Noting that

$$1 - \frac{\sin(e^t v)}{e^t v} > 0, \quad \forall v \neq 0$$

and $v \leq 1$ on $\partial\Omega$, we get from the Maximum Principle that $v(x,t) \leq 1$ in $\Omega \times (0, \infty)$. Hence $u(x,t) \leq e^t$ in $\Omega \times (0, \infty)$.

6315

Consider the Cauchy problem for Burger's equation:

$$u_t - u_{xx} + u u_x = 0, \mathbb{R} \times (0, \infty),$$
$$u(x,0) = g(x),$$

where $g \in L^\infty \cap L^1(-\infty, \infty)$. Derive the explicit formula of the solution for the Cauchy problem.

(*Northwestern*)

Solution.

Setting

$$v(x,t) = \exp \left(-\frac{1}{2} \int_0^x u(\xi, t) d\xi \right),$$

we have

$$u = -2\frac{v_x}{v},$$

$$u_t = 2\frac{v_t v_x}{v^2} - 2\frac{v_{xt}}{v}, \tag{*}$$

$$u_x = 2\frac{v_x^2}{v^2} - 2\frac{v_{xx}}{v}$$

and

$$u_{xx} = -4\frac{v_x^3}{v^3} + \frac{v_x v_{xx}}{v^2} - 2\frac{v_{xxx}}{v}.$$

Moreover,

$$u_t - u_{xx} + u u_x = \frac{2}{v^2}\frac{\partial}{\partial x}(v(v_t - v_{xx})).$$

Therefore, if v is a solution to the Cauchy problem

$$\begin{cases} v_t - v_{xx} = 0, & \mathbb{R} \times (0, \infty) \\ v|_{t=0} = \exp\left(-\frac{1}{2}\int_0^x g(\xi)d\xi\right), \end{cases}$$

then u given by $(*)$ is a solution to the original problem.

From the Poisson's formula for the Cauchy problem of heat equation, we have

$$v = \frac{1}{2\sqrt{\pi t}}\int_{-\infty}^{\infty}\exp\left(-\frac{1}{2}\int_0^\eta g(\xi)d\xi - \frac{(x-\eta)^2}{4t}\right)d\eta$$

and

$$v_x = -\frac{1}{4\sqrt{\pi t}}\int_{-\infty}^{\infty}g(\eta)\exp\left(-\frac{1}{2}\int_0^\eta g(\xi)d\xi - \frac{(x-\eta)^2}{4t}\right)d\eta.$$

And finally we obtain

$$u(x,t) = \frac{\int_{-\infty}^{\infty}g(\eta)\exp\left(-\frac{1}{2}\int_0^\eta g(\xi)d\xi - \frac{(x-\eta)^2}{4t}\right)d\eta}{\int_{-\infty}^{\infty}\exp\left(-\frac{1}{2}\int_0^\eta g(\xi)d\xi - \frac{(x-\eta)^2}{4t}\right)d\eta}.$$

6316

Consider the rectangle $Q \doteq [0, a] \times [0, b] \subset \mathbb{R}$.

(i) By separation of variables, compute the first eigenvalue λ_1 and the first eigenfunction ϕ_1 of the Laplacian operator $-\Delta$ on Q:

$$\Delta u + \lambda u = 0, \quad \text{on } Q, \quad u = 0 \text{ on } \partial\Omega.$$

(ii) Let g be a smooth function that vanishes on the boundary ∂Q. Assuming that $\kappa < \lambda_1$, prove that the solution of the parabolic problem

$$u_t = \Delta u + \kappa u \quad t > 0, \ x \in Q$$
$$u(0, x) = g(x), \quad x \in Q$$
$$u(t, x) = 0, \quad x \in \partial Q$$

converges to zero uniformly on Q, as $t \to \infty$.

(*Penn State*)

Solution.

(i) Set

$$u(x) = X(x_1)Y(x_2).$$

From the equation, we have

$$\frac{X''}{X} + \lambda = -\frac{Y''}{Y} = \mu,$$

where μ is a constant. Therefore we have two 1-dimensional eigenvalue problems:

$$X'' + \nu X = 0, \quad X(0) = X(a) = 0$$

and

$$Y'' + \mu Y = 0, \quad Y(0) = Y(b) = 0,$$

where $\nu + \mu = \lambda$. It is easy to verify that their non-trivial solutions are

$$\nu_m = \left(\frac{m\pi}{a}\right)^2, \; X_m(x_1) = \sin \frac{m\pi x_1}{a}$$

and

$$\mu_n = \left(\frac{n\pi}{b}\right)^2, \; Y_n(x_2) = \sin \frac{n\pi x_2}{b},$$

respectively.

Therefore the eigenvalues and the eigenfunctions of the Laplacian operator $-\Delta$ are

$$\lambda_{m,n} = \left(\frac{m\pi}{a}\right)^2 + \left(\frac{n\pi}{b}\right)^2$$

and

$$\phi_{m,n} = \sin \frac{m\pi x_1}{a} \sin \frac{n\pi x_2}{b},$$

respectively. The first eigenvalue and the first eigenfunction are

$$\lambda_{1,1} = \left(\frac{\pi}{a}\right)^2 + \left(\frac{\pi}{b}\right)^2$$

and

$$\phi_{1,1} = \sin \frac{\pi x_1}{a} \sin \frac{\pi x_2}{b},$$

respectively.

(ii) We now solve the parabolic problem also by the separation of variables. Set

$$u(t, x) = T(t)X(x).$$

Then the problem is reduced to a 2-dimensional eigenvalue problem

$$-\Delta U = \lambda U \quad \text{in } Q, \quad U = 0 \quad \text{on } \partial Q$$

and an ordinary differential equation

$$T' - (\kappa - \lambda)T = 0.$$

In (i), we have got the eigenvalues $\lambda_{m,n}$ and the eigenfunctions $\phi_{m,n}$ of the 2-dimensional elliptic problem. Therefore, for the ordinary differential equation, we have

$$T_{m,n} = e^{(\kappa - \lambda_{m,n})t}.$$

Hence we get the solution of the parabolic initial-boundary problem

$$u(t, x) = \sum_{m,n=1}^{\infty} A_{m,n} e^{(\kappa - \lambda_{m,n})t} \sin \frac{m\pi x_1}{a} \sin \frac{n\pi x_2}{b},$$

where

$$A_{m,n} = \frac{1}{\|\phi_{m,n}\|_{L^2(Q)}} \int_Q g(x) \sin \frac{m\pi x_1}{a} \sin \frac{n\pi x_2}{b} dx.$$

It is not difficult to verify that

$$|A_{m,n}| \leq M,$$

where M is a constant, and that

$$|u(t, x)| \leq M e^{(\kappa - \lambda_{1,1})t} \sum_{m,n=1}^{\infty} e^{(\lambda_{1,1} - \lambda_{m,n})t}.$$

Noting that

$$\sum_{m,n=1}^{\infty} e^{(\lambda_{1,1} - \lambda_{m,n})t} < +\infty,$$

we get immediately that $u(t, x)$ converges to zero uniformly on Q, as $t \to \infty$.

Section 4

Hyperbolic Equations

6401

Assume that g is smooth and in $L^1(\mathbb{R}^n)$, n arbitrary. Let $\hat{u}(\xi, t)$ be the Fourier transform (with respect to x) of the solution to the n-dimensional wave equation:

$$u_{tt} - c^2 \Delta u = 0,$$
$$u(x, 0) = 0,$$
$$u_t(x, 0) = g(x).$$

Show that the L^p norm of $\hat{u}(\cdot, t)$ at time t for $p > n$ is bounded by a constant times $t^{1-n/p}$.

(*Indiana*)

Solution.

Transforming the equation and the initial conditions, we have

$$\begin{cases} \hat{u}_{tt} + c^2 |\xi|^2 \hat{u} = 0, \\ \hat{u}(\xi, 0) = 0, \\ \hat{u}_t(\xi, 0) = \hat{g}(\xi), \end{cases}$$

and

$$\hat{u}(\xi, t) = \frac{1}{c|\xi|} \hat{g}(\xi) \sin(c|\xi|t).$$

Then

$$\int_{\mathbb{R}^n} |\hat{u}|^p d\xi = \frac{1}{c^p} \int_{\mathbb{R}^n} \frac{|\sin(c|\xi|t)|^p}{|\xi|^p} |\hat{g}(\xi)|^p d\xi.$$

$g \in L^1(\mathbb{R}^n)$ implies that $\hat{g}(\xi)$ is bounded in $\mathbb{R}^n$. Therefore, it holds with a positive constant M that

$$\|\hat{u}\|^p_{L^p(\mathbb{R}^n)} \leq M \int_{\mathbb{R}^n} \frac{|\sin(c|\xi|t)|^p}{|\xi|^p} d\xi.$$

Set $\eta = c\xi t$ in the integrand. Then we get

$$\|\hat{u}\|^p_{L^p(\mathbb{R}^n)} \leq M(ct)^{p-n} \int_{\mathbb{R}^n} \frac{|\sin \eta|^p}{|\eta|^p} d\eta.$$

$p > n$ ensures the convergence of the integral. The above estimation gives the conclusion of the problem.

6402

Consider the Cauchy problem for the wave equation

$$u_{tt} - \Delta u = 0, \quad u = u(x,t), \quad x \in \mathbb{R}^n, \quad t > 0$$
$$u(x,0) = u_0(x),$$
$$u_t(x,0) = u_1(x).$$

Assume $u_0, u_1 \in S$.

(a) Use the Fourier-transform to show that the total energy at time t

$$E(t) = \frac{1}{2} \int_{\mathbb{R}^n} u_t^2(x,t) + |\nabla_x u(x,t)|^2 dx$$

is constant in time.

(b) Let E_0 denote the constant energy in (a), i.e., $E(t) \equiv E_0$. Show that

$$\lim_{t \to \infty} \int_{\mathbb{R}^n} u_t^2(x,t) dx = \lim_{t \to \infty} \int_{\mathbb{R}^n} |\nabla_x u(x,t)|^2 dx = E_0.$$

(*Iowa*)

Solution.

(a) Taking the Fourier-transforms of the wave equation and of the initial conditions with respect to the variable $x = (x_1, \ldots, x_n)$, we have

$$u_{tt} + |\xi|^2 \hat{u}(\xi, t) = 0,$$
$$\hat{u}(\xi, 0) = \hat{u}_0(\xi),$$
$$\hat{u}_t(\xi, 0) = \hat{u}_1(\xi),$$

where $\xi = (\xi_1, \ldots, \xi_n)$. Solving the above initial problem of the ordinary differential equation, we get

$$\hat{u}(\xi, t) = \hat{u}_0(\xi) \cos(|\xi|t) + \frac{\hat{u}_1(\xi)}{|\xi|} \sin(|\xi|t)$$

and

$$\hat{u}_t(\xi, t) = -\hat{u}_0(\xi)|\xi| \sin(|\xi|t) + \hat{u}_1(\xi) \cos(|\xi|t). \tag{1}$$

Therefore

$$\int_{\mathbb{R}^n} (|\hat{u}_t(\xi, t)|^2 + |\xi|^2 |\hat{u}(\xi, t)|^2) d\xi = \int_{\mathbb{R}^n} (|\hat{u}_1(\xi)|^2 + |\xi|^2 |\hat{u}_0(\xi)|^2) d\xi.$$

Then we can complete the proof of (a) by the Plancherel Theorem.

(b) From (1), we have

$$
\begin{aligned}
|\hat{u}_t(\xi, t)|^2 &= |\hat{u}_0(\xi)|^2 |\xi|^2 \sin^2(|\xi|t) + |\hat{u}_1(\xi)|^2 \cos^2(|\xi|t) \\
&\quad - 2\mathrm{Re}(\hat{u}_0 \bar{\hat{u}}_1) \sin(|\xi|t) \cos(|\xi|t) \\
&= \frac{1}{2} |\hat{u}_0(\xi)|^2 |\xi|^2 (1 - \cos(2|\xi|t)) \\
&\quad + \frac{1}{2} |\hat{u}_1(\xi)|^2 (1 + \cos(2|\xi|t)) \\
&\quad - \mathrm{Re}(\hat{u}_0 \bar{\hat{u}}_1) \sin(2|\xi|t) \\
&= \frac{1}{2} (|\hat{u}_1(\xi)|^2 + |\xi|^2 |\hat{u}_0(\xi)|^2) - \frac{1}{2} |\hat{u}_0(\xi)|^2 \cos(2|\xi|t) \\
&\quad + \frac{1}{2} |\hat{u}_1(\xi)|^2 \cos(2|\xi|t) - \mathrm{Re}(\hat{u}_0 \bar{\hat{u}}_1) \sin(2|\xi|t).
\end{aligned}
$$

Integrating the above equation on $\mathbb{R}^n$ and noting that

$$\lim_{t\to\infty} \int_{\mathbb{R}^3} |\hat{u}_0(\xi)|^2 |\xi|^2 \cos(2|\xi|t)d\xi$$

$$= \lim_{t\to\infty} \int_{\mathbb{R}^n} |\hat{u}_1(\xi)|^2 \cos(2|\xi|t)d\xi$$

$$= \lim_{t\to\infty} \int_{\mathbb{R}^3} \mathrm{Re}(\hat{u}_0\bar{\hat{u}}_1) \sin(2|\xi|t)d\xi$$

$$= 0,$$

we obtain

$$\lim_{t\to\infty} \int_{\mathbb{R}^n} |\hat{u}_t(\xi,t)|^2 d\xi = \frac{1}{2}\int_{\mathbb{R}^n} (|\hat{u}_1(\xi)|^2 + |\xi|^2|\hat{u}_0(\xi)|^2)d\xi.$$

Then the Plancherel Theorem gives us

$$\lim_{t\to\infty} \int_{\mathbb{R}^n} u_t^2(x,t)dx = E_0.$$

Combining this with (a), we finish the proof of (b).

6403

Consider the initial problem for the wave equation in three space variables:

$$u_{x_1x_2} + u_{x_2x_2} + u_{x_3x_3} - u_{tt} = 0; \quad x \in \mathbb{R}^3, t > 0,$$
$$u(x,0) = \phi(x); \quad x \in \mathbb{R}^3,$$
$$u_t(x,0) = \psi(x); \quad x \in \mathbb{R}^3.$$

(a) Write down the formula for the solution of the problem.

(b) If ϕ and ψ vanish outside a ball of radius 3 centered at the origin, find the set of points in $\mathbb{R}^3$ where you are sure that u vanishes when $t = 10$.

(c) If ϕ vanishes everywhere in $\mathbb{R}^3$, and

$$\psi(x) = \begin{cases} 0, & \text{for } |x| < 1, \\ k, & \text{for } 1 \le |x| \le 2, \\ 0, & \text{for } 2 < |x|, \end{cases}$$

where k is a constant, find $u(0,t)$ for all $t \geq 0$. (Your answer should be explicit, no integrals.)

(*Indiana–Purdue*)

Solution.

(a)

$$u(x,t) = \frac{1}{4\pi t} \int_{|y-x|=t} \psi(y) dS + \frac{\partial}{\partial t} \left(\frac{1}{4\pi t} \int_{|y-x|=t} \phi(y) dS \right).$$

(b) By applying the domain of dependence for the solutions to the wave equation, we can verify that u vanishes when $t = 10$ for $|x| \leq 7$ or $|x| \geq 13$.

In fact, when $|x| \leq 7$, it holds that

$$|y| \geq 3 \quad \text{for all } y \in \{y : |y - x| = 10\}.$$

Therefore

$$\int_{|y-x|=10} \phi(y) ds = \int_{|y-x|=10} \psi(y) = 0.$$

Similarly, we can show that the above equality holds when $|x| \geq 13$.

(c) It is easy to verify that

$$u(0,t) = \begin{cases} 0, & 0 \leq t < 1, \\ kt, & 1 \leq t \leq 2, \\ 0, & 2 < t. \end{cases}$$

6404

Let Ω be the upper-half space in $\mathbb{R}^n$

$$\Omega = \{(x_1, x_2, x_3) \in \mathbb{R}^3; x_3 > 0\}.$$

Consider the initial-boundary value problem

$$\frac{\partial^2 u}{\partial x_1^2} + \frac{\partial^2 u}{\partial x_2^2} + \frac{\partial^2 u}{\partial x_3^2} - \frac{\partial^2 u}{\partial t^2} = 0; \quad x \in \Omega, t > 0,$$

$$u(x,0) = \phi(x), \quad u_t(x,0) = \psi(x); x \in \Omega,$$

$$u(x,t) = 0, \quad x \in \partial\Omega, t \geq 0.$$

(a) Write down a formula for the value $u(x,t)$ of the solution at (x,t), $x \in \Omega$, when

$$t < x_3.$$

(b) Find a formula for the value $u(x,t)$ of the solution at (x,t), $x \in \Omega$ for all $t > 0$.

(Indiana–Purdue)

Solution.

(a) When $x \in \Omega$ and $0 < t < x_3$, the domain of dependence for the value $u(x,t)$ at (x,t) is in Ω. The value $u(x,t)$ is given by the formula for the Cauchy problem of the equation:

$$u(x,t) = \frac{1}{4\pi t} \int_{|y-x|=t} \psi(y)dS + \frac{\partial}{\partial t}\left(\frac{1}{4\pi t}\int_{|y-x|=t}\phi(x)dS\right).$$

(b) Define $\Phi(x)$ and $\Psi(x)$ as both odd functions with respect to x_3 in $\mathbb{R}^3$ as follows:

$$\Phi(x) = \begin{cases} \phi(x_1, x_2, x_3), & x_3 \geq 0, \\ -\phi(x_1, x_2, -x_3), & x_3 < 0, \end{cases}$$

$$\Psi(x) = \begin{cases} \psi(x_1, x_2, x_3), & x_3 \geq 0, \\ -\psi(x_1, x_2, -x_3), & x_3 < 0, \end{cases}$$

respectively. Then the value $u(x,t)$ is given by

$$u(x,t) = \frac{1}{4\pi t} \int_{|y-x|=t} \Psi(y)dS + \frac{\partial}{\partial t}\left(\frac{1}{4\pi t}\int_{|y-x|=t}\Phi(y)dS\right).$$

If both ϕ and ψ satisfy certain compatibility conditions, it is easy to see that $u(x,t)$ given by the above formula satisfies the wave equation and the initial conditions. When $x_3 = 0$,

$$\int_{|y-x|=t} \Psi(y)dS = \int_{y_3=\sqrt{t^2-(y_1-x_1)^2-(y_2-x_2)^2}} \psi(y_1, y_2, y_3)dS$$

$$- \int_{y_3=-\sqrt{t^2-(y_1-x_1)^2-(y_2-x_2)^2}} \psi(y_1, y_2, -y_3)dS = 0.$$

In a similar way, we have

$$\int_{|y-x|=t} \Phi(y)ds = 0, \quad \text{when } x_3 = 0.$$

Therefore, $u(x,t)$ satisfies the boundary condition on $x_3 = 0$.

6405

Let $u(x,t)$ be a solution of class C^2 to the Cauchy problem

$$u_{tt} = \Delta u \ (x \in \mathbb{R}^n, t > 0),$$
$$u(x,0) = \phi(x), \ u_t(x,0) = \psi(x); \ \phi, \psi \in C^\infty,$$

with compact support. Define the local energy by

$$E_R(u(t)) = \frac{1}{2} \int_{|x|<R} (u_t^2 + |\nabla u|^2)dx.$$

(i) Use the energy indentity to show that

$$E_{R-T}(u(0)) \le E_R(u(t)) \le E_{R+T}(u(0)).$$

Conclude that the total energy $E_\infty(u(t))$ is conserved.

(ii) When $n = 3$, evaluate $\lim_{t\to\infty} E_R(u(t))$.

(iii) Do (ii) above when $n = 1$.

(iv) Let $n = 2$, $\phi \equiv 0$ and suppose that $\psi \equiv 0$ for $|x| \ge k$, where $x = (x_1, x_2)$. Show that u satisfies

$$|u(x,t)| \le c(t+r+2k)^{-1/2}(t-r+2k)^{-1/2} \int_{|x|\le k} |\psi(x)|dx$$

for some constant c, on the set $r < t - 4k$. (Here $r = |x|$.)

(Indiana)

Solution.

(i) For $t \in [0,T]$, we have

$$E_{R-T+t}(u(t)) = \frac{1}{2} \int_0^{R-T+t} \int_{|x|=\rho} (u_t^2 + |\nabla u|^2)dSd\rho.$$

By the above expression, we can calculate that

$$\frac{d}{dt}E_{R-T+t}(u(t)) = \frac{1}{2}\int_{|x|=R-T+t}(u_t^2 + |\nabla u|^2)dS$$

$$+ \int_{|x|\leq R-T+t}(u_t u_{tt} + \nabla u \cdot \nabla u_t)dx$$

$$= \frac{1}{2}\int_{|x|=R-T+t}(u_t^2 + |\nabla u|^2)dS$$

$$+ \int_{|x|\leq R-T+t}(u_t\Delta u + \nabla u \cdot \nabla u_t)dx$$

$$= \frac{1}{2}\int_{|x|=R-T+t}\left(u_t^2 + |\nabla u|^2 + 2u_t\frac{\partial u}{\partial n}\right)dS$$

$$= \frac{1}{2}\int_{|x|=R-T+t}\sum_{i=1}^{n}(u_t\cos(n, x_i) + u_{x_i})^2 dS \geq 0.$$

Hence

$$E_{R-T}(u(0)) \leq E_{R-T+t}(u(t)) \leq E_R(u(t)).$$

Similarly, we can show that for $t \in [0, T]$

$$\frac{d}{dt}E_{R+T-t}(u(t)) \leq 0.$$

Hence

$$E_{R+T-t}(u(t)) \leq E_{R+T}(u(0)), \quad \forall t \in [0, T].$$

Then the estimation

$$E_R(u(t)) \leq E_{R+T}(u(0))$$

is obtained from the clear inequality

$$E_R(u(t)) \leq E_{R+T-t}(u(t)), \quad \forall t \in [0, T].$$

It is clear that the total energy $E_\infty(u(t))$ equals to $E_\infty(u(0))$.

(ii) From the formula of the solution to the problem

$$u(x,t) = \frac{\partial}{\partial t}\left(\frac{1}{4\pi t}\int_{|y-x|=t}\phi(y)dS\right) + \frac{1}{4\pi t}\int_{|y-x|=t}\psi(y)dS,$$

it is easy to verify

$$E_R(u(t)) = 0 \quad \text{if } t > R + a,$$

where a is the radius of a ball in which supp ϕ $\cup$ supp ψ is contained. Hence

$$\lim_{t\to\infty} E_R(u(t)) = 0.$$

(iii) By the formula

$$u(x,t) = \frac{1}{2}(\phi(x-t) + \phi(x+t)) + \frac{1}{2}\int_{x-t}^{x+t}\psi(y)dy,$$

it is easy to evalute

$$\lim_{t\to\infty} E_R(u(t)) = 0.$$

(iv) When $n = 2$, the solution is given by

$$u(x,t) = \frac{1}{2\pi}\int_{|y-x|\le t}\frac{\psi(y)}{\sqrt{t^2 - |y-x|^2}}dy.$$

If $|x| < t - 4k$, it is easy to see that

$$\{y : |y| \le k\} \subset \{y : |y - x| \le t\}.$$

Hence the formula of the solution can be rewritten as

$$u(x,t) = \frac{1}{2\pi}\int_{|y|\le k}\frac{\psi(y)}{\sqrt{t^2 - |y-x|^2}}dy.$$

When $|x| < t - 4k$ and $|y| \le k$, we have

$$|y - x| \le |x| + 2k,$$

and hence

$$\frac{1}{\sqrt{t^2 - |y - x|^2}} \le \frac{1}{\sqrt{(t + |x| + 2k)(t - |x| - 2k)}}.$$

It is easy to verify that

$$\sqrt{\frac{t - |x| + 2k}{t - |x| - 2k}} \le \sqrt{3} \quad \forall t - |x| > 4k.$$

Therefore, we obtain

$$|u(x, t)|$$
$$\le \frac{\sqrt{3}}{2\pi}(t + |x| + 2k)^{-\frac{1}{2}}(t - |x| + 2k)^{-\frac{1}{2}} \int_{|y| \le k} |\psi(y)| dy$$
$$\forall t - |x| > 4k.$$

6406

Let $\Omega \subset \mathbb{R}^n$ be a bounded domain and assume $u(x, t)$ is a function with $u \in C^2(\overline{\Omega} \times [0, \infty))$, which solves the equation

$$u_{tt} - \Delta u = u$$

with the boundary condition $u|_{\partial\Omega} = 0$. With

$$E(t) = \frac{1}{2} \int_\Omega u_t^2(x, t) + |\nabla_x u(x, t)|^2 dx$$

prove that there is a $C < \infty$ such that

$$E(t) \le \exp(Ct)E(0)$$

for all $t \ge 0$.

(Iowa)

Solution.
Multiplying the equation by u_t and integrating the resulting equation with respect to x in Ω, we have

$$\frac{d}{dt}E(t) = \int_\Omega u u_t \, dx.$$

Here we have applied the Green's formula:

$$\int_\Omega u_t \Delta u \, dx = -\int_\Omega \nabla_x u_t \cdot \nabla_x u \, dx.$$

By the Poincaré inequality, we can conclude that

$$\left| \int_\Omega u u_t \, dx \right| \leq \frac{1}{2} \int_\Omega (u^2(x,t) + u_t^2(x,t)) dx$$
$$\leq CE(t).$$

Thus we obtain a differential inequality for $E(t)$

$$\frac{d}{dt} E(t) \leq CE(t).$$

Then $E(t) \leq \exp(Ct)E(0)$ is an immediate consequence of the above differential inequality.

6407

(a) Find the explicit form of the solution of the initial value problem

$$\nabla^2 u = u_{tt}, \quad x \in \mathbb{R}^3, t > 0,$$
$$u(x,0) = \frac{f(r)}{r}, \quad x \in \mathbb{R}^3,$$
$$u_t(x,0) = -\frac{f'(r)}{r}, \quad x \in \mathbb{R}^3,$$

where $r^2 = x_1^2 + x_2^2 + x_3^2$, and $f(\eta)$ is a C^3 function on the real line with compact support such that

$$f(\eta) = 0, \quad \eta \geq 0,$$
$$\int_0^\infty \left\{ (f')^2 + \frac{1}{\eta^2}(\eta f' - f)^2 \right\} d\eta = 1,$$

where $f'(\eta)$ is the derivative of f with respect to η.

(b) What is its energy at time $t = 10$.

(*Indiana–Purdue*)

Solution.

Find the spherically symmetric solution to the problem. In the symmetric case, the equation can be written as

$$\frac{\partial^2 u}{\partial r^2} + \frac{2}{r}\frac{\partial u}{\partial r} = u_{tt}.$$

Set $w = ru$. Then w solves the following problem:

$$\begin{cases} w_{rr} = w_{tt}, & r > 0, \ t > 0, \\ w = 0, & r = 0, \\ w = f(r), & w_t = -f'(r), \ t = 0. \end{cases}$$

Solving this problem, and then we can get

$$u(x, t) = \begin{cases} \dfrac{1}{r} f(r - t), & r \geq t, \\ 0, & 0 \leq r < t. \end{cases}$$

(b) The energy of the problem is given by

$$E(t) = \frac{1}{2} \int_{\mathbb{R}^n} (u_t^2 + |\nabla u|^2) dx.$$

It is easy to verify that

$$E(t) = E(0), \quad \forall t > 0.$$

And therefore

$$E(10) = 2\pi \int_0^\infty \left(\frac{f'(r)^2}{r^2} + \left| \nabla\left(\frac{f(r)}{r} \right) \right|^2 \right) r^2 dr$$

$$= 2\pi \int_0^\infty (f'(r)^2 + r^{-2}(f - rf')^2) dr$$

$$= 2\pi.$$

6408

Consider the wave equation in $\mathbb{R}^3$:

$$u_{tt} - \Delta u = 0 \quad \text{for } x \in \mathbb{R}^3, \ t > 0,$$
$$u(x,0) = 0,$$
$$u_t(x,0) = g(x),$$

where $g \in C_0^\infty(\mathbb{R}^3)$. Prove that there exists a constant C depending only on the given data such that

$$\sup_{x \in \mathbb{R}^3} |u(x,t)| \leq \frac{C}{t}, \quad t > 0.$$

(Indiana)

Solution.

Let R and M be positive constants such that

$$\operatorname{supp} g \subset B_R = \{x \in \mathbb{R}^3, |x| < R\}$$

and

$$|g(x)| \leq M, \quad \forall x \in \mathbb{R}^3.$$

From the Poisson's formula, we have

$$u(x,t) = \frac{1}{4\pi t} \int_{|y-x|=t} g(y) dS_y.$$

It is easy to verify that

The area of the intersection $\{y \in \mathbb{R}^3, |y - x| = t\} \cap B_R$ $\leq$ The area of ∂B_R.

Therefore, we obtain

$$|u(x,t)| \leq \frac{1}{4\pi t} \cdot M \cdot 4\pi R^2 = \frac{MR^2}{t}, \quad t > 0.$$

6409

Let $u \in C^2(\mathbb{R}^{n+1}(x,t))$ be a solution of the equation $u_{tt} - \Delta u = 0$. Suppose also that $u = u_t = 0$ on the ball $B = \{(x,0) :$

$|x| \leq t_0\}$ in the plane $t = 0$. Prove that u vanishes in the conical region

$$D = \{(x,t) : 0 \leq t \leq t_0 \text{ and } |x| \leq t_0 - t\}.$$

(Indiana)

Solution.
Let

$$\Omega_\tau = \{(x,t) \in \mathbb{R}^{n+1}, |x| \leq t_0 - t, t = \tau\}$$

for $0 \leq \tau \leq t_0$. Multiplying the wave equation by u_t, and integrating the resulting equation with respect to x in Ω_r, we can obtain

$$\frac{1}{2}\int_{\Omega_\tau} \frac{\partial}{\partial \tau}(u_t^2 + |\nabla u|^2)(x,\tau)dx - \int_{\partial\Omega_r} u_t \sum_{i=1}^{n} u_{x_i}\cos(n,x_i)dS = 0.$$

Set

$$E(\tau) = \frac{1}{2}\int_{\Omega_\tau}(u_t^2 + |\nabla u|^2)(x,\tau)dx.$$

It is not difficult to verify that

$$\frac{dE(\tau)}{d\tau} = \frac{1}{2}\int_{\Omega_\tau}\frac{\partial}{\partial\tau}(u_t^2 + |\nabla u|^2)(x,\tau)dx$$
$$- \frac{1}{2}\int_{\partial\Omega_\tau}(u_t^2 + |\nabla u|^2)(x,\tau)dS.$$

Therefore, we have

$$\frac{dE(\tau)}{d\tau} = -\frac{1}{2}\int_{\partial\Omega_\tau}\left(u_t^2 + |\nabla u|^2 - 2u_t\sum_{i=1}^{n}u_{x_i}\cos(n,x_i)\right)dS$$
$$= -\frac{1}{2}\int_{\partial\Omega_\tau}\sum_{i=1}^{n}(u_t\cos(n,x_i) - u_{x_i})^2dS \leq 0, \quad \forall 0 < \tau < t_0.$$

It is clear that $E(0) = 0$. Hence

$$E(\tau) = 0, \quad \forall 0 \leq \tau \leq t_0.$$

This implies that

$$u = 0 \text{ in } D.$$

6410

Let $u(x,t)$ be a solution of the Cauchy problem

$$u_{tt} = \Delta u \quad (x \in \mathbb{R}^3, t > 0),$$
$$u(x,0) = f(x), \quad u_t(x,0) = g(x).$$

Assume that $f, g \in C_0^\infty(\mathbb{R}^3)$.

Prove that if

$$\lim_{t\to\infty} \int_{\mathbb{R}^3} u^2 \, dx = 0,$$

then

$$\int_{\mathbb{R}^3} g \, dx = 0.$$

(Indiana)

Solution.

Let $R > 0$ such that

$$\text{supp} f, \text{ supp} g \subset B_R = \{x \in \mathbb{R}^3, |x| < R\}.$$

It is not difficult to verify that

$$\text{supp} \, u(\cdot, t) \subset \{x \in \mathbb{R}^3, t - R < |x| < t + R\}$$

for any $t > R$.

By using Green's formula, we get the following equality from the equation

$$\int_{\mathbb{R}^3} u_{tt} \, dx = \int_{\mathbb{R}^3} \Delta u \, dx = 0.$$

Therefore, we have

$$\frac{d}{dt} \int_{\mathbb{R}^3} u_t \, dx = 0,$$

$$\int_{\mathbb{R}^3} u_t dx = \int_{\mathbb{R}^3} g \, dx$$

and

$$\int_{\mathbb{R}^3} u \, dx = \int_{\mathbb{R}^3} f \, dx + t \int_{\mathbb{R}^3} g \, dx \qquad (1)$$

by the initial conditions.

In the other side, by Cauchy inequality it follows that

$$\int_{\mathbb{R}^3} u dx \le \left(\int_{\mathbb{R}^3} u^2 dx \right)^{\frac{1}{2}} \left(\int_{t-R<|x|<t+R} dx \right)^{\frac{1}{2}}$$

$$= \left(\frac{8}{3} \pi R \right)^{\frac{1}{2}} (3t^2 + R^2)^{\frac{1}{2}} \|u\|_{L^2(\mathbb{R}^3)}, \qquad (2)$$

for any $t > R$. This contradicts (1) provided that

$$\int_{\mathbb{R}^3} g \, dx \ne 0.$$

6411

Consider the solution of the wave equation

$$u_{tt} = \Delta u \ (x \in \mathbb{R}^3, t > 0)$$

with data $u(x,0) = 0$, $u_t(x,0) = \psi(x) \in C_0^\infty$.

(a) Use Fourier transform to show that there is a positive constant c such that

$$\lim_{t \to \infty} \int_{\mathbb{R}^3} u^2 \, dx = c.$$

(b) Show that there is a positive constant C_1 such that

$$t \cdot \max_x |u| \ge C_1$$

for all sufficiently large t.

(Indiana)

Solution.

(a) Let $\hat{u}$ denote the Fourier transform of u with respect to the variable $x = (x_1, \ldots, x_n)$. By taking the Fourier transforms of the wave equation and the initial conditions, we can get

$$\hat{u}(\xi) = \frac{|\hat{\psi}(\xi)|}{|\xi|} \sin(|\xi|t).$$

Therefore

$$
\begin{aligned}
\int_{\mathbb{R}^3} u^2(x,t)dx &= (2\pi)^{-3} \int_{\mathbb{R}^3} |\hat{u}(\xi,t)|^2 d\xi \\
&= (2\pi)^{-3} \int_{\mathbb{R}^3} |\hat{\psi}(\xi)|^2 \frac{\sin^2(|\xi|t)}{|\xi|^2} d\xi \\
&= \frac{1}{2(2\pi)^3} \int_{\mathbb{R}^3} |\xi|^{-2} |\hat{\psi}(\xi)|^2 d\xi \\
&\quad - \frac{1}{2(2\pi)^3} \int_{\mathbb{R}^3} |\xi|^{-2} |\hat{\psi}(\xi)|^2 \cos(2(|\xi|t)d\xi.
\end{aligned}
$$

Noting

$$|\xi|^{-2} |\hat{\psi}(\xi)|^2 \in L^1(\mathbb{R}^3),$$

we have

$$\lim_{t \to \infty} \int_{\mathbb{R}^3} |\xi|^{-2} |\hat{\psi}(\xi)|^2 \cos(2(|\xi|t)d\xi = 0.$$

Then we get

$$\lim_{t \to \infty} \int_{\mathbb{R}^3} u^2(x,t) = c,$$

where

$$c = \frac{1}{2(2\pi)^3} \int_{\mathbb{R}^3} |\xi|^{-2} |\hat{\psi}(\xi)|^2 d\xi > 0$$

if $\psi \not\equiv 0$.

(b) Let $\Omega(t)$ be the support of $u(x,t)$ with respect to the variable $x = (x_1, x_1, x_3)$ for $t > 0$. If the conclusion is false, then there exists a sequence $\{t_n\}$ such that

$$t_n \to \infty$$

and

$$\varepsilon_n = t_n \cdot \max_x |u(x, t_n)| \to 0, \quad \text{as } n \to \infty.$$

It is not difficult to verify that

$$\Omega(t_n) = O(t_n^2).$$

Then we have

$$\int_{\mathbb{R}^3} u^2(x, t_n)dx \le \int_{\Omega(t_n)} \varepsilon_n^2 t_n^{-2}dx \to 0, \quad \text{as } n \to \infty.$$

This contradicts (a). The proof of (b) is completed.

6412

Consider the Cauchy problem for the wave equation

$$u_{tt} - \Delta u = 0 \quad (x \in \mathbb{R}^n, t > 0),$$
$$u(x, 0) = \phi(x), \quad u_t(x, 0) = \psi(x).$$

(a) Let $n = 3$; $\phi, \psi \in C_0^\infty(\mathbb{R}^3)$. Let $\theta < 1$. Determine the large time behavior of the integral

$$\int_{|x| \le \theta t} u^4(x, t)dx.$$

(b) Let $n = 2$, $\phi \equiv 0$, $\psi \in C_0^\infty(\mathbb{R}^2)$. Show that for any $\theta < 1$, we have

$$\sup_{|x| \le \theta t} |u(x, t)| = O(t^{-1}) \quad \text{as } t \to \infty.$$

(Indiana)

Solution.

(a) Assume that $R > 0$ is so large that

$$\operatorname{supp} \phi \cup \operatorname{supp} \psi \subset \{x \in \mathbb{R}^3, |x| < R\}.$$

From the Huygens's principle for the 3-dimensional wave equation, we conclude that if $t > R$ and

$$|x| < t - R,$$

then

$$u(x,t) = 0.$$

Therefore, when $t > \frac{R}{1-\theta}$, it holds that

$$u(x,t) = 0, \quad \forall |x| < \theta t,$$

which implies that

$$\int_{|x| \le \theta t} u^4(x,t)dx = 0, \quad \text{when } t > \frac{R}{1-\theta}.$$

(b) From the Poisson's formula, the solution to the 2-dimensional problem is given by

$$u(x_1, x_2, t) = \frac{1}{2\pi} \int_{|y-x| \le t} \frac{\psi(y_1, y_2)}{\sqrt{t^2 - |y-x|^2}} dy.$$

Let $R > 0$ be so large that

$$\operatorname{supp} \psi \subset \{x \in \mathbb{R}^2, |x| < R\}.$$

When $t > \frac{R}{1-\theta}$, it is clear that if $|x| \le \theta t$ and $|y| \le R$ then

$$|y - x| \le \theta t + R$$

and

$$\sqrt{t^2 - |y-x|^2} \ge \sqrt{t^2 - (\theta t + R)^2}.$$

Therefore, we have

$$|u(x_1, x_2, t)| \leq \frac{1}{2\pi} \int_{|y| \leq R} \frac{|\psi(y_1, y_2)|}{\sqrt{t^2 - (\theta t + R)^2}} dy$$

$$\leq \frac{C}{\sqrt{t^2 - (\theta t + R)^2}}, \quad \forall t > \frac{R}{1 - \theta}, |x| \leq \theta t,$$

where C is a constant. The proof is completed.

6413

The linear transport equation is following PDE for a scalar function f of seven variables $x, v, t (x \in \mathbb{R}^3, v \in \mathbb{R}^3, t \in \mathbb{R}_+^1)$:

$$f_t + v \cdot \nabla_x f = 0.$$

Consider the initial-value problem for this; i.e., let

$$f(x, v, 0) = \overset{\circ}{f}(x, v); \quad \overset{\circ}{f} \text{ given}, \ \overset{\circ}{f} \in C^1.$$

(a) Show that, if $\overset{\circ}{f} \geq 0$ for all x, v, then the same is true for f at all later times.

(b) Define the local density ρ by

$$p(x, t) = \int_{\mathbb{R}^3} f(x, v, t) dv.$$

Suppose that $0 \leq \overset{\circ}{f}(x, v) \leq h(x)$, where $h \in L^1(\mathbb{R}^3)$. Show that

$$\sup_x \rho = O(t^{-3}) \quad \text{as } t \to \infty.$$

(*Indiana*)

Solution.

(a) The characteristic curves of the equation are defined by

$$\frac{dx}{dt} = v$$

and along which it holds that

$$\frac{df}{dt} = 0. \tag{1}$$

Therefore, the forward characteristic curve departing from $t = 0$, $x = \alpha$ is

$$x = \alpha + tv. \tag{2}$$

From (1) and (2), we get the solution to the initial-value problem

$$f(x, v, t) = \overset{\circ}{f}(x - tv, v). \tag{3}$$

It is clear that $f \geq 0$ if $\overset{\circ}{f} \geq 0$.
  (b) From (3), we have

$$
\begin{aligned}
\rho &= \int_{\mathbb{R}^3} f(x, v, t) dv \\
&= \int_{\mathbb{R}^3} \overset{\circ}{f}(x - tv, v) dv \\
&\leq \int_{\mathbb{R}^3} h(x - tv) dv \\
&= \frac{1}{t^3} \int_{\mathbb{R}^3} h(u) du.
\end{aligned}
$$

This completes the proof.

6414

Solve the Cauchy problem

$$xu_x + yuu_y = -xy \ (x > 0)$$
$$u = 5 \quad \text{on } xy = 1,$$

if a solution exists.

(*Indiana*)

Solution.

The system of the characteristic equations for the problem is given by

$$\frac{dx}{ds} = x, \quad \frac{dy}{ds} = yu, \quad \frac{du}{ds} = -xy$$

with the initial conditions

$$s = 0 : x = \alpha, \quad y = \frac{1}{\alpha}, \quad u = 5.$$

From the above Cauchy problem, we have immediately

$$x = \alpha e^s$$

and

$$u\frac{du}{ds} = -\alpha e^s \frac{du}{ds} = -\frac{d}{ds}(\alpha e^s y + u).$$

This implies that

$$\frac{1}{2}u^2 + u + xy = C,$$

where C is a constant. The initial data gives $C = \frac{37}{2}$. Then we obtain immediately

$$u = -1 + \sqrt{38 - 2xy}.$$

6415

Compute an explicit solution $u(x,t)$ to the initial-value problem

$$u_t + (u_x)^2 = 0,$$
$$u(x,0) = x^2.$$

$$(Indiana)$$

Solution.

Set $v = u_x$. Differentiating the equation and the initial condition with respect to x, we get

$$v_t + 2vv_x = 0,$$
$$v(x, 0) = 2x. \tag{1}$$

The characteristic curve of the quasilinear equation (1) is given by

$$\frac{dx}{dt} = 2v,$$

along which it holds that

$$\frac{dv}{dt} = 0.$$

From the above characteristic equations, we can obtain immediately

$$v = \frac{2x}{4t + 1}.$$

Hence

$$u = \frac{x^2}{4t + 1}.$$

6416

Solve the following initial value problem.

$$u_t + uu_x = 0, \quad x \in \mathbb{R}, \ t > 0,$$
$$u(x, 0) = \begin{cases} 1, & x < 0, \\ 1 - x, & 0 \le x \le 1, \\ 0, & x > 1. \end{cases}$$

Show that the continuous solution exists only for a finite time, and find the discontinuous entropy solution, giving explicitly the discontinuity curve and the Rankine–Hugoniot jump condition along it.

(Indiana–Purdue)

Solution.

The characteristic curve departing from $t = 0$, $x = \beta$ is given by

$$\frac{dx}{dt} = u, \quad t = 0 : x = \beta, \tag{1}$$

along which u is a constant, i.e.,

$$\frac{du}{dt} = 0, \quad t = 0 : u = u^0(\beta), \tag{2}$$

where $u^0(x) = u(x, 0)$. From (1) and (2), we have

$$x = u^0(\beta)t + \beta, \quad u = u^0(\beta),$$

and

$$x = \begin{cases} t + \beta, & \beta < 0, \\ (1 - \beta)t + \beta, & 0 \le \beta \le 1, \\ \beta, & \beta > 1, \end{cases}$$

$$u = \begin{cases} 1, & \beta < 0, \\ 1 - \beta, & 0 \le \beta \le 1, \\ 0, & \beta > 1. \end{cases}$$

For $0 < t < 1, 0 \le \beta \le 1$, we have

$$\beta = \frac{x - t}{1 - t}$$

and hence

$$u = 1 - \frac{x - t}{1 - t} = \frac{1 - x}{1 - t}.$$

Therefore we get the continuous solution to the problem

$$u = \begin{cases} 1, & x < t, \\ \dfrac{1 - x}{1 - t}, & t \le x \le 1, \\ 0, & x > 1, \end{cases}$$

for $0 \le t < 1$ (see Fig. 6.2).

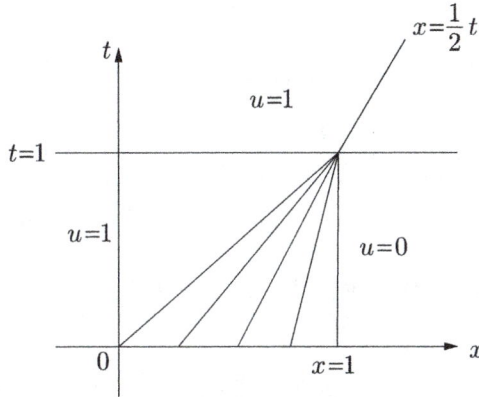

Fig. 6.2.

When $t = 1$, the continuous solution to the problem blows up.

For $t \geq 1$, the Rankine–Hugoniot condition along the discontinuous curve is given by

$$(u_+ - u_-)\frac{dx}{dt} - \left(\frac{u_+^2}{2} - \frac{u_-^2}{2}\right) = 0,$$

i.e.,

$$\frac{dx}{dt} = \frac{1}{2}(u_+ + u_-).$$

Noting $u_+ = 0, u_- = 1$, we have

$$\frac{dx}{dt} = \frac{1}{2}.$$

Therefore, the discontinuous curve is

$$x = \frac{1}{2}(t+1), \quad t \geq 1$$

and the entropy solution to the problem is given by

$$u = \begin{cases} 1, & x < \dfrac{1}{2}(t+1), \\[2mm] 0, & x > \dfrac{1}{2}(t+1). \end{cases}$$

6417

(a) Let $u(x,t)$ be a smooth solution of

$$\frac{\partial u}{\partial t} + a(x,u)\frac{\partial u}{\partial x} - bu = 0 \quad \text{for } 0 < t < T,$$

where b is a constant and a is smooth. Show that if

$$\alpha \le u(x,0) \le \beta \quad \text{for all } x,$$

then

$$\alpha e^{bt} \le u(x,t) \le \beta e^{bt}$$

for all x and all $t \in (0,T)$.

(b) Let $u(x,t)$ be smooth solution of

$$\frac{\partial u}{\partial t} + a(u)\frac{\partial u}{\partial x} = 0 \quad \text{for } 0 < t < T,$$
$$u(x,0) = u_0(x),$$

where u_0 and a are smooth. Show that, if $x(t,y)$ is the characteristic through $(y,0)$, then

$$u_x(x(t,y),t) = \frac{u_x(y,0)}{(\frac{d}{dx}a(u_0(x))|_{x=y})t + 1}.$$

(i) Use this to compute the largest T for which a smooth solution can exist up to time T.

(ii) Show that, if $0 \le t_1 < t_2 < T$, then

$$\int_{-\infty}^{\infty} |u_x(x,t_2)|dx = \int_{-\infty}^{\infty} |u_x(x,t_1)|dx.$$

(Indiana)

Solution.

(a) Let $u = e^{bt}v$. Then v satisfies the following equation

$$\frac{\partial v}{\partial t} + a(x, e^{bt}v)\frac{\partial v}{\partial x} = 0.$$

The solution v maintains a constant along a characteristic of the above equation, defined by the equation

$$\frac{dx}{dt} = a\big(x, e^{bt}v\big).$$

Therefore, it holds that

$$\inf_x v(x,0) \le v(x,t) \le \sup_x v(x,0).$$

This gives the estimation in the problem immediately.

(b) $u(x,t)$ is a constant along the characteristic $x = x(t,y)$. Hence

$$u_x(x(t,y),t) = u_0(y) \quad \forall t \in (0,T).$$

Differentiating above equality with respect to y, we have

$$u_x(x(t,y),t)\frac{\partial}{\partial y}x(t,y) = u_0'(y) \quad \forall t \in (0,T).$$

The characteristic $x(t,y)$ through $(y,0)$ is given by the following Cauchy problem:

$$\begin{cases} \dfrac{\partial}{\partial t}x(t,y) = a(u_0(y)), \\[2mm] x(0,y) = y. \end{cases}$$

Differentiating this problem with respect to y, we get

$$\begin{cases} \dfrac{\partial}{\partial t}\left(\dfrac{\partial}{\partial y}x(t,y)\right) = a'(u_0(y))u_0'(y), \\[2mm] \dfrac{\partial}{\partial y}x(0,y) = 1. \end{cases}$$

Then it follows from above problem that

$$\frac{\partial}{\partial y}x(t,y) = a'(u_0(y))u_0'(y)t + 1.$$

Therefore, we have

$$u_x(x(t,y),t) = \frac{u_0'(y)}{a'(u_0(y))u_0'(y)t + 1}.$$

This is just that we want to show.

(i) Let $m = \inf_y a'(u_0(y))u_0'(y)$. From the above expression, we can get the following conclusions easily.

If $m \geq 0$, the Cauchy problem admits a global smooth solution in $(0, \infty)$;

If $-\infty < m < 0$, the largest T for which the problem admits a smooth solution in $(0, T)$ is given by

$$mT + 1 = 0.$$

(ii) For any fixed $t \in (0, T)$, by the change of variables $x = x(t, y)$ in the integral, we have

$$\int_{-\infty}^{\infty} |u_x(x,t)|dx = \int_{-\infty}^{\infty} |u_x(x(t,y),t)| \frac{\partial}{\partial y} x(t,y)dy.$$

Taking the expressions of $u_x(x(t,y),t)$ and $\frac{\partial}{\partial y}x(t,y)$ into account, we obtain

$$\int_{-\infty}^{\infty} |u_x(x,t)|dx = \int_{-\infty}^{\infty} |u_0'(y)|dy.$$

This completes the proof of the problem.

6418

(Uniqueness of weak solutions) Consider the Cauchy problem

$$\begin{cases} u_t + u_x = 0, & \text{on } \mathbb{R} \times [0, T], \\ u(x, 0) = u_0(x), & \text{on } \mathbb{R} \times \{0\}. \end{cases} \tag{2}$$

A function $u \in L^p(\mathbb{R} \times [0, T])$ is defined to be a weak solution of (2) iff

$$\int_0^T \int_{-\infty}^{+\infty} u(y, s)(-\phi_t(y, s) - \phi_x(y, s))dyds$$

$$+ \int_{-\infty}^{\infty} u_0(y)\phi(y,0)dy = 0 \qquad (2')$$

for all test functions $\phi(x,t) \in C^{\infty}(\mathbb{R} \times [0,T])$ which are compactly supported and vanish on $\{t = T\}$.

If $u_0(x) \equiv 0$, prove that the only L^P solution of $(2')$ is $u \equiv 0$.

(*Indiana*)

Solution.

Consider the following Cauchy problem:

$$\begin{cases} \phi_t + \phi_x = f(t,x), & \text{on } \mathbb{R} \times [0,T], \\ \phi(x,T) = 0, & \text{on } \mathbb{R} \times \{T\}, \end{cases}$$

where $f \in C^{\infty}(\mathbb{R} \times [0,T])$ with a compact support.

It is not difficult to show that the above Cauchy problem admits a solution $\phi \in C^{\infty}(\mathbb{R} \times [0,T])$ which is compactly supported. Therefore, we have

$$\int_0^T \int_{-\infty}^{+\infty} u(y,s)f(y,s)dyds = 0 \qquad (*)$$

for all function $f \in C^{\infty}(\mathbb{R} \times [0,T])$ with a compact support, provided $u_0(x) \equiv 0$. As the space $C_0^{\infty}(\mathbb{R} \times [0,T])$, consisting of all C^{∞}-functions with compact support in $\mathbb{R} \times [0,T]$, is dense in $L^q(\mathbb{R} \times [0,T])$, where $q = p/(p-1)$, the equality $(*)$ is valid for all functions $f \in L^q(\mathbb{R} \times [0,T])$. Hence we have $u \equiv 0$.

6419

Consider the Cauchy problem for a function $u = u(x,t)$, where $x \in \mathbb{R}$ and $t > 0$.

$$\begin{cases} u_t + a(x,t)u_x = b(x,t), \\ u(x,0) = u_0(x) \in C(\mathbb{R}). \end{cases}$$

(a) Show that if $b \equiv 0$, then $TV(u(\cdot,t)) \leq TV(u_0(\cdot))$ for each $t > 0$ (where $TV = $ "total variation").

(b) Show that when $b \not\equiv 0$, one still has the bound

$$TV(u(\cdot,t)) \leq TV(u_0) + \int_0^t TV(b(\cdot,s))ds.$$

(Indiana)

Solution.

The characteristic curves of the equation are given by the equation

$$\frac{dx}{dt} = a(x,t),$$

along which the solutions to the equation satisfy

$$\frac{du}{dt} = b(x,t).$$

(a) Suppose that $b \equiv 0$. Then u is a constant along a characteristic curve. We divide the line $t = \tau$ into subintervals by the points $\ldots, x_{-n}, \ldots, x_{-1}, x_0, x_1, \ldots, x_n, \ldots$. Suppose that the backward characteristic curve through (x_i, τ) meets the line $t = 0$ at $(\xi_i, 0)$ $(i = \ldots, -n, \ldots, -1, 0, 1, \ldots, n, \ldots)$. Then we have

$$u(x_i, \tau) = u_0(\xi_i),$$

and

$$\sum_{i=-\infty}^{+\infty} |u(x_{i+1}, \tau) - u(x_i, \tau)| = \sum_{i=-\infty}^{+\infty} |u_0(\xi_{i+1}) - u_0(\xi_i)|.$$

Therefore, we get

$$TV(u(\cdot,\tau)) \leq TV(u_0(\cdot)), \quad \forall \tau > 0.$$

(b) Let $\{x_i\}$ and $\{\xi_i\}$ be given as in (a). When $b \not\equiv 0$, we have

$$u(x_i, \tau) = u_0(\xi_i) + \int_0^\tau b(\tilde{x}_i(s), s)ds,$$

where by $x = \tilde{x}_i(t)$ we denote the characteristic curve through (x_i, τ). From the above inequalities, we get

$$\sum_{i=-\infty}^{+\infty} |u(x_{i+1}, \tau) - u(x_i, \tau)|$$

$$\leq \sum_{i=-\infty}^{+\infty} |u_0(\xi_{i+1}) - u_0(\xi_i)|$$

$$+ \int_0^\tau \sum_{i=-\infty}^{+\infty} |b(\tilde{x}_{i+1}(s), s) - b(\tilde{x}_i(s), s)| ds$$

$$\leq TV(u_0) + \int_0^\tau TV(b(\cdot, s)) ds.$$

Then we can complete the proof easily.

6420

Suppose that $u(x, t)$ is a smooth solution of

$$u_t + u u_x = 0 \quad \text{for } x \in \mathbb{R}, t > 0$$
$$u(x, 0) = f(x) \quad \text{for } x \in \mathbb{R}.$$

Assume that f is a C^1 function such that

$$f(x) = \begin{cases} 0, & \text{for } x < 0 \\ 1, & \text{for } x > 1 \end{cases}$$

and $f'(x) > 0$, for $|x| < 1$. Show that for $t > 0$

$$\lim_{r - \infty} u(rx, rt) = \begin{cases} 0, & \text{for } x < 0 \\ x/t, & \text{for } 0 < x < t \\ 1, & \text{for } x > t. \end{cases}$$

(*Maryland*)

Solution.

The characteristic of the equation is defined by

$$\frac{dx}{dt} = u$$

and along which, the unknown function u satisfies

$$\frac{du}{dt} = 0.$$

So when $x > t + 1$, the characteristic through a at $t = 0$ is

$$x = t + a,$$

along which $u = 1$. Therefore

$$u \equiv 1 \quad \text{for } x > t + 1.$$

In the similar way, we can find that

$$u \equiv 0 \quad \text{for } x < -1.$$

Now we discuss the solution in the region $-1 < x < t + 1$. The characteristic through a at $t = 0$ is defined by

$$\frac{dx}{dt} = f(a),$$

that is

$$x = f(a)t + a,$$

along which $u = f(a)$. Because $f'(a)t + 1 > 0$, we can solve from the above equation

$$a = a(x, t).$$

Therefore the solution to the problem in this region is

$$u(x, t) = f(a(x, t)).$$

Now we discuss the asymptotic behavior of the solution. It is clear that we need only study the solution in the region $x < t + 1$. When $x > t$, we have

$$rx > rt + 1$$

and

$$u(rx, rt) = 1$$

for all large enough r. When $x < 0$, then $rx < -1$ for large r. Therefore

$$u(rx, rt) = 0$$

for all large enough r.

Now we study the situation in the region $0 < x < t$. It is not difficult to verify that $\alpha(rx, rt)$ is a monotone function of r (whether it is increasing or decreasing depends on that $\alpha(x, t)$ is bigger or smaller than 0). Set

$$\alpha(rx, rt) \to \bar{\alpha} \quad \text{as } r \to \infty.$$

Then

$$u(rx, rt) \to f(\bar{\alpha}) \quad \text{as } r \to \infty.$$

But

$$rx = f(\alpha(rx, rt))rt + \alpha(rx, rt)$$

and α is bounded. Let $r \to \infty$, from the above equation, we get $f(\bar{\alpha}) = x/t$. Therefore

$$u(rx, rt) \to x/t \quad \text{as } r \to \infty.$$

6421

Consider the Cauchy problem for conservation law

$$u_t + \left(\frac{u^2}{2}\right)_x = 0, \quad u(0, x) = g(x),$$

where g is a (nontrivial) smooth function with bounded support.

(i) Prove that the solution develops a shock in finite time.

(ii) At a fixed time t, let $x(t)$ be the position of a shock and let u^-, u^+ be the left and right states across the shock. Using the divergence theorem on a suitable triangle with vertex at $(t, x(t))$, prove the identity

$$\int_{x(t)-tu^-}^{x(t)-tu^+} g(x)dx = t\left(\frac{(u^-)^2}{2} - \frac{(u^+)^2}{2}\right).$$

(Penn State)

Solution.

(i) As g has bounded support, there exist $x_1, x_2 \in \mathbb{R}$ such that $x_1 < x_2$ and

$$g(x_1) > g(x_2).$$

The characteristic of the equation is defined by

$$\frac{dx}{d\tau} = u,$$

along which u satisfies

$$\frac{du}{d\tau} = 0.$$

So the characteristic lines through x_1 and x_2 at $\tau = 0$ are

$$x = g(x_1)\tau + x_1$$

and

$$x = g(x_2)\tau + x_2,$$

respectively. They must meet at some time $\tau = t > 0$. But along them the unknown functions are

$$u = g(x_1)$$

and

$$u = g(x_2),$$

respectively. The solution develops a shock.

(ii) Through $(t, x(t))$ we draw two backward characteristic lines

$$x = u^-(\tau - t) + x(t)$$

and

$$x = u^+(\tau - t) + x(t).$$

They meet the initial line $\tau = 0$ at

$$x = x(t) - u^- t$$

and

$$x = x(t) - u^+ t,$$

respectively.

Applying the divergence theorem of the conservation law to the equation we have

$$\oint u\,dx - \frac{u^2}{2}dt = 0,$$

where the integral curse is the triangle constructed by the two backward characteristics and the initial line. From the equation we obtain

$$\int_{x(t)-tu^-}^{x(t)-tu^+} g(x)dx = \int_0^t \frac{(u^-)^2}{2}d\tau - \int_0^t \frac{(u^+)^2}{2}d\tau$$

$$= t\left(\frac{(u^-)^2}{2} - \frac{(u^+)^2}{2} \right).$$

6422

Consider the first order system of equations

$$u_t + \sum_{j=1}^n A_j u_{x_j} = 0,$$

where $u(x,t) = (u_1(x,t), \ldots, u_m(x,t)), (x,t) \in \mathbb{R}^n \times \mathbb{R}$, and the A_j's are symmetric $m \times m$ matrices with constant real entries. Use an energy argument to show that the domain of dependence of $(x^0, t^0), t^0 > 0$, for a solution of the system is contained in the cone

$$\{|x - x^0| \le \Lambda(t^0 - t)\}$$

where $\Lambda = \max_{|\xi|=1, 1 \le l \le m} |\lambda_l(\xi)|$, and $\lambda_l(\xi), l = 1, \ldots, m$, are the eigenvalues of the matrix $A(\xi) = \sum_{j=1}^{n} \xi_j A_j$.

(UC, Los Angeles)

Solution.

In fact, we only need to show that the domain of dependence of (x^0, t^0) is in the ball

$$\Omega_0 : |x - x^0| \le \Lambda t^0$$

in the initial plane. By Ω_τ we denote the transversal domain of the cone by the plane $t = \tau$, that is

$$\Omega_\tau : |x - x^0| \le \Lambda(t^0 - t), \quad t = \tau.$$

We shall prove that

$$\int_{\Omega_\tau} |u(x, \tau)|^2 dx \le \int_{\Omega_0} |u(x, 0)|^2 dx,$$

which implies the conclusion on the domain of dependence of the solutions.

Multiplying both sides of the equation by u and then integrating the resulting equation in domain Q_τ, where Q_τ is the part of the cone between $t = 0$ and $t = \tau$, we have

$$\int_{Q_\tau} \frac{\partial}{\partial t} |u^2| dx + \sum_{j=1}^{n} \int_{Q_\tau} \frac{\partial}{\partial x_j} (A_j u \cdot u) dx = 0.$$

Using integration by parts (Green's formula), we get from the above equation that

$$\int_{\Omega_\tau} |u(x, \tau)|^2 dx - \int_{\Omega_0} |u(x, 0)|^2 dx$$

$$+ \int_{\Sigma_\tau} \left(\xi_0 |u|^2 + \sum_{j=1}^{n} \xi_j A_j u \cdot u \right) dS = 0,$$

where Σ_τ is the surface of the cone between $t = 0$ and $t = \tau$, and $(\xi_0, \xi_1, \ldots, \xi_n)$ is its outer normal, that is

$$(\xi_0, \xi_1, \ldots, \xi_n) = \frac{1}{\sqrt{1 + \Lambda^2}} \left(\Lambda, \frac{x_1 - x_1^0}{|x - x^0|}, \ldots, \frac{x_n - x_n^0}{|x - x^0|} \right).$$

As $\Lambda = \max_{|\xi|=1, 1 \leq l \leq m} |\lambda_l(\xi)|$, all the eigenvalues of the real symmetric matrix $\Lambda I + \Sigma_{j=1}^n \xi_j A_j$ are not negative. Therefore

$$\xi_0 |u|^2 + \sum_{j=1}^n \xi_j A_j u \cdot u \geq 0.$$

And hence

$$\int_{\Omega_\tau} |u(x, \tau)|^2 dx \leq \int_{\Omega_0} |u(x, 0)|^2 dx.$$

Abbreviations of Universities in this Book

Auburn	Auburn University
Cincinnati	University of Cincinnati
Columbia	Columbia University
Courant Inst.	Courant Institute of Mathematical Sciences, New York University
Harvard	Harvard University
Illinois	University of Illinois (Urbana)
Indiana	Indiana University (Bloomington)
Indiana–Purdue	Indiana University–Purdue University (Indianapolis)
Iowa	University of Iowa (Iowa City)
Maryland	University of Maryland
Michigan	University of Michigan (Ann Arbor)
Minnesota	University of Minnesota (Minneapolis)
Northwestern	Northwestern University
Penn State	Pennsylvania State University
Princeton	Princeton University
Purdue	Purdue University
Rutgers	Rutgers University
Stanford	Stanford University
SUNY, Albany	State University of New York (Albany)
SUNY, Stony Brook	State University of New York (Stony Brook)
Texas A&M	Texas A&M University
Temple	Temple University
Toronto	University of Toronto
UC, Berkerly	University of California (Berkerly)
UC, Irvine	University of California (Irvine)
UC, Los Angeles	University of California (Los Angeles)

Made in the USA
Monee, IL
08 January 2025